Microstructure-Sensitive Design for Performance Optimization

Brent L. Adams

Surya R. Kalidindi

David T. Fullwood

AMSTERDAM • BOSTON • HEIDELBERG • LONDON
NEW YORK • OXFORD • PARIS • SAN DIEGO
SAN FRANCISCO • SINGAPORE • SYDNEY • TOKYO

Butterworth-Heinemann is an imprint of Elsevier

Butterworth-Heinemann is an imprint of Elsevier
225 Wyman Street, Waltham, MA 02451, USA
The Boulevard, Langford Lane, Kidlington, Oxford, OX5 1GB, UK

Library of Congress Cataloging-in-Publication Data
Application submitted

British Library Cataloguing-in-Publication Data
A catalogue record for this book is available from the British Library.

ISBN: 978-0-12-396989-7

For information on all Butterworth-Heinemann publications
visit our Web site at http://store.elsevier.com

Contents

Preface

The prominent "grand challenge" in materials engineering for the twenty-first century is to effect a reversal of the paradigm by which new materials are developed, especially for highly constrained design (HCD) applications. Traditional methodologies for new materials development are driven mainly by innovations in processing; it follows that only a limited number of readily accessible microstructures are considered,[1] with attention focused on a small number of properties or performance objectives. For HCD applications, the designer faces increasingly complex requirements with multiple property objectives/constraints and material anisotropy affecting system performance. It is evident that the time- and resource-consumptive empiricism that has dominated materials development during the past century must give way to a greater dependence on modeling and simulation.[2]

We need to invert the current paradigm in new materials development from the present (deductive) cause-and-effect approach to a much more powerful and responsive (inductive) goal–means approach (Olson, 1997). This shift could substantially reduce system development time and cost for materials-sensitive HCD problems. There has existed a fundamental incompatibility during the past two decades between materials science and engineering and the engineered product design cycle. Current methodology for introducing new materials into engineered components requires up to 10 years of development time. This compares with design optimization methodologies that are presently capable of introducing sophisticated design evolvements (excluding materials considerations) in a matter of days or weeks. One consequence of this incompatibility is a fundamental weakness in the nexus that links materials science and engineering to the design enterprise, where the goal is to tailor a material's microstructure to meet the stringent properties and performance requirements of complex components and systems. Addressing this gap is the primary motivation for this book.

To the best of the authors' knowledge, this book presents the first mathematically rigorous framework for addressing the inverse problems of materials design and process design, while using a comprehensive set of hierarchical measures of the microstructure statistics and composite theories that are based on the same description of the microstructure. The framework presented in the book utilizes highly efficient spectral representations to arrive at invertible linkages between material structure, its properties, and the processing paths used to alter the material structure. Several recent high-profile reports ("Integrated Computational Materials Engineering (ICME)," The National Academies Press, 2008; "Materials Genome Initiative for Global Competitiveness," National Science and Technology Council, 2011; "A National Strategic Plan for Advanced Manufacturing," National Science and Technology Council, 2012) have all called for the creation of a new materials innovation infrastructure to facilitate the design, manufacture, and deployment of new high-performance materials at a dramatically accelerated pace in emerging advanced technologies. We believe that the framework presented in this book can serve as the core enabler for these strategic initiatives.

---

[1]It is known that the space of potential microstructures is vastly larger than the set that is typically characterized.

[2]This position has been strongly articulated in the report of findings of a National Science Foundation (NSF)-sponsored workshop entitled "New Directions in Materials Design Science and Engineering," edited by McDowell and Story (1998).

This book is primarily intended as a reference for specialists engaged in the emerging field of materials design. It can also be used as a textbook for a sequence of two courses offered for a high-level undergraduate class or a graduate class. Chapters 1 through 3 serve as background material and can be skipped (or assigned as self-reading) if students have familiarity with this material. Chapters 4 through 11 introduce the basic concepts of the first-order theories and illustrate their usage in first-order inverse solutions to materials and process design problems. These chapters could be the focus of a first course in MSDPO (microstructure-sensitive design for performance optimization). In a second follow-up course, the focus can be on the more difficult concepts associated with second-order theories presented in Chapters 12 through 15. Chapter 16 provides useful background material on microscopy techniques (with a strong focus on electron backscatter diffraction) that can be used in either course to give the student a solid introduction to at least one characterization technique that complements the computational approaches found in the rest of the text. Our past experience indicates that these courses are highly amenable to the incorporation of team projects by small groups of students as an integral part of the course.

Acknowledgments

The authors would like to acknowledge the many people who have contributed to the production of this book. Much of the groundwork for the book was undertaken during a sabbatical leave that Brent Adams took at Drexel University to collaborate with Surya Kalidindi in 2004–2005. The authors are thankful for the support of Brigham Young University and Drexel University that facilitated this fruitful period of idea development.

Innumerable colleagues contributed to the progress of this work, with particular thanks to Hamid Garmestani and Graeme Milton. Stuart Wright of Edax-TSL kindly provided various Orientation Imaging Microscopy (OIM™) images used in the book, and the overview of OIM that formed the basis for Chapter 16.

Furthermore, numerous students contributed directly or indirectly to this development, including Stephen Niezgoda, Sadegh Ahmadi, Max Binci, Hari Duvvuru, Brad Fromm, Carl Gao, Ben Henrie, Eric Homer, Josh Houskamp, Marko Knezevic, Colin Landon, Scott Lemmon, Ryan Larsen, Mark Lyon, Gwénaëlle Proust, Craig Przybyla, Joshua Shaffer, Xianping Wu, and Tony Fast. Bradford Singley produced many of the figures.

Partial funding was also provided by the Army Research Office, under program manager David Stepp, and various NSF grants. Earlier work was also made possible under a grant from the Air Force Office of Scientific Research with program manager Craig Hartley. Surya Kalidindi was partially funded for his effort under grants from the Office of Naval Research (ONR) with program managers Julie Christodolou and William Mullins. Generous supplemental funding, provided through the Warren N. and Wilson H. Dusenberry Professorship held by Brent Adams, was instrumental in providing support for undergraduate students engaged in developing the case studies, and for travel that has greatly facilitated this work.

Nomenclature

$a, b, \ldots$	(lowercase italic) scalar
$\mathbf{v}, \mathbf{n}, \ldots$	(lowercase, bold) vector
$v_i, \ldots$	(lowercase italics) vector component
$\mathbf{e}_1, \ldots$	basis vector
$\mathbf{T}, \mathbf{C}, \ldots$	(uppercase bold) tensor
$T_{12}, C_{ijkl}, \ldots$	(uppercase italics) tensor components
$\mathbf{u} \cdot \mathbf{v}$	dot product of vectors
$\lvert\mathbf{u}\rvert$	modulus of a vector
$\mathbf{u} \times \mathbf{v}$	cross-product of vectors
$\in_{ijk}$	permutation symbol
$\mathbf{e}_i \otimes \mathbf{e}_k$	dyadic (outer) product of vectors
δ_{ij}	delta symbol
$\mathbf{I} = \delta_{ij}\mathbf{e}_i \otimes \mathbf{e}_j = \mathbf{e}_i \otimes \mathbf{e}_i$	identity tensor
$\mathbf{S}^T$	transpose of a matrix/tensor
$tr(\mathbf{S})$	trace of a matrix/tensor
Q_{ij}	unitary matrix (transformation)
g_{ij}	coordinate transformation using a rotation
$\{\varphi_1, \Phi, \varphi_2\}$	Euler angles describing a rotation
$\det \mathbf{T}$	determinant of a matrix/tensor
$\mathbf{R}$	rotation matrix
∇	del operator
grad	grad operator
$\text{div } \mathbf{v} = \nabla \cdot \mathbf{v}$	divergence
$\text{curl } \mathbf{v} = \nabla \times \mathbf{v}$	curl
$\nabla^2\phi = \nabla \cdot \nabla\phi$	Laplacean operator
σ_{ji}	stress tensor component
$\sigma_{ji,j}$	comma denotes differentiation
ε_{ij}	strain tensor component
$\forall$	"for all"
$\dot{\varepsilon}_{ij}$	dot denotes differentiation by time (i.e., a rate)
C_{ijkl}	stiffness tensor component
S_{ijkl}	compliance tensor component
C_{ij}	stiffness tensor component in reduced notation (i.e., matrix form)
$\mathbf{b}$	Burgers vector
(111)	slip planes

$\langle 110 \rangle$	slip directions
$\mathbf{n}^{\alpha}$	slip plane normal
$\mathbf{m}^{\alpha}$	slip direction
τ_{RSS}^{α}	resolved shear stress
τ_{CRSS}	critical resolved shear stress
$\dot{\gamma}^{\alpha}$	effective shear rate
$\mathbf{L}^{app}$	applied velocity gradient tensor
$\mathbf{L}^{p}$	plastic velocity gradient tensor
$\mathbf{W}^{*}$	rigid-body spin
$\mathbf{W}^{p}$	anti-symmetric component of $\mathbf{L}^{p}$
$\mathbf{D}^{p}$	symmetric component of $\mathbf{L}^{p}$
$\mathrm{sgn}(a)$	sign of **a**
$\because$	"because"
$\bar{\mathbf{a}}$	average value of **a**
$\mathbf{C}^{*}, \mathbf{C}^{eff}$	effective (macroscopic) value of **C** (the asterisk is also used for 'complex conjugate' in various equations involving Fourier analysis)
$\mathbf{C}^{r}$	reference value for **C**
$\mathbf{C}'$	polarized value of C (i.e., $\mathbf{C}^{r} - \mathbf{C}$ or $\overline{\mathbf{C}} - \mathbf{C}$)
h	local state
$M(\vec{x}, h)$	microstructure function
dV/V	volume fraction
dh	the invariant measure
$\mathbf{H}$	the local state space
$\{\mathbf{a}_1^o, \mathbf{a}_2^o, \mathbf{a}_3^o\}$	lattice basis vectors
L^o	local lattice
$(a_1^o,\ a_2^o,\ a_3^o,\ \alpha^o,\ \beta^o,\ \gamma^o\ )$	lattice parameters (magnitudes and angles)
$\tilde{\mathbf{I}}$	signifies the inversion tensor
FZ	fundamental zone
$SO(3)$	special orthogonal group (rotation group)
$SO(3)/G$	left coset of G with $SO(3)$
$FZ(O)$	fundamental zone for cubic lattices
$FZ(D_6)$	fundamental zone for hexagonal lattices
${}^{(k)}(\bullet)$	kth element of an ensemble
$\langle \bullet \rangle = \frac{1}{\mathrm{K}} \sum_{\mathrm{k}=1}^{\mathrm{K}} {}^{(\mathrm{k})}(\bullet)$	ensemble average
$f_2(h, h' \vert \vec{r})$	two-point local state correlation function
Ω	sample region, representative volume element (RVE), etc.
$\Omega \vert \mathbf{r}$	subset of Ω such that adding the vector **r** to any point remains in Ω
$\Psi(\Omega)$	space of vectors, **r**, that can fit in Ω
∂H	local state space of interfaces (between grains)
∂h	local state at an interface

S_V	surface area per unit volume
$S_V(R_A, \hat{n}, R_B)$	grain boundary character distribution (GBCD)
$\Delta R = R_A^T R_B$	lattice misorientation
$S_V(\Delta R)$	misorientation distribution function (MDF)
$[0, D_k)$	real interval greater than or equal to zero, less than D_k
$\omega_{s_1 s_2 s_3} \leftrightarrow \omega_s$	subcell of Ω
$\chi_s(x) = \begin{cases} 1 \text{ if } x \in \omega_s \\ 0 \text{ otherwise} \end{cases}$	spatial indicator function for subcell s
γ_n	or ω_n subcell of local state space H, containing local state h_n
$\chi^n(h) = \begin{cases} 1 \text{ if } h \in \gamma_n \\ 0 \text{ otherwise} \end{cases}$	local state indicator function for subcell n
D_s^n	microstructure coefficients in primitive Fourier approximation
$F_t^{nn'} = H_{ss't}\, D_s^n D_{s'}^{n'}$	two-point correlation function Fourier coefficients
S^2	the set of all physically distinct unit vectors
$\mathbf{T}(R)$	complex representation of the special orthogonal group $SO(3)$
$L^2(S^2)$	set of square integrable functions on unit sphere
$k_l^m(\hat{n})$	surface spherical harmonic functions (SSHFs)
$T_l^{\mu n}(R)$	generalised spherical harmonic functions (GSHFs)
$\dot{T}_l^{\mu n}(R)$	GSHF with crystal symmetry
$\dot{T}_l^{m\nu}(R)$	GSHF with statistical symmetry
S	statistical symmetry group
$\ddot{T}_l^{\mu\nu}(R)$	GSHF with crystal and statistical symmetry
$W_p^q(x_k)$	Haar wavelet function
$W_{p_1 p_2 p_3}^{q_1 q_2 q_3}(x)$	3D Haar wavelet function
d_s	orientation distribution in cell s in primitive basis
$\hat{d}_s^{(i)}$	single orientation in cell s, or eigentexture
M_s	microstructure hull
$d_k^{k'}$	distance between microstructure functions k and k'
$Pr(\cdot)$	probability of
$Pr(T \uparrow d\beta)$	probability that T "hits" $d\beta$
$L(d\beta \cap T)$	line segment of intersection
$A_p(\bullet)$	area of projection
P_P, L_L, A_A, V_V	point, line, area, and volume fraction
L_A	line length per unit of area
$\{\hat{e}_i\}$	lab coordinate basis
$\{\hat{e}_i^S\}$	sectioning coordinate frame
$\overline{P}_r^t(\cos\nu)$	normalized associated Legendre functions
σ_f^2	variance of function, f
$\theta(x) = \begin{cases} 1 \text{ if } x \in \Omega \\ 0 \text{ otherwise} \end{cases}$	localization function
$H(\Omega)$	caliper length of Ω
$\partial M(\overrightarrow{x}, \partial h)$	interface function

CHAPTER

Introduction

1

CHAPTER OUTLINE

Design is that mysterious process, at the same time human and divine, that conceives the shaping of material into objects and systems of clever functionality, useful in leveraging and enhancing human activity. *Microstructure-sensitive design for performance optimization* (MSDPO) describes a new component of design activity in which the specific requirements in properties and functionality of the materials are realized only in specific preferred microstructures. MSDPO requires bridges that cross over two distinct length scales—the macroscopic scale discerned by the natural eye in which specified material properties are required to meet the needs of the designer, and a microscopic scale of the microstructure that usually requires the assistance of microscopy to examine. It is here in the details of the microstructure that the material can be designed to meet the macroscale properties design requirements.

It is difficult to imagine what our world would be like without microstructure-designed products—jet aircrafts that transport people across continents in a matter of hours; computers and telecommunications systems that rapidly perform calculations, store data, and communicate vast quantities of information around the world and into the solar system; modern pharmaceuticals specifically designed to arrest disease and improve health in plants, animals, and humans. In these and many other examples, matter is organized in particular ways that provide the designer with the properties and functionality essential for the conceived design. For example, in the hot section of jet engines nickel-based superalloys have been developed, containing precipitation-strengthening phases that are stable at high temperatures. These permit the designer to realize engines that operate at high levels of thermodynamic efficiency.

Shifting from aluminum to highly textured copper-based alloys in the metallic interconnects of computer chips facilitates smaller circuits with higher current densities and higher operating temperatures, suppressing the electromigration and void failures that limited the application of aluminum alloys. Secondary recrystallization has been exploited in iron-silicon alloys to obtain high levels of the {110}<001> Goss orientation that is ideal for the magnetic properties required in electric power transformers. Thin aluminum beverage cans became a technological reality when materials engineers learned to exactly balance the rolling and recrystallization texture components of the microstructure of the sheet product that is input into deep drawing and ironing operations. Thus, in each of these examples, and many others, particular characteristics of microstructure are sought that lead to combinations of properties that facilitate the desired functionality of the design.

As our ability to tailor materials to meet the functionality envisioned by designers increases, so does the range of possibilities and performance of designs. We say that the design space is "enhanced" or "opened up." For any augmentation of the design space that increases or improves functionality is welcome news to the designer. The purpose of this book is to introduce the engineer to rigorous methodology for specifically tailoring the microstructure of materials to meet the properties and functionality required by the designer. Where this is achieved, the design process transcends the traditional materials selection component of design by introducing material microstructure as a design variable.

1.1 CLASSIC MICROSTRUCTURE–PROPERTIES RELATIONSHIPS

Futurists envision the day when individual atoms might be moved into position, constructing, atom by atom, materials of specified chemistry, molecular arrangement, and internal structure, to achieve desirable combinations of properties and functionality, much like an architect would design a building one beam at a time. We are very far from realizing this futuristic vision in at least two ways. First, the concept of moving individual atoms into prescribed locations faces many practical limitations—one cannot simply push atoms around by mechanical devices.

The second limitation is in terms of predicting macroscopic behavior of materials on the basis of atomic-scale theory. Quantum theory is currently the best available physical theory of a solid state, but if one imagines the system of $3N$ wave equations that must be solved for a sample containing N electrons, each with three degrees of freedom (adding the protons and neutrons requires additional equations), one rapidly comes to the conclusion that modern computational resources fall far short of what would be required to simulate at the atomic level the properties of even a very small macroscopic system consisting of, for example, one mole of atoms. If, however, classic nineteenth-century physics can be used to model material properties, then significant progress is made toward the goal of designing microstructures suitable for specific designs.

The methodology of focus in this book embraces classic (pre–quantum mechanics) microstructure–properties theory. The basic building blocks of materials are taken to be small grains or crystallites, or small regions of homogeneous material phase. It is presumed that the local physical laws governing the properties of these building blocks are known, by previous experimental determination. Perhaps they are established by simulation with reliable theories at finer-length scales. In this context, local properties relationships can be thought of as the averaged or homogenized behavior of quantum mechanical relations that govern the atomic-scale behavior. The oscillations of that behavior over distance and time scales appropriate for atomic simulations are no longer present in the classic relations. For many properties of interest, atomic-scale physics can be ignored, and models incorporating mesoscale building blocks can be considered. Other properties may indeed require a quantum-level treatment, where averaging would not be appropriate. Microstructure at this intermediate- or mesoscopic-length scale (*mesoscale*) refers to the volume fractions, the shapes, sizes, and spatial arrangements of these building blocks that form an aggregate or composite structure.

Furthermore, for a theory of materials to be of practical use in engineering, the material states that govern the properties, as described in the theory, must be readily observable. With the advent of modern microscopy, the states of materials at the mesoscale are retrievable in a straightforward manner. An orientation imaging microscopy overview, which is the main tool for study of mesoscale microstructures in crystalline materials, is presented in Chapter 16. The data provided by such techniques has formed the experimental basis for much of the methodology described in this book.

It is certain that the approximations embodied in a mesoscale theory limit its range of applicability. Not all observable phenomena can be modeled. For example, a mesoscale theory may be inappropriate for modeling behaviors that are sensitive to the details of the spatial fields of point and line defects (dislocations). Limitations of the mesoscale theory, sometimes called the *generalized theory of composites*, are understandable in terms of a fundamental requirement that there exist three distinct length scales. The finest scale, say l_1, is associated with the largest size of homogeneous phase present in the structure. Homogeneous phases at this fine scale must be discernible to the microscopic probe. It is the myriad details of sizes, shapes, and spatial arrangement of these constituent homogeneous phases that constitute the microstructure of the material.

At an intermediate-length scale one defines l_2 to be the length at which the composite microstructure is statistically homogeneous. Sometimes referred to loosely as the representative volume element, regions of size l_2 contain sufficient grains or crystallites in their many configurations to be statistically representative of the material of interest. It is at this intermediate-length scale that physical theory and modeling take place.

The largest-length scale of interest, l_3, is characteristic of the macroscale, and must be chosen to be less than the defining dimensions of the part of interest. It is often at this macroscale that the properties and characteristics of the material are of interest to the designer. Modern theory of composites, based on statistical continuum theory, requires that the micro-, meso-, and macroscales must be well separated:

$$l_1 \ll l_2 \ll l_3 \tag{1.1}$$

Comprehensive treatment of these requirements in conjunction with the theory of composites is presented in other works (Milton, 2002; Torquato, 2002).

Some examples of material microstructure that are amenable to the generalized composite theory include polycrystalline metal and ceramic alloys, aligned and chopped fiber composites, particulate composites, interpenetrating multiphase composites, cellular materials, concrete, block copolymers, ice, stone, bone, soil, and many others. Figures 1.1 and 1.2 show some common examples of the microstructures of materials that have been modeled this way.

1.2 MICROSTRUCTURE-SENSITIVE DESIGN FOR PERFORMANCE OPTIMIZATION

Microstructure-sensitive design for performance optimization embodies a methodology for the systematic design of material microstructure to meet the requirements of design in optimal ways. It

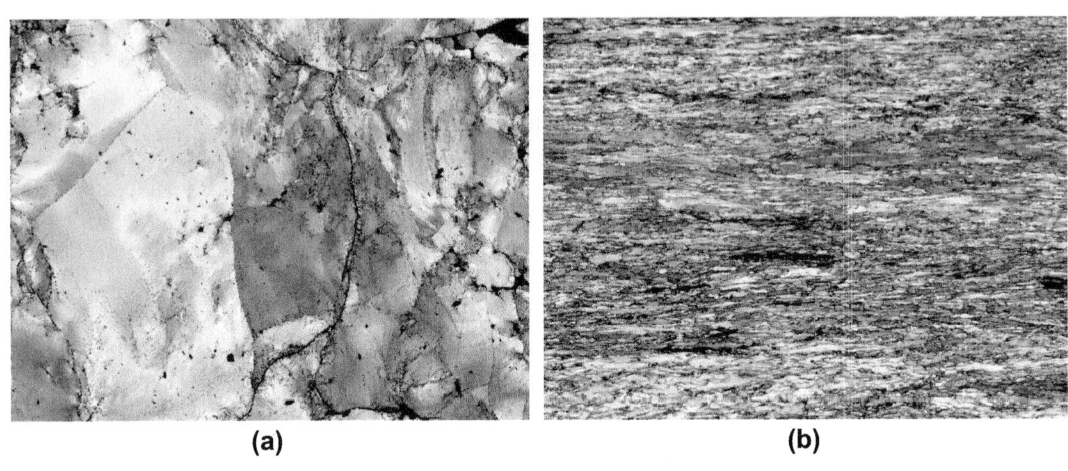

FIGURE 1.1

Images from orientation imaging microscopy: (a) iron with evidence of deformation and (b) rolled copper. *Images courtesy of EDAX-TSL.*

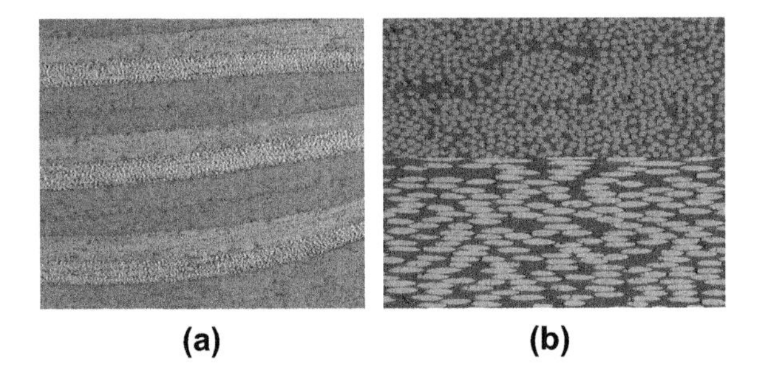

FIGURE 1.2

Typical laminate carbon epoxy lay-up showing (a) layers of fibers with different orientations and (b) close-ups of two layers. The filament diameter is approximately 7 μm.

embraces generalized composite theory as the means of relating macroscopic effective properties to microstructure. The methodology of MSDPO is conducted in selected Fourier spaces that facilitate two important constructs—the *microstructure hull* and the *properties closure.* It is important to understand the significance of these two central ideas, which are not presently included in standard works in materials science and engineering.

The microstructure hull is the set of all possible microstructures that could exist in the material of interest. Readers might imagine a material comprising only two distinct material phases, labeled α and β. Suppose these phases fill the region of three-dimensional (3D) space occupied by the material sample. One could imagine a microstructure in which the α-phase component occurs as spherical particles embedded in a matrix of the β phase. A hierarchy of descriptions of this microstructure can be considered. The simplest is just the volume fraction of α versus β phases. This description requires only one number: the volume fraction of the α phase in the sample. Volume fraction tells us nothing about how the two phases are spatially distributed in the sample. A more complex description might add to the volume fraction a description of the distribution of radii of the embedded spherical phase. This additional information about the microstructure would require the specification of a single-parameter distribution function describing the frequency of occurrence of the spherical α phase as a function of the sphere radius. This is substantially more demanding than just describing the volume fraction of the α phase.

To this could be added information about the relative placement of the centers of the spherical particles. This information requires correlation functions, which are functions of vectors; these are much more complex than the distribution function of the sphere radius. If this process is continued, one eventually arrives at a comprehensive description of this microstructure wherein the phase is specified at each point within the microstructure of the sample. This extensive description is called the *microstructure function*, and we will return to it later in the book. Obviously, this detailed description of the placement of the phase at each point in the sample requires a very large amount of information. And it would be limited in the sense that it would describe only one microstructure—only one element of the set of all possible microstructures that comprise the microstructure hull.

The most important question that must be asked is: How detailed must our description of material microstructure be to make reliable and accurate predictions of material properties? Fortunately, physical theory often provides at least a partial answer to this question; and we discover that the more precisely we must predict the properties from physical theory, the more we must know about the microstructure. If it is sufficient within our physical theory for properties to know only the volume fraction of the α phase in our imagined material, then the unit line interval [0,1] contains all possible volume fractions—this is the (limited) microstructure hull for this model approximation of the properties. If additional information about spatial placement of the phases is important to an accurate prediction of properties, then the microstructure hull can become much more complex, as we will see later. The important thing to remember about the (limited) microstructure hull is that it is the set of all possible microstructures within a selected physical microstructure–properties theory. If the physical theory requires only volume fractions, then in the two-phase example the unit line interval [0,1] is a proper mathematical model for the (limited) microstructure hull, since the real numbers lying upon the unit line interval can be taken to describe all possible volume fractions of the microstructure.

MSDPO requires microstructure hulls of significantly greater complexity than the unit line interval described here. The degree of complexity of the hull is a function of the type of material of interest, and the specific requirements of the physical theory required to predict properties. All of these hulls will be *limited hulls* because they will consider only partial information about the microstructure. Hereafter, we will not use the word *limited*, but readers should understand that all of the hulls we can consider offer only an incomplete description of the microstructure, and are therefore limited.

We shall find that the microstructure hull is always a convex geometrical body in a Euclidean representation space. This representation space may have only one dimension, as in the case of the unit line interval, or it may have many dimensions, as required by the physical theory. The set of points lying within and on the surface of the microstructure hull represent all possible (physically realizable) microstructures pertinent to the selected mesoscale property physics. (For the example of the unit line interval, any two points in the interval define a line on a subinterval contained within the unit interval; every point in the subinterval lies within the original hull, and therefore, by definition, the unit line interval is convex. The endpoints of the interval, 0 and 1, are the boundary of the hull of volume fractions specified by the unit interval.) The fact that the microstructure hull is convex is very important to applications; it implies that the hull can be readily searched for optimal solutions to design problems.

We will see that the mesoscale physics embodied in generalized composite theory for predicting properties can be represented as families of property surfaces in the Euclidean space (flat and curved) that intersects the microstructure hull. (In the example of the unit line interval hull, these "surfaces" are simply individual points lying in the line interval.) Loosely speaking, all points lying on that part of a property surface that intersects the microstructure hull are predicted by the theory to have the same property value. (In the example of the unit line interval hull, all microstructures having the same volume fraction, represented by the same point on the line interval, are predicted to have the same property.)

This is an important idea—namely that in most instances more than one microstructure is predicted to give rise to the same property. If specified property p is associated with a particular surface in the representation space, then property $p + \Delta p$ will belong to a nearby "parallel" surface in the representation space. And if we consider all possible properties predicted by the physical theory, these will belong to the *family* of surfaces associated with the property of interest. We must always remember, however, that only those portions of the property surfaces that intersect the microstructure hull in the representation space will be associated with real, physical microstructures. For two properties of interest, say p and q, each associated with its specific family of surfaces, it is the locus of intersection of pairs of surfaces with the microstructure hull that identifies microstructures predicted to have specific combinations of two properties (p, q). The same principles apply to combinations of more than two properties. By exploring all possible combinations of specified properties allowed by these intersections in the representation space, the *properties closure* is recovered.

Properties closure constitutes a powerful tool for the design of microstructures. Just knowing the range of all theoretically possible combinations of properties for a particular material can enable the designer to entertain new and novel design solutions using that material. The designer can implement a search over the properties closure, using various strategies that will be described later in this book, to find the set of properties that would be optimal for the design. These particular combinations of properties can then be readily mapped back into the microstructure hull to find the class of microstructures that best meets the design requirements. We use the term *class of microstructures* here because all of our microstructure hulls will be limited hulls, and a single point in the hull represents all microstructures that exhibit the same limited features identified by the coordinates of that point in the hull. *Back-mapping* is an important component of MSDPO. Microstructure design is fundamentally an inverse problem. The designer seeks material microstructures that are predicted to realize specified combinations of properties. Thus, back-mapping from the properties closure to the microstructure hull lies at the heart of MSDPO.

Realization of a particular material microstructure design, through back-mapping, is the final challenge of the problem. Conventional processes for manufacturing the material may or may not impress microstructures of the type identified by the back-mapping. Manufacturing constraints and the costs of processing must be considered. For various practical reasons it may be infeasible to impress/manufacture the optimal microstructure within the material. When this situation occurs it could be advantageous to restrict the MSDPO methodology to consider only processes and microstructures that are readily available. Applying restrictions to the processing/manufacture of materials microstructure is an additional challenge to MSDPO implementation. However, when these constraints are implemented in the methodology there will still be an opportunity to explore new and rich design concepts, with the additional degrees of freedom afforded by considering material microstructure to be an additional set of design variables.

1.3 ILLUSTRATION OF THE MAIN CONSTRUCTS OF MSDPO

For the purposes of illustration we will demonstrate the main ideas of MSDPO by considering the problem of designing microstructures comprising two distinct phases in an imaginary two-dimensional (2D) space. As in the previous section, the two phases are labeled α and β. Our focus will be on elastic properties, and it will be assumed that both phases are elastically isotropic. We will illustrate the main steps of MSDPO methodology in the sections that follow, in terms of a first-order treatment, involving only volume (area) fractions; subsequently, we will briefly touch on second-order analysis that brings the morphology of the microstructure into the problem. In this text, the term *first-order* generally refers to microstructure properties relations that depend only on volume fractions of the microstructure's constituents. An example of a two-phase material is shown in Figure 1.3 (along with an orientation map of the same sample to illustrate its multistate nature from an orientation perspective); readers should be aware that individual phases of typical materials are usually not elastically isotropic.

1.3.1 Identification of principal properties and candidate materials

The first step in the materials design process is to identify the *principal properties* of interest for a specific application (i.e., yield strength, ductility, thermal/electrical conductivity,) and one or more *candidate materials* that may achieve them. The number of potential suitable materials for the

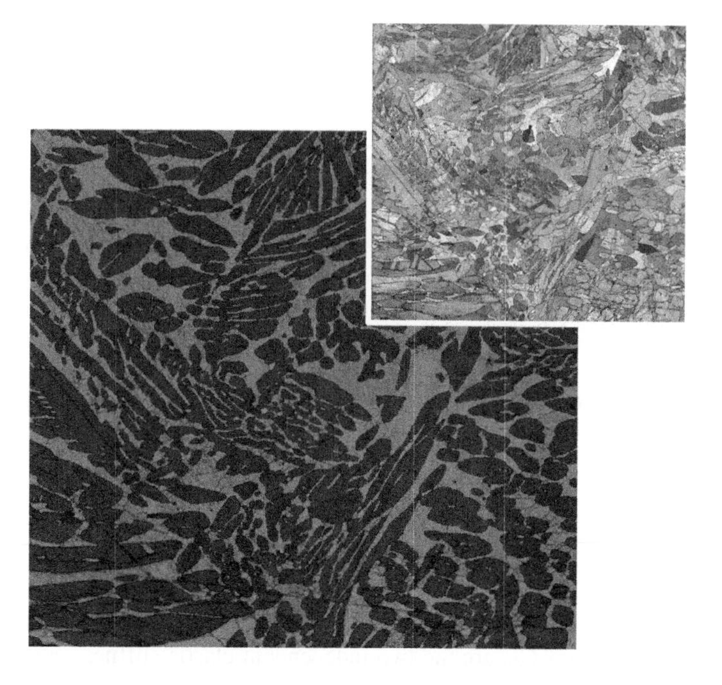

FIGURE 1.3

Two-phase stainless steel, showing phases and local crystal orientations (*inset*) of the same sample. *Images courtesy of EDAX-TSL.*

application is inversely related to the number of properties that must be achieved by the material. An intensely specific and demanding property specification often results in greater difficulty in identifying a suitable material, imposing greater limitations on design options. Because of this, in choosing the list of principal properties the relative importance of each property must be considered. If the initial properties list results in an overrestricted or nonexistent set of materials, less rigorous property requirements could be employed according to priority.

Once the principal properties of interest have been determined, candidate materials (phases) may be selected that could exhibit the desired combination of properties. Sometimes, when a particularly difficult set of properties is not amenable to a single material phase, composites of two or more phases can be considered. Often composites demonstrate properties that are intermediate between those of the constituent phases; thus, in combination these phases might achieve the desired combination of properties. Of course, not all combinations of materials are chemically suited to be combined to form composites; in some cases, however, composites have been constructed that exhibit properties that neither constituent phase possesses individually. In selecting potential materials, a greater number of candidate materials result in a higher likelihood of finding appropriate options and greater design freedom. However, if the possible options for a design become burdensomely many, it may be advantageous to augment the list of principal property requirements, thereby limiting the number of candidate materials. Often this can be accomplished simply by adding material cost to the list of principal properties.

For the purposes of illustration we suppose that our design calls for a thin membrane, extensive in the x-y plane but thin in the z-direction. The mechanical loading of this membrane is taken to be in-plane. Furthermore, we assume that it has been determined that the membrane must be constructed of a material possessing an elastic tensile stiffness in the x-direction of $C^*_{11} = 140$ GPa, and an in-plane shear stiffness of $C^*_{44} = 70$ GPa. A brief description of the technical meaning of these elastic properties is provided next.

Isotropic elastic properties modeling

In the case of elastic properties, suitable models have been established from which a limited set of local state variables can be related to the local elastic properties of the phases. Typically, the local state variables are the phase itself, which might be distinguished from other phases by its chemistry, and, for crystalline phases, the orientation of its crystal lattice. For noncrystalline phases lattice orientation is not required; here we assume that our two phases are noncrystalline and therefore only phase identity will be important.

Isotropic elastic properties must obey the following form of Hooke's law in 2D plane-stress problems:

$$\begin{bmatrix} \sigma_{11} \\ \sigma_{22} \\ \sigma_{12} \end{bmatrix} = \begin{bmatrix} C_{11} & (C_{11} - 2C_{44}) & 0 \\ (C_{11} - 2C_{44}) & C_{11} & 0 \\ 0 & 0 & C_{44} \end{bmatrix} \begin{bmatrix} \varepsilon_{11} \\ \varepsilon_{22} \\ \varepsilon_{12} \end{bmatrix} \tag{1.2}$$

Here σ_{ij} denotes the components of the 2-D Cauchy stress tensor, ε_{ij} are elements of the infinitesimal strain tensor in 2D, and C_{11} and C_{44} are the two independent elastic stiffness constants that must be determined for each of the two component phases. The subscripts i and j associated with the stress and strain components generally range from 1 to 3, but in this 2D example they range over 1 and 2 only. σ_{11} denotes the tensile stress (force per unit area) acting in the x-direction of the material on the y-z plane;

likewise, σ_{22} is the tensile stress acting in the y-direction on the x-z plane. σ_{12} is a shear stress acting in the x-direction on the x-z plane.

Typical units for the stress are MPa. Likewise, ε_{11} is the unitless strain (change in length normalized to the original length) of the material in the x-direction, ε_{22} is the strain in the y-direction, and ε_{12} is the (unitless) small half-change of angle between the x- and y-directions that occurs upon deformation. That the material must obey Hooke's law of the form of (1.2) is the first element of our physical theory. It is known from previous experiments that relation (1.2) is a good model that approximates, to a high degree of accuracy, the relationship between strain and stress for materials under quasistatic loading conditions. The coefficients in the matrix linking stress and strain are called *elastic stiffness*, and have units typically of GPa. Since each phase is taken to be isotropic, the form of Hooke's law shown in Eq. (1.2) is completely independent of our choice of coordinate frame. Whatever choice of x- and y-directions is selected in the membrane material, the form of Hooke's law remains the same.

The inverse form of Hooke's law is described in terms of the elastic compliances (S_{11}, S_{44}), in the following form:

$$\begin{bmatrix} \varepsilon_{11} \\ \varepsilon_{22} \\ \varepsilon_{12} \end{bmatrix} = \begin{bmatrix} S_{11} & \frac{1}{2}(2S_{11} - S_{44}) & 0 \\ \frac{1}{2}(2S_{11} - S_{44}) & S_{11} & 0 \\ 0 & 0 & S_{44} \end{bmatrix} \begin{bmatrix} \sigma_{11} \\ \sigma_{22} \\ \sigma_{12} \end{bmatrix} \tag{1.3}$$

Obviously, the unit of the elastic compliances is GPa^{-1}. The following relationships between the elastic stiffness and compliance components can be verified. Thus, only two independent elastic constants are required to specify 2D elastic properties for isotropic phases:

$$S_{11} = \frac{C_{11} - C_{44}}{C_{44}(3C_{11} - 4C_{44})}, \quad S_{44} = \frac{1}{C_{44}} \tag{1.4}$$

Since our microstructure comprises just two phases, we can easily distinguish the elastic properties of each phase by a left superscript, or ${}^{\alpha}C_{11}$, ${}^{\alpha}C_{44}$, ${}^{\alpha}S_{11}$, ${}^{\alpha}S_{44}$ for the α phase and ${}^{\beta}C_{11}$, ${}^{\beta}C_{44}$, ${}^{\beta}S_{11}$, ${}^{\beta}S_{44}$ for the β phase. If a material point in the membrane is identified with either phase, then the local elastic behavior of the material at that point is known through relations (1.2–1.4) if the basic elastic constants of the constituent phases are known. Indeed, if we could cut out a tiny piece of the material of size ~l_1, containing the material point, and subject it to a controlled strain (stress), then the stress (strain) response of the material would obey relations (1.2) and (1.3). Note that the macroscopic or effective properties of the membrane, which has dimensions ~l_3, will be related to the volume fraction of each phase present in the membrane, and to other details of how the phases are spatially distributed therein over regions of dimension ~l_2.

Suppose that our examination of available materials data, with other factors taken into consideration, leads us to consider a mixture of chemically compatible α and β phases, with each phase possessing isotropic elastic properties. These material phases have been selected because of the natural range of their elastic properties. That is, take the elastic properties of the two phases to be ${}^{\alpha}C_{11} = 110$ GPa, ${}^{\alpha}C_{44} = 30$ GPa (pure α) and ${}^{\beta}C_{11} = 240$ GPa, ${}^{\beta}C_{44} = 120$ GPa (pure β). Note that the desired macroscopic tensile modulus is ordered according to ${}^{\alpha}C_{11} < C_{11}^{*} < {}^{\beta}C_{11}$, and the needed in-plane shear modulus is ordered according to ${}^{\alpha}C_{44} < C_{44}^{*} < {}^{\beta}C_{44}$. Thus, we may expect that some composite mixture of these two phases might exhibit the desired combination of elastic properties. On this basis we have identified a candidate material system for the membrane design.

Readers should note that it is not unusual to be concerned about properties where appropriate models are not yet available to facilitate a design process. For example, the principal properties list for a particular design application and metal alloy candidate might include a property like pitting corrosion. If the relationship between the susceptibility to localized corrosion and measurable local state variables is unknown, then we are not in a position to include pitting corrosion as part of the materials design process. Examples of measurable local state variables, using modern methods of microscopy, include phase identity, chemistry, lattice orientation, and others; but microstructure comprises many features at differing length scales that may or may not be detected and quantified by our instruments. Thus, it is not surprising that some properties or behaviors of interest would not yet be linked to local state variables of the microstructure.

1.3.2 First-order homogenization relations

As previously indicated, our interest is in a thin material sample, or membrane. Suppose that this membrane is contained by a continuous rectangular region in 2D space, which we shall designate as Ω. Readers are invited to visualize region Ω and its microstructure like the red–blue (light–dark in the black-and-white version) mixture of phases in Figure 1.3. For our example, specifying microstructure within Ω is a matter of specifying local state, either phase α or β, at each material point $x \in \Omega$. Such a description ignores the atomic-scale disruption that occurs at the interfaces between the two phases that is known to exist. In real materials this disruption occurs over very small distances, typically not more than about 1 nm. For materials with a minimum phase feature size of $l_1 \gg 1$ nm, the influence of atomic disruption at interphase boundaries on elastic properties is negligible.

The next step of MSDPO requires us to obtain expressions for the effective (macroscopic) elastic properties, C^*_{ij}, S^*_{ij}, in terms of the known constants ${}^{\alpha}C_{ij}$, ${}^{\alpha}S_{ij}$ and ${}^{\beta}C_{ij}$, ${}^{\beta}S_{ij}$. Relationships of this kind are called *homogenization relations* because they "blend" together the properties of the constituent phases to obtain estimates of the effective properties. Among the earliest models for blending elastic properties were the Voigt and Reuss averages; these are of the form:

$$\begin{aligned} C^*_{ij} &= \left(1 - f_\beta\right) {}^{\alpha}C_{ij} + f_\beta {}^{\beta}C_{ij} = \langle C_{ij} \rangle \\ S^*_{ij} &= \left(1 - f_\beta\right) {}^{\alpha}S_{ij} + f_\beta {}^{\beta}S_{ij} = \langle S_{ij} \rangle \end{aligned} \tag{1.5}$$

where $f_\beta \in [0, 1]$ is the volume fraction (area fraction in the 2D case) associated with phase β, and $f_\alpha = (1 - f_\beta)$ is the fraction associated with phase α. The angular brackets $\langle \bullet \rangle$ are used to symbolize volume (area) averages. Evident in (1.5) is the express dependence of the effective properties on the volume fraction of the two phases present, f_β; no other characteristic of the microstructure is required for these models.

In selecting any homogenization relation it is always important to ask just how accurate and reliable the models will be in predicting the desired properties. An incomplete answer can be given here. For example, sophisticated mechanical considerations, to be derived in Chapter 8, will show that these simple volume averages are in reality *bounds* on the effective elastic properties for components lying on the diagonal of the elastic matrix:

$$C_{11}^{*} \le \langle C_{11} \rangle, \; C_{22}^{*} \le \langle C_{22} \rangle$$
$$S_{11}^{*} \le \langle S_{11} \rangle, \; S_{22}^{*} \le \langle S_{22} \rangle \tag{1.6}$$

These bounding relations are often called the Hill–Paul bounds, honoring the important work of two of the early pioneers in homogenization theory. Thus, the volume averages of terms lying on the diagonal of the stiffness and compliance matrices are greater than or equal to the effective values of the stiffness and compliance in the corresponding positions in the effective elastic matrices. The homogenization relations of (1.5) and (1.6) illustrate a second component of physical theory required by MSDPO. This component of modeling relates the microstructure of the material to macroscopic/effective properties.

Model relations for the off-diagonal effective elastic properties will be discussed in Chapter 8. They are significantly more complex than the bounding relations for the diagonal terms. Readers should note that the bounds discussed here only pertain to the diagonal components of the effective elastic stiffness and compliance relations. The off-diagonal parts of relations (1.5) and (1.6) are not available to us at this stage. Fortunately, for the design of the membrane currently under consideration, these off-diagonal terms are not required.

In Chapter 8, it will also be demonstrated that the two inequalities in relation (1.6) can be combined into a single expression of the form:

$$\langle S \rangle_{ii}^{-1} \le C_{ii}^{*} \le \langle C_{ii} \rangle \tag{1.7}$$

where $\langle S \rangle_{ii}^{-1}$ is the diagonal ii component of the inverse of the average compliance matrix. The salient effective elastic stiffness properties are, according to (1.7), bounded from above by the volume average stiffness of the two constituent phases, and from below by the inverse of the volume average compliance of the same. Combining Eqs. (1.4) and (1.7) we obtain

$$\begin{aligned}
&\frac{4A-B}{B(3A-B)} \le C_{11}^{*} \le (1-f_\beta)^{\alpha}C_{11} + f_\beta{}^{\beta}C_{11} \\
&\frac{1}{B} \le C_{44}^{*} \le (1-f_\beta)^{\alpha}C_{44} + f_\beta{}^{\beta}C_{44} \\
&A = \frac{(1-f_\beta)({}^{\alpha}C_{11} - {}^{\alpha}C_{44})}{{}^{\alpha}C_{44}(3{}^{\alpha}C_{11} - 4{}^{\alpha}C_{44})} + \frac{f_\beta({}^{\beta}C_{11} - {}^{\beta}C_{44})}{{}^{\beta}C_{44}(3{}^{\beta}C_{11} - 4{}^{\beta}C_{44})} \\
&B = \frac{(1-f_\beta)}{{}^{\alpha}C_{44}} + \frac{f_\beta}{{}^{\beta}C_{44}}
\end{aligned} \tag{1.8}$$

As an aid to visualize our theoretical results, let's plot the two bounds given by Eq. (1.8). Notice that the hypothetical values of the basic elastic properties of the two selected phases are represented by the values associated with the termini of the moon-shaped domains in Figure 1.4. The termini in the lower left of these figures are the properties of the α phase, corresponding to $f_\beta = 0$; the termini in the upper right of the diagrams indicate the properties of the β phase, with $f_\beta = 1$. The basic shape of these figures shows the functional form of the upper and lower bounds predicted for C_{11}^{*} and C_{44}^{*} as a function of f_β, the volume (area) fraction of phase β. The theory predicts that for any particular choice of f_β the upper line is

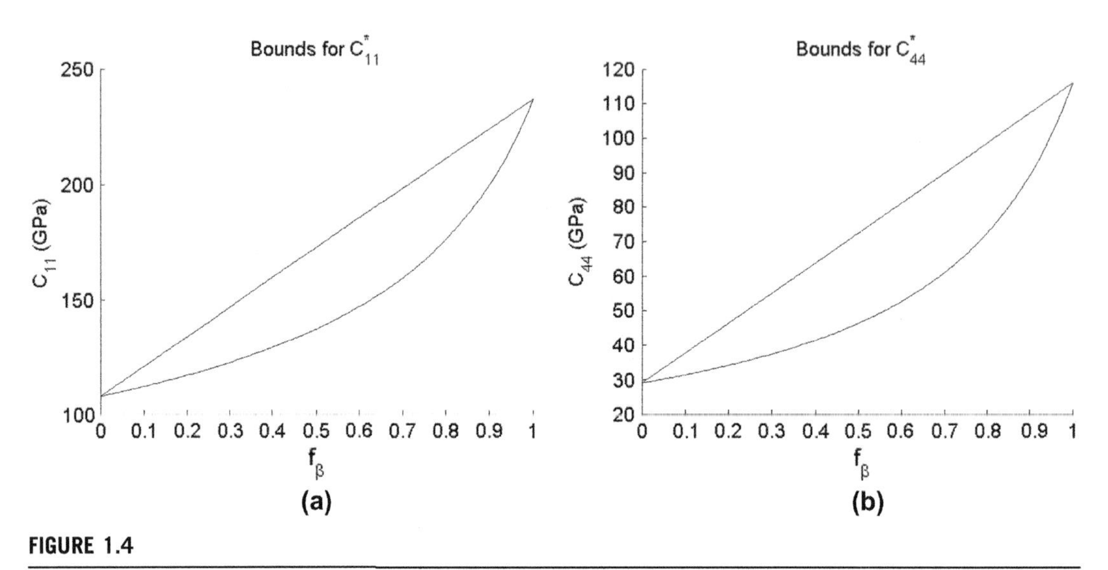

FIGURE 1.4

Plot of bounds for (a) C^*_{11} and (b) C^*_{44} versus fraction of β phase.

the largest value of the elastic stiffness that may be achieved, and the lower line is the lowest value. All possible properties lie between the two lines—hence the lines are called "bounds."

1.3.3 Microstructure hull

As illustrated in the previous section, the required microstructure representation can be very simple. Often, however, our homogenization models require as input one or more distribution functions of the microstructure. Quantification and visualization of these distributions' functions, as we will see later in the book, are achieved most conveniently by the method of Fourier series. The distribution function itself is approximated by a finite series of basis functions (the Fourier basis), weighted by appropriate coefficients. Selecting this Fourier basis is an important part of the third step in the design process. For our present purposes we can ignore the Fourier representation of microstructures, but we will return to it in later examples where more complex distribution functions of the microstructure are required.

Relation (1.8), illustrated in Figure 1.4, depends only on the volume fraction of the phases present in the microstructure. Thus, it is a very simple matter to identify the set of all possible microstructures that could exist within the region Ω. All possible microstructures are simply represented by all possible volume fractions f_β extending between pure α ($f_\beta = 0$), through all possible mixtures of the two phases, to the case of pure β phase ($f_\beta = 1$). This set of all microstructures is called the *microstructure hull*, and we will use the symbol M to represent it. For the simple homogenization relations pertinent to our problem, M is just the set of all possible f_β, or, mathematically,

$$M = \{f_\beta | f_\beta \in [0, 1]\} \tag{1.9}$$

Note that [0,1] here denotes the interval of real numbers lying between 0 and 1 (including both 0 and 1). The symbol $\in$ means "belongs to." This equation can be expressed in words as "M is the set of all possible volume fractions of β, f_β, such that this volume fraction lies in the real interval extending from 0 to 1." Clearly, we could have instead specified f_α on the same interval, since $f_\alpha + f_\beta = 1$.

1.3.4 Properties closure

The next step of MSDPO is to determine the *properties closure* from the homogenization relations. Readers should note that for each of the two physical theories (the lower and upper bounds) only one property is predicted for a specified f_β. For example, consider the upper-bound theory. For a specified effective property, C^*_{11} or C^*_{44}, only one volume fraction f_β is predicted to achieve that upper bound on the property. Now taking this upper-bound value for each volume fraction $f_\beta \in [0, 1]$, we obtain a single pair of properties, (C^*_{11}, C^*_{44}). By varying f_β, these pairs form a continuous curve (a straight line in this case) in the property space: C^*_{11} versus C^*_{44}. This curve, formed by either theory, is an example of a properties closure—it is a predicted set of possible properties based on theory.

Taking a somewhat more advanced approach that includes the additional physical principle of bounding, consider the range of properties, C^*_{11} and C^*_{44}, that are predicted by both bounds to be accessible upon fixing f_β. Fix a vertical line at f_β in Figure 1.4, and recover $C^*_{11}(f_\beta)^+$ and $C^*_{11}(f_\beta)^-$ and

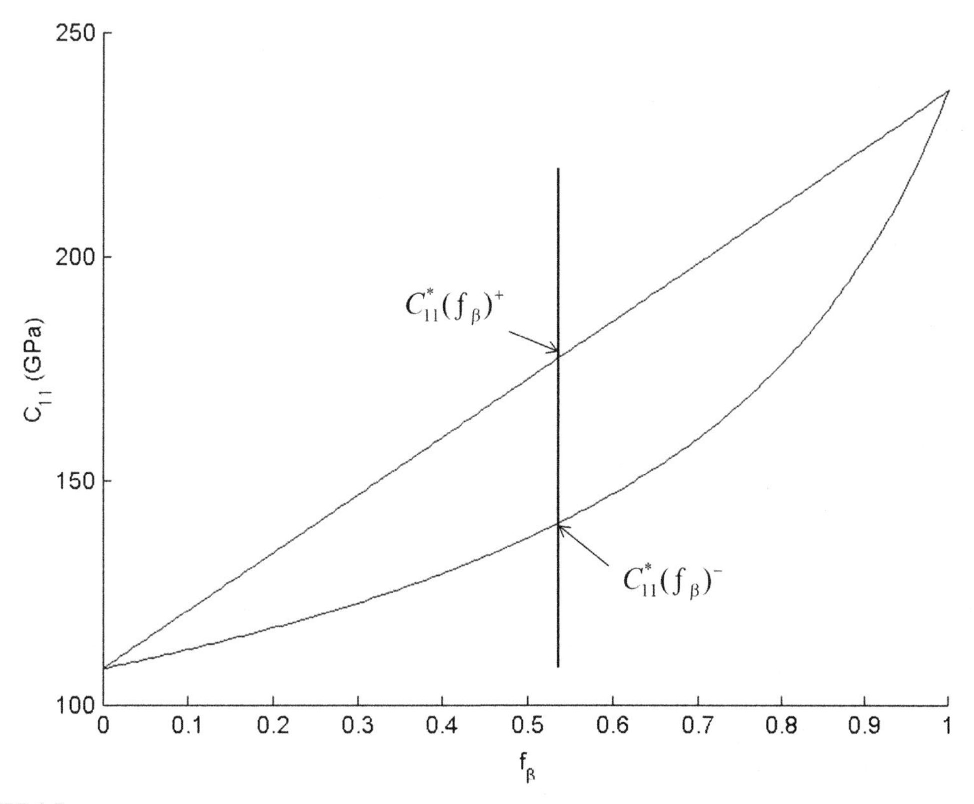

FIGURE 1.5

Recovery of the upper and lower bounds of C^*_{11} for a specified value of f_β.

$C^*_{44}(f_\beta)^+$ and $C^*_{44}(f_\beta)^-$ as illustrated in Figure 1.5 for C^*_{11}. The superscripts $+$, $-$ indicate the upper and lower bounds, respectively.

Then exercise the homogenization relations in this manner over the full microstructure hull to obtain the set of all theoretically possible properties combinations for C^*_{11} and C^*_{44}. This is a more complete approximation of the properties closure. More precisely, in the 2D properties space, C^*_{11} versus C^*_{44}, we plot the set of all pairs of points $(C^*_{44}(f_\beta)^+, C^*_{11}(f_\beta)^-)$ and $(C^*_{44}(f_\beta)^-, C^*_{11}(f_\beta)^+)$ that occur over the set of all $f_\beta \in [0, 1]$. This process is illustrated in Figure 1.6 for a selected set of f_β (0.1, 0.5, and 0.9). The recovered upper and lower bounds, at corresponding values of f_β, give rise to the "boxes" from which the elastic properties closure can be constructed, as shown in Figure 1.7. It is useful to note that these "boxes" get progressively smaller as $f_\beta \to 0$ or 1.

Note that the effective properties required by the design, $C^*_{11} = 140$ GPa and $C^*_{44} = 70$ GPa, lie very near the upper surface of the properties closure shown in Figure 1.7. This fact validates our decision to consider a mixture of these two particular phases.

Readers should note that when the upper and lower bounds are considered in this way the effective properties closure becomes a 2D region. When guided only by the Hill–Paul bounds we predict that all properties combinations lying within the closure *may be possible*; but we are not guaranteed that they can be realized. Only higher-order homogenization theory, incorporating some aspects of the morphology of the constituent phases, can make a more precise statement. For a specified f_β, the Hill–Paul theory tells us only that $(C^*_{11}(f_\beta)^- \leq C^*_{11} \leq C^*_{11}(f_\beta)^+)$, and $(C^*_{44}(f_\beta)^- \leq C^*_{44} \leq C^*_{44}(f_\beta)^+)$, which is the "box" in properties space that we have used to construct the closure. When f_β approaches either 0 from the right or 1 from the left on the unit interval, this properties box approaches a point—the point representing the properties of phase α or phase β, respectively. This is evident from the construction illustrated in Figure 1.7. Note that when f_β lies near ½, the box is quite large, which suggests large uncertainties in predicting the properties combinations. If greater precision of properties prediction is required at these intermediate volume fractions, then additional

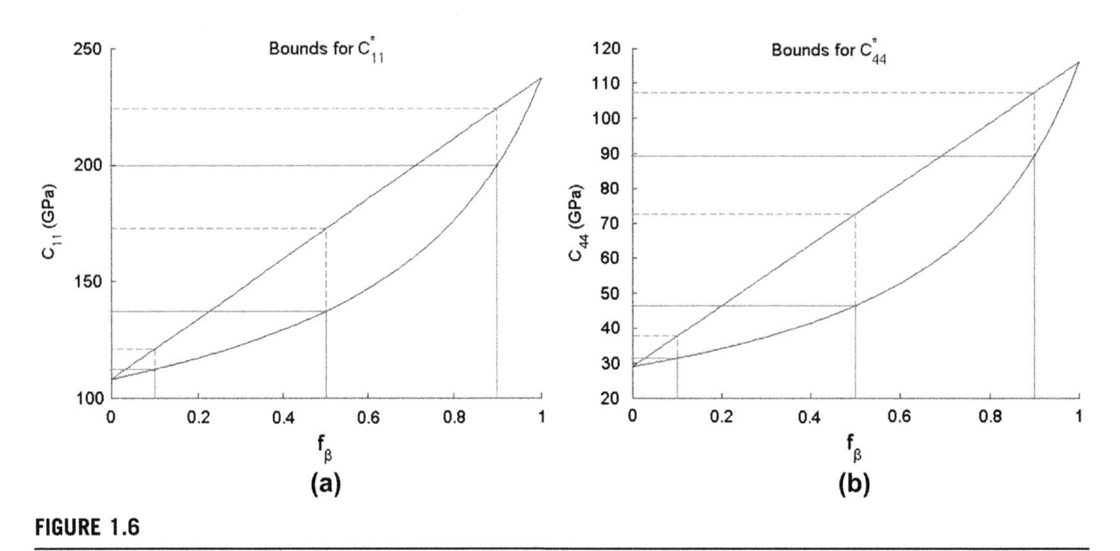

FIGURE 1.6

Recovery of upper bounds (*dashed lines*) and lower bounds (*solid lines*) for (a) C^*_{11} and (b) C^*_{44} at selected f_β.

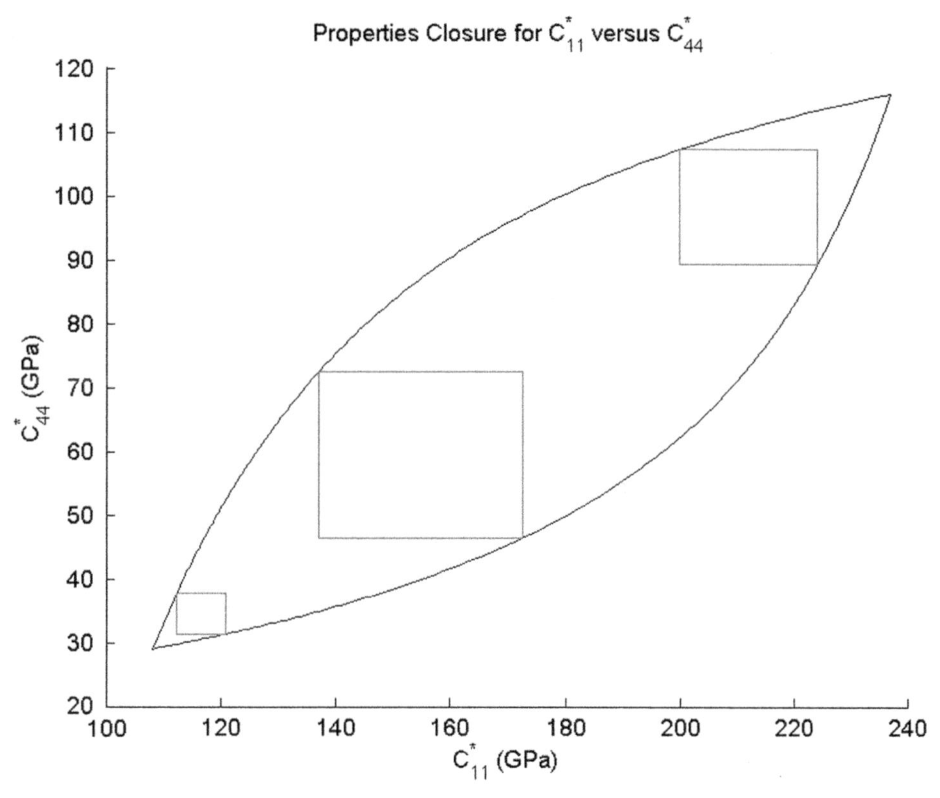

FIGURE 1.7

Recovered properties closure for C_{11}^{*} versus C_{44}^{*}. Each box is associated with a particular choice of f_{β}, corresponding to those presented in Figure 1.6.

information about the microstructure must be specified. But this topic requires higher-order homogenization theory.

1.3.5 Back-mapping to discover optimized microstructures

The final step of MSDPO is to apply *back-mapping to discover optimized microstructures*. The properties closure can be regarded as the principal result of the MSDPO methodology. It is this closure that is provided to the designer. It consists of the predicted properties combinations allowed by all theoretically possible microstructures (i.e., belonging to the microstructure hull). However, after the designer has used the properties closure in conjunction with the design, it is necessary to recover the class or type of microstructures that are predicted to achieve the properties. For our simple example, plots of the bounding relations as a function of f_{β}, such as seen earlier in Figure 1.4, may be directly used to arrive at appropriate choices for f_{β}, given the properties requirements.

To illustrate this, just ask the question: What range of f_{β}, between the upper- and lower-bound theories, is predicted to achieve $C_{11}^{*} = 140$ GPa? We can calculate this range to be $0.23 \leq f_{\beta} \leq 0.52$

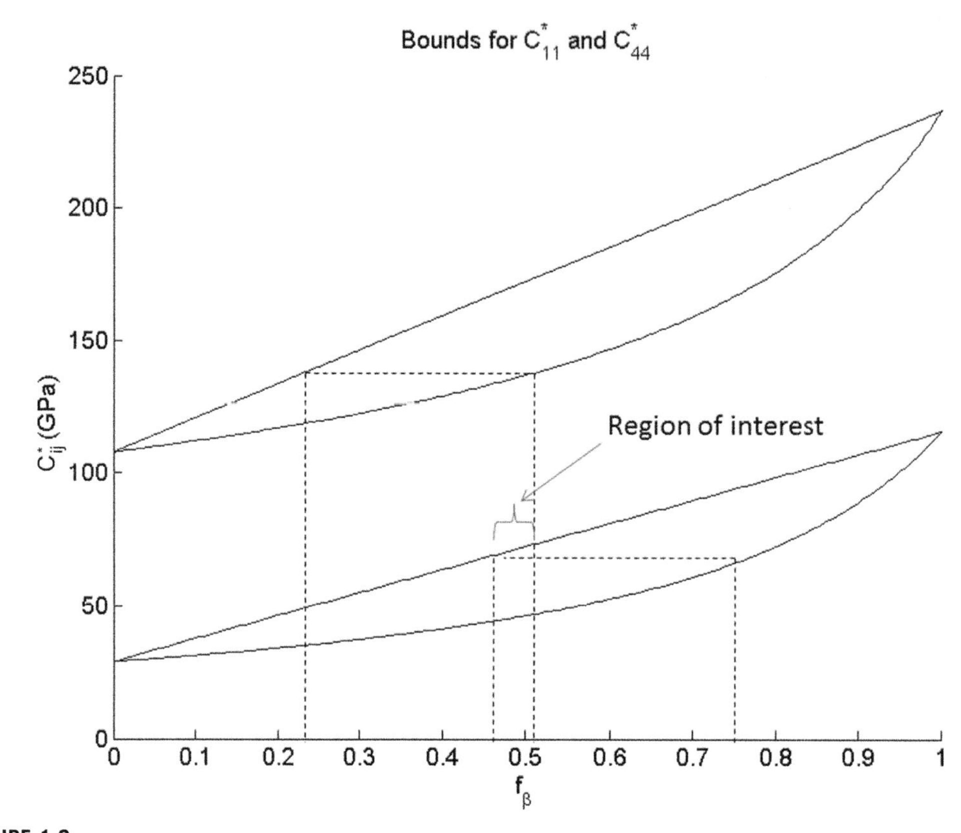

FIGURE 1.8

Demonstration of the back-mapping process. The interval is the predicted class or set of microstructures that may achieve the desired properties.

by solving the bounding expression of Eq. (1.8) for f_β, setting $C^*_{11} = 140$ GPa. Solving the upper-bound equation recovers $f_\beta = 0.23$, and solving the lower-bound relation gives $f_\beta = 0.52$. Our theory says that any f_β lying between these two extremes might be able to deliver $C^*_{11} = 140$ GPa. Similarly, we ask: What range of f_β is predicted to achieve $C^*_{44} = 70$ GPa? And we find $0.44 \leq f_\beta \leq 0.76$ by solving the bounding expression of Eq. (1.8) for f_β, setting $C^*_{44} = 70$ GPa. Note that these ranges of f_β overlap over the interval $0.44 \leq f_\beta \leq 0.52$. The process for finding this overlap is illustrated in Figure 1.8. In this overlapping region we expect the volume fraction to be such that we *may* or *might* achieve the desired combination of effective properties.

If no overlap were to occur then the theory would predict that no physically possible microstructure would give the desired properties combination. This latter statement is stronger than the first, because exclusion from the intervals dictated by the first-order bounds asserts that the effective properties cannot be obtained, no matter what the additional features of microstructure might be specified. If, as is the case here, a class of microstructures is predicted to exist such that the desired combination of properties *may exist*, we do not have a guarantee that a real microstructure *does exist* that will achieve the desired properties. If we must know with greater certainty that our set of recovered microstructures will in fact achieve the required properties, then we must construct examples of the microstructures predicted to achieve them and, by testing, verify they are achieved. Alternatively, we must implement a higher-order homogenization theory that will be more accurate and reliable. Higher-order homogenization will generally require information about the microstructure beyond the simple volume fraction information used in these basic Hill–Paul theories for elasticity.

This procedure illustrates the process of back-mapping. We began with the properties desired by the designer, and we back-mapped from those properties to the predicted optimal class of microstructures—in this case identified by the volume fraction parameter f_β. The next section illustrates how considerations of microstructure morphology can lead to additional understanding of homogenization.

1.3.6 Second-order homogenization relations

First-order methodology embraces only volume fraction information about microstructure, and it comprises the central focus of this book. More advanced treatments of homogenization indicate that a very broad set of local state spaces can be treated within the MSDPO framework. Information about the spatial placement of the component phases is not included in the first-order treatment. And yet it is known that significant differences in properties can be affected by alterations in the morphology of the constituents. A simple example of this is commonly treated in elementary discussions of composite systems. Consider two distinct (3D) placements of the elastically stiffer β phase, as shown in Figure 1.9.

It is known that the microstructure shown Figure 1.9(a) yields an effective C_{11}^* that would be close to the upper bound $C_{11}^*(f_\beta)^+$. On the other hand, the microstructure shown in Figure 1.9(b) yields an effective C_{11}^* that would be closer to the lower bound $C_{11}^*(f_\beta)^-$. Similarly, elementary morphologies that produce effective properties close to the bounds for the other elastic stiffness components can be established. Thus, we conclude that the bounds on effective elastic properties are only realized when the morphology of constituent phases is prescribed. This is the province of second-order MSDPO discussed in the later chapters of this book. Readers should also note that it is also true that even a partial description of microstructure morphology adds significantly to the dimensionality of the representation space, with commensurate increase in the numerical difficulties in implementing MSDPO methodology.

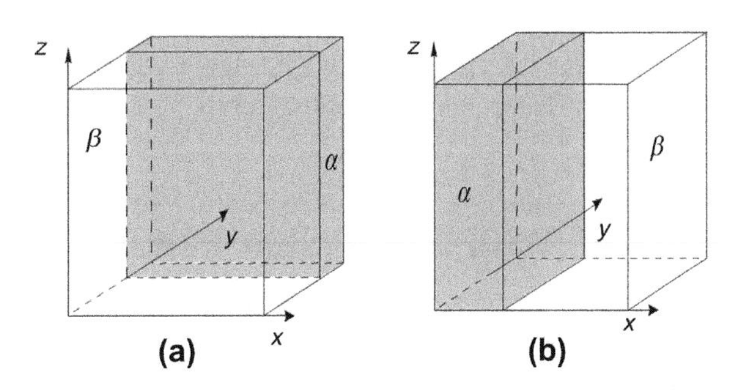

FIGURE 1.9

Sketches of the placement of phases α and β in a 3D region Ω in geometries that approximately give the (a) upper bound and (b) lower bound for C_{11}^*.

1.3.7 Summary of the MSDPO process

The simplified two-phase, 2D example we just described in this section has identified a simplified process for the design of a material microstructure to meet the various performance and properties demands of the design. The five steps are summarized briefly:

1. In the first step we must identify the *principal properties* of interest in the design, and the *candidate materials* that will be considered, as we strive to meet those properties. For the example given it was assumed that two particular elastic properties of a thin membrane component were of primary interest. Furthermore, a two-phase material system was selected as a candidate for the design on the basis that the required properties were found to lie in the range between the two components of the composite mixture. Further analysis led us to conclude that this was an acceptable choice, but had we not been successful with our first choice, we could have looked for other candidates.
2. Our second step is described as identifying *homogenization relations.* These take the local state variables of the constituent material phases, and their local properties, and couple them with quantitative representations of the microstructure to obtain predictions of the macroscopic or effective properties. Readers should always remember that usable homogenization relations do not exist for all of the properties that may be of interest. When this is the case, microstructure design can only consider those properties for which valid and rigorous homogenization relations exist.
3. To conduct MSDPO, having identified the appropriate homogenization relation(s), it is necessary to enumerate all possible microstructures that may occur. This set is called the *microstructure hull.* Although in the simplified example we chose this was a very simple set, isomorphic to the real line interval [0,1], we will see later in the book that it can become considerably more complex in more realistic design problems.
4. Having identified all possible microstructures pertinent to our chosen homogenization relation(s), the next step is to exercise those relationships over the entire microstructure hull to discover the set of all possible combinations of the principal properties. This set is called the *properties closure*; it becomes the main interface between the designer and the materials specialist.
5. Having identified the properties combination of interest in the design, the final step of MSDPO is *back-mapping to discover the optimal material microstructure.* Having identified the most desirable properties for the design, these properties can be inserted into the homogenization relationships where they can be solved to identify the most desirable microstructures—those that are predicted to achieve optimal properties combinations.

1.4 IMPLEMENTATION OF MSDPO IN DESIGN PRACTICE

MSDPO methodology offers distinctive advantages over conventional design practice. The treatment is fully *anisotropic* at both the macroscale and the microscale. Although much of design practice in the past depended on the assumption of isotropy of properties in materials, modern tools of design are not limited by this assumption. For example, most finite-element analysis tools permit the user to introduce anisotropic (effective) elastic and plastic properties. Also, readers are reminded that most natural and engineered materials in application today exhibit some degree of anisotropy; hence, design conducted with the assumption of isotropy can be costly, especially when the degree of anisotropy and its nature are not understood by the design team. MSDPO considers the full set of anisotropic properties combinations that are possible within the selected physical theory.

Two distinct levels of MSDPO can be applied in design. The simplest is the design of a single homogeneous microstructure that best meets the requirements of the design. Thus, for example, if a design embodies regions of material that are particularly vulnerable to failure by elevated stress concentrations, MSDPO enables the designer to find a class of microstructures that optimally mitigate against these stress concentrations at the local "hot spots." It is envisioned that the material to be used throughout the component will have this same microstructure. We call this *homogeneous design*. The objective is to define the class of material microstructure with anisotropic properties combinations predicted to best meet the most demanding requirements of the design. Homogeneous design is the primary focus of this book.

A second, more comprehensive design strategy permits the microstructure to vary from point to point in the component. This is called *heterogeneous design*. Thus, in each specific spatial region a distinctive microstructure is obtained that, in combination with the performance of all other regions of the component, is predicted to give rise to a global optimized solution. Figure 1.10 illustrates the fact that materials often demonstrate heterogeneity in the microstructure that may lead to heterogeneous distributions of properties. These characteristics may need to be taken into account whether we choose to actively pursue a heterogeneous design or not.

An example of where heterogeneous design can be very beneficial is shown in Figure 1.11; this type of orthoplanar-compliant mechanism has been used in pressure sensors. Note that each of the legs of the mechanism is composed of two compliant beams. Displacement of the central region of the mechanism (in or out) relative to its periphery will cause these compliant beams to flex. From

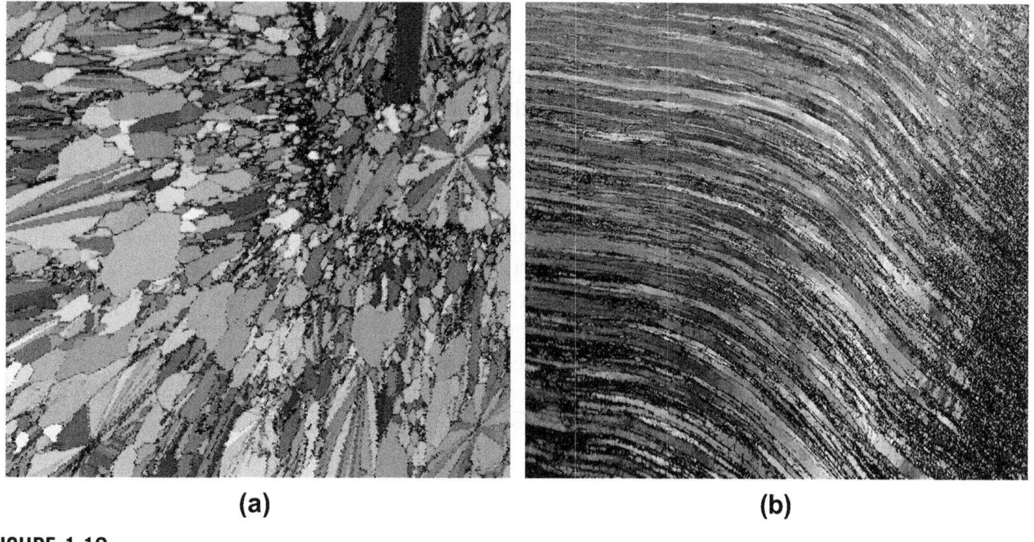

FIGURE 1.10

Polycrystalline silicon (a) and aluminum after friction stir welding (b), demonstrating significant natural heterogeneity. *Images courtesy of EDAX-TSL.*

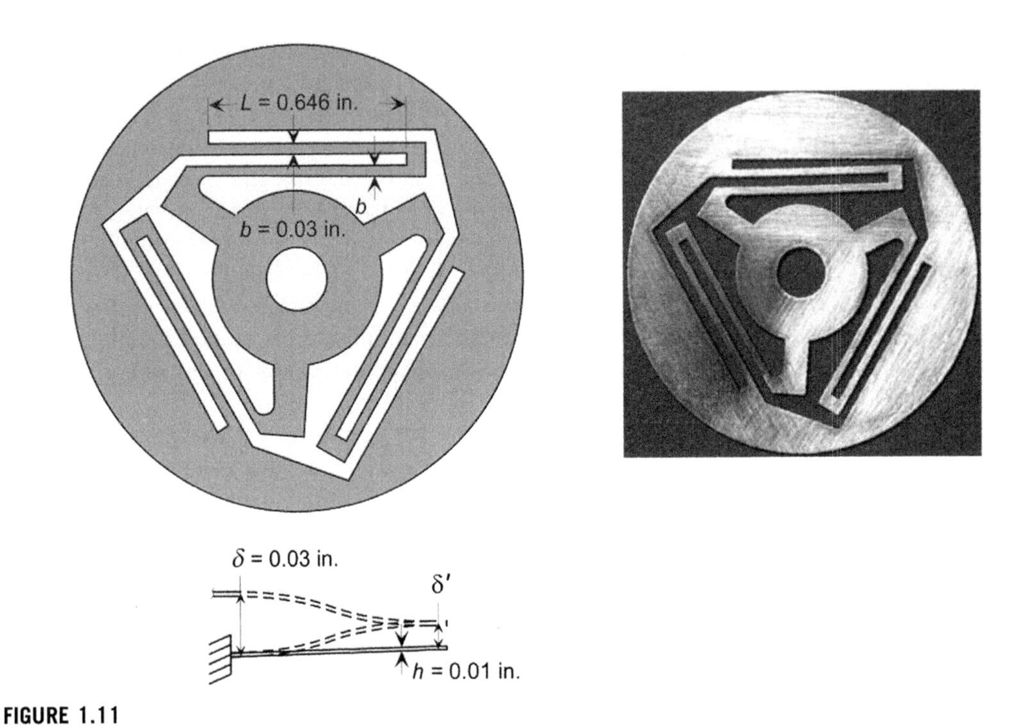

FIGURE 1.11

Orthoplanar spring mechanism. *Courtesy of L. Howell, Brigham Young University.*

a mechanical design perspective, the performance of these beams is limited, nominally, by the ratio of the yield strength to Young's modulus of elasticity of the material along the axis of the beam. Thus, in a design for the illustrated orthoplanar sample, three different directions of the compliant beams require consideration of yielding and stiffness properties in three directions.

At the post and base of the mechanism are certain additional requirements for elevated stiffness of the material, given that these must attach to the two sides of the sensor. These requirements obviously conflict with the high compliance that is optimal along the compliant beams. Thus, for this example heterogeneous design will be ideal if the heterogeneous microstructure (and the concomitant heterogeneous properties) can be realized by clever processing. The implementation of heterogeneous design optimization can also suggest other geometries that will result in improved performance. Heterogeneous design generally requires the repeated implementation of homogeneous design at several key locations in the component. Manufacturing constraints on heterogeneous design may, naturally, be significantly more pronounced than in homogeneous design.

1.5 THE CENTRAL CHALLENGE OF MSDPO

Posing the problem of microstructure design, as will be described in this book, has many advantages. Perhaps chief among these is the existence of a convex microstructure hull that can be searched in connection with the homogenization relations to find optimal microstructures and properties. But with this development there also comes a significant challenge. Although the hull of the elementary two-phase problem treated in Section 1.3 is very simple, most materials of practical interest require microstructure hulls with representations in many dimensions. And the number of dimensions can become very large. This high dimensionality is called the *large numbers problem.*

Often, this large number of dimensions precludes the use of an ordinary 3D geometrical depiction of the problem. Although results can be described in 2D or 3D projections, the large numbers involved demand a very large number of these projections to communicate graphically the solutions to the problems. From the standpoint of pedagogy, the large numbers problem poses a significant challenge to visualization of the MSDPO methodology. However, mathematically, the problem of searching the microstructure hull for the best solutions to a particular design problem is relatively straightforward. For this reason, the methods of MSDPO can be applied to a host of practical materials microstructures to achieve improvements in performance in highly constrained design problems.

1.6 ORGANIZATION OF THE BOOK

This book is organized into 16 chapters. Chapters 2 and 3 give essential mathematical and mechanics background in tensors and Fourier series methods. Quantitative representation of microstructure, using distribution functions and their Fourier approximations, are described in Chapters 4 and 5. Chapter 6 describes mathematical treatments to exploit known symmetry in the quantitative description of the microstructure functions. Chapter 7 begins the representation of properties. Our focus will be on mechanical and thermal properties, but these basic constructs can be extended to a larger set of properties. The formal treatment of first-order homogenization theory is introduced in Chapter 8; in this text the focus is entirely on volume fractions, with no consideration given to phase morphology. Methods for constructing microstructure hulls and properties closures are discussed in Chapter 9.

Linking these two constructs together to achieve the back-mapping required for microstructure design is discussed in Chapter 10 with several relevant case studies of the methodology. A brief description of microstructure evolution is presented in Chapter 11. This treatment will necessarily be quite limited, but it will introduce readers to the basic idea of tailoring processing to get at special microstructures with optimized properties and performance. Chapter 12 then introduces higher-order representation of structure, with a higher-order homogenization framework following in Chapter 13. Second-order hulls and closures are covered in Chapter 14, followed by a short section on higher-order methods as applied to processing in Chapter 15. Chapter 16 gives a brief introduction to the experimental methods and microscopy upon which the MSDPO methodology is dependent for its practical implementation.

SUMMARY

This chapter introduced the main steps involved in microstructure-sensitive design for performance optimization. It also introduced the novel concepts of microstructure hulls and properties closures that are central to MSDPO. Although the main steps involved in MSDPO are demonstrated in this chapter through a grossly simplified case study, it is hoped that readers have now gained a deep appreciation for the tremendous challenge that lies ahead. In particular, it is hoped that readers appreciate the difficulties associated with exploring the unimaginably large design space of potential material microstructures and with making invertible connections from this space to the properties and the processing spaces.

EXERCISES

1. **(short project)** Imagine that the 2D material described in this chapter is given a 3D nature by giving it a thickness t. Further, by rolling and joining, a tube is formed with diameter D. Assume that the joining does not alter the volume fractions of the two phases that comprise the microstructure, and assume that the microstructure does not vary from the inside to the outside of the tube wall. Consider an application that requires a tube of this material, of length L, to be loaded in tension in the axial direction of the tube (corresponding to the one direction in the material) with a force F, without increasing the length by more than ΔL. For this application, when loaded in torsion giving rise to a specified twist of $\gamma = D\theta/2L$, it is desired that the stored elastic energy be as little as possible. You can assume that the yield strength in tension for each phase of the composite is known (σ_y^α, σ_y^β), and that the composite yield strength will be well approximated by weighting the yield strengths of the phases by their volume fraction. You can also assume that the yield in shear is 50% of the yield in tension.

   Describe how you would proceed to design the microstructure of the material for this application. You can consider L, F, θ, C_{11}^α, C_{11}^β, C_{44}^α, C_{44}^β, σ_y^α, σ_y^β to be known. Clearly identify the objective for the design, and list each of the constraints. Describe how you would use the elastic closure to obtain a final design for the microstructure.

2. **(short essay)** Write a one-page essay about why the length scales must be well separated as described in relation (1.1.1). Your discussion should be based on information obtained from sources external to this chapter. You may wish to consider St. Venant's principle. (Be adventurous! Free up your mind to explore ideas that may prove to be unsound later. Be as creative as possible.)

CHAPTER

Tensors and Rotations

2

CHAPTER OUTLINE

In this chapter, we provide a concise review of the important mathematical concepts needed for later chapters. Readers who are already familiar with these basic concepts of tensors and their applications should be able to skip this chapter and follow the discussion in subsequent chapters without any difficulty. It is also not our intent to provide a comprehensive review of tensors in this chapter, but merely to collect together all of the tensor concepts needed for the later chapters and present them here in one place for the convenience of those readers who need a brief review. It is also assumed here that readers are already familiar with basic concepts in matrix algebra.

2.1 DEFINITIONS AND CONVENTIONS

The familiar 3D physical space is referred to as the *Euclidean 3-space*. Quantitative descriptions of physical phenomenon in this space often require selection of a reference point and reference directions, the combination of which is referred to as a *reference frame*. However, the phenomena themselves and the physical quantities involved are oblivious to the existence of the reference frame. For example, one may describe the location of a point A as being 5 m north and 6 m west of another point B. In this description, point B and the directions north and west constitute a reference frame, and the location of point A is uniquely described in this reference frame. However, point A exists completely independent of point B and the selection of the reference directions.

Furthermore, it is also evident that the selection of the reference frame is not unique. Since an infinite number of choices exist for the selection of a reference frame (both the reference point and the reference directions), it is necessary to develop a mathematical framework to reconcile the multiple descriptions of point A arising from all of the possible choices of a reference frame. This is accomplished by the use of the mathematical concept of *tensors*. All physical quantities that need to be expressed in the Euclidean space can be classified as tensors of different ranks, with the *rank* of the tensor identifying precisely the relationship between the multiple quantitative descriptions of the given physical quantity in any of the selected reference frames.

The simplest physical entities to describe are the tensors of rank zero, also referred to as *scalars*. The quantitative description of these physical quantities is completely independent of our choice for the reference frame. Energy, temperature, and density are all examples of physical quantities that can vary spatially and temporally in the Euclidean space, but can be represented as scalar functions.

Tensors of rank one are also called *vectors*. Vectors are defined as the difference between an ordered pair of points. Figure 2.1 shows schematically several points in the Euclidean space labeled as ***a*** through ***g***. As an example, a vector **v** has been defined as the difference of points b and a, that is, $\mathbf{v} = b - a$. Note that the definition of vector **v** does not need the specification of a reference frame. However, the full quantitative description of **v** requires the selection of a reference frame. Vectors may be used to describe physical quantities, such as displacement and its derivatives, with respect to time, such as velocity and acceleration. Note that derivatives with respect to time do not change the rank of a tensor because they are not affected by the choice of a reference frame in the Euclidean 3-space. A number of other physical quantities can then be defined as vectors by exploring their relationships to those that have already been previously defined as vectors. For example, force is classically defined as the product of mass and acceleration. Since mass is a scalar (which is itself defined as a product of two scalars—density and volume) and acceleration is a vector, force has to be defined as a vector.

Tensors of rank two and higher are often simply referred to as tensors. Higher-ranked tensors are needed to describe more complicated physical quantities. For example, one might need to describe a change in the shape of a given body. Figure 2.2 shows a schematic of the stretching of a circular object into an elliptical object. In this process, we have transformed a very large set of vectors that could be defined in the circular object to their new positions in the elliptical object. One simple way to visualize this is to consider a set of radial vectors in the circular object and see how they transform to the elliptical object, as shown in the figure. Clearly, the information we want to capture is significantly more complicated than can be captured by a simple vector. A convenient way of visualizing higher-ranked tensors is to treat them as linear transformation mappings from a lower-ranked tensor space into another lower-ranked tensor space. For example, a tensor of rank two can serve as a linear mapping of all the vectors (tensors of rank one) in the Euclidean space into itself. The mapping defined in Figure 2.2 can therefore be described as a tensor of rank two. Likewise, a fourth-rank tensor can be defined as a linear transformation mapping of all second-rank tensors in the Euclidean space into itself. These ideas will be explored in more detail in later chapters.

As noted earlier, it becomes necessary to define a reference frame to specify quantitatively any tensors of interest except scalars. In mechanics studies, a reference frame is often defined by

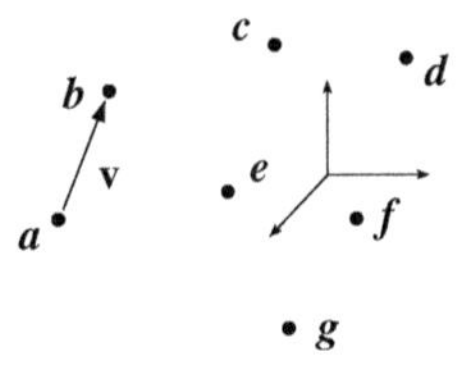

FIGURE 2.1

Schematic description of points and vectors in Euclidean 3-space. The *black* circles labeled **a** through **g** represent points and the *arrow* from **a** to **b** represents the vector $\mathbf{v}=\mathbf{b}-\mathbf{a}$.

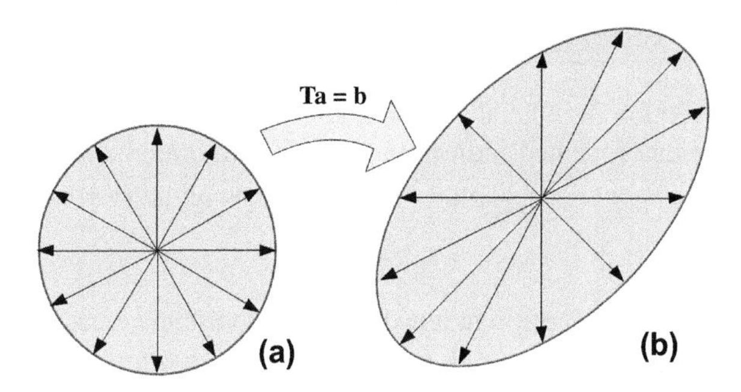

FIGURE 2.2

Schematic description of a tensor of rank two, **T**, as a linear transformation mapping of vectors in the Euclidean space into itself.

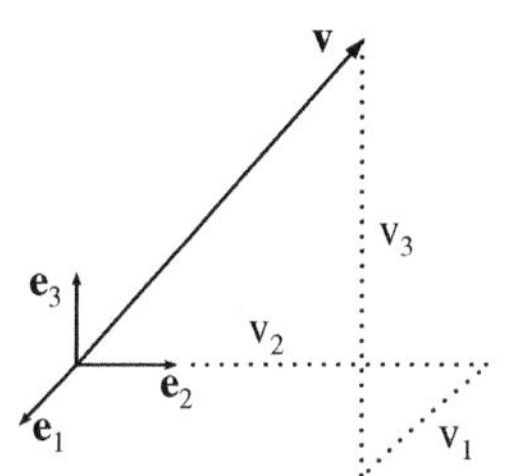

FIGURE 2.3

A Cartesian reference frame defined by a set of three orthonormal unit vectors $\{\mathbf{e}_1, \mathbf{e}_2, \mathbf{e}_3\}$, and the decomposition of a vector **v** into its components in this selected reference frame.

introducing a set of three orthonormal unit vectors $\{\mathbf{e}_1, \mathbf{e}_2, \mathbf{e}_3\}$, also called an *orthonormal basis*. The unit vectors in the basis are labeled following the right-hand convention (Figure 2.3). This definition yields the familiar rectangular Cartesian coordinate system and allows us to express any vector in the Euclidean 3-space as a weighted sum of the basis vectors. For example, vector **v** in Figure 2.3 can be expressed as

$$\mathbf{v} = \mathrm{v}_1\mathbf{e}_1 + \mathrm{v}_2\mathbf{e}_2 + \mathrm{v}_3\mathbf{e}_3 = \sum_{i=1}^{3} \mathrm{v}_i\mathbf{e}_i \tag{2.1}$$

The quantities v_1, v_2, and v_3 are referred to as the *components* of vector **v** in the selected Cartesian reference frame.

At this point, it is important to recognize some of the main conventions used with tensors. Note that all tensors of rank higher than zero are shown with bold font style (e.g., **v** in Eq. 2.1). The components of the tensor in a specific reference frame are scalars and are shown using regular (not bold) font style. We have also introduced *indicial notation* in Eq. (2.1). The subscript i in v_i may refer to any one of the components of **v**. Note also that we have used indicial notation in Eq. (2.1) to refer to the components of the orthonormal basis $\{\mathbf{e}_i\}$. It is important to recognize that while v_i refers to a scalar quantity, $\mathbf{e}_i$

refers to a vector quantity (note the bold style) and most important, $\mathbf{e}_i$ does not refer to the ith component of vector $\mathbf{e}$.

Another standard convention in indicial notation, known as the *summation convention* or Einstein summation convention, is that repeated indices in a single term automatically imply summation. In other words, one can drop the summation sign in Eq. (2.1) and simply express the equation as

$$\mathbf{v} = v_i \mathbf{e}_i \tag{2.2}$$

Finally, it is also a standard convention in tensor notation to use lowercase symbols for vectors and uppercase symbols for tensors of higher rank. All of these conventions will be used throughout this book.

2.2 TENSOR OPERATIONS

A variety of operations can be defined on tensors. Once again, our aim is not to provide a comprehensive treatment but to present a concise review of some of the essential tensor operations typically used in the study of mechanical behavior of crystalline solids. In understanding the various operations on tensors, detailed attention has to be given to the rank of the tensor and how that changes with the specific operation.

The most basic operation on tensors is the addition of tensors of equal rank. If $\mathbf{u}$ and $\mathbf{v}$ are vectors defined in the Euclidean 3-space, their addition is defined as

$$\mathbf{u} + \mathbf{v} = (u_i + v_i)\mathbf{e}_i \tag{2.3}$$

Equation (2.3) is also called the *parallelogram law of vector addition.* In fact, this operation was already invoked in our earlier discussion when we defined a vector as a sum of its components along the directions of the orthonormal basis (Eq. 2.1).

Next, we introduce the concept of a scalar product (also called the *dot product* or an *inner product*). The *scalar product* of two vectors, $\mathbf{u}$ and $\mathbf{v}$, is defined as

$$\alpha = \mathbf{u} \cdot \mathbf{v} = u_i v_i \tag{2.4}$$

Scalar products may be visualized as projections. For example, Eq. (2.1) can be recast using the concept of scalar product as

$$v_i = \mathbf{v} \cdot \mathbf{e}_i \tag{2.5}$$

In other words, the component of vector $\mathbf{v}$ along $\mathbf{e}_i$ may be visualized as the projection of vector $\mathbf{v}$ on the unit vector $\mathbf{e}_i$, and can be obtained by a simple scalar product of $\mathbf{v}$ with $\mathbf{e}_i$. The definition of the scalar product allows us to define the *magnitude* or the *norm* of a vector as

$$|\mathbf{u}| = \sqrt{\mathbf{u} \cdot \mathbf{u}} = (u_i u_i)^{1/2} \tag{2.6}$$

It is also sometimes convenient to think of the scalar product as

$$\mathbf{u} \cdot \mathbf{v} = |\mathbf{u}||\mathbf{v}|\cos\theta \tag{2.7}$$

where θ is the angle between the vectors **u** and **v**. The scalar product possesses the following properties:

$$\mathbf{v}\cdot\mathbf{u} = \mathbf{u}\cdot\mathbf{v} \tag{2.8}$$

$$(\alpha\mathbf{u}+\beta\mathbf{v})\cdot\mathbf{w} = \alpha\mathbf{u}\cdot\mathbf{w}+\beta\mathbf{v}\cdot\mathbf{w} \tag{2.9}$$

$$\mathbf{u}\cdot\mathbf{u} \geq 0 \tag{2.10}$$

$$\mathbf{u}\cdot\mathbf{u} = 0, \quad \textit{iff} \;\; \mathbf{u} = 0 \tag{2.11}$$

The orthonormal basis of unit vectors (the magnitude or norm of these vectors is unity) described in Section 2.1 has the following property:

$$\mathbf{e}_i\cdot\mathbf{e}_j = \begin{Bmatrix} 1 \;\; \textit{if} \;\; i=j \\ 0 \;\; \textit{if} \;\; i\neq j \end{Bmatrix} = \delta_{ij} \tag{2.12}$$

δ_{ij} defined in Eq. (2.12) is called the *Kronecker delta*. It is often used to operate on tensor components to extract the specific components of interest—for example, $v_j\delta_{ij} = v_i$ and $T_{ij}\delta_{jk} = T_{ik}$.

A *vector product* (also called a *cross product*) of two vectors **u** and **v** results in a vector and is defined as

$$\mathbf{u}\times\mathbf{v} = u_i v_j \in_{ijk} \mathbf{e}_k \tag{2.13}$$

where $\in_{ijk}$ is called the *permutation symbol* and is defined as

$$\in_{ijk} = \begin{cases} 1 \;\; \textit{if} \;\; (i,j,k) \text{ is an even permutation of } (1,\;2,\;3) \\ -1 \;\; \textit{if} \;\; (i,j,k) \text{ is an odd permutation of } (1,\;2,\;3) \\ = 0 \;\; \textit{if} \;\; (i,j,k) \text{ is not a permutation of } (1,\;2,\;3) \end{cases} \tag{2.14}$$

According to Eq. (2.14), $\in_{123} = \in_{231} = \in_{312} = 1$, $\in_{132} = \in_{321} = \in_{213} = -1$, while all other $\in_{ijk}$ are equal to 0.

Note that the scalar product resulted in a scalar outcome and the vector product resulted in a vector outcome. Now let us look into an operation between two vectors that will result in a second-rank tensor. However, we first need to establish a suitable representation for a second-rank tensor. We have seen that a vector can be represented by its components along the basis vectors of the reference frame, $\{\mathbf{e}_i\}$ (see Eq. 2.2). In Euclidean 3-space, this means that a vector is represented by a set of three components. Because we expect second-rank tensors to hold much more information than a vector, it stands to reason that it will have more components than a vector. The appropriate basis to describe second-rank tensors is denoted by $\{\mathbf{e}_i\otimes\mathbf{e}_j\}$, where $\otimes$ denotes a *dyadic product* (also called the *outer product*). In Euclidean 3-space, the total number of basis dyads is nine, since both indices i and j can take values 1, 2, or 3.

It is sometimes convenient to arrange the components of a vector as a column. For example, unit vectors $\{\mathbf{e}_i\}$ in Euclidean 3-space can be depicted as

$$\mathbf{e}_1 = \begin{Bmatrix} 1 \\ 0 \\ 0 \end{Bmatrix}, \quad \mathbf{e}_2 = \begin{Bmatrix} 0 \\ 1 \\ 0 \end{Bmatrix}, \quad \mathbf{e}_3 = \begin{Bmatrix} 0 \\ 0 \\ 1 \end{Bmatrix} \tag{2.15}$$

Extending this representation to dyads, one may depict all of the basis dyads in Euclidean 3-space as

$$\begin{aligned}
\mathbf{e}_1 \otimes \mathbf{e}_1 &= \begin{bmatrix} 1 & 0 & 0 \\ 0 & 0 & 0 \\ 0 & 0 & 0 \end{bmatrix}, \quad \mathbf{e}_1 \otimes \mathbf{e}_2 = \begin{bmatrix} 0 & 1 & 0 \\ 0 & 0 & 0 \\ 0 & 0 & 0 \end{bmatrix}, \quad \mathbf{e}_1 \otimes \mathbf{e}_3 = \begin{bmatrix} 0 & 0 & 1 \\ 0 & 0 & 0 \\ 0 & 0 & 0 \end{bmatrix} \\
\mathbf{e}_2 \otimes \mathbf{e}_1 &= \begin{bmatrix} 0 & 0 & 0 \\ 1 & 0 & 0 \\ 0 & 0 & 0 \end{bmatrix}, \quad \mathbf{e}_2 \otimes \mathbf{e}_2 = \begin{bmatrix} 0 & 0 & 0 \\ 0 & 1 & 0 \\ 0 & 0 & 0 \end{bmatrix}, \quad \mathbf{e}_2 \otimes \mathbf{e}_3 = \begin{bmatrix} 0 & 0 & 0 \\ 0 & 0 & 1 \\ 0 & 0 & 0 \end{bmatrix} \\
\mathbf{e}_3 \otimes \mathbf{e}_1 &= \begin{bmatrix} 0 & 0 & 0 \\ 0 & 0 & 0 \\ 1 & 0 & 0 \end{bmatrix}, \quad \mathbf{e}_3 \otimes \mathbf{e}_2 = \begin{bmatrix} 0 & 0 & 0 \\ 0 & 0 & 0 \\ 0 & 1 & 0 \end{bmatrix}, \quad \mathbf{e}_3 \otimes \mathbf{e}_3 = \begin{bmatrix} 0 & 0 & 0 \\ 0 & 0 & 0 \\ 0 & 0 & 1 \end{bmatrix}
\end{aligned} \tag{2.16}$$

Using these basis dyads, it is possible to define any second-rank tensor, **T**, as (note that summation is implied on repeated indices as per the conventions adopted earlier)

$$\mathbf{T} = T_{ij}\mathbf{e}_i \otimes \mathbf{e}_j \tag{2.17}$$

where T_{ij} denotes the components of the tensor in the coordinate reference frame defined by $\{\mathbf{e}_i\}$. Note the analogy between Eqs. (2.17) and (2.1). We hope that readers are able to see how this rationale can be extended to define a tensor of any rank. For example, the appropriate basis to define a tensor of rank three is $\{\mathbf{e}_i \otimes \mathbf{e}_j \otimes \mathbf{e}_k\}$.

The dyadic product defined earlier can be performed on any two vectors (not just the unit basis vectors) to produce a second-rank tensor. Let **U** denote the dyadic product of vectors **u** and **v**. **U** can then be expressed as

$$\mathbf{U} = \mathbf{u} \otimes \mathbf{v} = u_i\mathbf{e}_i \otimes v_j\mathbf{e}_j = u_i v_j \mathbf{e}_i \otimes \mathbf{e}_j = U_{ij}\mathbf{e}_i \otimes \mathbf{e}_j \tag{2.18}$$

It is, therefore, easy to see that the components of the second-rank tensor **U** are simply defined by products of the corresponding components of vectors **u** and **v**, that is, $U_{ij} = u_i v_j$. By considering the maximum number of independent terms for tensors defined in Eqs. (2.17) and (2.18), it is readily seen that it is not generally possible to describe an arbitrary second-rank tensor as a single dyadic product of two vectors.

These definitions allow us to define an important operation called the *tensor product*. In general, it refers to an operation where the product of two tensors is of a reduced rank compared to the sum of the ranks of the tensors being multiplied. For example, the tensor product between a second-rank tensor, $(\mathbf{u} \otimes \mathbf{v})$, and a vector, **c**, may be defined as

$$(\mathbf{u} \otimes \mathbf{v})\mathbf{c} = \mathbf{u}(\mathbf{v} \cdot \mathbf{c}) \tag{2.19}$$

Using the interpretation given in Figure 2.2, it can be seen that the second-rank tensor $(\mathbf{u} \otimes \mathbf{v})$ can be visualized as a linear mapping that transforms any vector **c** in the Euclidean space to a vector parallel to **u**, the length of which is defined by $(\mathbf{v} \cdot \mathbf{c})|\mathbf{u}|$.

In direct analogy to Eq. (2.5), it can be shown that

$$T_{ij} = \mathbf{e}_i \cdot \mathbf{T}\mathbf{e}_j \tag{2.20}$$

Example 2.1

Given $\mathbf{Ta} = \mathbf{b}$, prove that $T_{ij}a_j = b_i$.

Solution

$$\begin{aligned}\mathbf{Ta} &= (T_{ij}\mathbf{e}_i \otimes \mathbf{e}_j)(a_k\mathbf{e}_k) = T_{ij}a_k(\mathbf{e}_j \cdot \mathbf{e}_k)\mathbf{e}_i = T_{ij}a_k\delta_{jk}\mathbf{e}_i = T_{ij}a_j\mathbf{e}_i \\ \mathbf{b} &= b_i\mathbf{e}_i \\ \therefore T_{ij}a_j &= b_i\end{aligned}$$

Note the similarity in the expression derived in Example 2.1, $T_{ij}a_j = b_i$, and the definition of the scalar product in Eq. (2.4), $\alpha = u_iv_i$. In both these operations, the rank of the product tensor is reduced by two compared to the sum of the ranks of the tensors being multiplied. This is also evident from the fact that the expressions for the tensor product have one repeated index, which essentially reduces the overall rank by two. Similarly, it is also possible to define a tensor product between two tensors of rank two such that the product tensor is also of rank two:

$$\mathbf{TS} = T_{ij}S_{jk}\mathbf{e}_i \otimes \mathbf{e}_k \tag{2.21}$$

Note that the rank of the product is once again reduced by two compared to the sum of the ranks of the tensors being multiplied. Equation (2.21) further implies that the tensor product between two tensors is not commutative, that is, $\mathbf{TS} \neq \mathbf{ST}$. For future use, we introduce the notation $\mathbf{S}^2 = \mathbf{SS}$, $\mathbf{S}^3 = \mathbf{SSS}$, etc.

A *scalar product* of two second-rank tensors, $\mathbf{S}$ and $\mathbf{T}$, is simply defined by an extension of Eq. (2.4) as

$$\alpha = \mathbf{S} \cdot \mathbf{T} = S_{ij}T_{ij} \tag{2.22}$$

As with vectors, the scalar product allows us to define a *magnitude* or *norm* of a second-rank tensor as

$$|\mathbf{S}| = \sqrt{\mathbf{S} \cdot \mathbf{S}} = \sqrt{S_{ij}S_{ij}} \tag{2.23}$$

A unit (spherical) tensor of rank two (also called an *identity tensor*) is defined as

$$\mathbf{I} = \delta_{ij}\mathbf{e}_i \otimes \mathbf{e}_j = \mathbf{e}_i \otimes \mathbf{e}_i \tag{2.24}$$

This second-rank identity tensor has the expected property that when it operates on other vectors or tensors it leaves their values unaltered, that is, $\mathbf{Iv} = \mathbf{v}$, $\mathbf{IA} = \mathbf{A}$.

The *transpose* of a second-rank tensor, $\mathbf{S}$, is denoted by $\mathbf{S}^T$ and is defined such that

$$\mathbf{S} = S_{ij}\mathbf{e}_i \otimes \mathbf{e}_j, \quad \mathbf{S}^T = S_{ji}\mathbf{e}_i \otimes \mathbf{e}_j \tag{2.25}$$

Based on this definition, it can be shown that

$$\mathbf{Su} \cdot \mathbf{v} = \mathbf{u} \cdot \mathbf{S}^T\mathbf{v} = S_{ij}u_jv_i = u_jS_{ij}v_i = u_jS^T_{ji}v_i \tag{2.26}$$

holds true for all vectors **u** and **v** in the Euclidean space. The transpose operation also exhibits the following very useful properties:

$$(\mathbf{S}+\mathbf{T})^T = \mathbf{S}^T + \mathbf{T}^T \tag{2.27}$$

$$(\mathbf{S}\mathbf{T})^T = \mathbf{T}^T\mathbf{S}^T \tag{2.28}$$

$$\left(\mathbf{S}^T\right)^T = \mathbf{S} \tag{2.29}$$

If $\mathbf{T}$ denotes a *symmetric* second-rank tensor, then $\mathbf{T} = \mathbf{T}^T$. If $\mathbf{W}$ denotes a tensor with the property $\mathbf{W} = -\mathbf{W}^T$, then $\mathbf{W}$ is referred to as a *skew-symmetric* or *anti-symmetric* tensor. It then follows that an arbitrary second-rank tensor $\mathbf{L}$ can be decomposed uniquely into its symmetric and skew symmetric components as $\mathbf{L} = \mathbf{T} + \mathbf{W}$, where

$$\mathbf{T} = \frac{1}{2}\left(\mathbf{L}+\mathbf{L}^T\right) \qquad \mathbf{W} = \frac{1}{2}\left(\mathbf{L}-\mathbf{L}^T\right) \tag{2.30}$$

2.3 COORDINATE TRANSFORMATIONS

As described in the previous sections, the concept of tensors is invoked primarily to address our need to reconcile the multiple quantitative representations of a given physical quantity or phenomena that arise due to the use of different reference frames by different observers. In this section, we take a detailed look at the coordinate transformation laws associated with a change of reference frames. Let $\{\mathbf{e}_1, \mathbf{e}_2, \mathbf{e}_3\}$ and $\{\mathbf{e}'_1, \mathbf{e}'_2, \mathbf{e}'_3\}$ represent two different Cartesian reference frames that share the same origin. Let v_i and v'_i denote the components of a given vector $\mathbf{v}$ in these two reference frames. The relationship between these components can be expressed as

$$\mathbf{v} = v_j\mathbf{e}_j = v'_j\mathbf{e}'_j \tag{2.31}$$

The following derivation converts the relationship in Eq. (2.31) into more useful forms:

$$v'_j\mathbf{e}'_j\cdot\mathbf{e}'_i = v_j\mathbf{e}_j\cdot\mathbf{e}'_i$$

$$v'_j\delta_{ij} = v'_i = \mathbf{e}'_i\cdot\mathbf{e}_j v_j$$

$$v'_i = Q_{ij}v_j, \qquad Q_{ij} = \mathbf{e}'_i\cdot\mathbf{e}_j \tag{2.32}$$

$$Q_{ij}\mathbf{e}_j = \left(\mathbf{e}'_i\cdot\mathbf{e}_j\right)\mathbf{e}_j = \left(\mathbf{e}_j\otimes\mathbf{e}_j\right)\mathbf{e}'_i = \mathbf{e}'_i \tag{2.33}$$

Equation (2.32) describes the coordinate transformation law for vectors. Q_{ij} represents the components of a transformation matrix, which are given by the dot products (direction cosines) of various basis vectors in one reference frame with the basis vectors in the other reference frame.

Example 2.2

Derive the components of the transformation matrix for two Cartesian reference frames that are related to each other by a rotation about the $\mathbf{e}_3$-axis by an angle θ.

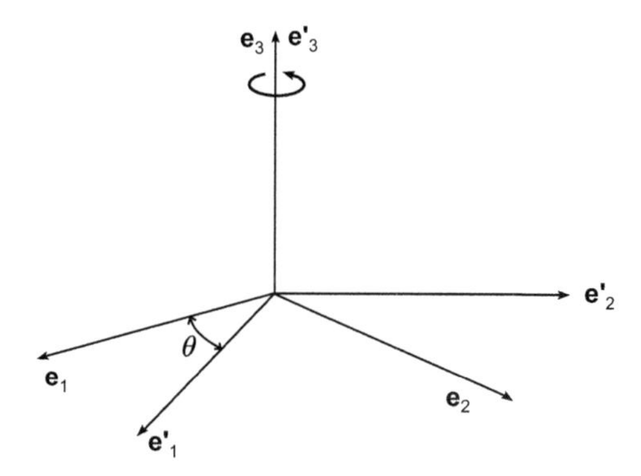

FIGURE 2.4

Two Cartesian reference frames that are related by a rotation of θ about the 3-axis.

Solution

Figure 2.4 shows two Cartesian reference frames that are related by a rotation of θ about $\mathbf{e}_3$. Evaluating the coefficients of Q_{ij} using Eq. (2.32) and arranging them into a matrix, we obtain

$$Q_{ij} = \begin{bmatrix} \cos\theta & \sin\theta & 0 \\ -\sin\theta & \cos\theta & 0 \\ 0 & 0 & 1 \end{bmatrix}$$

Let us now address what would be needed to describe the transformation matrix between two arbitrarily selected Cartesian reference frames in the Euclidean 3-space. According to a formalism developed by Euler, one can describe this transformation uniquely by a sequence of three rotations taken about specific directions. These rotation angles are usually referred to as *Euler angles*. There are a number of variants in the definitions of the Euler angles corresponding to different choices made in the selection of the three directions about which the rotations are defined. The particular definition of the Euler angles used extensively in the study of crystal mechanics is referred to as the Bunge–Euler angles.

Bunge–Euler angles are defined to describe a transformation from a sample Cartesian reference frame $\{\mathbf{e}_1, \mathbf{e}_2, \mathbf{e}_3\}$ to a local Cartesian crystal frame $\{\mathbf{e}_1^c, \mathbf{e}_2^c, \mathbf{e}_3^c\}$, as shown in Figure 2.5. The goal is to define a sequence of transformations that would relate the sample reference frame to the crystal reference frame. In other words, our objective is to express the complete transformation matrix $Q_{ij} = \mathbf{e}_i^c \cdot \mathbf{e}_j$ in terms of a sequence of three (simplified) transformations, say $Q_{il} = Q_{ij}(\phi_2)Q_{jk}(\Phi)Q_{kl}(\phi_1)$.

Physically, one way to accomplish this task would be to rotate the sample frame about $\mathbf{e}_3$ by an angle of ϕ_1 so that the $\mathbf{e}_1$-axis after the rotation is aligned with a direction $\mathbf{e}_1'$ that is perpendicular to

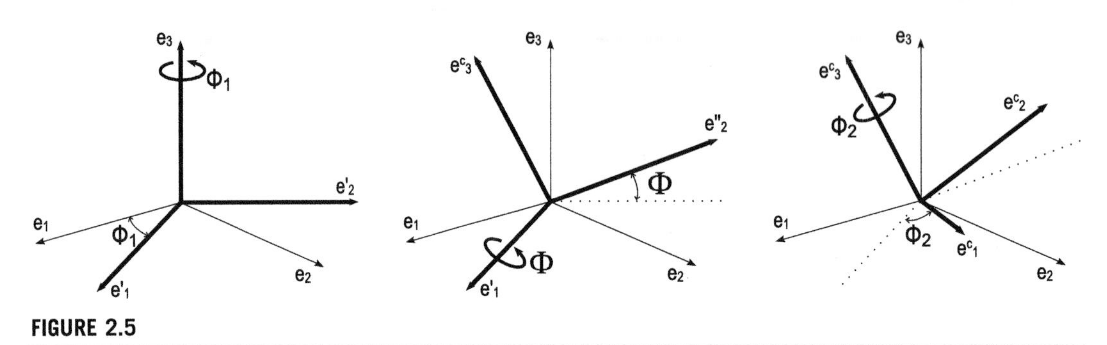

FIGURE 2.5

Schematic description of the Bunge–Euler angles used to establish the relationship between two arbitrarily defined Cartesian reference frames as a sequence of three rotations.

both $\mathbf{e}_3$ and $\mathbf{e}_3^c$. After this first rotation, the sample reference frame $\{\mathbf{e}_1, \mathbf{e}_2, \mathbf{e}_3\}$ will form the intermediate reference frame$\{\mathbf{e}_1', \mathbf{e}_2', \mathbf{e}_3\}$, as shown in Figure 2.5. This new reference frame can now be rotated about the $\mathbf{e}_1'$-axis by an angle Φ to bring $\mathbf{e}_3$ into perfect alignment with $\mathbf{e}_3^c$, forming another intermediate reference frame $\{\mathbf{e}_1', \mathbf{e}_2'', \mathbf{e}_3^c\}$, which is also shown in Figure 2.5. A final rotation by an angle ϕ_2 can then be applied about $\mathbf{e}_3^c$ to bring $\{\mathbf{e}_1', \mathbf{e}_2'', \mathbf{e}_3^c\}$ in perfect alignment with$\{\mathbf{e}_1^c, \mathbf{e}_2^c, \mathbf{e}_3^c\}$.

Mathematically, to construct $Q_{kl}(\phi_1)$, consider the transformation from $\{\mathbf{e}_1, \mathbf{e}_2, \mathbf{e}_3\}$ to $\{\mathbf{e}_1', \mathbf{e}_2', \mathbf{e}_3\}$ depicted on the left in Figure 2.5. The components of this transformation are readily seen to be

$$Q_{kl}(\phi_1) = \begin{bmatrix} \cos\phi_1 & \sin\phi_1 & 0 \\ -\sin\phi_1 & \cos\phi_1 & 0 \\ 0 & 0 & 1 \end{bmatrix}$$

Next, construct the transformation from $\{\mathbf{e}_1', \mathbf{e}_2', \mathbf{e}_3\}$ to $\{\mathbf{e}_1', \mathbf{e}_2'', \mathbf{e}_3^c\}$ as depicted in the central portion of Figure 2.5. As with the first transformation, the geometry of the direction cosines is easily seen to give the following components:

$$Q_{jk}(\Phi) = \begin{bmatrix} 1 & 0 & 0 \\ 0 & \cos\Phi & \sin\Phi \\ 0 & -\sin\Phi & \cos\Phi \end{bmatrix}$$

The product of these two transformations, $Q_{jk}(\Phi)Q_{kl}(\phi_1)$, relates $\{\mathbf{e}_1, \mathbf{e}_2, \mathbf{e}_3\}$ to $\{\mathbf{e}_1', \mathbf{e}_2'', \mathbf{e}_3^c\}$. The third transformation relates $\{\mathbf{e}_1', \mathbf{e}_2'', \mathbf{e}_3^c\}$ to the final crystal frame $\{\mathbf{e}_1^c, \mathbf{e}_2^c, \mathbf{e}_3^c\}$. Depicted on the right in Figure 2.5, the direction cosines are seen to be

$$Q_{ij}(\phi_2) = \begin{bmatrix} \cos\phi_2 & \sin\phi_2 & 0 \\ -\sin\phi_2 & \cos\phi_2 & 0 \\ 0 & 0 & 1 \end{bmatrix}$$

The definition of the Bunge–Euler angles allows us to evaluate the overall coordinate transformation matrix between the sample frame, $\{\mathbf{e}_1, \mathbf{e}_2, \mathbf{e}_3\}$, and the crystal frame, $\{\mathbf{e}_1^c, \mathbf{e}_2^c, \mathbf{e}_3^c\}$, as a product of the three transformation matrices. The overall transformation matrix from the sample reference frame to the local crystal frame, with components $Q_{il}(\phi_1, \Phi, \phi_2) = Q_{ij}(\phi_2)Q_{jk}(\Phi)Q_{kl}(\phi_1)$, is denoted by $g^{s\rightarrow c}$ and is expressed as

$$g_{il}^{s\rightarrow c} = \begin{bmatrix} \cos\phi_2 & \sin\phi_2 & 0 \\ -\sin\phi_2 & \cos\phi_2 & 0 \\ 0 & 0 & 1 \end{bmatrix} \begin{bmatrix} 1 & 0 & 0 \\ 0 & \cos\Phi & \sin\Phi \\ 0 & -\sin\Phi & \cos\Phi \end{bmatrix} \begin{bmatrix} \cos\phi_1 & \sin\phi_1 & 0 \\ -\sin\phi_1 & \cos\phi_1 & 0 \\ 0 & 0 & 1 \end{bmatrix}$$

$$= \begin{bmatrix} \cos\phi_1\cos\phi_2 - \sin\phi_1\cos\Phi\sin\phi_2 & \sin\phi_1\cos\phi_2 + \cos\phi_1\cos\Phi\sin\phi_2 & \sin\Phi\sin\phi_2 \\ -\cos\phi_1\sin\phi_2 - \sin\phi_1\cos\Phi\cos\phi_2 & -\sin\phi_1\sin\phi_2 + \cos\phi_1\cos\Phi\cos\phi_2 & \sin\Phi\cos\phi_2 \\ \sin\phi_1\sin\Phi & -\cos\phi_1\sin\Phi & \cos\Phi \end{bmatrix} \tag{2.34}$$

A closer look at the definition of the Bunge–Euler angles reveals a redundancy in the definition. There are indeed two possible choices for the selection of $\mathbf{e}_1'$ while satisfying the requirement that this direction be perpendicular to both $\mathbf{e}_3$ and $\mathbf{e}_3^c$. The choices are $\mathbf{e}_1'$ and $-\mathbf{e}_1'$. To address this redundancy, the following limits are set for the three Bunge–Euler angles:

$$0 \le \phi_1 < 2\pi, \quad 0 \le \Phi \le \pi, \quad 0 \le \phi_2 < 2\pi \tag{2.35}$$

Even with the restricted range specified in Eq. (2.35) for the Bunge–Euler angles, there continues to be a redundancy in their definition. This occurs in situations when $\Phi = 0$, for which we can specify only $(\varphi_1 + \varphi_2)$, and not φ_1 and φ_2 individually. Similarly, for $\Phi = 180$, we can only specify $(\phi_1 - \phi_2)$ uniquely. Except for the situations just noted, it is possible to select a unique set of Bunge–Euler angles to describe the relationship between any two arbitrarily selected Cartesian reference frames in Euclidean 3-space. The transformation described in Eq. (2.34) will be used extensively in later chapters. For now, we return to a discussion of the coordinate transformation matrix and the properties expected of this matrix.

The following derivation reveals an extremely important property associated with the transformation matrix Q_{ij} associated with two arbitrarily selected Cartesian reference frames defined by $\{\mathbf{e}_1, \mathbf{e}_2, \mathbf{e}_3\}$ and $\{\mathbf{e}_1', \mathbf{e}_2', \mathbf{e}_3'\}$:

$$\begin{aligned} \mathbf{e}_i' \cdot \mathbf{e}_j' &= Q_{ik}\mathbf{e}_k \cdot Q_{jl}\mathbf{e}_l = Q_{ik}Q_{jl}\delta_{kl} = Q_{ik}Q_{jk} \\ \mathbf{e}_i' \cdot \mathbf{e}_j' &= \delta_{ij} \\ \Rightarrow Q_{ik}Q_{jk} &= \delta_{ij} \end{aligned} \tag{2.36}$$

In matrix notation, Eq. (2.36) is expressed as

$$[Q]\,[Q]^T = [Q]^T[Q] = [I], \qquad [Q]^T = [Q]^{-1} \tag{2.37}$$

where $[I]$ represents an identity matrix (diagonal terms are equal to 1 and off-diagonal terms are equal to 0) and $[Q]^{-1}$ represents the matrix inverse of $[Q]$. Because of the property described in Eq. (2.36), the coordinate transformation matrix Q_{ij} is referred to as an *orthogonal* matrix. A set of six equations is embedded in Eq. (2.36); these impose the conditions that all columns and rows of the Q_{ij} matrix be all normal to each other. It is important to recognize that although the Q_{ij} matrix contains a total of nine terms, the conditions set in Eq. (2.36) reduce the number of independent variables needed to prescribe the Q_{ij} matrix to only three. Note that we have already shown in Eq. (2.34) that it is possible to specify uniquely the Q_{ij} matrix by specification of three independent Bunge–Euler angles.

There are indeed other forms of specification of a set of three independent variables to describe the relationship between two arbitrarily selected Cartesian reference frames. In addition to other variants in the selection of Euler angles (e.g., Roe's angles), one can also specify an axis–angle pair to uniquely describe this relationship. In this description, one specifies an axis and a rotation angle about this axis that would bring the two selected Cartesian frames in complete alignment with each other. It should be noted that this description again involves specification of three independent variables (two associated with the direction of the axis and the third being the rotation angle). The matrix describing this transformation will be discussed in Example 2.3.

Equation (2.36) further implies that the inverse transformation between the two selected reference frames is simply given by the transpose of the transformation matrix. For example, if $[Q]$ denotes the transformation matrix for quantities from $\{\mathbf{e}_1, \mathbf{e}_2, \mathbf{e}_3\}$ to $\{\mathbf{e}'_1, \mathbf{e}'_2, \mathbf{e}'_3\}$, then $[Q]^T$ will provide the transformation matrix for transformation of tensors from $\{\mathbf{e}'_1, \mathbf{e}'_2, \mathbf{e}'_3\}$ to $\{\mathbf{e}_1, \mathbf{e}_2, \mathbf{e}_3\}$. Particularly useful in later chapters is the fact that transformation from the crystal reference frame to the sample reference frame in the study of crystal mechanics is simply given by $(g^{s\to c})^T$, with $g^{s\to c}$ being defined as in Eq. (2.34). In other words, the overall transformation matrix from the crystal reference frame to the sample reference frame can be expressed as

$$g_{ij}^{c\to s} = \begin{bmatrix} \cos\phi_1\cos\phi_2 - \sin\phi_1\cos\Phi\sin\phi_2 & -\cos\phi_1\sin\phi_2 - \sin\phi_1\cos\Phi\cos\phi_2 & \sin\phi_1\sin\Phi \\ \sin\phi_1\cos\phi_2 + \cos\phi_1\cos\Phi\sin\phi_2 & -\sin\phi_1\sin\phi_2 + \cos\phi_1\cos\Phi\cos\phi_2 & -\cos\phi_1\sin\Phi \\ \sin\Phi\sin\phi_2 & \sin\Phi\cos\phi_2 & \cos\Phi \end{bmatrix} \tag{2.38}$$

The coordinate transformation laws for second-rank tensors can be derived following the same approach that was used earlier for vectors. The following derivation provides the desired result:

$$\begin{aligned} \mathbf{T} &= T_{ij}\mathbf{e}_i \otimes \mathbf{e}_j = T'_{rs}\mathbf{e}'_r \otimes \mathbf{e}'_s \\ T_{ij}(Q_{ri}\mathbf{e}'_r) \otimes (Q_{sj}\mathbf{e}'_s) &= Q_{ri}Q_{sj}T_{ij}\mathbf{e}'_r \otimes \mathbf{e}'_s = T'_{rs}\mathbf{e}'_r \otimes \mathbf{e}'_s \\ \Rightarrow T'_{rs} &= Q_{ri}Q_{sj}T_{ij} \end{aligned} \tag{2.39}$$

Note the similarity between Eqs. (2.32) and (2.39). Indeed, the coordinate transformation laws for higher-rank tensors are simply extensions of these equations. For example, for a fourth-rank tensor, the coordinate transformation rule is expressed as

$$c'_{ijkl} = Q_{ip}Q_{jq}Q_{kr}Q_{ls}c_{pqrs} \tag{2.40}$$

In closing this section, we reiterate that the concept of coordinate transformation essentially deals with reconciling the different descriptions of a given tensor in two different reference frames. This concept has broad usage in mechanics. In this book, this concept will be applied largely in transforming the physical tensors of interest between the material (or crystal) reference frame and the sample (or global) reference frame. This is often very useful because the material behavior is most easy to describe in the natural material reference frame, whereas the design performance of a component is most conveniently expressed in the global reference frame.

2.4 ROTATIONS

An alternative, but completely equivalent, approach to connecting the crystal reference frame and the sample reference frame is through the use of rotation tensors. Here, instead of creating two different reference frames, we stick to a single reference frame, which is typically selected as the sample reference frame. We then start with a hypothetical strain-free crystal in a reference orientation that is perfectly aligned with the sample reference frame, and then rotate the reference crystal in its entirety in such a way that the crystal ends up in its actual current orientation. Since only rigid-body rotation is applied, the crystal remains distortion-free. Therefore, in the description of a rotation tensor, there is often only a single reference frame and a rotation of all material points in the crystal from its reference configuration to its current configuration. As noted earlier, such a linear mapping of points (or vectors) can be described by a second-rank tensor, **R**, as

$$\mathbf{x} = \mathbf{R}\mathbf{x}_o \tag{2.41}$$

where $\mathbf{x}_o$ and $\mathbf{x}$ denote a material point in the crystal in its reference configuration and its current configuration, respectively. When there is no distortion of lattice in the crystal between its reference configuration and its current configuration, it can be shown that **R** is an orthogonal tensor:

$$\begin{aligned} \mathbf{x}\cdot\mathbf{x} &= \mathbf{R}\mathbf{x}_o\cdot\mathbf{R}\mathbf{x}_o = \mathbf{x}_o\cdot\mathbf{x}_o \\ &\Rightarrow \mathbf{R}^T\mathbf{R} = \mathbf{R}\mathbf{R}^T = \mathbf{I} \end{aligned} \tag{2.42}$$

A more rigorous treatment of this mapping, including lattice distortions, is presented later in Chapter 7.

Let us now focus on the similarities and differences between the rotation tensor **R** and the coordinate transformation matrix Q_{ij} introduced in the previous section. The similarities between Eqs. (2.32) and (2.41), and between Eqs. (2.37) and (2.42), have caused a great deal of confusion in the interdisciplinary collaborations between practitioners in the mechanics and materials communities. The following notes are offered to help understand the precise differences between these two concepts.

1. The coordinate transformation matrix Q_{ij} is not a tensor. This matrix is defined essentially through a selection of two specific reference frames, and its components have a physical meaning only for those two selected reference frames. The rotation tensor **R**, on the other hand, describes a physical rotation of the crystal, and in principle its definition is not predicated on the selection of a specific reference frame (see Eq. 2.41). However, in practice, the components of the rotation tensor **R** are most often described in the sample reference frame.
2. In dealing with coordinate transformations, one typically rotates the reference frame while keeping the physical quantities of interest stationary, whereas, in dealing with rotation tensors, we keep the reference frame stationary and rotate the physical bodies of interest (crystals in the present discussion). Because of this distinction, coordinate transformations are generally described as *passive* transformations, while the rotation tensors are described as *active* transformations. Furthermore, it should be clear that the two descriptions are exact inverses of each other.
3. To precisely identify the connections between the coordinate transformation matrices and the rotation tensors, let us consider a simple example. As discussed in the previous section, let $g^{s\rightarrow c}$

denote the coordinate transformation matrix from the sample reference frame to the crystal reference frame (see Eq. 2.34). Let $R_{ij}^{s\to c}$ denote the components of the rotation tensor in the sample reference frame describing the rotation of a crystal from its reference orientation (aligned with the sample reference frame) to its current orientation. Because these transformations are exact inverses of each other, and keeping in mind that these are orthogonal matrices and tensors, it is easy to see that

$$\mathbf{g}^{s\to c} = (\mathbf{R}^{s\to c})^{-1} = (\mathbf{R}^{s\to c})^{T} \tag{2.43}$$

By way of proof, consider the action of rotation $\mathbf{R}$ on the unit reference vectors $\{\mathbf{e}_1, \mathbf{e}_2, \mathbf{e}_3\}$ to yield new unit vectors $\{\mathbf{e}_1', \mathbf{e}_2', \mathbf{e}_3'\}$, that is, $\mathbf{e}_i' = \mathbf{R}\mathbf{e}_i$. Note that the $(j,i)^{\text{th}}$ component of rotation $\mathbf{R}$ is expressed as $R_{ji} = \mathbf{R}\mathbf{e}_i \cdot \mathbf{e}_j$ (Eq. 2.20). Using this definition, one may write $R_{ji} = \mathbf{R}\mathbf{e}_i \cdot \mathbf{e}_j = \mathbf{e}_i' \cdot \mathbf{e}_j = g_{ij}^{s\to c}$. Hence, we see that (2.43) is correct.

Note that a coordinate transformation matrix can also be expressed from the crystal reference frame to the sample reference frame as $g^{c\to s}$ (see Eq. 2.38). It can be seen that this matrix can be related to the rotation tensor as

$$\mathbf{g}^{c\to s} = (\mathbf{g}^{s\to c})^{-1} = \mathbf{R}^{s\to c} \tag{2.44}$$

As mentioned earlier, the apparent similarities between the components of the coordinate transformation matrix and the rotation tensor have caused substantial confusion in the minds of students entering the interdisciplinary field of mechanics and materials. Although the two descriptions are completely equivalent, because of the long history and conventions adopted in their respective fields, it is simply impractical to completely abandon one of the concepts. We urge students to always ascertain the precise definition of a coordinate transformation matrix or a rotation tensor before using it in their calculations. For example, if a coordinate transformation matrix is specified, it is important to establish if it is defined as a transformation from the sample reference frame to the crystal reference frame or vice versa (of course, these two are exact transposes of each other, but knowing which definition is being used with a given matrix often has a major impact on the solutions obtained to a given problem; one cannot and should not guess this information).

Likewise, for a quantitative description of a rotation tensor, it is important to know if the rotation tensor is being expressed in the sample reference frame or a crystal reference frame (remember that rotation is a tensor and can in principle be expressed in any reference frame of choice), and also if it is a rotation from the reference configuration to the current configuration or the exact opposite rotation from the current configuration to the reference configuration. Once a precise definition of a given transformation is noted, it should be easy to establish the correct rules for its usage in subsequent calculations.

It is also worth noting that the concept of rotation tensor can also be used to transport other physical tensors, other than points and vectors described in Eq. (2.41), from the hypothetical reference configuration of the crystal to its current configuration. The rules for the transport of higher-ranked tensors during a rotation of a crystal indeed correspond exactly with the coordinate transformation laws for the higher-ranked tensors described earlier (see Eqs. 2.39 and 2.40), keeping in mind the connections between these two alternative approaches (as described, for example, in Eqs. 2.43 and 2.44).

Example 2.3

It is often convenient to describe rotations using an axis of rotation $\mathbf{n}$ and a rotation angle θ, taken in the right-hand rule sense about $\mathbf{n}$. The vector $\theta\mathbf{n}$ is then a complete description of the rotation (i.e., the length of the vector, θ, gives the rotation angle); hereafter, we refer to this as $\mathbf{R}(\theta\mathbf{n})$. Prove that

$$R_{ij}(\theta\mathbf{n}) = \delta_{ij}\cos(\theta) - \in_{ijk} n_k \sin(\theta) + (1-\cos(\theta))n_i n_j \tag{2.45}$$

Solution

Consider an arbitrary vector $\mathbf{a}$ subjected to rotation $\mathbf{R}(\theta\mathbf{n})$ resulting in vector $\mathbf{b}$. Decompose vector a into a component along direction $\mathbf{n}$ and normal to direction $\mathbf{n}$ as

$$\mathbf{a} = (\mathbf{a}\cdot\mathbf{n})\mathbf{n} + \mathbf{a} - (\mathbf{a}\cdot\mathbf{n})\mathbf{n} = (\mathbf{a}\cdot\mathbf{n})\mathbf{n} + (\mathbf{I} - \mathbf{n}\otimes\mathbf{n})\mathbf{a}$$

In the rotation of vector $\mathbf{a}$ by $\mathbf{R}(\theta\mathbf{n})$ only the component normal to $\mathbf{n}$ experiences the rotation. Therefore, the $(\mathbf{I} - \mathbf{n}\otimes\mathbf{n})\mathbf{a}$ component produces a new vector in the plane normal to $\mathbf{n}$ without a change in its magnitude. A simple consideration of the geometry reveals that any vector $\mathbf{c}$ that is normal to vector $\mathbf{n}$, when subjected to $\mathbf{R}(\theta\mathbf{n})$, maps into the vector $\mathbf{c}\ \cos\theta - \mathbf{c}\times\mathbf{n}\sin\theta$. Applying this result to the $(\mathbf{I} - \mathbf{n}\otimes\mathbf{n})\mathbf{a}$ component, vector $\mathbf{b}$ may be expressed as

$$\begin{aligned}
\mathbf{b} &= (\mathbf{a}\cdot\mathbf{n})\mathbf{n} + \cos\theta(\mathbf{I} - \mathbf{n}\otimes\mathbf{n})\mathbf{a} - \sin\theta(\mathbf{I} - \mathbf{n}\otimes\mathbf{n})\mathbf{a}\times\mathbf{n} \\
&= \cos\theta\ \mathbf{a} + \sin\theta\ \mathbf{n}\times\mathbf{a}\ + (1-\cos\theta)(\mathbf{n}\otimes\mathbf{n})\mathbf{a} \\
&= (\cos\theta\ \mathbf{I} + \sin\theta\ \mathbf{N}\ + (1-\cos\theta)(\mathbf{n}\otimes\mathbf{n}))\mathbf{a} \\
\Rightarrow \mathbf{R} &= (\cos\theta\ \mathbf{I} + \sin\theta\ \mathbf{N}\ + (1-\cos\theta)(\mathbf{n}\otimes\mathbf{n})),
\end{aligned}$$

where $\mathbf{n}$ is the dual vector of the anti-symmetric tensor $\mathbf{N}$ defined such that $\mathbf{N}\mathbf{v} = \mathbf{n}\times\mathbf{v}$ for any arbitrary vector $\mathbf{v}$. The expression for the rotation tensor can also be conveniently expressed in the indicial notation as

$$R_{ij}(\theta\mathbf{n}) = \delta_{ij}\cos(\theta) - \in_{ijk} n_k \sin(\theta) + (1-\cos(\theta))n_i n_j$$

Note that $\mathbf{R}(\pi\mathbf{n}) = \mathbf{R}(-\pi\mathbf{n})$, indicating that these two rotations have identical physical meaning. Note also that

$$R(\theta\mathbf{e}_3) = \begin{bmatrix} \cos\theta & -\sin\theta & 0 \\ \sin\theta & \cos\theta & 0 \\ 0 & 0 & 1 \end{bmatrix}$$

indicating that the components of the rotation tensor are given by the transpose of the transformation coordinate matrix (derived in Example 2.2).

Example 2.4

A crystalline sample is being studied in a microscope. The crystal reference frame is denoted as $\{\mathbf{e}_1^c, \mathbf{e}_2^c, \mathbf{e}_3^c\}$, the sample reference frame as $\{\mathbf{e}_1^s, \mathbf{e}_2^s, \mathbf{e}_3^s\}$, and the microscope reference frame as $\{\mathbf{e}_1^m, \mathbf{e}_2^m, \mathbf{e}_3^m\}$. It was noted that the sample preparation procedures resulted in the sample reference frame being rotated by 15° about $\mathbf{e}_1^m$

with respect to the microscope reference frame. The crystal orientation was previously measured and described by a 45° rotation about $\mathbf{e}_3^s$ with respect to the sample reference frame. Find the coordinate transformation matrix between the microscope reference frame and the crystal reference frame. Find also the rotation tensor describing the crystal orientation in the microscope reference frame and verify if it is indeed the transpose of the coordinate transformation matrix obtained.

Solution

In coordinate transformations, we start with the microscope reference frame, transform to the sample reference frame, and then to the crystal reference frame as (see also Example 2.2 and Eq. 2.34)

$$g_{ij}^{m\to c} = g_{ik}^{s\to c} g_{kj}^{m\to s} = \begin{bmatrix} \cos(45) & \sin(45) & 0 \\ -\sin(45) & \cos(45) & 0 \\ 0 & 0 & 1 \end{bmatrix} \begin{bmatrix} 1 & 0 & 0 \\ 0 & \cos(15) & \sin(15) \\ 0 & -\sin(15) & \cos(15) \end{bmatrix}$$

In rotations, we start with a hypothetical crystal oriented in perfect alignment with the microscope reference frame, rotate it 15° about $\mathbf{e}_1^m$, and then rotate it 45° about $\mathbf{e}_3^s$. The overall rotation is then expressed as

$$\mathbf{R} = \mathbf{R}(45\mathbf{e}_3^s)\mathbf{R}(15\mathbf{e}_1^m)$$

Note that the rotation tensor $\mathbf{R}$ can be expressed in any frame as desired. However, in this example, we have been asked to express it in the microscope reference frame. This means that both $\mathrm{R}(15\mathrm{e}_1^m)$ and $\mathrm{R}(45\mathrm{e}_3^s)$ $\mathbf{R}(45\mathbf{e}_3^s)$should be expressed in the microscope reference frame to perform the product operation previously described. Noting the inverse relationship between rotation tensors and the coordinate transformation matrices, the first rotation can be expressed in the microscope reference frame as

$$[\mathbf{R}(15\mathbf{e}_1^m)]^m = \begin{bmatrix} 1 & 0 & 0 \\ 0 & \cos(15) & -\sin(15) \\ 0 & \sin(15) & \cos(15) \end{bmatrix}$$

The second rotation $\mathbf{R}(45\mathbf{e}_3^s)$ is more conveniently expressed in the sample reference frame and then transformed into the microscope reference frame using a second-rank coordinate transformation law as

$$[\mathbf{R}(45\mathbf{e}_3^s)]^s = \begin{bmatrix} \cos(45) & -\sin(45) & 0 \\ \sin(45) & \cos(45) & 0 \\ 0 & 0 & 1 \end{bmatrix}, \text{ or}$$

$$[\mathbf{R}(45\mathbf{e}_3^s)]^m = \begin{bmatrix} 1 & 0 & 0 \\ 0 & \cos(15) & -\sin(15) \\ 0 & \sin(15) & \cos(15) \end{bmatrix} \begin{bmatrix} \cos(45) & -\sin(45) & 0 \\ \sin(45) & \cos(45) & 0 \\ 0 & 0 & 1 \end{bmatrix} \begin{bmatrix} 1 & 0 & 0 \\ 0 & \cos(15) & \sin(15) \\ 0 & -\sin(15) & \cos(15) \end{bmatrix}$$

Finally, we evaluate $\mathbf{R}$ in the microscope reference frame as

$$[\mathbf{R}]^m = [\mathbf{R}(45\mathbf{e}_3^s)]^m[\mathbf{R}(15\mathbf{e}_1^m)]^m = \begin{bmatrix} 1 & 0 & 0 \\ 0 & \cos(15) & -\sin(15) \\ 0 & \sin(15) & \cos(15) \end{bmatrix} \begin{bmatrix} \cos(45) & -\sin(45) & 0 \\ \sin(45) & \cos(45) & 0 \\ 0 & 0 & 1 \end{bmatrix}$$

Note that $[\mathbf{R}]^m = [g_{ij}^{m\to c}]^T$.

For completeness we write the rotation tensor in terms of Euler angles:

$$\mathbf{R}(\phi_1, \Phi, \phi_2) = \mathbf{R}(\phi_1\mathbf{e}_3)\mathbf{R}(\Phi\mathbf{e}_1)\mathbf{R}(\phi_2\mathbf{e}_3) \tag{2.46}$$

$$\mathbf{R} = \begin{bmatrix} \cos\phi_1\cos\phi_2 - \sin\phi_1\cos\Phi\sin\phi_2 & -\cos\phi_1\sin\phi_2 - \sin\phi_1\cos\Phi\cos\phi_2 & \sin\phi_1\sin\Phi \\ \sin\phi_1\cos\phi_2 + \cos\phi_1\cos\Phi\sin\phi_2 & -\sin\phi_1\sin\phi_2 + \cos\phi_1\cos\Phi\cos\phi_2 & -\cos\phi_1\sin\Phi \\ \sin\Phi\sin\phi_2 & \sin\Phi\cos\phi_2 & \cos\Phi \end{bmatrix} \tag{2.47}$$

2.5 EIGENVALUES AND EIGENVECTORS

We introduced in Section 1.1 the notion that higher-rank tensors can be interpreted as linear transformation mappings from a lower-ranked tensor space into another lower-ranked tensor space. Such mappings are often associated with principal unit directions in the lower-ranked tensor space that usually carry important physical meaning for the specific mapping being studied. As an example, let $\mathbf{T}$ denote a second-rank tensor. Then, it becomes possible to seek special vectors $\mathbf{a}$ such that

$$\mathbf{T}\mathbf{a} = \lambda\mathbf{a} \tag{2.48}$$

The solutions to Eq. (2.48) identify the special directions, $\mathbf{a}$, that remain unchanged in direction in the linear transformation mapping associated with $\mathbf{T}$. However, vector $\mathbf{a}$ does undergo a change in magnitude, described by the scalar multiplier λ, as a consequence of the linear transformation. The solutions of Eq. (2.48) in λ and $\mathbf{a}$ are referred to as its *eigenvalues* (also called principal values) and *eigenvectors* (also called principal directions), respectively. Note that there can be multiple solutions to Eq. (2.48) for a prescribed tensor.

Equation (2.48) can be recast as

$$(\mathbf{T} - \lambda\mathbf{I})\mathbf{a} = \mathbf{0} \tag{2.49}$$

where $\mathbf{0}$ denotes the null vector. Equation (2.49) has a nontrivial solution only if

$$\det(\mathbf{T} - \lambda\mathbf{I}) = 0 \tag{2.50}$$

The determinant function in Eq. (2.50) is defined as is usually done in matrix algebra and is applied here on the components of $(\mathbf{T} - \lambda\mathbf{I})$ in any selected Cartesian reference frame. Expansion of Eq. (2.50) results in the following *characteristic equation*:

$$\lambda^3 - i_1(\mathbf{T})\lambda^2 + i_2(\mathbf{T})\lambda - i_3(\mathbf{T}) = 0 \tag{2.51}$$

with

$$i_1(\mathbf{T}) = tr\,\mathbf{T} = T_{ii} \tag{2.52}$$

$$i_2(\mathbf{T}) = \frac{1}{2}\left[(tr\ \mathbf{T})^2 - tr\ (\mathbf{T}^2)\right] = \frac{1}{2}\left[(T_{ii})^2 - T_{ij}T_{ji}\right] \tag{2.53}$$

$$i_3(\mathbf{T}) = \det\ \mathbf{T} = \epsilon_{ijk}T_{i1}T_{j2}T_{k3} \tag{2.54}$$

Recalling that $\mathbf{T}$ represents a physical quantity or a physical phenomenon that exists independent of a reference frame selection, we expect the eigenvalues associated with $\mathbf{T}$ to also be independent of the reference frame selection. In other words, we expect the solutions of the characteristic equation (2.51) to be completely independent of the selection of the Cartesian reference frame selected to evaluate the coefficients $(i_1(\mathbf{T}), i_2(\mathbf{T}), i_3(\mathbf{T}))$. This is only possible if the coefficients $(i_1(\mathbf{T}), i_2(\mathbf{T}), i_3(\mathbf{T}))$ themselves are independent of the selection of the reference frame.

Consequently, the list $(i_1(\mathbf{T}), i_2(\mathbf{T}), i_3(\mathbf{T}))$ is called the list of the *invariants* of tensor $\mathbf{T}$, with the members of the list referred to as the first, second, and third invariants. The values of these invariants for a given tensor are completely independent of the coordinate reference frame selection used to quantify the tensor components. Since Eq. (2.51) is a cubic equation, it is expected to produce three roots, although not all are expected to be real or distinct. The *Cayley–Hamilton theorem* states that every tensor will obey its own characteristic equation:

$$\mathbf{T}^3 - i_1(\mathbf{T})\mathbf{T}^2 + i_2(\mathbf{T})\mathbf{T} - i_3(\mathbf{T})\mathbf{I} = 0 \tag{2.55}$$

A *symmetric tensor* $\mathbf{S}$ possesses three real eigenvalues (λ_1, λ_2, λ_3) and at least one orthonormal set of corresponding eigenvectors $\{\mathbf{p}_1, \mathbf{p}_2, \mathbf{p}_3\}$. Consequently, symmetric tensors in Euclidean 3-space admit the following representation (from applying Eqs. 2.18 and 2.48):

$$\mathbf{S} = \sum_{i=1}^{3} \lambda_i \mathbf{p}_i \otimes \mathbf{p}_i \tag{2.56}$$

Note that we have encountered in Eq. (2.56) a situation where an index is repeated three times. Under these special circumstances, to avoid any possible confusion, the summation sign will be shown explicitly throughout this book. Equation (2.56) allows us to recast Eqs. (2.52) through (2.54) as

$$\begin{aligned} i_1(\mathbf{S}) &= \lambda_1 + \lambda_2 + \lambda_3 \\ i_2(\mathbf{S}) &= \lambda_1\lambda_2 + \lambda_2\lambda_3 + \lambda_3\lambda_1 \\ i_3(\mathbf{S}) &= \lambda_1\lambda_2\lambda_3 \end{aligned} \tag{2.57}$$

A tensor $\mathbf{T}$ is said to be *positive definite* if $\mathbf{v}\cdot\mathbf{Tv} > 0$ holds true for all vectors $\mathbf{v}$ in the Euclidean space, except the null vector (zero vector). Likewise, the tensor is said to be *negative definite* if $\mathbf{v}\cdot\mathbf{Tv} < 0$ for all vectors other than the null vector. However, if $\mathbf{v}\cdot\mathbf{Tv} \geq 0$, it is said to be positive semi-definite, and if $\mathbf{v}\cdot\mathbf{Tv} \leq 0$, it is said to be negative semi-definite. For a positive definite symmetric tensor, the eigenvalues are real and positive. If $\mathbf{C}$ is a positive definite symmetric tensor, then it admits the representation

$$\mathbf{C} = \sum_{i=1}^{3} \lambda_i \mathbf{e}_i \otimes \mathbf{e}_i, \quad \lambda_i > 0 \tag{2.58}$$

where λ_i and $\mathbf{e}_i$ denote the eigenvalues and eigenvectors, respectively, of $\mathbf{C}$. If $\mathbf{U} = \sqrt{\mathbf{C}}$, then $\mathbf{U}$ is also positive definite and admits the representation

$$\mathbf{U} = \sum_{i=1}^{3} \sqrt{\lambda_i}\mathbf{e}_i \otimes \mathbf{e}_i \tag{2.59}$$

The *inverse* of $\mathbf{C}$ is denoted by $\mathbf{C}^{-1}$ and admits the representation

$$\mathbf{C}^{-1} = \sum_{i=1}^{3} \lambda_i^{-1}\mathbf{e}_i \otimes \mathbf{e}_i \tag{2.60}$$

2.6 POLAR DECOMPOSITION THEOREM

A particularly useful theorem in the study of finite deformations on solids is the *polar decomposition theorem*. According to this theorem, if $\mathbf{F}$ is a second-rank tensor with a positive determinant, it admits the following representation:

$$\mathbf{F} = \mathbf{RU} = \mathbf{VR} \tag{2.61}$$

where $\mathbf{R}$ represents a proper orthogonal tensor with

$$\mathbf{RR}^T = \mathbf{R}^T\mathbf{R} = \mathbf{I},\ \det(\mathbf{R}) = 1 \tag{2.62}$$

and $\mathbf{U}$ and $\mathbf{V}$ denote positive definite symmetric tensors defined as

$$\mathbf{U} = \sqrt{\mathbf{F}^T\mathbf{F}}, \quad \mathbf{V} = \sqrt{\mathbf{FF}^T} \tag{2.63}$$

Let $\{\mathbf{r}_1, \mathbf{r}_2, \mathbf{r}_3\}$ and $\{\mathbf{l}_1, \mathbf{l}_2, \mathbf{l}_3\}$ denote the sets of orthonormal eigenvector bases for $\mathbf{U}$ and $\mathbf{V}$, respectively. Then the various quantities previously described admit the following representations:

$$\begin{aligned} &\mathbf{U} = \sum_{i=1}^{3} \lambda_i \mathbf{r}_i \otimes \mathbf{r}_i, \quad \mathbf{V} = \sum_{i=1}^{3} \lambda_i \mathbf{l}_i \otimes \mathbf{l}_i, \quad \mathbf{R} = \sum_{i=1}^{3} \mathbf{l}_i \otimes \mathbf{r}_i, \\ &\mathbf{F} = \sum_{i=1}^{3} \lambda_i \mathbf{l}_i \otimes \mathbf{r}_i, \quad \mathbf{F}^{-1} = \sum_{i=1}^{3} \lambda_i^{-1} \mathbf{r}_i \otimes \mathbf{l}_i \end{aligned} \tag{2.64}$$

2.7 TENSOR GRADIENTS

In mechanics, we routinely encounter spatial gradients of various tensors. For example, strain is derived from displacement gradients. We will find the following "del" operator quite useful in describing a variety of spatial gradients:

$$\nabla = \frac{\partial}{\partial x_1}\mathbf{e}_1 + \frac{\partial}{\partial x_2}\mathbf{e}_2 + \frac{\partial}{\partial x_3}\mathbf{e}_3 = \frac{\partial}{\partial x_i}\mathbf{e}_i \tag{2.65}$$

where x_i denotes the spatial coordinates. Gradients of scalars and tensors are then defined by

$$\text{grad } \varphi = \nabla\varphi = \frac{\partial \varphi}{\partial x_i}\, \mathbf{e}_i \tag{2.66}$$

$$\text{grad } \mathbf{u} = \nabla\mathbf{u} = \frac{\partial u_i}{\partial x_j}\, \mathbf{e}_i \otimes \mathbf{e}_j \tag{2.67}$$

There are a number of operations that can be defined using the del operator. Some of these are as follows:

$$\text{div } \mathbf{v} = \nabla\cdot\mathbf{v} = \frac{\partial v_i}{\partial x_i} \tag{2.68}$$

$$\text{div } \mathbf{T} = \nabla\cdot\mathbf{T} = \frac{\partial T_{ij}}{\partial x_j}\, \mathbf{e}_i \tag{2.69}$$

$$\text{curl } \mathbf{v} = \nabla\times\mathbf{v} = \in_{ijk}\frac{\partial v_k}{\partial x_j}\, \mathbf{e}_i \tag{2.70}$$

$$\nabla^2\varphi = \nabla\cdot\nabla\varphi = \frac{\partial^2\varphi}{\partial x_i \partial x_i} = \varphi,_{ii} \tag{2.71}$$

In Eq. (2.71) we introduced another standard convention in tensor notation where we *represent the partial derivatives with a comma*. This convention will be used extensively in later chapters.

For smooth tensor fields (i.e., the gradients exist and are unique), the following identities, known as *Gauss theorems* or *divergence theorems*, will be quite useful in converting integrals over volumetric domains, D, to integrals over bounding surfaces, ∂D, where $\mathbf{n}$ denotes the surface normal on ∂D:

$$\int_D \text{div } \mathbf{T}\, dV = \int_{\partial D} \mathbf{T}\,\mathbf{n}\, dA, \text{ that is, } \int_D T_{ij\ldots k,q}\, dV = \int_{\partial D} T_{ij\ldots k} n_q\, dA \tag{2.72}$$

$$\int_D \text{div } \mathbf{v}\, dV = \int_{\partial D} \mathbf{v}\cdot\mathbf{n}\, dA \tag{2.73}$$

$$\int_D \text{grad } \varphi\, dV = \int_{\partial D} \varphi\,\mathbf{n}\, dA \tag{2.74}$$

SUMMARY

This chapter was designed to serve as a “crash” refresher for the important concepts needed for later chapters. The concept of tensors is central to physics-based modeling of materials phenomena. As needed, students may refer to numerous other standard texts covering this area. Likewise, the representation of crystal lattice orientation is central to many of the considerations in later chapters. That is why this specific topic received a lot of attention in this chapter. It is hoped that readers are now familiar and comfortable with the multiple equivalent representations used in the literature for crystal orientations.

Exercises

Section 2.2

1. Using the concepts described in Eqs. (2.4) and (2.6), show that

$$\mathbf{u}\cdot\mathbf{v} = |\mathbf{u}||\mathbf{v}|\cos\theta$$

where θ is the angle between the vectors **u** and **v**.

Prove the following identities:

2. $\mathbf{e}_i \times \mathbf{e}_j = \in_{ijk}\mathbf{e}_k$.
3. If $\mathbf{w} = \mathbf{u}\times\mathbf{v}$, then $|\mathbf{w}| = |\mathbf{u}||\mathbf{v}|\sin\theta$, where θ is the angle between **u** and **v**, $0 \le \theta \le \pi$.
4. $\mathbf{u}\times\mathbf{v} = -\mathbf{v}\times\mathbf{u}$.
5. $(\alpha\mathbf{u}+\beta\mathbf{v})\times\mathbf{w} = \alpha(\mathbf{u}\times\mathbf{w})+\beta(\mathbf{v}\times\mathbf{w})$.
6. $\mathbf{u}\cdot(\mathbf{u}\times\mathbf{v}) = 0$.
7. $(\mathbf{u}\times\mathbf{v})\cdot(\mathbf{u}\times\mathbf{v}) = (\mathbf{u}\cdot\mathbf{u})(\mathbf{v}\cdot\mathbf{v})-(\mathbf{u}\cdot\mathbf{v})^2$.
8. Prove Eq. (2.20).
9. Prove Eq. (2.21).
10. Prove $(\mathbf{TS})\,\mathbf{a} = \mathbf{T}(\mathbf{Sa}) = T_{ij}S_{jk}a_k\mathbf{e}_i$.
11. Prove $(\mathbf{a}\otimes\mathbf{b})\cdot(\mathbf{c}\otimes\mathbf{d}) = (\mathbf{a}\cdot\mathbf{c})(\mathbf{b}\cdot\mathbf{d}) = a_ib_jc_id_j$.
12. If $\mathbf{S} = \mathbf{S}^T$, then $\mathbf{S}\cdot\mathbf{A} = \mathbf{S}\cdot\mathbf{A}^T$ for all **A**.
13. If $\mathbf{W} = -\mathbf{W}^T$, then $\mathbf{W}\cdot\mathbf{A} = -\mathbf{W}\cdot\mathbf{A}^T$ for all **A**.
14. Prove $\mathbf{R}\cdot\mathbf{ST} = (\mathbf{S}^\mathrm{T}\mathbf{R})\cdot\mathbf{T} = \mathbf{RT}^\mathrm{T}\cdot\mathbf{S}$.
15. Prove $\mathbf{u}\cdot\mathbf{Sv} = \mathbf{S}\cdot(\mathbf{u}\otimes\mathbf{v})$.
16. If **W** is a skew-symmetric tensor, show that $\mathbf{Wv} = \mathbf{w}\times\mathbf{v}$ for all vectors **v** in the Euclidean 3-space. Express the components of **w** in terms of the components of **W**. Note that **w** is referred to as the dual vector of **W**.

Section 2.3

17. It is a standard practice in physical metallurgy to represent the lattice orientation of a crystal in terms of Miller indices as (*hkl*)[*uvw*], where it is understood that the (*hkl*) plane is perpendicular to the sample $\mathbf{e}_3$ axis and the [*uvw*] direction is parallel to the sample $\mathbf{e}_1$ axis. Prove that these Miller indices can be obtained from the Bunge—Euler angles using the following relationships, where n and n' denote constants that have been selected to yield the lowest integer values for h, k, l, and u, v, w, respectively.

$$h = n\sin\Phi\sin\varphi_2, \quad k = n\sin\Phi\cos\varphi_2, \quad l = n\cos\Phi$$

$$u = n'(\cos\varphi_1\cos\varphi_2 - \sin\varphi_1\sin\varphi_2\cos\Phi), \quad v = n'(-\cos\varphi_1\sin\varphi_2 - \sin\varphi_1\cos\varphi_2\cos\Phi),$$
$$w = n'\sin\varphi_1\sin\Phi$$

18. Invert the relationships described in Exercise 2.17 and prove

$$\Phi = \cos^{-1}\left(\frac{l}{\sqrt{h^2+k^2+l^2}}\right)$$

$$\varphi_2 = \cos^{-1}\left(\frac{k}{\sqrt{h^2+k^2}}\right) = \sin^{-1}\left(\frac{h}{\sqrt{h^2+k^2}}\right)$$

$$\varphi_1 = \sin^{-1}\left(\frac{w}{\sqrt{u^2+v^2+w^2}}\sqrt{\frac{h^2+k^2+l^2}{h^2+k^2}}\right)$$

Section 2.4

19. Derive Eq. (2.45) from the geometry of rotations.

20. Verify relations (2.47) from relations (2.45) and (2.46).

21. For cubic phases it is customary to define the orientation as the set of direction cosines $g_{ij} = \frac{\mathbf{a}_i}{|\mathbf{a}_i|}\cdot\mathbf{e}_j$, where $\mathbf{e}_j$ comprise the right-hand orthonormal coordinate system in the laboratory or sample frame, and where it is presumed that the set of unit vectors $\frac{\mathbf{a}_i}{|\mathbf{a}_i|}$ also comprises a right-hand orthonormal set in the local crystal. Show that for this case the direction cosines are related to components of the orientation tensor by transposition: $g_{ij} = R_{ji}$.

Section 2.5

22. Prove Eq. (2.51).

23. Prove that the magnitude or the norm of a vector is invariant during a coordinate transformation.

24. Prove $tr(\mathbf{a}\otimes\mathbf{b}) = \mathbf{a}\cdot\mathbf{b}$.

25. Prove $\mathbf{S}\cdot\mathbf{T} = tr(\mathbf{S}^T\mathbf{T})$. *Hint:* The dot product of two second-rank tensors is analogous to the dot product of two vectors and is expressed as $\mathbf{S}\cdot\mathbf{T} = S_{ij}T_{ij}$.

26. Prove Eq. (2.56).

27. Prove Eq. (2.57).

28. Starting with Eq. (2.58), prove Eqs. (2.59) and (2.60).

29. If $\mathbf{W}$ is a skew symmetric tensor, show that it has only one real eigenvalue. What is its real eigenvalue?

Section 2.6

30. Prove Eq. (2.64).

31. Let $\mathbf{F} = \mathbf{I} + \mathbf{e}_1\otimes\mathbf{e}_2$. Using the polar decomposition theorem, evaluate the corresponding $\mathbf{R}$ and $\mathbf{U}$.

CHAPTER

3

Spectral Representation

Generalized Fourier Series

CHAPTER OUTLINE

Fourier series are an essential tool in many compact descriptions of complex functions. Splitting a given function into harmonic components often allows manipulation of the individual parts and subsequent solution of otherwise difficult problems. Furthermore, in many instances, a small number of terms in the Fourier series adequately describe a complex function, resulting in an efficient approximation that reduces computing time and facilitates visualization in a low number of dimensions. In this chapter, we remind readers of some of the fundamental properties of Fourier analysis, beginning with a reminder of Fourier series and moving to Fourier transforms and fast Fourier transforms. Our goal is to efficiently and rapidly capture functions that underlie microstructure description and design. We also review some of the basics of the generalized spherical harmonics and the surface spherical harmonics used as Fourier bases in representations of functions over orientation spaces.

3.1 PRIMITIVE BASIS

In the computer age there is a rapid trend toward capturing naturally continuous phenomena in digital format. As music, pictures, radio, and TV go digital there is a need to ensure that the information is captured in the most efficient manner possible. Take, for example, a black and white photograph that we wish to capture digitally. In continuous space the image may be described by $P(\mathbf{x}, \mathbf{c})$, where $\mathbf{x}$ is a vector describing a point in real space (i.e., a point on the photo), $\mathbf{c}$ is a vector in greyscale space, and P is a distribution function over these spaces. Thus, $P(\mathbf{x}, \mathbf{c})d\mathbf{c}$ is the "volume fraction" of grey $\mathbf{c}$ at point $\mathbf{x}$ for some measure $d\mathbf{c}$ in greyscale space. To arrive at a digital version of the picture we must first decide on a desired resolution—for example, the standard 640×480 size used on many computer screens. The original photograph is split into a grid of this size. We must subsequently decide on the resolution that we will adopt in the greyscale space—for example, the common 256-greyscale scheme. Then each cell in our grid is assigned the shade from the original 256 shades of grey that best matches the average shade in that grid.

One way to capture this process mathematically is to make use of the so-called indicator functions. These are discrete functions that are used to "pick out" cells of the grid that we have formed. They take the value of 1 in the cell to which they refer and 0 everywhere else. Thus, for the example picture format we would introduce two types of indicator functions—one for the spatial resolution and one for the greyscale resolution:

$$\begin{aligned} \chi_n(\mathbf{c}) &= \begin{cases} 1 \text{ if } \mathbf{c} \text{ is closest to shade in cell } \# \, n \text{ in greyscale space} \\ 0 \text{ otherwise} \end{cases} \\ \chi^s(\mathbf{x}) &= \begin{cases} 1 \text{ if } \mathbf{x} \in \text{ cell } \# \text{ sin real space} \\ 0 \text{ otherwise} \end{cases} \end{aligned} \tag{3.1}$$

where n takes a value from 1 to 256, and s takes a value from 1 to 307,200 ($= 640 \times 480$).

We may now fully describe the digitized picture using the parameter D_s^n, where for a given (n,s), the parameter gives the fraction of cell s that is covered by shade n. Since we are only allowing a single shade in each cell, D_s^n takes the value of 1 if shade n is in cell s, and the value 0 otherwise. Using the variables $\mathbf{x}$ and $\mathbf{c}$ over the two spaces (real and greyscale spaces), the picture may now be described in terms of the indicator functions (summation implied over repeated indices):

$$P(\mathbf{x}, \mathbf{c}) \approx D_s^n \chi_s(\mathbf{x}) \chi^n(\mathbf{c}) \tag{3.2}$$

To formalize the mathematics further, we formulate an inner product on any functions that might be defined over these spaces. For a real space, Ω (the area of the picture), the inner product for functions is defined by

$$f \cdot g = \int_{\Omega} f(\mathbf{x}) g(\mathbf{x}) d\mathbf{x} \tag{3.3}$$

Using this definition, the indicator functions on real space can be seen to be orthogonal to each other—the inner product of two distinct indicator functions is 0. By changing the definition of the inner product slightly we may ensure that the indicators' functions form an orthonormal basis:

$$\chi_s \cdot \chi_{s'} = \frac{1}{V_c} \int_{\Omega} \chi_s(\mathbf{x}) \chi_{s'}(\mathbf{x}) d\mathbf{x} = \delta(s, s') = \begin{cases} 1 \text{ if } s = s' \\ 0 \text{ if } s \neq s' \end{cases} \tag{3.4}$$

where V_c is the area of a cell. Using a similar definition for greyscale space, we may now determine the parameters D_s^n in a procedure commonly used in Fourier series. First, multiply Eq. (3.2) by $\chi_s(\mathbf{x})\chi^n(\mathbf{c})$ and integrate over the real and greyscale spaces:

$$\begin{aligned} \int_{\Omega}\int_{C} \chi_s(\mathbf{x})\chi^n(\mathbf{c})P(\mathbf{x}, \mathbf{c}) d\mathbf{c} d\mathbf{x} &\approx \int_{\Omega}\int_{C} \chi_s(\mathbf{x})\chi^n(\mathbf{c}) D_{s'}^{n'} \chi_{s'}(\mathbf{x})\chi^{n'}(\mathbf{c}) d\mathbf{c} d\mathbf{x} \\ &= D_{s'}^{n'} \delta(n, n')\delta(s, s') = D_s^n \end{aligned} \tag{3.5}$$

for the required coefficient.

It should be noted that this process simply "averages" or homogenizes the shade over each cell on the photograph. The result is an image similar to a bitmap. Note that we might have chosen to allow mixtures of shades in a given cell rather than only a single shade. The single color approach corresponds to the eigen-microstructures introduced later, and the mixture method corresponds to noneigen.

The indicator functions described in the preceding constitute a *primitive basis*. When we employ this basis (for both real and orientation spaces) we are in essence breaking up the function space into a series of "bins" or cells, within each of which we approximate the value of the function. In this manner, integration may be undertaken using summation over the bins, leading to a numerical scheme that converts some difficult problems into more tractable ones. The same basis is used next to present a further simple example of its use.

Example 3.1

Let $f(x) = x^2$. Take the Fourier decomposition of f using the primitive basis functions.

Solution

We decompose f into a series of functions that only have a nonzero value on a given integer interval of the real line:

$$f(x) \approx \sum_{n=-\infty}^{\infty} F_n \chi_n(x) \tag{3.6}$$

where

$$\chi_n(x) = \begin{cases} 1 \text{ if } n \leq x < n+1 \\ 0 \text{ otherwise} \end{cases}$$

are the basis functions (called *indicator functions*). Note that this decomposition does not lead to an exact value for *f(x)* irrespective of how many terms in the series are summed. This emphasizes the fact that the decomposition is simply a vehicle for simplifying certain problems by using numerical techniques.

We calculate F_n using the fact that

$$\int_{-\infty}^{\infty} \chi_n(x)\chi_m(x) dx = \begin{cases} 1 \text{ if } n = m \\ 0 \text{ otherwise} \end{cases}$$

That is, the basis functions are orthonormal; if we had chosen an interval of length other than 1, clearly the normality would not occur. We proceed by multiplying both sides of (3.6) by $\chi_m(x)$ and integrating:

$$F_m = \int_{-\infty}^{\infty} f(x)\chi_m(x)dx = \int_{m}^{m+1} x^2 dx = \left((m+1)^3 - m^3\right)/3$$

The primitive basis is in fact not a complete basis for the function space; hence, a decomposition using this basis is not strictly speaking a generalized Fourier series. The requirements for a complete basis are defined in the next section. In short, a complete basis should be able to represent any function within a specified error, ε (see Eq. 3.10). For a given primitive basis over a finite space (e.g., the space of greyscale shades described earlier) the number of cells, or indicator functions, is fixed (256 in the case of the greyscale space). Therefore, the resolution is fixed. If we choose a ε that is beyond this resolution we cannot add terms to the series to improve the approximation—we have to define a whole new set of indicator functions with the finer resolution.

On the other hand, the finite length of the series allows one to use matrix algebra to manipulate equations captured in the primitive basis, leading to inversion of some otherwise intractable problems. If a complete series is required with properties similar to the primitive basis, then *Haar wavelets* might be used. These are more sophisticated functions on real space than the indicator functions, and can thus reflect more of the structure of functions on the space. Furthermore, a complete orthogonal basis of Haar wavelets can be defined. Hence, some of the intuitive simplicity of the primitive basis is retained, while also enabling decomposition into a true generalized Fourier series. This basis will not be used in this text, but it is applicable to some of the problems we will tackle and therefore may be of interest to the inquisitive reader.

3.2 FOURIER SERIES

Bitmap image storage, mentioned in the previous section, is a highly inefficient method for storing photographic data. A much more efficient procedure uses the JPEG format. Each section of the picture is captured with a discrete cosine transformation. In effect, the picture is compared to 2D cosine functions of varying frequencies, and is then approximated by a linear combination of those that best match the picture (Figure 3.1).

One advantage of this process is that the cosine functions pick out any natural pattern (or correlation) that exists in the image; patterned areas will require fewer terms in the cosine transformation to describe them accurately. Furthermore, it turns out that the eye cannot detect rapid variations in color very well; hence, high-frequency components of the image can be deemphasized without loss of visual quality. Since the cosine functions each have a specific spatial frequency, we may simply reduce the information contained in the higher-frequency components (right and bottom areas of Figure 3.1) to reduce the amount of data required.

When we introduce spherical harmonic functions to capture information about the microstructure of materials (which may be thought of as a snapshot of the material), the idea is very similar to the use of the cosine functions used to capture JPEG image data in an efficient manner. But first we will review

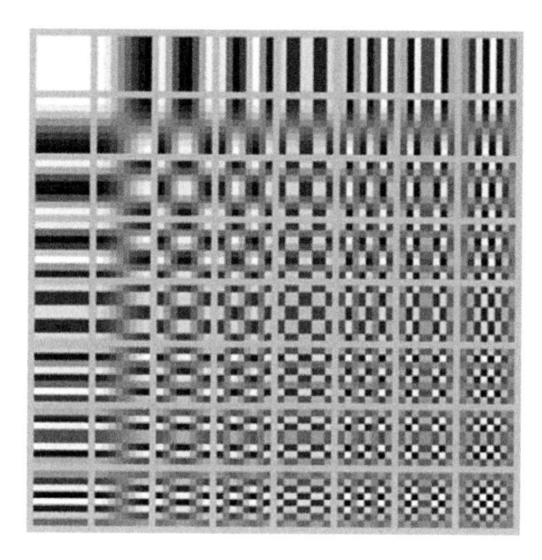

FIGURE 3.1

Visualization of the 64 cosine functions matched to a small area of the picture. *Source*: http:\\en.wikipedia.org/wiki/jpeg.

standard Fourier series techniques a little further. It is assumed readers are familiar with the standard Fourier series. A function is written as an infinite series of trigonometric functions in a technique known as *harmonic analysis*. If a different set of basis functions (i.e. not sine and cosine) is chosen such that the functions are orthogonal and complete, then a *generalized Fourier series* results.

A simple application of a Fourier series might be to decompose an electrical signal in terms of a sine series. Let the voltage of the signal, V, vary with time, t; then it may be decomposed over the interval $(0,T)$ as

$$V(t) = \sum_{n=1}^{\infty} a_n \sin(n\pi t/T) \tag{3.7}$$

The values of the coefficients, a_n, are determined by exploiting the fact that the sine functions of different orders are "perpendicular" (orthogonal) to each other for a chosen inner product. The standard inner product in function space (over the real numbers) is the integral of the function product over a given interval. In our example,

$$\int_0^T \sin(n\pi t/T)\sin(m\pi t/T)dt = \begin{cases} \frac{T}{2} \text{for } n = m \\ 0 \text{ for } n \neq m \end{cases} \tag{3.8}$$

Thus, multiplying Eq. (3.7) by $\sin(m\pi t/T)$ and integrating, we obtain

$$a_m = \frac{2}{T}\int_0^T V(t)\sin(m\pi t/T)dt \tag{3.9}$$

Clearly, the value of a_m is only valid within the interval over which integration occurs.

Another point worth mentioning is that if the signal under consideration is generated by a harmonic generator (e.g., suppose that the voltage is directly proportional to the position of an object undergoing simple harmonic motion), then the function V may be described in terms of a limited number of sine functions; that is, all of the a_n will be 0 except for those representing harmonic frequencies present in the motion. This makes the sine function a very good basis for the spectral decomposition (meaning a decomposition of the original function into a spectrum of harmonics that contribute to the signal—much the same as splitting light into colors of different frequencies). In this case, the basis functions will transform in the same way as the original function V under a given function. Thus, transformations of the function V may be studied in spectral form. This would not be possible if the basis functions were not naturally related to the original function.

Apart from mutual orthogonality, the other fundamental property of the Fourier series basis is *completeness*. Intuitively, completeness means there are "enough" functions in the basis, such that the infinite series will eventually converge to any target function. Take target function $f(t)$, defined on domain $[t_0, t_1]$; then, for given $\varepsilon > 0$, there exists N such that

$$\int_{t_0}^{t_1} \left| f(t) - \sum_{n=1}^{N} a_n \sin(n\pi t/L) \right| \langle \varepsilon \tag{3.10}$$

Clearly, the information contained in the function $f(t)$ is transferred to the Fourier coefficients a_n. In a continuous space there is an infinite number of points at which $f(t)$ is defined, and an infinite number of coefficients a_n required to fully represent the function. In practice, we often truncate the series of coefficients at some finite number, thus enabling us to accurately capture the main information for a function at an infinite number of points with a finite number of coefficients.

To make the relationship between the original function and the Fourier series coefficients more transparent we now introduce the concept of the Fourier transform.

3.3 FOURIER TRANSFORM

Harmonic analysis enables us to take a different view of a physical phenomenon. In the example of a voltage signal that may be written as a Fourier sine series, both the original signal (in the space of time) and the series view (in the space of frequency) contain the full information of the signal. However, there are different advantages to the two views. If we wish to simulate the effect of the voltage on an electrical circuit over time, we will work with the original signal. If, on the other hand, we wish to look at how the frequency of the signal changes with time, then we will consider the data in frequency space. Furthermore, in this space we can cut out unwanted noise by cutting out data for frequencies that are of no interest to the main phenomena. Before demonstrating this we introduce the concept of the Fourier transform as a convenient way to jump between these two views.

In general, information regarding physical phenomena is captured in digital form at discrete points. Hence, we will move directly to the discrete Fourier transform. Furthermore, we remind readers of the relationship between the exponential function and the standard trigonometric functions:

$$e^{ix} = \cos(x) + i\sin(x) \tag{3.11}$$

Thus, harmonic analysis using the exponential function combines the advantages of both a Fourier sine series and a Fourier cosine series.

In real situations data is generally available at a discrete set of points in space rather than on a continuous domain. In this case the resultant transform is termed a discrete Fourier transform (DFT). Suppose that we have a function, f, defined on a discrete grid of points, x_n, where n takes values from 0 to $N - 1$. Let the value of the function at x_n be f_n; then the points in the transformed space are given by

$$F_k = \sum_{n=0}^{N-1} f_n e^{-2\pi ikn/N} \tag{3.12}$$

The transform from frequency space back to real space is given by

$$f_n = \frac{1}{N} \sum_{k=0}^{N-1} F_k e^{2\pi ikn/N} \tag{3.13}$$

There are alternative definitions for Fourier transforms; hence, it is important that the actual definition that is being used by a particular computer algorithm be determined. The preceding definition is consistent with MATLAB and various other numerical analysis software tools.

While the discrete Fourier transforms may be calculated exactly as given in Eqs. (3.12) and (3.13), this is not the most efficient method. In Eq. (3.12) there are N terms that must be calculated for each F_k, and N values of k. Hence, the number of operations is of order N^2. However, algorithms have been developed that exploit redundancies in these calculations to reduce the number of operations to order $N \log N$. Such algorithms are generically termed fast Fourier transforms (FFTs). We will often refer to DFTs as FFTs, although this terminology is slightly inaccurate since the term FFT more specifically refers to the algorithm rather than the transform. As an example, if $N = 1000$, then the number of operations for the complete DFT is of order 1,000,000, while the FFT requires operations of order 7000—a dramatic reduction in time and effort (see Figure 3.2).

3.3.1 Function representation

Readers should be familiar with the standard Cartesian vector space. Any vector in $\Re^3$ can be represented as a linear combination of the three basis vectors $\mathbf{e}_1, \mathbf{e}_2, \mathbf{e}_3$. Furthermore, an inner product is defined on the vector space such that $\mathbf{v} \cdot \mathbf{w} = \sum_{i=1}^{3} v_i w_i$ and $\mathbf{e}_i \cdot \mathbf{e}_j = \delta_{ij}$, where δ_{ij} is the Kronecker delta. The representation of general functions by harmonic functions is completely analogous to the Cartesian vector space. In the case of a Fourier sine series, the sine functions provide an infinite set of basis functions (analogous to the Cartesian unit vectors) with the Fourier coefficients being analogous to the components of a vector in Cartesian space. As mentioned before, an inner product is defined in terms of integrals (Eq. 3.8), and the basis functions are orthogonal under the action of this inner product.

In the case of FFTs, any function on a grid of N points may be written in terms of N exponential basis functions. The inner product on this space is given by

$$f \cdot g = \sum_{n=0}^{N-1} f_n g_n \tag{3.14}$$

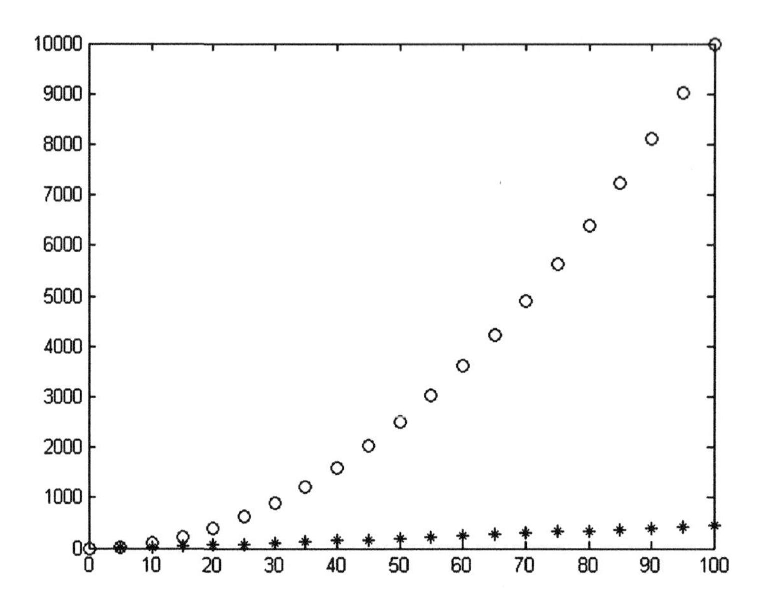

FIGURE 3.2

Schematic of the number of calculations required to take a DFT without FFTs (*circles*) and with FFTs (*asterisks*).

Then the exponential basis functions are orthogonal under this inner product:

$$\sum_{n=0}^{N-1} e^{2\pi ikn/N} e^{-2\pi ik'n/N} = N\delta_{kk'} \tag{3.15}$$

This result forms the basis for Plancherel's theorem, which relates the inner product in real space to the inner product in the frequency domain (see the following).

To illustrate the idea of a function in real space being comprised of a series of harmonic functions, suppose that we generate a 60-Hz sine "signal" on the domain [0,1] with a randomly generated noise that masks the original wave. This is illustrated in Figure 3.3(a). If we take the FFT of this signal, it is clear that the dominant harmonic component is 60 Hz (Figure 3.3b). If we were to remove the other components from the FFT and then invert the FFT we would arrive at a clean signal without the noise (Figure 3.3c).

3.3.2 Sampling

If a function is to be faithfully captured by a transform then the sampling frequency (in our case the grid spacing) must be adequate to capture the highest-frequency component of the data. The Nyquist frequency is half the sample frequency, and must be greater than or equal to the highest-frequency component if this frequency is to be accurately represented by the transform without aliasing. Two functions are aliases of each other if they are indistinguishable for the given sampling rate. For example, if we were to sample a function at integer intervals then $\sin(\pi x)$ would look exactly the same as $\sin(2\pi x)$.

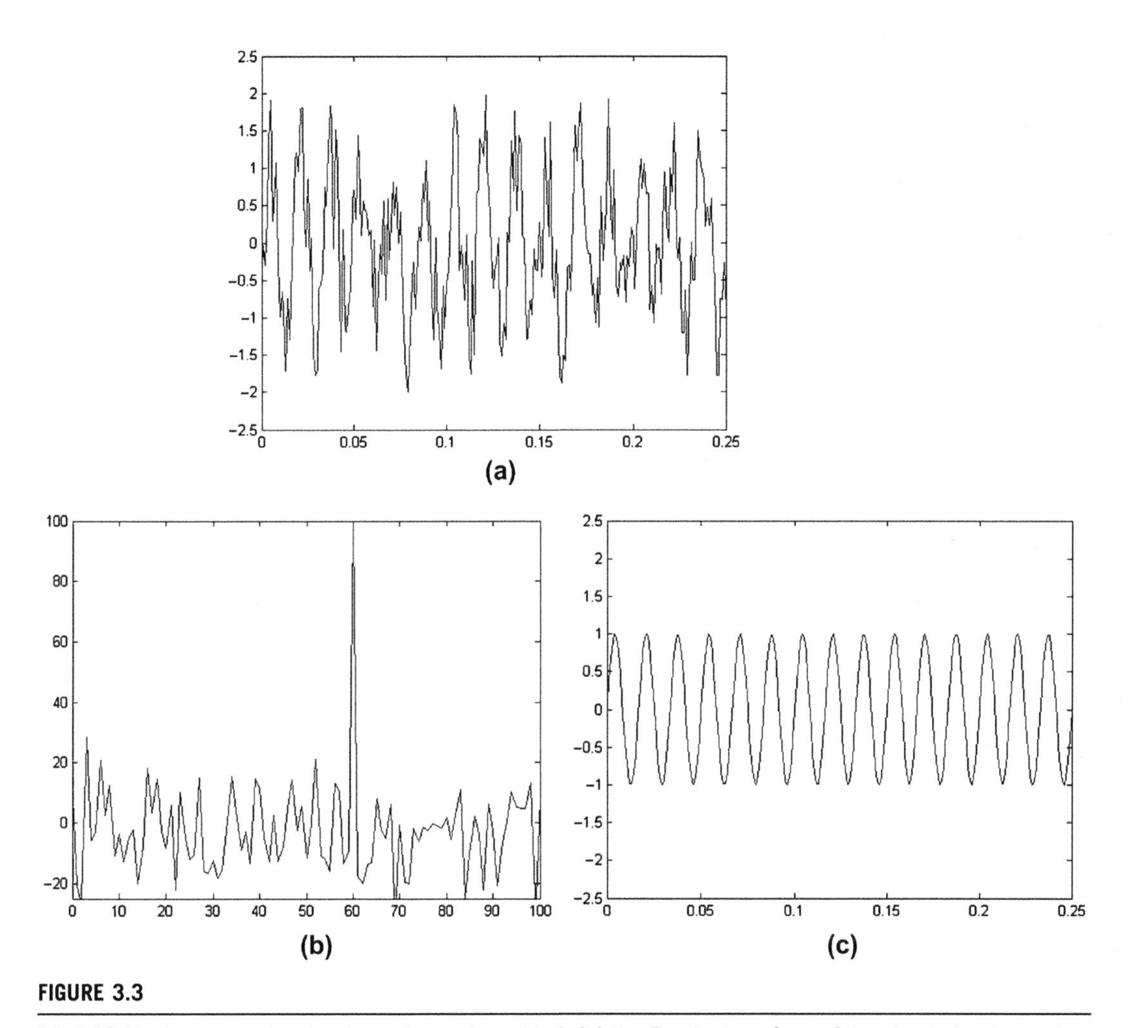

FIGURE 3.3

(a) A 60-Hz sine wave signal with random noise added; (b) the Fourier transform of the signal showing a clear 60-Hz peak; and (c) the 60-Hz component extracted from the original signal.

For the case of a grid and associated FFT that must capture the features of an image, the grid spacing must be half of the smallest feature that we wish to investigate/represent. Sampling can be an extremely important issue in the analysis of data using Fourier transforms, but we will generally not need more than the simple rule of thumb presented previously for our purposes.

3.3.3 FFTs in higher dimensions

In the next section, it will be convenient to demonstrate the power of harmonic analysis using 2-D images. To work in more than one dimension we must generalize the definition of FFTs to higher dimensions. This is easily done. Suppose that we have a function defined in two dimensions on a discrete rectangular grid enumerated by $n = 0 : N - 1$ in one direction and $m = 0 : M - 1$ in the other direction. Note that there is no requirement that the spacing be the same in the two directions, although the spacing between the points in a single direction should be the same. Assume that a function, f, is defined on the grid, with values f_{mn} at the grid points. Then the 2D Fourier transform is defined as

$$F_{jk} = \Im(f_{mn}) = \sum_{m=0}^{M-1} \sum_{n=0}^{N-1} f_{mn} e^{-2\pi imj/M} e^{-2\pi ink/N} \tag{3.16}$$

The inverse Fourier Transform is given by

$$f_{mn} = \Im^{-1}\left(F_{jk}\right) = \frac{1}{MN} \sum_{j=0}^{M-1} \sum_{k=0}^{N-1} F_{jk} e^{2\pi imj/M} e^{2\pi ink/N} \tag{3.17}$$

In this representation, the original function is being broken down into a set of harmonic waves in 2D instead of 1D. The extension to higher dimensions is similar.

3.3.4 Properties of the discrete Fourier transform

We now review some basic properties of the discrete Fourier transform. First, if f is a real-valued function (which is generally the case for the operations in this book), then clearly the imaginary components of F_k must cancel in Eq. (3.13), leading to the condition that

$$imag(F_{N-k}) = -imag(F_k) \quad \text{for} \quad k > 0 \tag{3.18}$$

Similarly (again for real-valued f),

$$real(F_{N-k}) = real(F_k) \quad \text{for} \quad k > 0 \tag{3.19}$$

Intuitively, it makes sense that only half of the Fourier coefficients are independent since each coefficient has a real and imaginary part, giving a total of N independent variables using just half of the coefficients; this is all we need to represent a function of N real values.

Periodicity and the shift theorem

While the definition of DFT given before does not rely on periodic properties of the original function, f, it is clear that the exponential function is periodic ($e^{i(x+2\pi)} = e^{ix}$). It can therefore be useful to extend f to a periodic function that repeats itself every N points. This is done by simply allowing n and k to be defined on the complete set of integers; then it is easy to see that

$$f_{n+N} = \frac{1}{N} \sum_{k=0}^{N-1} F_k e^{2\pi ik(n+N)/N} = \frac{1}{N} \sum_{k=0}^{N-1} F_k e^{2\pi ikn/N} e^{2\pi ikN/N} = \frac{1}{N} \sum_{k=0}^{N-1} F_k e^{2\pi ikn/N} = f_n \tag{3.20}$$

where kN/N being an integer ensures $e^{2\pi ikN/N} = 1$.

It is then easy to verify the shift theorem:

$$\Im\left(f_n \cdot e^{2\pi inm/N}\right)_k = F_{k-m} \tag{3.21}$$

where $\Im$ indicates the discrete Fourier transform, and the product in brackets indicates that, for each n, f_n is multiplied by the corresponding value of $e^{2\pi inm/N}$. Basically, multiplication of the original function by an exponential can be used to translate (shift) the Fourier transform of the function (in the equation the size of the translation is m). Similarly, one may perform the same trick in reverse:

$$\Im\left(F_k \cdot e^{-2\pi ikm/N}\right)_n = f_{n-m} \tag{3.22}$$

Convolution theorem

Suppose that we have two functions, f and g. In discrete 1D space the convolution of these functions is given by

$$(f * g)_m = \sum_{n=0}^{N-1} f_n g_{m-n} = \Im^{-1}(F_k.G_k)_m \tag{3.23}$$

where the product $F_k.G_k$ is taken pointwise (i.e., F_k is multiplied by the corresponding value of G_k for each k). Convolutions are used extensively in many areas of numerical analysis. For example, let f and g represent the grayscale shading on two similar images, such that image g is translated by a vector $\mathbf{t}$ with respect to image f. The convolution effectively offsets image g by a vector value $\mathbf{r}_m$ with respect to image f, subsequently takes the product of each of the grayscale values of the points that now lie on top of each other, and finally takes the sum of these products. Thus, the maximum value of $(f * g)$ will occur when the vector $\mathbf{r}_m$ realigns the images; that is, $\mathbf{r}_m = -\mathbf{t}$. Hence, the technique provides a rapid algorithm for correcting problems such as camera shake, and so on.

It may be helpful to work through a proof of this theorem to become more familiar with the discrete Fourier transform. As described previously, the convolution of the two functions is given by

$$(f * g)_m = \sum_{n=0}^{N-1} f_n g_{m-n} \tag{3.24}$$

Take the DFT of both sides:

$$\begin{aligned} \Im(f * g) &= \sum_{m=0}^{N-1} \sum_{n=0}^{N-1} f_n g_{m-n} e^{-2\pi imk/N} \\ &= \sum_{n=0}^{N-1} f_n \sum_{m=0}^{N-1} g_{m-n} e^{-2\pi imk/N} \end{aligned} \tag{3.25}$$

Now let $m - n = p$. Then

$$\begin{aligned} \Im(f * g) &= \sum_{n=0}^{N-1} f_n \sum_{p=-n}^{N-1-n} g_p e^{-2\pi i(n+p)k/N} \\ &= \sum_{n=0}^{N-1} f_n e^{-2\pi ink/N} \sum_{p=-n}^{N-1-n} g_p e^{-2\pi ipk/N} \end{aligned} \tag{3.26}$$

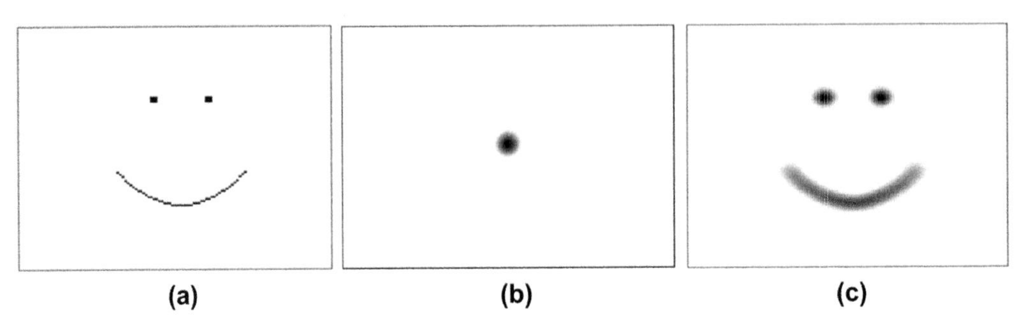

FIGURE 3.4

(a) A simple image where each pixel is either black or white; (b) a circular "pulse" of width 5 pixels that smoothly transitions between 0 and 1; and (c) the convolution of the previous two functions.

Now let us assume periodicity (cyclicity); that is, in a single dimension we have $g_m = g_{m+N}$. Since $e^{-2\pi izk/S}$ is also cyclic with the same period, we may renumber the second summation to obtain

$$\begin{aligned}\Im(f*g) &= \sum_{n=0}^{N-1} f_n e^{-2\pi ink/N} \sum_{p=0}^{N-1} g_p e^{-2\pi ipk/N} \\ &= F_k \cdot G_k\end{aligned} \tag{3.27}$$

as required. The result is that in practice we may efficiently obtain the convolution of two functions using pointwise multiplication of the FFTs, which is dramatically more efficient than the brute-force method.

As an example, take a figure with very sharp features such as the one shown in Figure 3.4(a). This image is captured as a 100×100 matrix with each entry in the matrix representing a pixel. The black pixels have a value of 1 and the white pixels have a value of 0 (we have inverted the scale of a typical picture that has white as 1 and black as 0). If we take a second matrix that represents a pulse of radius 5 pixels (as illustrated in Figure 3.4b) and take the convolution of the image and the pulse, the result is to smear the image to give much softer features (as illustrated in Figure 3.4c). The operation is much faster than other methods of producing a similar result due to the efficiency of both the convolution theorem and the FFT.

Parseval's theorem

A final useful theorem is Parseval's theorem, which relates the "amplitude" of a function in real space to the amplitude in Fourier space. It arises from Plancherel's theorem, which states that

$$\sum_{n=0}^{N-1} f_n \cdot g_n^* = \frac{1}{N} \sum_{k=0}^{N-1} F_k \cdot G_k^* \tag{3.28}$$

where the asterisk indicates a complex conjugate. If we replace g by f in this formula we arrive at Parseval's theorem:

$$\sum_{n=0}^{N-1} |f_n|^2 = \frac{1}{N} \sum_{k=0}^{N-1} |F_k|^2 \tag{3.29}$$

since $f_n \cdot f_n^* = |f_n|^2$.

Similarly, if we define $g_n = 1$ for all n, then $G_0 = N$ and $G_k = 0$ (for $k \neq 0$); likewise, if we define $G_k = 1$ for all k, then $g_0 = 1$ and $g_n = 0$ (for $n \neq 0$). Hence,

$$\sum_{n=0}^{N-1} f_n = F_0 \text{ and } \sum_{k=0}^{N-1} F_k = Nf_0 \tag{3.30}$$

3.3.5 Filtering and data compression

As indicated before, an important advantage of the Fourier transform is the ability to filter out unwanted "frequencies" from the original data, thus enabling data compression. This approach is used extensively in acoustics and imaging. For example, when the music industry moved from analog format (such as that captured on a vinyl record) to digital format, it was clear that an infinite number of bits would be required to fully capture a musical experience. However, the higher frequencies contained in music (e.g., higher harmonics from different instruments) are not audible to the human ear and therefore need not be replicated. In modern highly compressed music formats these high frequencies are eliminated without significant loss of music quality. Such filtering of the music can be achieved by taking the Fourier transform of the music signal and ignoring the coefficients F_k for all k greater than a certain cut-off frequency.

This smaller number of coefficients can be efficiently stored in digital format, and the inverse transform can be applied to them when the music is listened to. Similarly, when a camera takes a photograph with a high-resolution digital camera, it is memory intensive to store the color and intensity of each pixel. Furthermore, in the same way that the ear does not decode high-frequency sound for the brain, the eye does not decode high frequencies (i.e., rapid changes in color or intensity in a small area) in an image. This fact is exploited in the JPEG image format, which uses harmonic analysis to remove high frequencies from the image, and stores only those Fourier coefficients needed to reproduce an image that is almost equivalent to the original image when viewed by the human eye.

We demonstrate this approach using the black-and-white image in Figure 3.5. The original high-resolution image is given on the left. In mathematical form this image is given as a matrix of size 1084×811, where each entry in the matrix gives the grayscale value (from 0 to 250) of the pixel that it represents. Since these entries are real, the symmetry given by Eqs. (3.18) and (3.19) means that only the first 543×406 values of the Fourier transform are required to fully represent the original function. Suppose that we wish to remove 50% of the frequencies in each dimension, starting with the high frequencies (while maintaining the relations (3.18) and (3.19) to ensure that the resultant transform relates to a real image); we may remove these higher-frequency terms by setting all F_k to 0 for all k's in the range $272 \leq k_1 \leq 814$ in the first dimension and $204 \leq k_2 \leq 609$ in the second dimension. If we then take the inverse transform we arrive at the image in Figure 3.5(b). To the human eye it looks almost identical to the original image. If we remove even more frequencies, the image will eventually become blurred, as is shown in Figure 3.5(c), where 90% of the high frequencies in each dimension are removed; in other words, we have removed 81% of the information regarding the picture (effectively compressing the information to 19% of the original data), and yet we still retain a reasonable resemblance to the original!

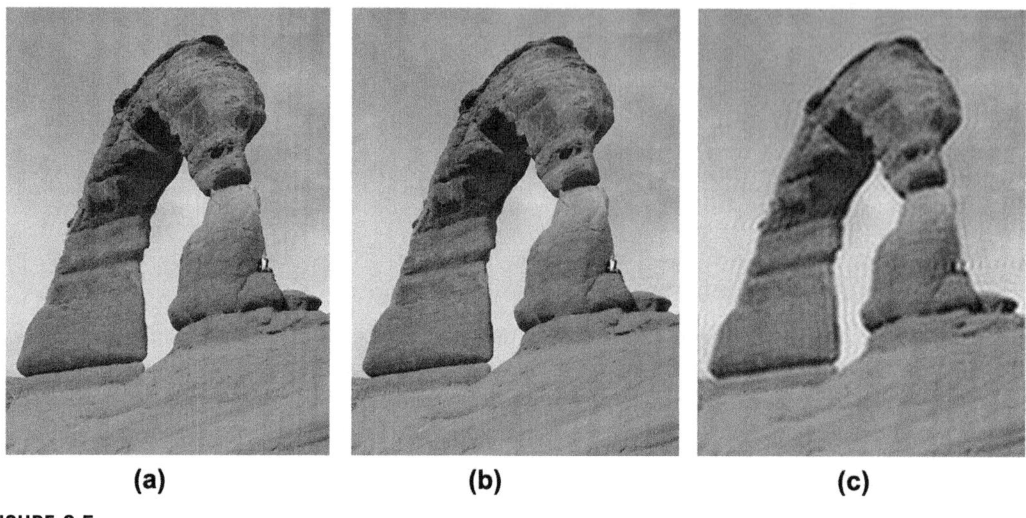

FIGURE 3.5

(a) Original grayscale image; (b) the same image after taking an FFT, removing 50% of the high frequencies in each dimension and recovering the resulting image; and (c) the same image after removing 90% of the high frequencies in each dimension in Fourier space.

Readers should note that the JPEG format regularly compresses a bitmap file by more than 90%, with minimal loss in quality, by using more sophisticated compression/filtering algorithms than the simple example that we have shown here.

3.3.6 Interpolation

In the previous section we discussed data compression, where minimal information is stored for adequate reproduction of a function. In a similar way, we may wish to reproduce a function starting from insufficient information. This is the idea behind interpolation. Readers should be familiar with simple linear interpolation. If the value of a function is given at two points, then the value midway between the points may be estimated by drawing a straight line between the known points and interpolating the new value at the midpoint of the line. If the original function were linear, then this method would obviously be a good approach. However, most functions of use are not linear. In this case one may use higher-order polynomials, for example. Another option is to use harmonic functions to estimate the missing data.

Suppose that we know the values of a function on a certain grid of points in real space, and we wish to estimate the values the function would take on a grid with smaller spacing between the points. In an identical analogy to the decompression of compressed data, Fourier transforms may be used to estimate the function values at the new points. One major advantage of this approach is that all of the interpolated data points on a refined grid may be calculated at once, and very efficiently, using FFTs.

As an example, suppose that $y = x^2$, with x taking the values [0,1,2,3]. Let Y be the FFT of y, $Y = [Y_0, Y_1, Y_2, Y_3]$. Now suppose that we wish to estimate y on a grid of 8 points instead of 4

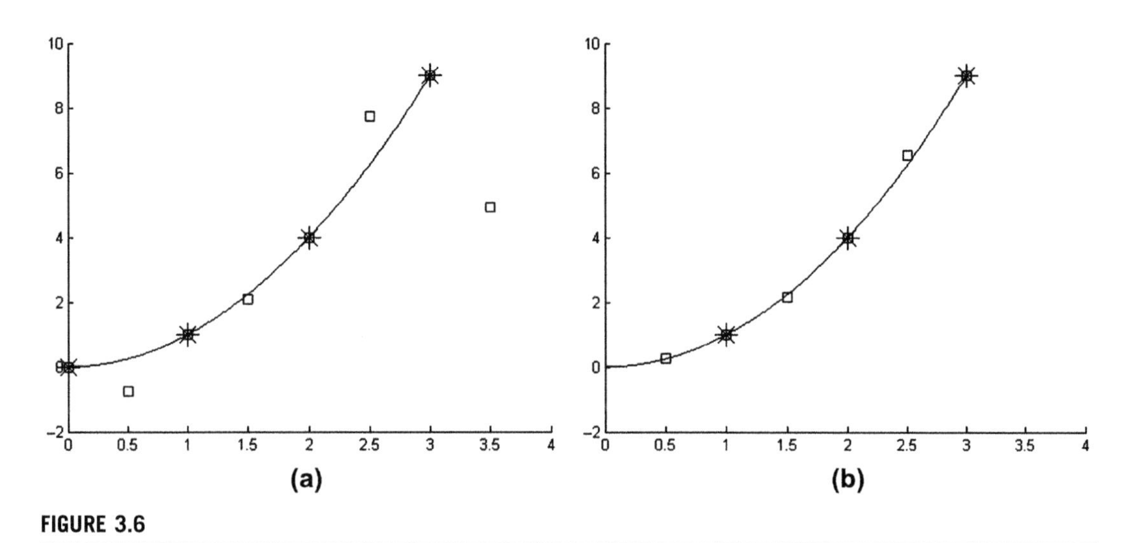

FIGURE 3.6

(a) Original points on a parabola (*asterisks*) and interpolated points using FFTs (*squares*). (b) The same example, but the FFT is taken of the original image plus its mirror image (not shown) to make it more harmonic; the result is greatly improved interpolation (*squares*).

(in fact, one of the new points will be $x = 3.5$, which we will ignore, so the useful domain will contain 7 points). We "pad" Y—that is, we insert higher-frequency components with zero amplitude: $Y' = [Y_0, Y_1, Y_2/2, 0, 0, 0, Y_2/2, Y_3]$. Note that the values of Y' from 1 to N must be symmetric about the center point to arrive back at the real-valued function when the inverse transform is applied (see Eq. 3.18); hence, if N is even, there is an odd number of points from 1 to N, and the value of Y at the Nyquist frequency of the original function must be halved and split to provide the symmetry.

If we now take the inverse FFT, the result is a function, y, that does not interpolate the x^2 function very well. This is shown in Figure 3.6(a). The reason for the poor interpolation is that we are trying to interpolate a nonharmonic function, x^2, with a harmonic function, e^x. However, if we take the values of x from -3 to 3 instead of 0 to 3, the function looks more like a harmonic function, and we get a much better interpolation (Figure 3.6b; negative values of x are not plotted). Similarly, interpolation of any nonharmonic function may sometimes be improved by taking its mirror image and appending it to the original function to make it look more harmonic.

3.4 GENERALIZED SPHERICAL HARMONIC FUNCTIONS

Readers will appreciate that different basis functions for Fourier series suit different problems. For example, the sine, cosine, Bessel, and Legendre functions are used wherever they most naturally, and efficiently, capture the nature of the problem. *Generalized spherical harmonic functions* (GSHFs) and *surface spherical harmonic functions* (SSHFs) are natural bases for functions involving the lattice orientation, R, of crystals and the interface inclination, $\hat{n}$, respectively. Specific subsets of the full set of basis functions may be chosen that naturally incorporate the symmetry of a particular crystal lattice (e.g., a cubic or hexagonal lattice), thus limiting the number of spectral functions required in the representation. A basis chosen in this manner also transforms in the same natural way as the crystal orientation, thus maintaining the essential crystal attributes throughout the solution of a problem. These bases are often referred to as the classic bases in the description of crystal orientations.

A complete treatment of the classic basis functions lies beyond the scope of this work. However, a brief description of their origin, their notation, and the central properties that enable them to be used in Fourier representation is given here. Before embarking on a more technical description, it should be borne in mind that conceptually the GSHFs are identical to the cosine functions used to capture a picture as described in Section 3.3, except that we are now describing the orientation of materials rather than color. The graphs of GSHFs given in Appendix 2 may be compared with the cosine functions shown in Figure 3.1. A typical GSHF is denoted as $T_l^{mn}(g)$. The three indices might be combined into a single index, but the separation lends several advantages that will become apparent later as they relate to the symmetry of the particular basis function.

The method of introduction parallels the description provided by Van den Boogaart (Van den Boogaart, 2002). The original theory of these functions, as defined in terms of Lie groups and the theory of representation, is described by Gelfand et al. (Gelfand et al., 1963). Since commercial algorithms are available for use by the research community, based on the work of Bunge (1993) and others, the focus here is on understanding some general characteristics of the set of basis functions. A comprehensive listing of the functions themselves, useful algorithms, and various tools for their manipulation are found in the work of Bunge (1993). Note that a full understanding of the derivation of these functions is not necessary for actually using them.

3.4.1 Formulation of the GSHFs

The *complex representation* $\mathbf{T}(R)$ of the special orthogonal group $SO(3)$ is an injective mapping from $R \in SO(3)$ to the set of square matrices with real or complex entries,

$$\mathbf{T} : SO(3) \rightarrow C^{d \times d} \tag{3.31}$$

that fulfill the property

$$\mathbf{T}(R_1)\mathbf{T}(R_2) = \mathbf{T}(R_1 R_2) \tag{3.32}$$

Note that in (3.31) $C^{d \times d}$ represents the square $d \times d$ matrices with complex entries. The only common example of this mapping is to the set of orthogonal 3×3 matrices described in relation (2.4.7) of Chapter 2. However, in representation theory it has been shown that an infinite number of possible representations satisfying (3.31) and (3.32) exist. Most of the representations of higher dimension are related to those of smaller dimension. Some representations are related to others by transformation of the coordinate basis. For these it can be shown that for every R, the two representations are related by

$$\mathbf{A}\mathbf{T}_a(R)\mathbf{A}^{-1} = \mathbf{T}_b(R) \tag{3.33}$$

where $\mathbf{A} \in O(d)$ belongs to the set of orthogonal coordinate transformations in dimension d. Representation theory also shows that there exists an infinite sequence $\mathbf{T}_0, \mathbf{T}_1, \ldots$ of representations of increasing dimension d that cannot be obtained by combining those of lower dimension. These are called the *characteristic representations of* $SO(3)$, and they are unique up to the equivalence relation expressed in (3.33).

The characteristic representations of $SO(3)$ are defined in the following way (Gelfand et al., 1963; Bunge, 1993):

$$\mathbf{T}_l\left(R\right) = \left(T_l^{mn}\left(R\right)\right)_{m=-l,\ldots,+l,n=-l,\ldots,+l} \in C^{(2l+1)\times(2l+1)} \tag{3.34}$$

with

$$T_l^{mn}\left(R\right) = e^{im\phi_2} P_l^{mn}\left(\cos\Phi\right) e^{in\phi_1} \tag{3.35}$$

in terms of the Euler angles introduced in Section 2.3. The functions $P_l^{mn}(\cos\Phi)$ are defined by the relation

$$P_l^{mn}(x) = \frac{(-1)^{l-m} i^{n-m}\sqrt{\dfrac{(l-m)!(l+n)!}{(l+m)!(l-n)!}}}{2^l (l-m)!\sqrt{(1-x)^{n-m}(1+x)^{n+m}}} \frac{d^{l-n}}{dx^{l-n}}\left((1-x)^{l-m}(1+x)^{l+m}\right) \tag{3.36}$$

For $-1 \le \cos\Phi \le +1$ they are purely real if $m+n$ is even, or purely imaginary if $m+n$ is odd. For calculation purposes, we use the following equations with coefficients a_l^{mns} taken from a data file (note that for the graphs in Appendix 2 the data was scaled to be between -1 and 1):

$$\begin{aligned} P_l^{mn}(x) &= \sum_{s=0}^{l} a_l^{mns} \cos sx \quad \text{for} \quad m+n \text{ even} \\ P_l^{mn}(x) &= \sum_{s=0}^{l} a_l^{mns} \sin sx \quad \text{for} \quad m+n \text{ odd} \end{aligned} \tag{3.37}$$

The functions $T_l^{mn}(R) = T_l^{mn}(\phi_1, \Phi, \phi_2)$ are called the GSHFs. They form a complete basis for $L^2(SO(3))$, which is the set of all square integrable real-valued functions on $SO(3)$. This is a Hilbert space (infinite dimensional), with the *scalar product*

$$(f_1, f_2) = \iiint_{SO(3)} f_1(R) f_2(R) dR \tag{3.38}$$

Thus, any real-valued function $f(R) \in L^2(SO(3))$ can be expressed in Fourier series with complex coefficients F_l^{mn}:

$$f(R) = \sum_{l=0}^{\infty} \sum_{m=-l}^{+l} \sum_{n=-l}^{+l} F_l^{mn} T_l^{mn}(R) \tag{3.39}$$

$$F_l^{mn} = (2l+1) \int_{SO(3)} f(R) T_l^{*mn}(R) dR \tag{3.40}$$

where the asterisk indicates a complex conjugate. In particular, for single-phase polycrystals where $H = SO(3)$, $f(R)$ can represent the *orientation distribution function* of the material.

3.4.2 Orthogonality relations among the GSHFs

The GSHFs comprise an orthogonal basis, with normalization taken such that the following relations hold:

$$\iiint_{SO(3)} T_l^{mn}(R) T_{l'}^{*m'n'}(R) dR = \frac{1}{2l+1} \delta_{ll'} \delta_{mm'} \delta_{nn'} \tag{3.41}$$

where the asterisk denotes a complex conjugate. dR is the invariant measure, which in the Euler-angle parameterization is given as

$$dR = \frac{1}{8\pi^2} \sin\Phi d\phi_1 d\Phi d\phi_2 \tag{3.42}$$

As an example of the usefulness of this relation, take a single crystal of a certain material. The orientation distribution function describes a single orientation, R', and is given by

$$f(R) = \delta(R' - R) \tag{3.43}$$

where δ is the usual Dirac function. For such a distribution, the coefficients in Eq. (3.39) are readily found using the standard technique of taking the integral inner product of the equation with a particular basis function:

$$\begin{aligned} F_l^{mn} &= (2l+1) \int_{SO(3)} f(R) T_l^{*mn}(R) dR = (2l+1) \int_{SO(3)} \delta(R' - R) T_l^{*mn}(R) dR \\ &= (2l+1) T_l^{*mn}(R') \end{aligned} \tag{3.44}$$

Since a general orientation distribution function can be expressed as a convex combination of all the distribution functions of the single crystals in the sample, we have

$$f\left(R\right) = \sum v_i \delta(R - R_i), \qquad \sum v_i = 1, \quad v_i > 0 \tag{3.45}$$

where v_i denotes the volume fraction of the material associated with orientation R_i. Thus, the hull of all possible distribution functions (the texture hull) may be obtained by finding the Fourier coefficients for individual crystals (Eq. 3.44) and taking convex combinations of the resulting distributions. In practice, a finite number of orientations are chosen from the fundamental zone, in a similar way to the division of orientation space for the primitive basis, and the hull is obtained from convex combinations of these "eigenstates" (see the examples in Chapter 10).

3.4.3 Addition theorem and other useful relations

Products of GSHFs satisfy the following relationship, called the *addition theorem* (addition of two rotations R and R'):

$$T_l^{mn}(R \cdot R') = \sum_{s=-l}^{+l} T_l^{ms}(R) T_l^{sn}(R') \tag{3.46}$$

If $R = I$, the identity element, it is seen that

$$T_l^{mn}(I) = \delta_{mn} \tag{3.47}$$

Since the elements of the characteristic representation $\mathbf{T}_l(R)$ are unitary matrices, it follows that (again, the asterisk denotes a complex conjugate)

$$T_l^{mn}(R^{-1}) = T_l^{*nm}(R) \tag{3.48}$$

3.5 SURFACE SPHERICAL HARMONIC FUNCTIONS

Representation of microstructures also requires considerations of real-valued functions of the type $f(\hat{\mathbf{n}}) \in L^2(S^2)$. Here $L^2(S^2)$ is the set of square-integrable functions on the unit sphere in three dimensions, S^2, and $\hat{\mathbf{n}} \in S^2$ denotes a vector from the origin of the unit sphere to a particular location on the surface of the unit sphere. Representations of functions of this type are obtained from similar considerations to those described for the GSHFs. Theoretical considerations lead to characteristic representations in terms of SSHFs, which are here labeled $k_l^m(\hat{\mathbf{n}})$, where the components of $\hat{\mathbf{n}}$ in *spherical coordinates* are

$$[\hat{\mathbf{n}}] \equiv \begin{bmatrix} \sin\alpha \cos\beta \\ \sin\alpha \sin\beta \\ \cos\alpha \end{bmatrix} \tag{3.49}$$

where α is the polar angle and β the colatitude angle in a specified orthonormal coordinate frame (Figure 3.7).

Thus, any real-valued function $f(\hat{\mathbf{n}}) \in L^2(S^2)$ can be expressed in Fourier series with complex coefficients k_l^m:

$$f(\hat{\mathbf{n}}) = \sum_{l=0}^{\infty} \sum_{m=-l}^{+l} F_l^m k_l^m(\hat{\mathbf{n}}) \tag{3.50}$$

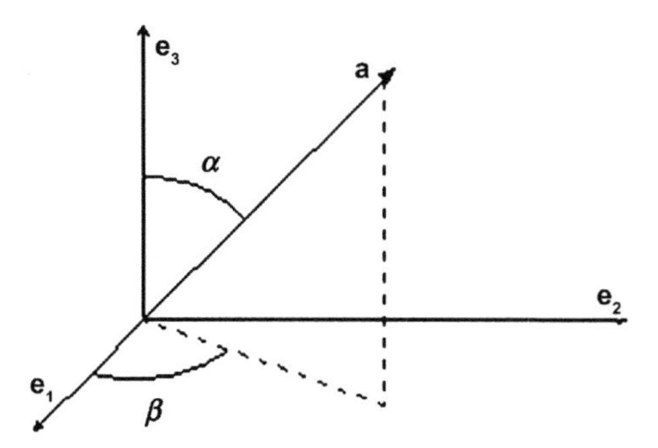

FIGURE 3.7

Schematic displaying spherical polar coordinates of the vector **a**.

$$F_l^m = \int_{S^2} f(\hat{\mathbf{n}}) k_l^{*m}(\hat{\mathbf{n}}) d\hat{\mathbf{n}} \tag{3.51}$$

When suitably normalized, the SSHFs can be expressed as

$$k_l^m(\alpha, \beta) = \frac{1}{\sqrt{2\pi}} e^{im\beta} \overline{P}_l^m(\cos\alpha) \tag{3.52}$$

where $\overline{P}_l^m(\cos\alpha)$ are real-valued functions for $-1 \leq \cos\alpha \leq 1$, called normalized associated Legendre functions. These are defined by

$$\overline{P}_l^m(x) = \sqrt{\frac{(l+m)!}{(l-m)!}} \sqrt{\frac{2l+1}{2}} \frac{(-1)^{l-m}}{2^l l!} \left(1-x^2\right)^{-m/2} \frac{d^{l-m}}{dx^{l-m}} \left(1-x^2\right)^l \tag{3.53}$$

They are related to the $P_l^{mn}(x)$ functions of (3.36) by expressions

$$P_l^{0m}(x) = P_l^{m0}(x) = i^{-m} \sqrt{\frac{2}{2l+1}} \overline{P}_l^m(x) \tag{3.54}$$

It can also be shown that this leads to a correspondence between the GSHFs and the SSHFs of the form

$$T_l^{m0}(\Phi, \phi_2) = \sqrt{\frac{4\pi}{2l+1}} k_l^m(\Phi, \phi_2 - \pi/2), T_l^{0n}(\Phi, \phi_2) = \sqrt{\frac{4\pi}{2l+1}} k_l^n(\Phi, \phi_1 - \pi/2) \tag{3.55}$$

3.5.1 Orthogonality relations among the SSHFs

The SSHFs fulfill the orthonormalization relations

$$\iiint_{S^2} k_l^m(\hat{\mathbf{n}}) k_{l'}^{*m'}(\hat{\mathbf{n}}) d\hat{\mathbf{n}} = \delta_{ll'} \delta_{mm'} \tag{3.56}$$

where $d\hat{\mathbf{n}} = \frac{1}{4\pi} \sin\alpha \, d\alpha \, d\beta$ is the invariant measure in S^2.

SUMMARY

This chapter reviewed five different spectral representations that can be applied to microstructure–property and structure–property relations:

1. Primitive basis
2. Fourier series
3. Discrete Fourier transform
4. Generalized spherical harmonic functions
5. Surface spherical harmonic functions

In the following chapters different representations are applied to different concepts of structure and to homogenization relations to facilitate inverse design approaches in microstructure-sensitive design. In Chapter 4 we discuss various methods of quantifying the underlying structure in materials, then in Chapter 5 the spectral techniques covered in this chapter are applied to specific structure metrics.

Exercises

Section 3.3.4

1. Verify Eqs. (3.18) and (3.19) on a simple example by taking the DFT of $y = x^2$ for $x \in [-10 : 10]$. Plot the real and imaginary parts of the DFT.
2. Prove the shift theorem, Eq. (3.21).
3. Use the convolution theorem to realign two images that have been offset by finding the peak value of the convolution of the images. Here is some example code for setting up the original images using MATLAB:

```
m1 = round(rand(50,50)); % first image

m2 = circshift(m1, [-13,27]); % create second image by offsetting the
first by (−13,27) figure(1); pcolor(m1); figure(2); pcolor(m2); % view
the images
```

4. Plancherel thought his theorem was self-evident. Prove the theorem.
5. Prove that $f_n \cdot f_n^* = |f_n|^2$ as required for Parseval's theorem.

Section 3.3.5

6. Take any grayscale image and plot the original image and the image obtained by removing 50% and 90% of the information in the Fourier transform as described for Figure 3.5.

CHAPTER

Description of the Microstructure

4

CHAPTER OUTLINE

In the previous two chapters we reviewed general mathematical concepts and tools that will be used in a microstructure-design framework. In this chapter we take our first look inside real materials by considering how structure is defined in composite and polycrystalline materials. Of particular importance will be the orientation distribution function, which will be combined with spectral methods in Chapter 5 to form our most commonly used structure descriptor for the rest of the book.

The word "microstructure" refers to the myriad features of the internal structure of heterogeneous materials at a variety of length scales. At coarse-length scales it is common for microscopes to reveal a material partitioned into regions of nearly continuous phase or properties, such as grains, fibers, or precipitates. Deeper inquiry reveals the atomic structure of these constituents, including lattice constants and lattice orientation of the crystalline components that are found in many natural and engineered materials.

Still deeper inquiry leads to the study of defects in the atomic structure of constituent phases, including the state of vacancy and interstitial concentrations, the presence of fine-scale precipitates, voids, grain boundaries, and the state of dislocation in the material. Beyond the atomic structure of the material lies the realm of electronic and nuclear states of the atoms and their aggregates. All of these features of the internal structure impact the properties and the performance of the material.

A complete description of microstructure in materials is well beyond our current capabilities. Fortunately, a complete description of internal structure is not required to successfully model many important microstructure–properties relationships of materials. What is required is an unambiguous understanding of how the local properties of a material point depend on a small set of local state parameters, and how these are distributed, by volume fraction and by spatial placement, in the material. Typically, not all local state parameters are of equivalent significance to properties. It is known that vacancy concentration, for example, has negligible effect on the elasticity of most materials under ordinary conditions. The same is true of the dislocation state. What is of primary importance to elasticity is the distribution by volume fraction of phases (and their orientations in crystalline phases) and correlations in their spatial placement within the microstructure.

Taking another example, initial yield strength of crystalline materials depends on the volume distribution of phases and orientations, but it also depends on grain size. Considerations of flow stress must add to these considerations the history-dependent state of dislocation structure in the constituent phases. Those who aspire to model microstructure–properties relationships must be guided by physical theory and by empirical evidence to select those local state parameters and features of microstructure that have first-order impact on properties. This selection will be the first, and perhaps the most important, consideration affecting the success of modeling.

Random, heterogeneous materials will be our principal concern throughout this book. Examples of engineered materials belonging to this class are polycrystalline metal and ceramic alloys, fiber- and particle-reinforced composites, polyphase amorphous materials, and some classes of polymers (see Figure 4.1). Examples of natural materials in this class include rocks, sandstone, sea ice, wood, bone, and others. Representing the statistical or stochastic geometry of phase and orientation is the emphasis throughout the book. It is known that in most engineered materials these aspects of microstructure govern the anisotropy of effective (macroscopic) properties, and are therefore of central importance in highly constrained materials design.

This chapter discusses the coarser features of microstructure (often described as the *mesoscale*), focusing only on its first-order description. The first-order description of microstructure is often referred to as one-point statistics, and captures only the volume fraction information on the distinct local states present in the material. In later chapters, we extend the treatment to include a higher-order

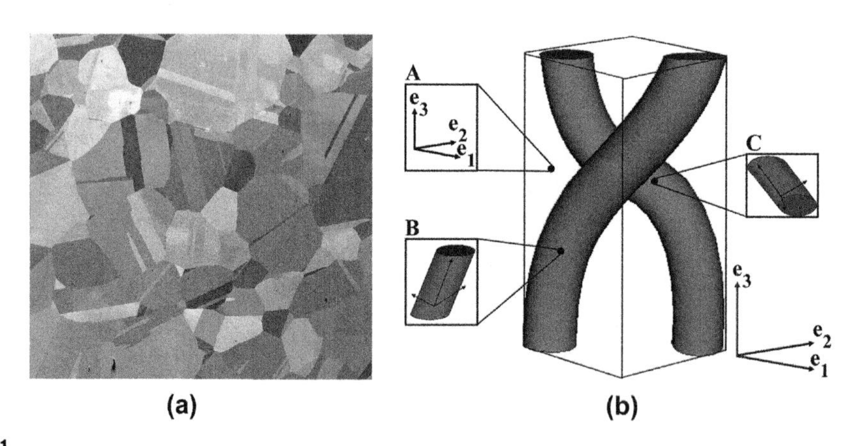

FIGURE 4.1

(a) OIM image of polycrystalline stainless steel sample. Image courtesy of Edax-TSL. (b) Schematic illustrating placement of fibers in the matrix of a reinforced composite. The Ti sample exhibits significantly lower contrast in the local elastic properties in comparison to the carbon fiber–reinforced epoxy composite.

description of the microstructure. We begin this chapter with the fundamental notion of local state and local state space that enables the rigorous definition of a local state distribution function representing the first-order statistics of the microstructure. These concepts will then be illustrated through two specific examples—a grain size distribution and a grain orientation distribution. We also show how these two distinct characteristics of local state can be coupled to facilitate a simultaneous consideration of both grain size and grain orientation distributions.

4.1 LOCAL STATES AND LOCAL STATE SPACE

A microstructure constituent that can be assigned distinct local properties can be considered to possess a distinct local state. It is implicitly assumed that this local state is either measurable at the length scale of interest or distinctly identifiable in digital representations of the microstructure, and exhibits the same local properties independent of its spatial location in the microstructure. However, these local properties are expected to depend on local state parameters. An example of one such state parameter would be temperature, T.

4.1.1 Multiphase composites

If the microstructure of interest comprises two or more phases[1] (labeled α, β, ...), the local state space could be described by the set listing the phases: $H = \{\alpha, \beta, \ldots\}$. In Figure 4.1, the carbon fiber–reinforced epoxy composite can be treated as a two-phase composite.

Beyond simply listing the phases that are present, other scalar variables might be important to the specification of the local state. Consider the example of composition in a two-component, two-phase material, assumed to be in equilibrium. Let c denote the mass fraction of component B in either phase. Let α denote the phase rich in component A, with $c_\alpha(T)$ the solubility limit for component B at temperature T (Figure 4.2). Similarly, let β indicate the phase rich in component B, with $c_\beta(T)$ its

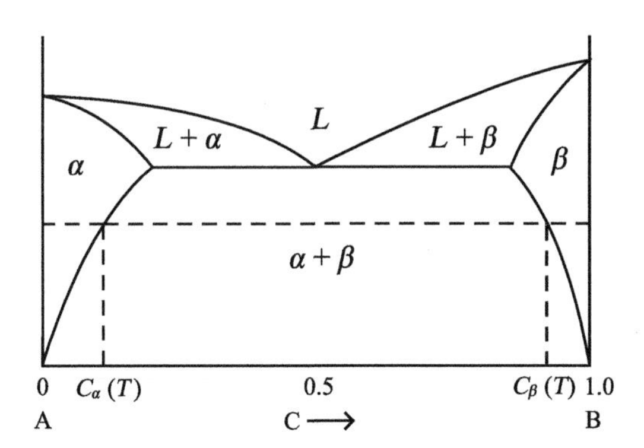

FIGURE 4.2

Sketch of two-component binary phase diagram, defining the two sets given in relation (4.3).

[1] Here the word "phase" is defined to indicate a material of distinct chemical composition and structure.

solubility limit. Considering equilibrium to be established, if $0 \le c \le c_\alpha(T)$ the system is entirely phase α, and if $c_\beta(T) \le c \le 1$ it is entirely phase β. If the overall composition of the alloy, say $\bar{c}$, lies in the range $c_\alpha(T) < \bar{c} < c_\beta(T)$, the *lever principle* requires that the system must contain mass fraction $(c_\beta(T) - \bar{c})/(c_\beta(T) - c_\alpha(T))$ of phase α at composition $c_\alpha(T)$, and mass fraction $(\bar{c} - c_\alpha(T))/(c_\beta(T) - c_\alpha(T))$ of phase β at composition $c_\beta(T)$.

For this problem, composition c belongs to the set of local state variables. Taking a set-theoretical approach, consider two sets on the domain of the (local) composition variable:

$$C_\alpha = \{c | c \in [0, c_\alpha(T))\},\ C_\beta = \{c | c \in (c_\beta(T),\ 1]\} \tag{4.1}$$

For the two-component, two-phase material, specification of the local state requires two variables, one stipulating the phase of the material point, and the second the composition. Let the ordered pair $(\rho, c) = h$ define the local state, where ρ represents the phase. Define two sets of ordered pairs:

$$H_\alpha = \{(\rho, c) | \rho = \alpha,\ c \in C_\alpha\},\ H_\beta = \{(\rho, c) | \rho = \beta,\ c \in C_\beta\} \tag{4.2}$$

It follows that the complete local state space, H, is defined by

$$H = \{(\rho, c) \in H_\alpha \cup H_\beta\} \tag{4.3}$$

Notice that compositions $c_\alpha(T) < c < c_\beta(T)$ never occur in the local state space H because they cannot occur at any material point in the system (although they occur in an average sense over larger length scales). Under conditions of equilibrium, and for average compositions lying in the range $c_\alpha(T) < c < c_\beta(T)$, as described previously the lever principle can be used to obtain mass/volume fractions of the phases present at concentrations at the solubility limits.

4.1.2 Polycrystalline microstructures

Many materials of practical importance contain crystalline phases. Usually, local properties of crystalline phases cannot be considered to be isotropic. Anisotropic local properties depend not only on phase but also on the lattice orientation of the crystalline phase relative to the sample coordinate system in which the macroscale properties of the sample are to be specified.[2]

As noted in Chapter 2, the crystal lattice orientation can be related to the sample reference frame through a set of three rotation angles called the Bunge–Euler angles, denoted by $g = (\phi_1, \Phi, \phi_2)$. The coordinate transformation matrix transforming tensors defined in the crystal reference frame to the sample reference frame was defined in terms of these three angles in Eq. (2.38). It was also noted that there was a redundancy in the definition of these angles. It can be shown that (ϕ_1, Φ, ϕ_2) and $(\phi_1 + \pi,\ 2\pi - \Phi,\ \phi_2 + \pi)$ produce the exact same transformation matrix. Consequently, the complete Bunge–Euler space defined by

$$FZ_T = \{g = (\phi_1, \Phi, \phi_2) | 0 \le \phi_1 < 2\pi,\quad 0 \le \Phi \le \pi,\quad 0 \le \phi_2 < 2\pi\} \tag{4.4}$$

[2]Of course, the local tensorial properties of the differently oriented crystals are the same in their own local crystal reference frames. However, our interest here is in defining the macroscale properties in the fixed sample reference frame. The local tensorial properties of the differently oriented crystals are different in the sample reference frame and are related by appropriate coordinate transformation laws.

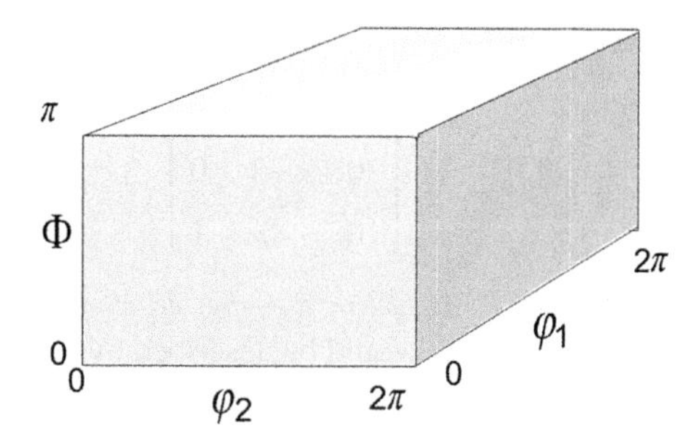

FIGURE 4.3

Geometrical description of the triclinic fundamental zone.

refers to the set of all physically distinct crystal lattice orientations that can theoretically occur in nature. This space is referred to as the *asymmetrical domain* or the *fundamental zone*, without any additional considerations of symmetry in either the crystal reference frame or the sample reference frame. The lack of additional crystal or sample symmetry is noted by the subscript T (denotes triclinic symmetry). In a similar vein, C and H refers to cubic and hexagonal crystal symmetries, respectively. Therefore, the Bunge–Euler space identified by Eq. (4.4) is referred to as the *triclinic fundamental zone*. In many written works, this space is also referred to as $SO(3)$, the space of all proper orthogonal matrices in 3D space. The triclinic fundamental zone defined in Eq. (4.4) is depicted in Figure 4.3.

Consideration of crystal symmetry will significantly reduce the size of the fundamental zones. Figure 4.4 depicts a sample reference frame, $\{\mathbf{e}_1, \mathbf{e}_2, \mathbf{e}_3\}$, and two equivalent crystal reference frames, $\{\mathbf{e}_1^c, \mathbf{e}_2^c, \mathbf{e}_3^c\}$ and $\{\mathbf{e}_1^a, \mathbf{e}_2^a, \mathbf{e}_3^a\}$. Let $Q_{ij}^{c\to s}$, $Q_{ij}^{a\to s}$, and $Q_{ij}^{a\to c}$ denote the related transformation matrices, where s denotes the sample reference frame. Using the concepts described earlier in Chapter 2, it is easy to see that

$$Q_{ij}^{a\to s} = Q_{ik}^{c\to s} Q_{kj}^{a\to c} \tag{4.5}$$

Let (ϕ_1, Φ, ϕ_2) denote the Bunge–Euler angles describing the relationship between $\{\mathbf{e}_1, \mathbf{e}_2, \mathbf{e}_3\}$ and $\{\mathbf{e}_1^c, \mathbf{e}_2^c, \mathbf{e}_3^c\}$. In other words, $Q_{ij}^{c\to s}$ can be expressed in terms of the angles (ϕ_1, Φ, ϕ_2) using Eq. (2.38). Similarly, let $(\phi_1^a, \Phi^a, \phi_2^a)$ denote the Bunge–Euler angles describing the relationship between $\{\mathbf{e}_1, \mathbf{e}_2, \mathbf{e}_3\}$ and $\{\mathbf{e}_1^a, \mathbf{e}_2^a, \mathbf{e}_3^a\}$. Then, it should be possible to find the relationship between (ϕ_1, Φ, ϕ_2) and $(\phi_1^a, \Phi^a, \phi_2^a)$ using Eq. (4.5).

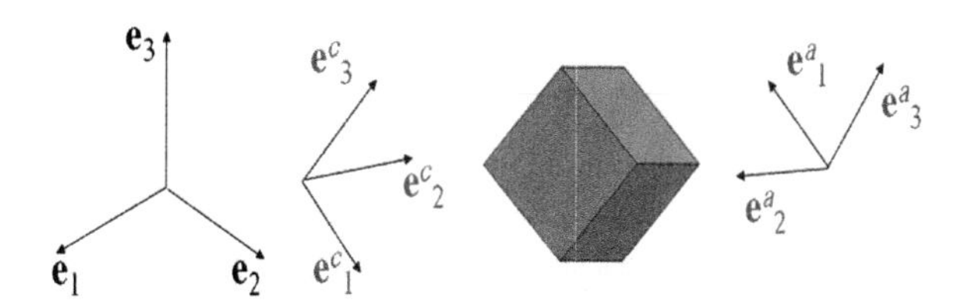

FIGURE 4.4

Illustration of crystal symmetry in the definition of lattice orientation.

As a simple example, consider the following crystal symmetry:

$$Q_{ij}^{a\to c} = \begin{bmatrix} -1 & 0 & 0 \\ 0 & -1 & 0 \\ 0 & 0 & 1 \end{bmatrix} \tag{4.6}$$

For this example, we can show that $\Phi^a = \Phi$, $\phi_2^a = \pi + \phi_2$, $\phi_1^a = \phi_1$. If this symmetry were to be imposed on FZ_T, the reduced fundamental zone would be described by$(0 \leq \phi_1 < 2\pi,\ 0 \leq \Phi \leq \pi,\ 0 \leq \phi_2 < \pi)$, since for each orientation $(\varphi_1, \Phi, \varphi_2)$ satisfying the condition $\pi \leq \varphi_2 < 2\pi$, the orientation $(\varphi_1, \Phi, \varphi_2 - \pi)$ is physically indistinguishable. Thus, only one of the two sets of Euler angles should be included in the fundamental zone.

Cubic crystal lattices belong to the symmetry group O_h. There are 24 rotational symmetry elements in that group, which means that there are 24 equivalent descriptions in FZ_T for any selected lattice orientation of a cubic crystal. These can be easily visualized by recognizing that there are six independent choices for selecting $\mathbf{e}_1^c$ (corresponding to positive and negative directions along the edges of the cube), and for each of these choices there are four independent choices for the selection of $\mathbf{e}_2^c$. Note that $\mathbf{e}_3^c$ is uniquely defined by the right-hand rule (i.e., $\mathbf{e}_3^c = \mathbf{e}_1^c \times \mathbf{e}_2^c$), after the selection of $\mathbf{e}_1^c$ and $\mathbf{e}_2^c$. The fundamental zone for cubic crystals, denoted by FZ_C, would therefore constitute only 1/24 of FZ_T.

Consideration of eight of the lattice symmetry operations associated with a cubic crystal lattice, following the basic approach described earlier, allows us to confirm that orientations corresponding to $(\varphi_1 + \pi, \pi - \Phi, 2\pi - \varphi_2)$, $(\varphi_1 + \pi, \pi - \Phi, \pi - \varphi_2)$, $(\varphi_1 + \pi, \pi - \Phi, \pi/2 - \varphi_2)$, $(\varphi_1 + \pi, \pi - \Phi,\ 3\pi/2 - \varphi_2)$, $(\varphi_1, \Phi, \varphi_2 + \pi/2)$, $(\varphi_1, \Phi, \varphi_2 + \pi)$, and $(\varphi_1, \Phi, \varphi_2 + 3\pi/2)$ correspond to the exact same crystal lattice orientation defined by $(\varphi_1, \Phi, \varphi_2)$ in the case of cubic crystal lattices. As a result of these considerations, we can reduce the fundmanetal zone of interest in cubic crystals initially to 1/8 of FZ_T, or three times that of FZ_C. The other symmetry operations not accounted for at this point do not produce simple mapping rules, and therefore will be addressed later. We shall find this specific fundamental zone to be very useful later in performing fast Fourier transforms of functions defined in the Bunge–Euler space for cubic crystals. Therefore, for future reference, we shall denote this space as $FZ3$ (noting that this space is three times larger than FZ_C) and define it as

$$FZ3 = \left\{ g = (\varphi_1, \Phi, \varphi_2) \middle| \begin{matrix} \left(0 \leq \phi_1 < 2\pi,\ 0 \leq \Phi < \frac{\pi}{2},\ 0 \leq \phi_2 < \frac{\pi}{2}\right) \\ \left(0 \leq \phi_1 < \pi,\ \Phi = \frac{\pi}{2},\ 0 \leq \phi_2 < \frac{\pi}{2}\right) \end{matrix} \right\} \tag{4.7}$$

A rigorous consideration of all 24 symmetry operations associated with the cubic crystal lattice reduces the fundamental zone to

$$FZ_C = \left\{ g = (\varphi_1, \Phi, \varphi_2) | 0 \leq \varphi_1 < 2\pi,\ \cos^{-1}\left(\frac{\cos\varphi_2}{\sqrt{1 + \cos^2\varphi_2}} \right) \leq \Phi \leq \pi/2,\ 0 \leq \varphi_2 \leq \pi/4 \right\} \tag{4.8}$$

This space is depicted in Figure 4.5. Readers should note that FZ_C is not rectangular in shape; it is this complexity that leads to the use of $FZ3$ in various calculations involving Fourier transforms.

Consideration of all 12 of the lattice symmetry operations associated with a hexagonal crystal lattice, following the basic approach described earlier, allows us to confirm that in the case of the hexagonal

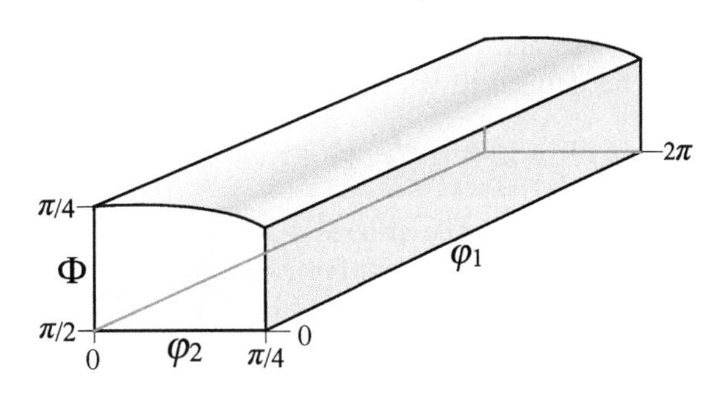

FIGURE 4.5

Fundamental zone for cubic crystal lattices.

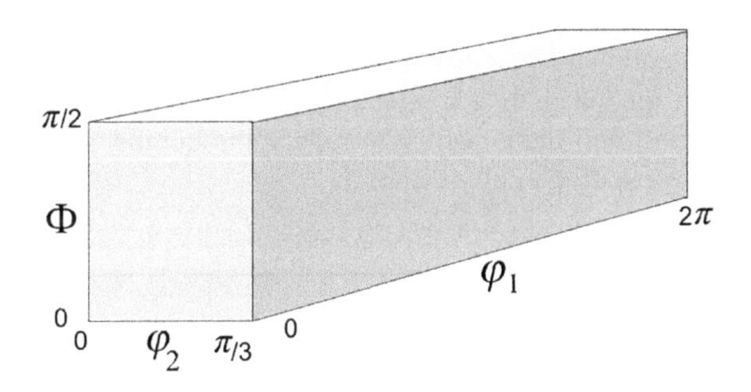

FIGURE 4.6

Fundamental zone for hexagonal crystal lattices.

crystal lattices, orientations corresponding to $(\phi_1 + \pi, \pi - \Phi, 2\pi - \phi_2)$, $(\phi_1 + \pi, \pi - \Phi, 5\pi/3 - \phi_2)$, $(\phi_1 + \pi, \pi - \Phi, 4\pi/3 - \phi_2)$, $(\phi_1 + \pi, \pi - \Phi, \pi - \phi_2)$, $(\phi_1 + \pi, \pi - \Phi, 2\pi/3 - \phi_2)$, $(\phi_1 + \pi, \pi - \Phi, \pi/3 - \phi_2)$, $(\phi_1, \Phi, \phi_2 + \pi/3)$, $(\phi_1, \Phi, \phi_2 + 2\pi/3)$, $(\varphi_1, \Phi, \varphi_2 + \pi)$, $(\phi_1, \Phi, \phi_2 + 4\pi/3)$, and $(\phi_1, \Phi, \phi_2 + 5\pi/3)$ correspond to exactly the same crystal lattice' orientation defined by $(\varphi_1, \Phi, \varphi_2)$. This leads to the following fundamental zone for hexagonal crystals:

$$FZ_H = \{g = (\phi_1, \Phi, \phi_2) | 0 \leq \phi_1 < 2\pi,\ 0 \leq \Phi \leq \pi/2,\ 0 \leq \phi_2 \leq \pi/3\} \tag{4.9}$$

This fundamental zone is depicted in Figure 4.6. It is rectangular, and presents no difficulties in terms of Fourier representations.

4.1.3 Multiphase polycrystals

For single-phase polycrystals, in the absence of any consideration other than orientation, $H \equiv FZ$. For mixtures of crystalline phases a more sophisticated description must be provided. Since any given material point may associate with any of the phases that are present in the material, H must contain all possible combinations of phase and orientation. More precisely, h consists of in the ordered pair $h = (\rho, g)$, and H can be defined to be

$$H = \{(\rho, g) | (\rho, g) \in \bigcup_{\rho} H_\rho,\ H_\rho = \{(\rho, g) | g \in FZ(\rho)\}\} \tag{4.10}$$

If additional state variables are needed to describe local properties, they can be readily appended in the manner illustrated in relations (4.3) and (4.10). For example, if the grain size l is important for a specified set of properties (e.g., yield strength), then l can be appended as follows. Suppose we are dealing with a single-phase polycrystalline material with variable grain size. Then the local state space can be expressed as

$$H = \{(l, g) \mid 0 < l \leq l_{\max}, g \in FZ\} \tag{4.11}$$

where $l_{\max}$ denotes the maximum average grain size to be considered. Geometrically, appending grain size to the fundamental zone as defined in (4.11) now requires a 4D representation, which is not easily visualized. With each added variable of local state there comes a significant increase in numerical (and experimental) complexity in solving problems, and in terms of visualization. Thus, there will always be the tension between the tendency to add more local state variables to achieve more precise descriptions of local properties, and the tendency to reduce the number to render the problem more tractable from a numerical or experimental viewpoint.

4.2 MEASURE OF LOCAL STATE SPACE

If we are to use calculus to explore the operation of functions on the local state space, H, then we must introduce a measure on the space. For discrete local state spaces as well as some continuous local state spaces, the invariant measure is trivial. However, for local state spaces containing the fundamental zone, the invariant measure is more complex. To establish a measure of the orientation space, we need to first address how many lattice orientations actually exist in the orientation space. One of the most convenient ways to think about the overall size of the orientation space is to think of how many ways we might be able to orient one of the crystal reference axes (e.g., $\mathbf{e}_3^c$) and then compound it with the number of ways we can orient one of the other remaining axes (e.g., $\mathbf{e}_1^c$). We do not need to worry about the third crystal reference axis, as it is automatically defined by the right-hand rule based on the selection of the other two axes. The orientation of a single axis in 3D space is best defined by its intersection on a unit sphere. Therefore, it is easy to see that the domain size for the selection of $\mathbf{e}_3^c$ is 4π, the surface area of a unit sphere. One can rationalize this by noting that every point on this unit sphere corresponds to a distinct choice of $\mathbf{e}_3^c$, and that all points on the unit sphere should be afforded equal weight in the selection of $\mathbf{e}_3^c$. Likewise, it is also easy to see that the domain size for the choice of $\mathbf{e}_1^c$ is 2π. Compounding these two domains, it is easy to see that the orientation space size is $8\pi^2$. Note that this definition of the orientation space size is completely independent of how we chose to represent the orientation; that is, it is independent of whether or not the orientation is represented using Bunge–Euler angles.

When the orientation is represented by Bunge–Euler angles, the relationship between $\mathbf{e}_3^c$ and $\mathbf{e}_3^s$ is fully prescribed by angles (ϕ_1, Φ). If these angles are given small increments $(d\phi_1, d\Phi)$, based on the measure of orientations defined previously, the area on the unit sphere covered by these increments is

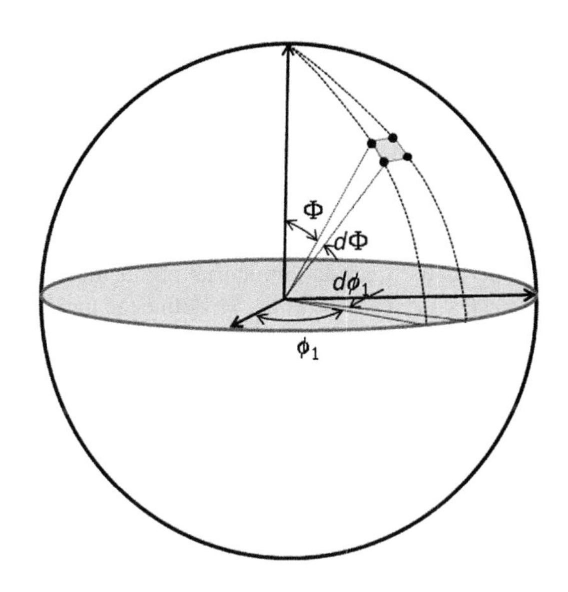

FIGURE 4.7

Illustration of the invariant measure of the orientation space.

actually sin $\Phi d\Phi d\phi_1$ (see Figure 4.7). As described earlier, the size of this areal element controls directly the number of invariant orientations associated with the imposed increments. Note also that the size of this areal element on the unit sphere is strongly influenced by the value of Φ. For example, for Φ approaching zero degrees, the area of this surface element approaches 0, and it is maximum at $\Phi = \pi/2$. It is, therefore, easy to see that the invariant number of orientations in a volume element of the Bunge–Euler space identified by $d\phi_1 d\Phi d\phi_2$ is actually $\sin\Phi d\Phi d\phi_1 d\phi_2$. In defining distribution functions, it is also convenient to normalize the measure of the local state space such that its overall size is unity. Consequently, we define an invariant incremental measure of the orientation space as

$$dg = \frac{1}{8\pi^2}\sin\Phi d\Phi d\phi_1 d\phi_2 \tag{4.12}$$

It is also important to note that this definition is consistent with our expectation that $\int\limits_{FZ_T} dg = 1$.

The normalization factors needs to be adjusted appropriately for fundamental zones associated with other crystal symmetries. For a rigorous treatment of the invariant measure for the orientation space, see Gelfand et al. (1963). A standard approach to the invariant measure, based on differential geometry, is presented by Morawiec (2004); Bunge provides a statistical motivation for the same (1993).

4.3 LOCAL STATE DISTRIBUTION FUNCTIONS

Given the random, statistical nature of microstructures, it is convenient to describe the various features they contain as distribution functions. These will be identified later as the appropriate input into microstructure–properties relations for estimating the effective properties of the material.

Having introduced the local state space, fundamental zones, and the concept of invariant measures, we are now in a position to describe the local state distribution function. When considering only the orientation characteristics of single-phase polycrystals, this function is the well-known orientation distribution function (ODF). Readers are reminded that the local state space describes the set of all possible local states of the local material points. Specifying the local state for a material point implies that we know everything that is required to assign local properties to that point. As we later construct microstructure–properties relationships we discover that the simplest of these requires that we know the volume fraction of material points that associate with each possible local state h belonging to the local state space H. Thus, we define the local state distribution function, $f(h)$, to be the volume density of materials points in the microstructure that associate with local state h. Formally,

$$f(h)dh = V_{h \pm dh/2}/V \tag{4.13}$$

where $V_{h \pm dh/2}/V$ denotes the volume fraction of the microstructure that associates with local states lying within a region of invariant measure dh of local state h. If the local state distribution function is integrated over the entire local state space H, one must obtain, by definition,

$$\int_H f(h)dh = 1 \tag{4.14}$$

Historically, one of the earliest descriptors of a heterogeneous microstructure is grain size. Materials of equivalent chemical composition, but with differing grain size, can exhibit very different properties. Remarkably, both yield strength and toughness tend to increase as grain size decreases; thus, grain size refinement is often the preferred route to improved mechanical properties.

4.3.1 Grain size distribution function

The simplest approach to measuring average grain size is to construct randomly placed (and randomly oriented) test lines across the exposed section plane, or micrograph. (Note that microstructure is naturally restricted to observations on 2D sections through the 3D material due to the opacity of materials to penetrating radiation. This limitation is explored by the methods of stereology, by which certain rigorous information on 3D microstructure is recovered from 2D sectioning. A brief introduction to stereology, with emphasis on the stereology of grain boundaries, is presented in Chapter 16 of this book.

To ensure statistical randomness, several (oblique) section planes should be examined. The number of points on each test line that intersects grain boundaries is recorded. Dividing this number by the length of the test line, one obtains the number of points per unit length of test line P_L. Averaging over all available test lines obtains $\overline{P}_L$, which can be shown to equal precisely one-half of the total surface area per unit volume of grain boundaries or interfaces, or $2\overline{P}_L = S_V$. This result is entirely independent of the shape of grains or phases as long as the sampling by test lines is random. It is convenient to define the average grain size, $\overline{l}$, to equal one over the average number of points of intersection per unit of length, $\overline{l} = 1/\overline{P}_L$. Note that this description of the grain size must be physically interpreted as the average length of intersection of a random test line with the grains or phases present in the microstructure.

Other estimates of grain size have become common because of the prevalence of digital image processing applied to micrographs. For example, methods for separating microstructural images into aggregates of distinct grains are readily obtained from noisy data using the methods of *mathematical morphology*. The image is probed using an array of small dots of various arrangements, called *structuring elements*. Thereby, fully contiguous aggregates of grains are recovered, from which the estimated grain or phase area for each grain in the micrograph is obtained. Taking the average area of the exposed grains to be $\overline{A}$, and relating this to the average grain size $\overline{l}$ through a *circular equivalent grain size* approximation, $\overline{A} = \pi \overline{l}^2/4$, we can express the average grain size in terms of the average exposed grain area, $\overline{l} = 2\sqrt{\overline{A}/\pi} \approx \sqrt{\overline{A}}$.

While the stereological results just presented deal with averages over many grains, the grain size of individual grains, l, can also be considered. Thus, in a single-phase polycrystal the local state could be the set of Bunge–Euler angles and the local grain size, $(\varphi_1, \Phi, \varphi_2, l)$. We will return later to considerations of the distribution of these four-parameter local states.

Average grain size is an example of a scalar parameter representation of microstructure. It is a single real number that represents a characteristic length associated with the microstructure of the material. Local grain size is an example of a spatially specific representation of microstructure. These two are obviously related by volume-averaging procedures. Other examples of scalar representations include average density, free energy, and others. Readers should note at this time that if the probing radiation is changed (e.g., the accelerating voltage of an electron microscope), new and different features of the microstructure might be observed, leading to new characteristic length scales. Different microscopes or conditions often reveal different characteristics of the microstructure. Given the vast amount of information contained in any heterogeneous microstructure, it is understood that physical theory must be invoked to determine the relative significance of the various components of the complete microstructure data, as they relate to the properties of interest. This is the domain of microstructure–properties (or homogenization) relations, which constitute the central focus of the discipline of materials science and engineering. We will examine several of the most important homogenization relations later in this book.

The grain size distribution is an example of a local state distribution function where the local state variable is a scalar and the local state space is a continuous subset of the space of positive real numbers. A typical (false color, but shown in greyscale here) orientation imaging microscopy (OIM) micrograph is shown in Figure 4.8. Various elements of mathematical morphology have been applied to secure this image of the nugget zone of friction-stirred 7075-T7 aluminum alloy. Some 6377 individual grain area measurements (including 321 edge grains) were recovered. Imagine constructing a table of individual area measurements on this set of grains. Label the ith individual area measurement A_i where $i = 1, 2, \ldots, I$. Each individual grain area can then be assigned its circular equivalent grain size, $l_i = 2\sqrt{A_i/\pi}$. When I is sufficiently large one can define a grain size distribution function, $f(l)$, such that $f(l)dl$ is the probability that the microstructure contains grains of equivalent grain size in the range $l - dl/2 \leq l < l + dl/2$. More precisely, the number fraction of grains that fall in a grain of size $l - dl/2 \leq l < l + dl/2$ is given by $f(l)dl$. From this definition it follows that the probability of finding grains of size between $a \leq l \leq b$ is given by

$$p(a \leq l \leq b) = \int_a^b f(l)dl \tag{4.15}$$

FIGURE 4.8

Grain map for the nugget zone of friction-stirred 7075-T7 aluminum, comprising 6377 individual grain area measurements, including 321 edge grains.

In particular,

$$\int_0^{\infty} f(l)dl = \int_0^{l_{max}} f(l)dl = 1 \tag{4.16}$$

where l_{max} is the size of the largest grain in the data set.

Figure 4.9 shows the recovered grain size distribution function, associated with Figure 4.8, plotted as a function of the equivalent grain size l. The extended tail at large grain size is typical of most grain size distributions. (Experience has shown that grain size distributions can be described in an approximate way by the log-normal distribution.)

The nth moment of the grain size distribution, also named the nth *expected value*, $\mu'_n(l)$, is readily constructed from the expression

$$\mu'_n(l) = \int_0^{l_{max}} l^n f(l)dl \tag{4.17}$$

The average grain size $\bar{l}$ is obtained from the first moment: $\bar{l} = \mu'_1(l)$. Central moments may also be constructed, of the form

$$\mu_n(l) = \int_0^{l_{max}} (l - \bar{l})^n f(l)dl \tag{4.18}$$

Note that $\mu_1(l) = 0$, and it can be easily shown that $\mu_2(l) = \sigma^2(l) = \mu'_2(l) - \bar{l}^2$, which is the familiar *variance* of the distribution.

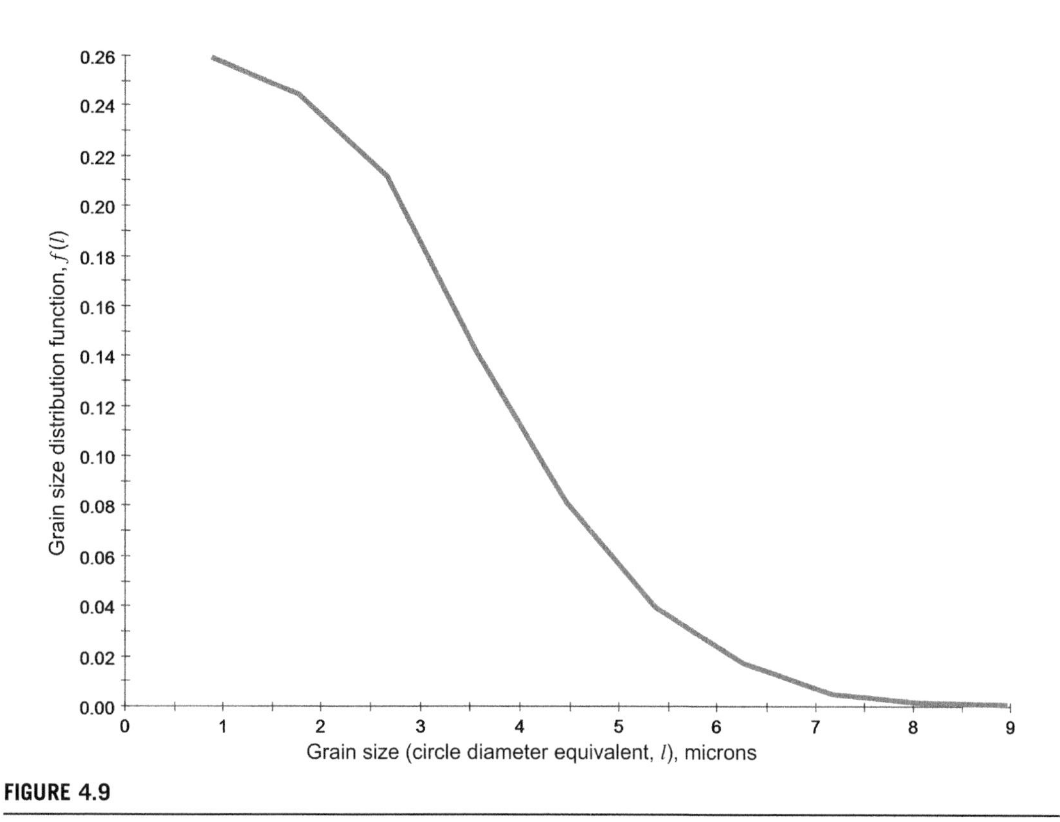

FIGURE 4.9

Grain size distribution function, $f(l)$, as a function of l for the microstructure shown in Figure 4.8. For this distribution l_{max} is approximately 9 microns.

Other single-variable distribution functions of importance to the materials engineer include molecular weight distributions (by mass and by length) for polymers, particle size distributions for precipitation-hardened alloys, and others. As with all representations of microstructure, these and other distribution functions are only useful when their relationship to the properties and performance of the materials is known, either by theory or by empirical knowledge.

4.3.2 Grain boundary inclination distribution function

As an example of a distribution function of two variables, consider grain boundary inclination. Inclination refers to the tangent normal direction associated with a point on the grain boundary network. Assuming that grain boundaries are sufficiently thin to be approximated by smooth 2D surfaces, then we may consider a point lying on the network of grain boundaries, and its associated tangent plane. The tangent plane is just the unique plane that touches the point in question on the grain boundary. This tangent plane is characterized by its unit normal $\mathbf{n}$. (Note that $\pm\mathbf{n}$ are equivalent if we do not distinguish the sides of the grain boundary.) Thus, each point belonging to the grain boundary network associates with a particular normal $\mathbf{n}$, and our interest here is in the amount of grain boundary of specified inclination that exists within the microstructure. Commonly, the unit normal $\mathbf{n}$ is parameterized relative to a selected orthonormal coordinate frame, $\{\mathbf{e}_1, \mathbf{e}_2, \mathbf{e}_3\}$, using spherical polar angles $\nu,\ \omega$. The polar angle is defined to be $\nu = \cos^{-1}(\mathbf{e}_3 \cdot \mathbf{n})$. Once ν is known, the colatitude angle ω can be determined from the expression $\mathbf{n} \cdot \mathbf{e}_1 = \sin\nu\cos\omega$. The range of possible boundary inclinations is the unit hemisphere, which is bounded by $0 \le \nu \le \pi/2,\ 0 \le \omega < 2\pi$.

Our interest here will be the grain boundary character distribution (GBCD) function, where for our present purpose "character" is defined to be the inclination of the boundary, $\mathbf{n}$. (Other features of a grain boundary, such as crystallographic features, can be added to the inclination to give a more precise definition of grain boundary character. This has been done in Figure 4.10.) Traditional notation labels the GBCD function as $S_V(\mathbf{n}) = S_V(\nu, \omega)$, which means the density (over the hemisphere) of "surface area per unit volume" of grain boundaries that have inclination $\mathbf{n}$. Note that if we define an infinitesimal patch on the surface of the hemisphere of possible $\mathbf{n}$, say the patch lying within the four points of inclination (ν, ω), $(\nu + d\nu, \omega)$, $(\nu, \omega + d\omega)$, $(\nu + d\nu, \omega + d\omega)$, we know that the surface area of this small patch is $\sin\nu d\nu d\omega$, which when integrated over the hemisphere of inclinations gives 2π, as expected. The term $d\mathbf{n} = \sin\nu d\nu d\omega$ is called the *invariant measure* associated with the space of inclinations, which is the surface of the unit hemisphere (refer to the earlier discussion of invariant measure in Section 4.2). By definition the GBCD function is defined such that $S_V(\nu, \omega)\sin\nu d\nu d\omega$ equals the total surface area of grain boundaries in the microstructure that have their inclinations lying within the patch defined previously by (ν, ω), $(\nu + d\nu, \omega)$, $(\nu, \omega + d\omega)$, $(\nu + d\nu, \omega + d\omega)$. If we integrate the GBCD function over the complete hemisphere of inclinations, we recover the total surface area per unit volume of grain boundaries, $\overline{S}_V$:

$$\int_0^{2\pi}\int_0^{\pi/2} S_V(\nu, \omega)\sin\nu d\nu d\omega = \overline{S}_V = 2\overline{P}_L \tag{4.19}$$

The GBCD function is an example of a distribution function in two variables. Distribution functions, in the context of microstructures, always tell us how much of the microstructure has a specified characteristic. In terms of the grain size distribution it was grain size, l. With the GBCD function, as defined here, it is the amount of surface area, per unit volume, of inclination $\mathbf{n}$. Examples of the GBCD function are shown in Figure 4.10. They represent the same 7075-T7 aluminum material shown in Figures 4.8 and 4.9. They have been obtained by multiple (oblique) sectioning methods using principles of stereology, described later in Section 16.7. Readers should note that the GBCD functions in Figure 4.10 are for particular types of grain boundaries, the crystallography of which has been precisely defined. CSL boundaries have a special geometrical structure due to an overlapping lattice structure. The Σ-number indicates the inverse fraction of overlapping lattice sites. These figures depict the ν, ω dependence of the GBCD function in the crystal frame for each particular CSL boundary type in the unit stereographic triangle bounded by $\{100\}$, $\{110\}$, $\{111\}$ crystallographic plane normals. These GBCD functions were recovered by stereology, similar to methods described later in Section 16.7.

For some distribution functions, such as the grain size distribution, moments and central moments are readily formed that have well-understood physical meaning, as we have seen. For other distribution functions, like the GBCD function, moments can also be formed, but their physical significance requires greater insight. In the sections that follow we pursue other, more complex distribution functions on the microstructure. Doing so requires careful definition of the domain of the function (the *local state space*), and any physical symmetry that may pertain to it.

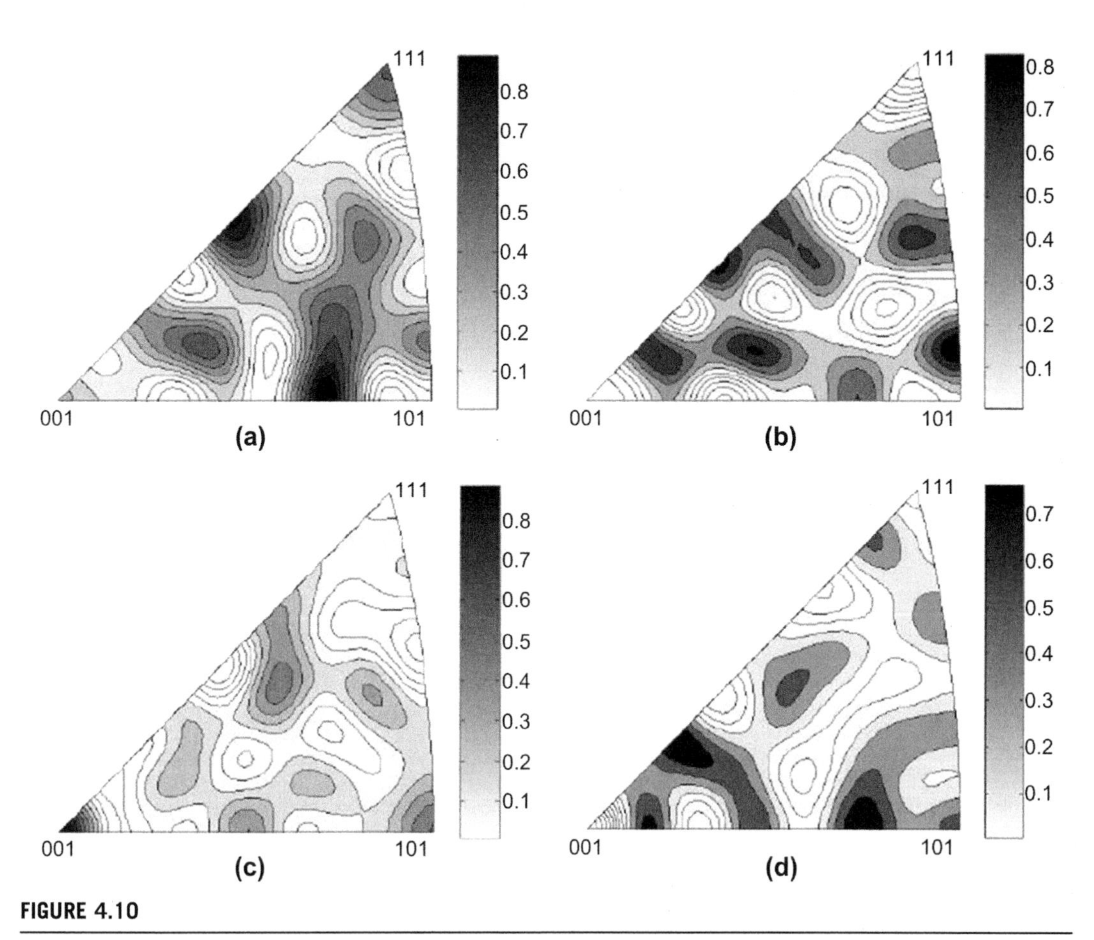

FIGURE 4.10

Recovered SV for four CSL boundaries in friction stir processed 7075-T7 aluminum: (a) Σ3, (b) Σ5, (c) Σ7, and (d) Σ9.

4.3.3 Orientation distribution function

When only the orientation characteristics of single-phase polycrystals are of interest, h reduces to the orientation of the single phase, g, and the local state distribution function reduces to the ODF, $f(h) = f(g) = f(\phi_1, \Phi, \phi_2)$. However, as noted earlier, the invariant measure for the orientation space is a little more complex, resulting in the following definition for the ODF:

$$f(g)dg = V_{g \pm dg/2}/V \tag{4.20}$$

When the ODF is expressed in the Bunge–Euler space, it takes on the following form:

$$f(\varphi_1, \Phi, \varphi_2)\frac{1}{C}\sin \Phi d\varphi_1 d\Phi d\varphi_2 = V_{g \pm dg/2}/V \tag{4.21}$$

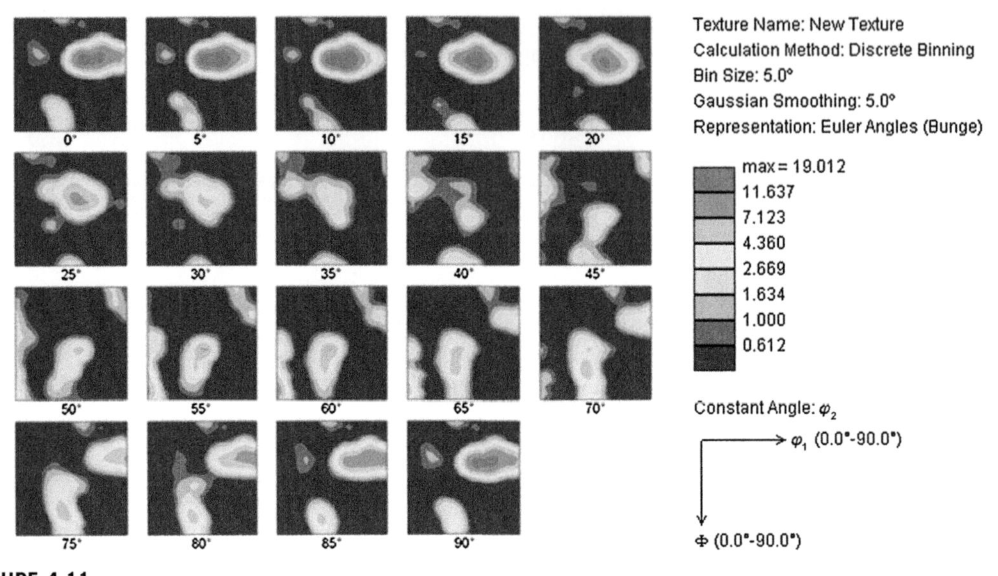

FIGURE 4.11

Partial ODF for aluminum processed by ECAP (see also Figure 4.12).

where the value of C depends on the specific fundamental zone being used in the problem. For example, the value of C is $8\pi^2$ for FZ_T (see Eq. 4.21), while it is π^2 for FZ_{3C}, $\pi^2/3$ for FZ_C, and $2\pi^2/3$ for FZ_H.

Note that the ODF is an example of a microstructure distribution function in three variables. Common graphical representations of the ODF consist of a set of 2D plots, holding one of the three Euler angles constant, while varying the other two continuously over their range. Details of the computation of the ODF will be presented following a detailed discussion of the spectral representation of the ODF using fast Fourier transforms.

Figure 4.11 shows, as an example, a portion of the ODF for heavily processed aluminum (using an equi-channel angular press, ECAP, to a strain of 10; the aluminum is high-purity PO815). In Figure 4.12 the φ_2 variable is given discretely in five-degree increments, while the Φ and φ_1 variables are described over the range of 0° to 90°. Greyscale shade indicate the strength of the ODF for any particular orientation.

4.3.4 Grain size and ODF-combined distribution function

When both lattice orientation and local grain size are considered to be important local state variables,[3] a four-parameter distribution function, called the grain size and ODF-combined distribution function (GSODF) can be defined. This function can be described as

$$f(g, l) = dV_{g\pm dg,\, l\pm dl}/V \tag{4.22}$$

[3]Steels and titanium alloys that exhibit large Hall–Petch slopes are examples of materials that may require the four-parameter GSODF.

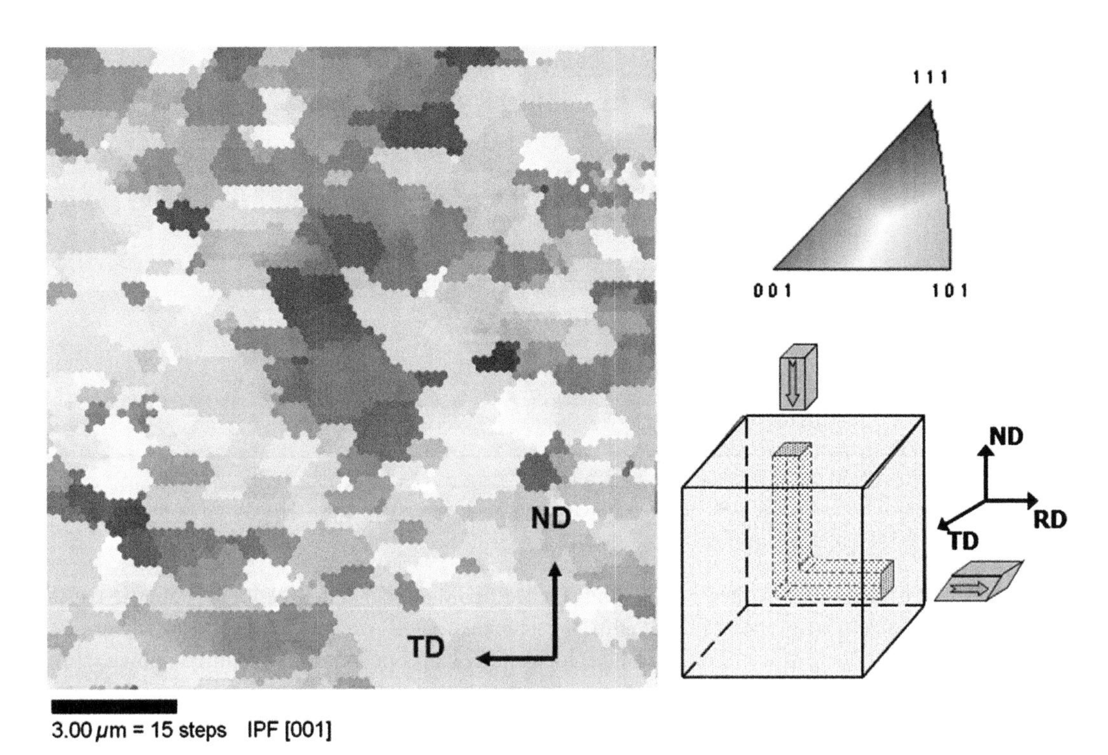

FIGURE 4.12

OIM image and [001] inverse pole figure for ECAP-processed aluminum (ODF shown in Figure 4.11).

If we let L denote the range of possible grain sizes, $L = \{l \mid l \in (0, l_{\max}]\}$, then the local state space of interest will be $FZ \times L$, or

$$FZ \times L = ((g, l) \mid g \in FZ;\ l \in L) \tag{4.23}$$

If orientation is described with Bunge–Euler angles, the GSODF can be expressed as

$$f(\varphi_1, \Phi, \varphi_2, l)\frac{1}{C \cdot l_{\max}}\sin\Phi d\varphi_1 d\Phi d\varphi_2 dl = dV_{g \pm dg/2; l \pm dl/2}/V \tag{4.24}$$

Graphical presentation of four- or higher-parameter distribution functions is obviously a significant challenge. Figure 4.13 shows a particular set of 2D graphical plots of the GSODF obtained for a rolled alpha titanium material. The local state space is $FZ_H \times L$.

4.4 DEFINITION OF THE MICROSTRUCTURE FUNCTION

At the outset we assume that a small list of parameters—that is, the local state variables or order parameters—can be specified (perhaps by experimental measurement) at each material point in the

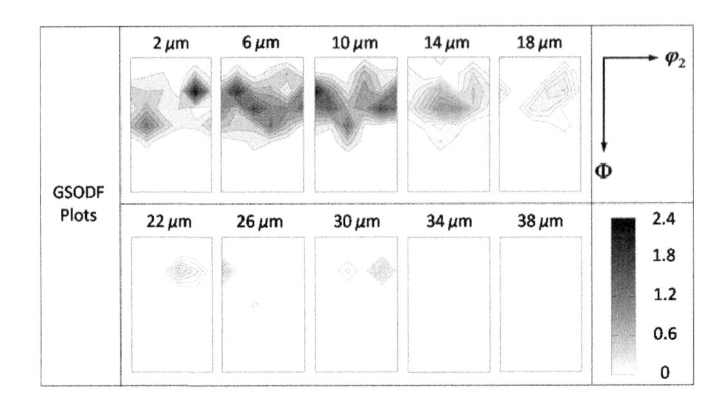

FIGURE 4.13

GSODF plots for alpha titanium. The plots shown are for constant $\varphi_1 = 0°$ sections of the Bunge–Euler angles at grain size values of 2, 6, 10, 14, 18, 22, 26, 30, 34, and 38 μm, respectively. The range of Φ and φ_2 was 0° to 90° and 0° to 60°, respectively, corresponding to FZ_H.

microstructure. Further, it is assumed that having so specified the local state, the local properties of the local material point are also specified. By this it is meant that if we had a sample of material of sufficient extent, consisting of a single (specified) local state, its properties are presumed known, either by theory or by measurement. Typical microstructure–properties models relate macroscopic effective properties, say P^*, to information on the spatial distribution of local properties in space, say $P(\mathbf{x})$. If local properties are known when the local state h is known, say $P(\mathbf{x}) = P(h(\mathbf{x}))$, then it is sufficient to specify the spatial distribution of the local state variables. To this end we introduce the *microstructure function*, $M(\mathbf{x}, h)$.

Let the function $M(\mathbf{x}, h)$ represent the mesoscale microstructure in region Ω of the material. $M(\mathbf{x}, h)$ is a real-valued function on the product space $f : \Omega \times H \rightarrow \Re$, where H denotes the comprehensive *local state space* of material points comprising the material. It is assumed that every material point $\mathbf{x} \in \Omega$ can be associated with one local state $h \in H$. However, within any finite region there could exist a distribution of local states. For example, for finite neighborhoods of a material point lying near an interface boundary, two local states may belong to the neighborhood.[4]

The *microstructure function*, $M(\mathbf{x}, h)$, associated with region Ω is formally defined by the relation

$$M(\mathbf{x}, h)dh = dV/V \tag{4.25}$$

dV/V is shorthand notation that indicates the volume fraction in an infinitesimally-small neighborhood of volume V containing position $\mathbf{x}$ that associates with local state lying within dh of local state h. dh denotes the invariant measure. Readers should compare the microstructure function with the local state

[4]Of course the interface itself is typically of finite thickness, and exhibits atomic and subatomic structure quite distinct from the structure at interior points in the phases. This structure has been studied, but a comprehensive description is still lacking. Furthermore, relatively little is known about local microstructure–properties relations of interfaces. For these reasons, the local state variables of interfaces are not included in the mesoscale representations of interest in this book. It is tacitly assumed that the Euclidean measure of the interfacial component of the microstructure is negligibly small, both in terms of its volume fraction and in terms of its effects on macroscopic properties. This assumption becomes increasingly tenuous as the intrinsic scale of microstructure declines.

distribution function described earlier in Section 4.3. The microstructure function is essentially a spatially specific local state distribution function.

The definition of the microstructure function (MF) provided previously is also motivated by certain realities of experimental microscopy, where the probe size (or information volume) of the instrument is of finite dimension. For length scales lying below the inherent probe size of the instrument, one may no longer determine spatial placement of h, only (by intensity) its distribution in the probe volume. Thus, experimental estimates of the microstructure function are limited in spatial resolution; in particular, phase boundaries, triple and quadruple junctions, etc., cannot be positioned more precisely than the probe size. Taken in the context of microscopy, definition (4.4.1) can be expressed in the following way: "In a neighborhood of point $\mathbf{x}$ of volume V, the subvolume dV occupied by material of local state lying in a neighborhood of (invariant) measure dh of local state h is $M(\mathbf{x}, h)Vdh$. Thus, $M(\mathbf{x}, h)$ is the volume density of local state h at location $\mathbf{x}$."

It is useful to establish the normalization of relation (4.25). When M is integrated over the entire pertinent region of space Ω, and over the entire local state space H, it should normalize to the value of 1:

$$\frac{1}{vol(\Omega)} \iiint\limits_{\Omega} \iiint\limits_{H} M(\mathbf{x}, h) dh d\mathbf{x} = 1 \tag{4.26}$$

This signifies that the entire region is filled with material from the local state space.

4.4.1 The microstructure ensemble

$M(\mathbf{x}, h)$ is defined for a particular region Ω in the material sample. A full characterization of the microstructure function could be destructive, as it is in the case of calibrated serial sectioning, where the sample is essentially reduced to a pile of dust when the characterization is complete. Newer experimental methodologies, such as those associated with high-energy x-ray diffraction, may ultimately result in nondestructive measurements of $M(\mathbf{x}, h)$. However, it is often not the individual microstructure function but its *ensemble average* that is of greatest importance in microstructure properties relationships.

The *microstructure ensemble* is a theoretical construct of considerable utility in modeling. Consider a very large set of regions of the same size and same shape as the original, Ω. These regions are presumed to have experienced the same macroscopic processing conditions as every other element of the ensemble. Thus, they have all been deformed alike, seen comparable temperature history, have the same chemistry, etc. And yet we do not expect the elements of the ensemble to have identical microstructure. Stochastic processes occurring at a microscopic level render each microstructure unique. These elements of the ensemble are sometimes referred to as *statistical volume elements* (SVEs).

We shall use the superscript k to identify distinct elements of the ensemble, with $k = 1, 2, 3, \ldots, K$ enumerating the entire set of elements (Figure 4.14). Let ${}^{(k)}\Omega$ indicate the regions occupied by elements of the ensemble. The ensemble average of some entity $(\cdot)$ is defined as

$$\langle(\cdot)\rangle = \frac{1}{K} \sum_{k=1}^{K} {}^{(k)}(\cdot) \tag{4.27}$$

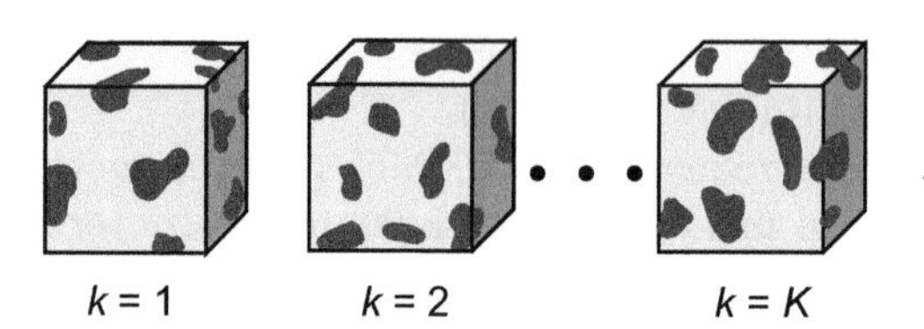

FIGURE 4.14

Sketch representing elements of the ensemble. These might be a series of samples taken from a bulk material.

In particular, define the ensemble average of the microstructure function to be

$$\langle M(\mathbf{x}, h)\rangle = \frac{1}{K}\sum_{k=1}^{K} {}^{(k)}M(\mathbf{x}, h) \tag{4.28}$$

In that which follows, the need will arise for the use of many indices to enumerate the basis functions and their coefficients. It will be useful to contract the number of indices needed for each component of the representation into a single index representing the entire set of indices pertaining to that component. The particular contractions to be adopted will be defined as they occur naturally in the development of the subject.

Also, we adopt the summation convention, that is, that repeated indices indicate summation over that index, when the repetition occurs within a single term of an equation. We shall use the convention, wherever it is possible, that upper indices are associated with local state variables and lower indices with spatial ones.

Furthermore, since there is little possibility for confusion, we will omit reference to any particular sample of the ensemble. For example, we use the symbol Ω to denote the space occupied by any particular sample of the ensemble, rather than the full notation ${}^{(k)}\Omega$, which was introduced in Section 4.4.1 for the kth element of the ensemble. In exceptional cases, where clarity demands it, the sample index will be shown as a left superscript, in parentheses: ${}^{(k)}(\cdot)$.

4.4.2 Other distribution functions from the microstructure function

In this section the relationship between the local state distribution function, introduced in Section 4.3, and the microstructure function is described. It will be emphasized that there are two distinct kinds of distributions: those that pertain to an individual microstructure and those that pertain to the ensemble of microstructures. Both are important.

The *local state distribution* for a selected microstructure function is formed by integration:

$${}^{(k)}f(h) = \frac{1}{vol(\Omega)}\iiint_{x\in\Omega} {}^{(k)}M(x, h)dx \tag{4.29}$$

${}^{(k)}f(h)$ is the volume fraction of the microstructure in the kth element of the ensemble that associates with local state h. The ensemble averaged local state distribution function is obtained by taking the ensemble average of both sides of (4.29):

$$f(h) = \langle {}^{(k)}f(h) \rangle = \frac{1}{vol(\Omega)} \iiint_{x \in \Omega} \langle {}^{(k)}M(x,h) \rangle dx \tag{4.30}$$

If $\langle {}^{(k)}M(x,h) \rangle$ is independent of position x we say that the microstructure ensemble is *statistically homogeneous* with respect to the microstructure function; for this case relation (4.30) can be expressed, for every position $x \in \Omega$, as

$$f(h) = \langle {}^{(k)}M(x,h) \rangle \tag{4.31}$$

SUMMARY

This chapter framed the concept of microstructure in terms of the local state of the material and the distribution of states that make up various distribution functions. These functions include the grain size and orientation distribution functions as typical examples. The microstructure function was introduced to capture the local distribution function at a neighborhood level. In Chapter 5 these concepts are couched in the framework of spectral methods to arrive at an efficient description of structure that forms the basis for microstructure design approaches.

Exercises

Section 4.1

1. Prove that orientations (ϕ_1, Φ, ϕ_2) and $(\phi_1 + \pi,\ 2\pi - \Phi,\ \phi_2 + \pi)$ produce the exact same transformation matrix.

Section 4.2

2. Verify that the invariant measure defined in Eq. (4.12) satisfies the requirement on the local state space that $\oint_H dh = 1$.

Section 4.3

3. Derive the relation $\mu_2(l) = \sigma^2(l) = \mu_2'(l) - \bar{l}^2$. Use it in conjunction with Figure 4.9 to estimate the variance for the 7075-T7 aluminum material.

4. Using Figure 4.9, find the approximate average grain size and variance of the grain size distribution by binning the distribution into a reasonable number of bins (~10), recovering the average grain size in each bin, and then averaging.

CHAPTER

Spectral Representation of Microstructure

5

CHAPTER OUTLINE

In this chapter we describe Fourier representation of microstructure distribution functions of one, two, and three variables. By "Fourier representation" we mean that a given function will be described as a series of special basis functions, weighted by appropriate coefficients. This series could be finite or infinite in extent. The basis functions of interest here were described in Chapter 3. When a distribution function has been expressed in a series of basis functions, we say the distribution has been *transformed into a Fourier space*. The basis functions, which are always chosen to be orthogonal to one another, form the coordinate-basis of a function space, and the coefficients weighting these basis functions are the coordinates in this Fourier space.

Readers might wonder why we would go to the trouble of representing our microstructure distributions in the Fourier space. Not only do these representations provide compact representations of local state distribution functions, they also facilitate efficient computations of the property closures that will be described in later chapters. Furthermore, transformation into the Fourier space often enables profound insights into the particular microstructure components that affect various properties. For example, one can see that first-order homogenization theory for effective elastic constants depends only on those components of microstructure that occupy the first few terms in the Fourier series (e.g., $l \leq 4$ in terms of generalized spherical harmonic functions). Careful consideration of the Fourier space region that associates with physical microstructures, within a selected framework of representation, gives rise to the concept of the microstructure hull—that convex region in Fourier space that encompasses all possible microstructures. Microstructure–properties relationships are found to be families of surfaces of various orders that intersect the microstructure hull. Examination of the full range of these intersections enables the prediction of all possible combinations of properties within the selected set of homogenization relationships and the selected framework of representation. These, in turn, lead to properties closures and microstructure design. These topics will be taken up in subsequent chapters.

The most basic Fourier series that we deal with is referred to as the *primitive basis*. The construct uses the *indicator functions* as a basis for both local state and spatial variables (see Section 3.1). (Readers are reminded that the indicator functions that form the primitive basis are not, strictly speaking, a Fourier basis, although they satisfy orthogonality conditions.) We will also illustrate the use of continuous and discrete Fourier expansions, as well as the surface spherical harmonic functions (SSHFs) and the generalized spherical harmonic functions (GSHFs) in Fourier representation of the microstructure distributions defined in this chapter.

5.1 PRIMITIVE BASIS

We begin with a primitive representation of the single-parameter grain size distribution. We follow with the common three-parameter orientation distribution function (ODF), which is commonly used to describe polycrystalline metal and ceramic alloys. The example will be for cubic polycrystals. Finally, we couple these spectral representations together to treat the GSODF.

5.1.1 Primitive basis representation of grain size distributions

Consider first the grain size distribution function, $f(l)$. The distribution shown in Figure 5.1 spans the range $0 < l \leq 9$ μm, with $l_{\max} = 9\ \mu\text{m}$. Split the domain of the distribution into a finite number of bins, say N. Define N primitive indicator functions ($n = 1, 2, \ldots, N$) of the form

$$\chi_n(l) = \begin{cases} 1 \text{ if } \left(\frac{n-1}{N}\right) l_{\max} < l \leq \left(\frac{n}{N}\right) l_{\max} \\ 0 \text{ otherwise} \end{cases} \tag{5.1}$$

Note that this particular definition of basis functions obeys the following orthogonality property:

$$\int_0^{l_{\max}} \chi_n(l)\chi_m(l)dl = \begin{cases} \frac{l_{\max}}{N} \text{ if } n = m \\ 0 \text{ otherwise} \end{cases} = \delta_{nm} l_{\max}/N \tag{5.2}$$

The (approximate) representation of $f(l)$ in terms of the indicator basis is just the finite series

$$f(l) \approx \sum_{n=1}^{N} F_n \chi_n(l) \tag{5.3}$$

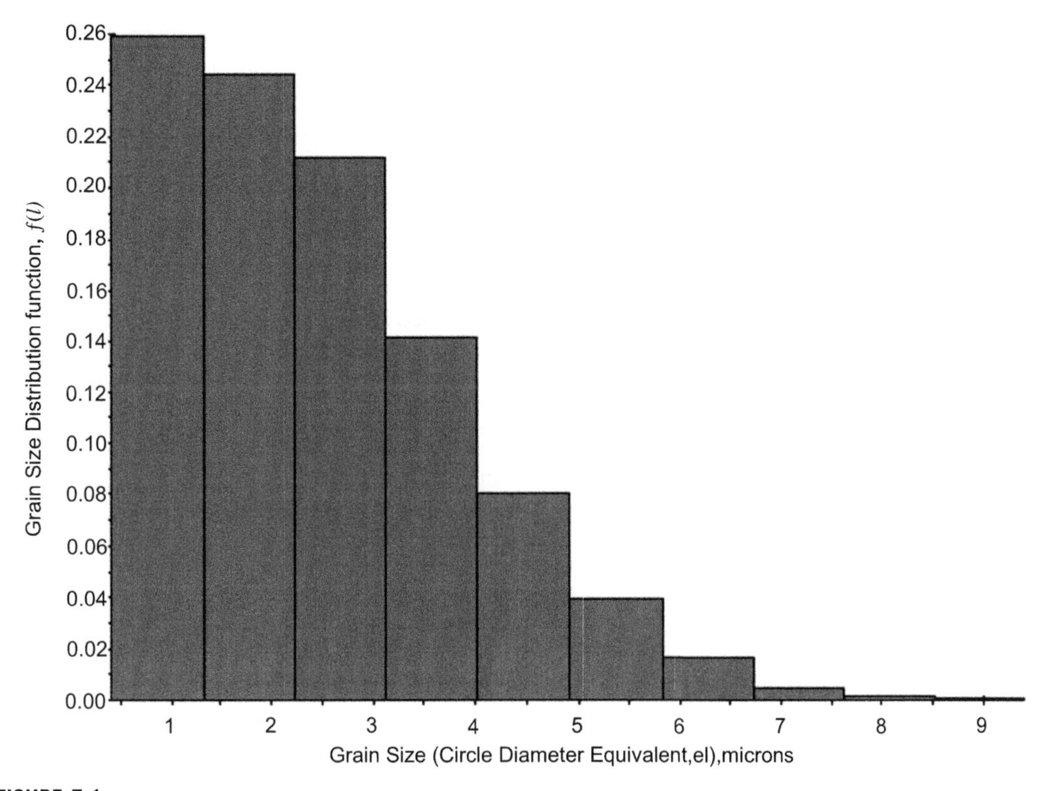

FIGURE 5.1

Tessellated grain size distribution function, $f(l)$, as a function of l for the microstructure shown in Figure 4.8. For this representation $N = 10$. (Compare with Figure 4.9.)

Multiplying both sides of this relation by $\chi_m(l)$, integrating over the complete domain, and applying (5.2) leads to the following exact and approximate expressions for the coefficients of the series:

$$F_m = \frac{N}{l_{\max}} \int_0^{l_{\max}} f(l)\chi_m(l)dl \cong \frac{\left(f\left(\left(\frac{m-1}{N}\right)l_{\max}\right) + f\left(\left(\frac{m}{N}\right)l_{\max}\right)\right)}{2} \tag{5.4}$$

Thus, the coefficients of the series expansion, F_m, are approximated by the average value of $f(l)$ over the interval spanned by the mth indicator function. Figure 5.1 illustrates a tessellation of the grain size distribution for the material shown in Figure 4.8.

5.1.2 Primitive basis representation of the orientation distributions

One can also use the primitive basis in the ODF representation defined in Section 4.3, although we will show later that the orientation distribution function is more efficiently described using spherical harmonics as the Fourier basis. We find that it is often convenient and much more intuitive to understand the fundamental concepts of microstructure-sensitive design (MSD) using the primitive basis, and to later switch the basis to spherical harmonics for significant economy in computations.

To use primitive basis for the orientation distribution function, we need to discretize the orientation space. Since we will often be dealing with discrete cells in the orientation space, we demonstrate one method of discretizing the fundamental zone for material with cubic symmetry. From these discretized subcells of the Euler-angle space, indicator functions can be defined, which comprise the primitive basis for an approximate representation of ODF. Representations for hexagonal materials are simpler, and they are readily constructed following the same procedure outlined here.

The tesserae (cells), B_n, in the fundamental zone are defined as illustrated in Figure 5.2, where it is assumed that each tesserae can be expressed as a rectangular parallelepiped. The boundaries for each side of the cube are defined according to $\varphi_{1,i}$, Φ_{jk}, and $\varphi_{2,k}$, where $i = 1, 2,\ldots, I, j = 1, 2,\ldots, J$, and $k = 1, 2, \ldots, K$ are used to enumerate the range of φ_1, Φ, and φ_2 associated with the individual tesserae. The total invariant measure of the fundamental zone is $\pi^2/3$ for lattices with cubic symmetry.

The boundary of each discretization $\varphi_{2,k}$ is determined by defining the boundaries of $\varphi_{2,k}$ at $k = 0$ and $k = K$ and then solving for the internal points according to the following equation.

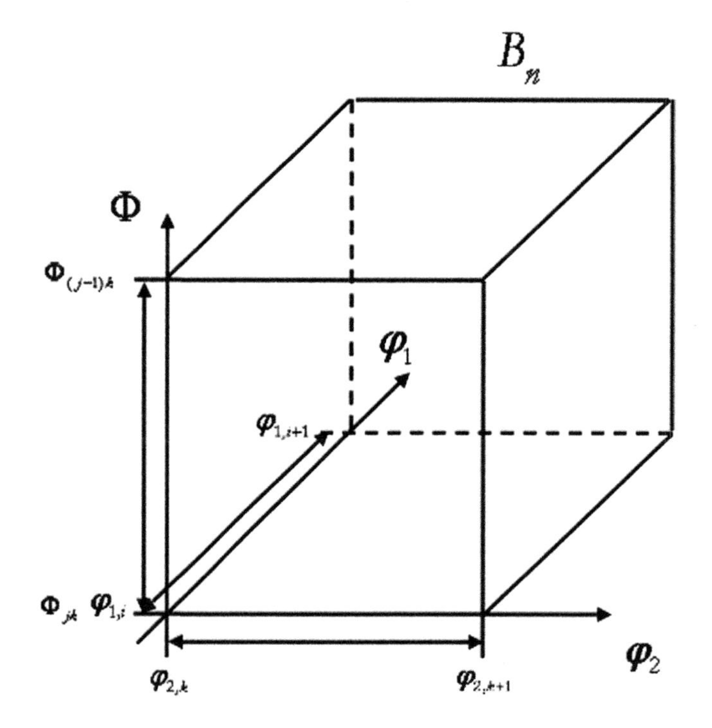

FIGURE 5.2

Definition of a generic "volume element" B_n in $SO(3)/G$, in terms of the Euler angles.

$$\int_{\varphi_{2,k}}^{\varphi_{2,k+1}} \xi d\varphi_2 = \frac{\pi}{6K}, \text{ for } k = 1..K-1 \tag{5.5}$$

where

$$\xi = \frac{1}{\sqrt{\frac{\cos(\varphi_2)^2+1}{\cos(\varphi_2)^2}}} \tag{5.6}$$

is obtained from the integration over Φ and $\pi/6K$ is the invariant measure for each discretization in φ_2.

The interior boundaries for the discretization in Φ are found similarly according to

$$\Phi_{(j-1)k} = \arccos\left(\cos\Phi_{jk} + \gamma_k\right) \text{ for } j = 1..J \text{ and for } k = 0..K-1 \tag{5.7}$$

where

$$\gamma_k = \left(\frac{\pi}{6JK}\right) \frac{1}{\left(\varphi_{2,k+1} - \varphi_{2,k}\right)} \quad \text{for } k = 0..K-1 \tag{5.8}$$

The boundaries of the discretizations along φ_1 are easily determined according to

$$\varphi_{1,i} = \frac{2\pi}{I} i \text{ for } i = 0..I \tag{5.9}$$

As an example using these methods, the fundamental zone of orientations was discretized into 512 tesserae where φ_1 is divided into 32 subdivisions and Φ and φ_2 are each divided into 4 subdivisions. The boundaries of the $\Phi - \varphi_2$ plane are given in Table 5.1, and illustrated in Figure 5.3. It is noted that for $k = 4$, Φ is not defined because the boundaries of Φ only need to be defined between each pair $\varphi_{2,k}$ for $k = 0...K$.

Having defined the boundaries of subcells in the *FZ(O)* of cubic orientation space, we are prepared to define indicator functions that will comprise the primitive basis for representation of the ODF. Thus, we assume the boundaries of all subcells B_n, for $n = 1, 2, \ldots, N$, have been defined by some procedure like that described before. Furthermore, consistent with the development in the preceding, we assume that each subcell is of equivalent invariant measure, $\pi^2/3N$.

Table 5.1 Boundaries of Tesserae in $\Phi - \varphi_2$ Plane Where $k = 0...K$

	$k = 0$	$k = 1$	$k = 2$	$k = 3$	$k = 4$
$\varphi_{2,k}$	0	0.1857	0.3747	0.5719	0.7854
Φ_{0k}	0.7854	0.7941	0.8213	0.8716	—
Φ_{1k}	1.0136	1.0249	1.0495	1.0931	—
Φ_{2k}	1.2105	1.2173	1.2324	1.2593	—
Φ_{3k}	1.3936	1.3968	1.404	1.4169	—
Φ_{4k}	1.5708	1.5708	1.5708	1.5708	—

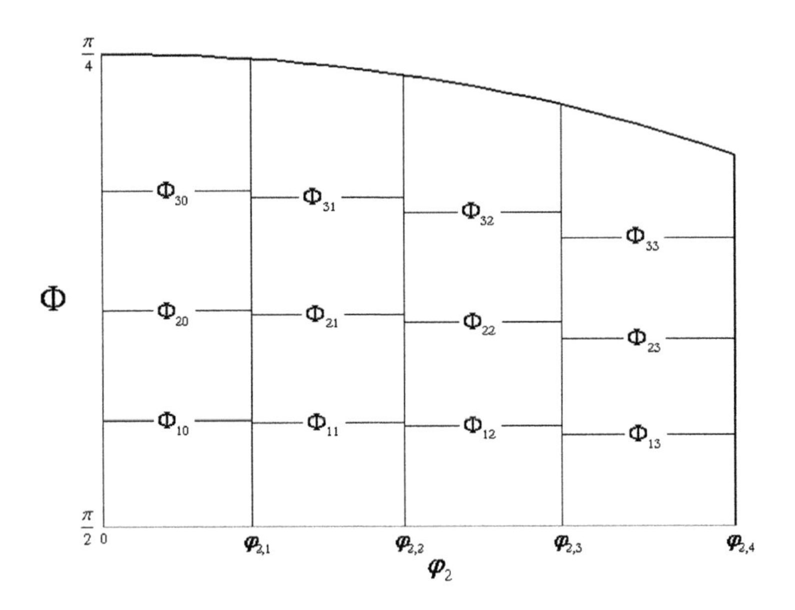

FIGURE 5.3

Boundaries of tesserae in $\Phi - \varphi_2$ plane for $k = 1...K$.

For each subcell we may then define an indicator function:

$$\chi_n(g) = \chi_n(\varphi_1, \Phi, \varphi_2) = \begin{cases} 1 \text{ if } g = (\varphi_1, \Phi, \varphi_2) \in B_n \\ 0 \text{ otherwise} \end{cases} \tag{5.10}$$

These indicator functions satisfy the following orthogonality condition:

$$\iiint\limits_{FZ(O)} \chi_n(g)\chi_m(g)dg = \pi\delta_{nm}/3N \tag{5.11}$$

The ODF can then be approximated by series representation of the form

$$f(g) \approx \sum_{n=1}^{N} F_n \chi_n(g) \tag{5.12}$$

where the primitive coefficients are defined by

$$F_m \approx \frac{3N}{\pi} \iiint\limits_{FZ(O)} f(g)\chi_m(g)dg \tag{5.13}$$

Modern methods of sampling the ODF often yield discrete orientation measurements. If J such measurements are obtained, each defining an orientation g_j, then the ODF is approximated by the expression

$$f(g) = \frac{1}{J}\sum_{j=1}^{J} \delta(g - g_j) \tag{5.14}$$

where $\delta(g - g_j)$ is the Dirac-delta function centered on orientation g_j. This function, when integrated over any region subdomain $\omega \subset FZ(O)$, must satisfy the relation

$$\iiint\limits_{\omega \subset FZ(O)} \delta(g - g_j)dg = \begin{cases} 1 \text{ if } g_j \in \omega \\ 0 \text{ otherwise} \end{cases} \tag{5.15}$$

Introducing (5.13) into (5.12), and making use of (5.14), we obtain

$$F_m \approx \frac{3N}{\pi}\left(\frac{J_m}{J}\right) \tag{5.16}$$

where J_m/J is the fraction of measurements that have orientation lying within subcell B_m.

5.1.3 Spectral representation of grain size and orientation distributions

Spectral representation of the GSODF follows by combining the techniques of representation for grain size and orientation variables. We will not give a complete description of this extension here, but indicate some of the key steps that are required. The example will be single-phase cubic polycrystals.

Binning must now cover the 4D space $FZ3 \times L_2$, where $FZ3$ and L_2 are augmented state spaces for orientation and grain size. $FZ3$ was introduced in Section 4.1.2, and the benefits of using this rectangular-shaped space over the usual irregular-shaped fundamental zone for cubic material will be elaborated on in Section 5.2.2. The augmented range for grain size results in a smoother discrete transform of the grain size distribution, as explained in Section 5.2.1. Following the preceding notation, each measurement point will fall in a bin designated by four numbers representing the discrete local state parameters, h_b, as associated with the finite set of bins enumerated by the index b: $(b_1, b_2, b_3, n) \leftrightarrow b$. The total number of bins (distinctive discrete local states) will be $B = B_1 B_2 B_3 N_l$. For single-phase cubic polycrystals there will be six crystallographically equivalent bins associated with each local state parameter. For each of the three equivalent sets of orientation parameters, there will be two equivalent grain size parameters, associated with the augmentation scheme.

The GSODF for a single-crystal/single-grain size microstructure[1] is conveniently described by a set of six Dirac-delta functions, $\delta(h - h^{(i)})$, where $i = 1, 2, \ldots, 6$ enumerates across the six equivalent local state bins, $b^{(i)}$. The single-crystal/single-grain size microstructure in the $FZ3 \times L_2$ space described here is, by analogy to relation (5.14),

[1]The term *single-crystal/single-grain size* does not correspond to any real microstructure, since the presence of grain boundaries implies that more than a single-crystal state must be present. However, from the point of view of polycrystals, these can be thought of as comprising many single-crystal/single-grain size states, represented by Dirac-delta functions and weighted by their occurrence in the microstructure.

$$f(h) = \frac{1}{6}\sum_{i=1}^{6}\delta\left(h - h^{(i)}\right) \tag{5.17}$$

The corresponding values of the discrete GSODF are given by

$$f^b = \begin{cases} \left(\int\limits_b \frac{1}{2\pi^2 l_{\max}} \sin\Phi d\phi_1 d\Phi d\phi_2 dl\right)^{-1} & \text{if } b = b^{(1)}, \text{ or } b^{(2)}, \text{ or}\ldots, b^{(6)} \\ 0, \text{ for all other values of } b & \end{cases} \tag{5.18}$$

Extensions to higher dimensions, if warranted, will be guided by the same principles.

5.2 FOURIER SERIES AND FOURIER TRANSFORM REPRESENTATIONS

We now consider spectral representations via Fourier series and discrete Fourier transforms, which were both described briefly in Chapter 3.

5.2.1 Fourier representation of grain size distributions

Although various choices are available we will illustrate the spectral representation of the grain size distribution using the classic Fourier basis that consists of the exponential functions. The natural periodicity of these functions is in multiples of 2π. It is tempting to alter the grain size variable $l \in [0, l_{\max})$ to a new variable, say $l' = 2\pi l / l_{\max} \in [0, 2\pi)$, that matches up to the natural periodicity of the exponential functions. Taking this alteration, the new grain size distribution function, $f'(l')$, is simply related to the measured one by the expression $f'(l') = f(l' \cdot l_{\max}/2\pi)$.

The Fourier representation of the new grain size distribution function, taking this approach, would have the form

$$f'(l') = \sum_{m=-\infty}^{+\infty} F_m e^{iml'} \tag{5.19}$$

We can see, on further reflection, however, that these choices would lead to a large number of significant Fourier coefficients, F_m, because the derivatives of this function, $df'(l')/dl'$, when evaluated at $l' = 0,\ 2\pi$, will typically be discontinuous. This can be seen to be the case with the grain size distribution shown in Figure 4.9.

But if we are willing to further alter or augment the function in ways that result in continuous derivatives at all points in the new function, then we can anticipate a significant reduction in the required number of Fourier coefficients. This process is called *Fourier continuation.* For the distribution plotted in Figure 4.9 a particular augmentation of L can be suggested. Since $df(l)/dl$ approaches 0 at $l = 0,\ l_{\max}$, define a new, augmented range of grain size that is two times the size of the present one, or $2l_{\max}$. We first alter our grain size function to become $f'(l)$ where

$$f'(l) = \begin{vmatrix} f(l) \text{ if } l \in [0, l_{\max}) \\ f(2l_{\max} - l) \text{ if } l \in [l_{\max}, 2l_{\max}) \end{vmatrix} \tag{5.20}$$

It should be evident that this first augmentation of the grain size function differs in its range, and it comprises a mirroring of the original grain size distribution about the $l = l_{max}$ position, extending it to $2l_{max}$. The second change we seek is the same we used to obtain Eq. (5.19): We want our new function to contain a natural periodicity over 2π. This can be accomplished with a change of variable of the form $l' = \pi\, l/l_{max} \in [0, 2\pi)$. The new function, with the desired properties, is defined by

$$f''(l') = f'(l' \cdot l_{max}/\pi) \tag{5.21}$$

This new function, $f''(l')$, will have the following desired properties: (1) it will be periodic in 2π, $f''(l' + 2m\pi) = f''(l')$, and (2) it will be efficient in representing grain size data like that shown in Figure 4.9, where to a first approximation $df''/dl' = 0$ at $l' = \pi$.

Of course grain size distributions with other characteristics may suggest other types of augmentation by Fourier continuation to reach suitable efficiencies in representation. Readers should also note that from a practical point of view our interest in the augmented grain size distribution function, $f''(l')$, is only over the first half of its domain $l' \in [0,\ \pi)$, which corresponds to grain size $l \in [0,\ l_{max})$. The remainder of the function is just "augmentation" that is useful for efficiency of representation and for numerical purposes.

The corresponding Fourier representation of the augmented grain size function is just

$$f''(l') = \sum_{m=-\infty}^{+\infty} F''_m e^{iml'} \tag{5.22}$$

A discrete representation of the grain size distribution function follows easily by splitting up the grain size domain into a specified number of bins. Suppose the number of bins on the augmented domain is to be N_l. (For our purposes here, N_l will be selected to be an even, positive integer.) Then the bins can be defined by line segments ν_n, given by

$$\nu_n = \{l\,|2\pi(n-1)/N_l \le l < 2\pi n/N_l\} \tag{5.23}$$

The grain size distribution function can then be defined by function values at the left endpoint of the line segments. Let these function values be $f''^n = f''(2\pi(n-1)/N_l)$ for each $n = 1, 2, \ldots, N_l$. The set $\{f''^1, f''^2, \ldots, f''^{N_l}\}$ will be the discrete representation of the augmented grain size distribution function.

How would we use measurements of grain size to obtain numerical values for $f''(l')$? Suppose we have M measurements of grain size coming from our microscopy. Let the measured values be enumerated by l_m, $m = 1, 2, \ldots, M$. These measurements can each be placed in the bins described in relation (5.23). Each measurement will be entered into two bins, to be consistent with our augmentation scheme. Let M_n be the number of measurements that fall in bin ν_n. If l_m naturally falls in bin ν_n (note that $1 \le n \le N_l/2$), then the number 1 is added to both ν_n and ν_{N_l-n}. After completing this process, $2M$ total points will have been distributed and $\sum_{n=1}^{N_l} M_n = 2M$. Taking the normalization of the grain size distribution to be that given in relation (5.23), we must have

$$\int_0^{l_{max}} f(l)dl = \int_0^{\pi} f''(l')dl' = 1 \approx \sum_{n=1}^{N_l/2} f''^n \int_{\nu_n} dl' = \frac{2\pi}{N_l}\sum_{n=1}^{N_l/2} f''^n = \sum_{n=1}^{N_l/2} M_n/M \tag{5.24}$$

Comparing terms on the right side of (5.24), we see that $f''^n = M_n N_l / 2\pi M$, which is just the fraction of measurement points falling in bin ν_n, divided by the measure of each bin $2\pi/N_l$.

In the frequency domain, the Fourier transform of the discrete Fourier representation is given by

$$F''k = \sum_{n=0}^{N_l-1} f''^n e^{-i2\pi nk/N_l} \tag{5.25}$$

and transformation back into the real domain is

$$f''n = \frac{1}{N_l} \sum_{k=0}^{N_l-1} F''^k e^{i2\pi nk/N_l} \tag{5.26}$$

Representing the grain size distribution function with the set of F''^k is completely equivalent to representing it with the set of f''^n. If good choices have been made in the augmentation scheme, the number of significant F''^k coefficients will be much smaller than N_l, and significant numerical advantage can be achieved in representing the grain size distribution.

5.2.2 Discrete Fourier representation of orientation distributions

Next we turn our attention to spectral representations of the ODF. Historically, the generalized spherical harmonics (GSH) in Euler space have been the most developed and successful of the various alternatives. The main advantages of the GSH representation are that they are highly efficient in representing the orientation dependence of several macroscale tensorial properties of interest, and they can be customized to automatically reflect the various symmetries associated with the selection of the crystal and sample reference frames. However, the main difficulty with GSH representations is the fact that they are computationally expensive. Here, we will explore the use of discrete Fourier transforms (DFTs) for spectral representation of ODF. A major advantage of using DFTs in place of GSH representations is that we can compute the transforms using the much more computationally efficient algorithms.

Discrete Fourier transforms (also sometimes referred to as fast Fourier transforms, or FFTs, because of the algorithms used in their computation) are typically computed using the function values on a uniform grid. In a purely mathematical sense, they are not a complete basis because of the inherent discretization involved, but they can be computed several orders of magnitude faster than the GSH coefficients. It is worth noting that although GSH constitute a complete basis, the computation of the GSH coefficients for most practical applications also reduces to evaluating the values of functions on a discrete grid and using them in an appropriate numerical integration scheme. Therefore, for all practical applications, there is only a minimal difference in the accuracy of representations between DFTs and the more rigorous GSH coefficients.

To represent the ODF using DFTs, we would have to bin the Bunge–Euler space. An important choice to consider here is whether we should define the orientation space in $(\varphi_1, \Phi, \varphi_2)$ or in a transformed distortion-free space defined by $(\varphi_1, \cos\Phi, \varphi_2)$. The main advantage of the latter choice is that the invariant measure of any region in the transformed orientation space would simply be its volume and the distortion inherent to Bunge–Euler space is completely remedied. However, in the process of transforming the orientation space to $(\varphi_1, \cos\Phi, \varphi_2)$, we lose the natural periodicity associated with the Bunge–Euler space, which has to be reintroduced artificially if we are to use DFTs effectively in the transformed space. For example, it will be implicitly assumed that any function defined in the $(\varphi_1, \cos\Phi, \varphi_2)$ space is periodic in $\cos\Phi$ with a period of two $(-1 \leq \cos\Phi \leq 1)$. The discontinuities associated with any such ad hoc introduction of periodicity produce additional Fourier transforms that are not present in the original function of interest. In the example presented here, we have, therefore, decided to stay with the original Bunge–Euler space. In all the theory developed here, we acknowledge and account explicitly for the distortion inherent to the Bunge–Euler space.

Let the 3D Bunge–Euler space of interest be discretized uniformly into $B_1 \times B_2 \times B_3$ bins, and let (b_1, b_2, b_3) enumerate these bins. The DFT representation of the ODF is defined as

$$f^{b_1b_2b_3} = \frac{1}{B_1B_2B_3} \sum_{k_1=0}^{B_1-1} \sum_{k_2=0}^{B_2-1} \sum_{k_3=0}^{B_3-1} F^{k_1k_2k_3} e^{\frac{i2\pi k_1 b_1}{B_1}} e^{\frac{i2\pi k_2 b_2}{B_2}} e^{\frac{i2\pi k_3 b_3}{B_3}} \tag{5.27}$$

where $f^{b_1b_2b_3}$ denotes the value of the ODF in the bin identified by (b_1, b_2, b_3). In the remainder of this book, for simplicity of notation, equations such as Eq. (5.27) will be expressed in a condensed notation as

$$f^b = \frac{1}{B} \sum_{k=0}^{B-1} F^k e^{\frac{i2\pi kb}{B}} \tag{5.28}$$

where F^k denotes the DFTs for ODF. For given values of the ODF on a uniform grid in the Bunge–Euler space, its DFTs are defined as (but are typically computed using FFT algorithms)

$$F^k = \sum_{b=0}^{B-1} f^b e^{\frac{-i2\pi kb}{B}} \tag{5.29}$$

The requirement that the values of the ODF be real translates to the requirement that $F^k = F^{*B-k}$ in the DFTs (where the asterisk denotes a complex conjugate).

Next, we address the range of the Bunge–Euler space to be considered while taking into account some of the symmetries associated with crystal and sample reference frames. It is ideal to define the space of interest as $(\phi_1 \in (0, 2\pi], \Phi \in (0, 2\pi], \phi_2 \in (0, 2\pi])$, because all functions of interest in our study are naturally periodic in this space. However, there are several redundancies within this space that can be exploited in the computations. For crystals of any symmetry, the definitions of the Bunge–Euler angles require that locations (ϕ_1, Φ, ϕ_2) and $(\phi_1 + \pi,\ 2\pi - \Phi,\ \phi_2 + \pi)$ correspond to the exact same crystal lattice orientation. Furthermore, consideration of eight of the lattice symmetry operations associated with a cubic crystal lattice (e.g., fcc, bcc) require that the locations corresponding to $(\phi_1 + \pi, \pi - \Phi, 2\pi - \phi_2)$, $(\phi_1 + \pi, \pi - \Phi, \pi - \phi_2)$, $(\phi_1 + \pi, \pi - \Phi, \pi/2 - \phi_2)$, $(\phi_1 + \pi, \pi - \Phi, 3\pi/2 - \phi_2)$, $(\phi_1, \Phi, \phi_2 + \pi/2)$, $(\phi_1, \Phi, \phi_2 + \pi)$, and $(\phi_1, \Phi, \phi_2 + 3\pi/2)$ also correspond to the exact same crystal lattice orientation defined by $(\phi_1, \Phi, \varphi_2)$. These equivalencies indicate that the natural periodic unit cell for all functions defined for cubic crystals in the Bunge–Euler space can be defined as $(\phi_1 \in (0, 2\pi], \Phi \in (0, 2\pi], \phi_2 \in (0, \pi/2])$

As a result of all of the considerations of symmetry for cubic crystals, it can be easily seen that we can focus our attention on the space defined by (see Section 4.1.2)

$$FZ3 = \left\{ (\phi_1, \Phi, \phi_2) \middle| \ \phi_1 \in [0,\ 2\pi), \Phi \in \left[0,\ \frac{\pi}{2}\right), \phi_2 \in \left[0,\ \frac{\pi}{2}\right) \right\} \tag{5.30}$$

It is important to note here that if the values of any desired function are obtained on a uniform grid in *FZ*3, then it is fairly simple to assign values on a uniform grid over ($\phi_1 \in [0, 2\pi)$, $\Phi \in [0, 2\pi)$, $\phi_2 \in [0, \pi/2)$) using the relationships mentioned earlier.

The space defined here as *FZ*3 is actually three times the *FZ* defined in conventional texture analyses for cubic-triclinic (i.e., cubic symmetry in the crystal reference frame and triclinic symmetry in the sample reference frame) functions. In other words, we expect three equivalent locations in *FZ*3 for every crystal orientation with a cubic lattice. Note that we have used only eight of the symmetry operators in defining *FZ*3. The main reason for using *FZ*3 instead of *FZ* is that a uniform grid in *FZ* does not map to uniform grids in the remainder of the *FZ*3. Since the use of DFTs requires a uniform grid (for maximum computational efficiency), we could establish the values of all desired functions in a uniform grid on *FZ*3, expand the grid to ($\phi_1 \in (0, 2\pi]$, $\Phi \in (0, 2\pi]$, $\phi_2 \in (0, \pi/2]$) using the relationships mentioned earlier, and then compute DFTs.

The ODF of a single crystal microstructure is conveniently described by a Dirac-delta function, $\delta(g - \tilde{g})$, in the continuous Bunge–Euler space. In the *FZ*3 space considered here, the single-crystal microstructure should be described as

$$f(g) = \frac{1}{3}\left[\delta\left(g - \tilde{g}\right) + \delta\left(g - \tilde{\tilde{g}}\right) + \delta\left(g - \tilde{\tilde{\tilde{g}}}\right)\right] \tag{5.31}$$

where $(\tilde{g}, \tilde{\tilde{g}}, \tilde{\tilde{\tilde{g}}})$ denote the three equivalent orientations of the single crystal in the *FZ*3. In the discretized *FZ*3 space used in computing the DFTs here, the values of the ODF for a single crystal of which the orientation falls in bins $(\tilde{b}, \tilde{\tilde{b}}, \tilde{\tilde{\tilde{b}}})$ are given by

$$f^b = \begin{cases} \left(\int\limits_b \frac{1}{\pi^2} \sin\Phi \, d\phi_1 d\Phi d\,\phi_2 \right)^{-1}, & \text{if } b = \tilde{b}, \text{ or } b = \tilde{\tilde{b}}, \text{ or } b = \tilde{\tilde{\tilde{b}}} \\ 0 \text{ for all other values of } b \end{cases} \tag{5.32}$$

Note that Eq. (5.32) implies that the values of $f^{\tilde{b}}, f^{\tilde{\tilde{b}}}$, and $f^{\tilde{\tilde{\tilde{b}}}}$ can all be different from each other. All polycrystal ODFs are simply visualized as a weighted sum of single-crystal ODFs, where the weights are essentially the volume fractions associated with the occurrence of the single-crystal orientation in the polycrystalline sample.

In terms of plotting ODFs, almost all commercial codes (see EDAX (2010)., OIM 6.0, EDAX-TSL; Adams, 1993; Wright, 1993) perform some form of smoothing of the ODF. Although this practice of smoothing the ODF has its origin in the X-ray methods of texture measurement and analyses (mainly attributed to the fact that the goniometer is constantly rotating in these measurements), it continues to be widely employed with modern methods of orientation measurements such as orientation image microscopy (OIM) that are significantly more accurate in their ability to measure crystal orientations. The smoothing of measurements acquired by the OIM technique is largely justified by the fact that the number of measurements performed in a typical scan is much lower than the number of measurements needed to establish the ODF accurately, as revealed by appropriate statistical analyses of the measured data. In the DFT representation of ODF, this smoothing can be accomplished quite easily by convolution with an appropriate filter.

5.3 SPHERICAL HARMONIC FUNCTION REPRESENTATIONS

Next we consider representation of the grain boundary character distribution function (GBCD) described in Section 4.3.2, using the SSHFs described in Section 3.5.

5.3.1 Representation of GBCD using SSHFs

The SSHFs defined in Eq. (3.52), and satisfying the orthogonality relations (3.56), form a complete orthonormal basis for functions on the hemispherical domain of all normal directions: $0 \leq \nu \leq \pi/2,\ 0 \leq \omega < 2\pi$. It is known that all square-integrable functions on the hemisphere (such as the GBCD) can be expressed as an infinite series in the SSHFs. The exact expression is

$$S_V(\nu, \omega) = \sum_{l=0,2}^{\infty} \sum_{m=-l}^{+l} Q_l^m k_l^m(\nu, \omega) \tag{5.33}$$

where $k_l^m(\nu, \omega)$ are the SSHFs, and the associated Fourier coefficients are Q_l^m. Only even values of the integers l are required in the series because we cannot distinguish sides of the grain boundaries, and therefore $S_V(\nu, \omega) = S_V(\pi - \nu, \omega + \pi)$. Using the orthogonality relations of (3.56), multiplying both sides of (5.33) by $k_{l'}^{*m'}(\nu, \omega,)$ (where the asterisk denotes a complex conjugate), and integrating over the hemisphere, we obtain the following expression for the Fourier coefficients Q_l^m:

$$Q_l^m = \frac{1}{2\pi} \int_0^{2\pi} \int_0^{\pi/2} S_V(\nu, \omega) k_l^{*m}(\nu, \omega) \sin \nu \, d\nu \, d\omega \tag{5.34}$$

Evaluation of the Fourier coefficients for the GBCD is not just a matter of direct sampling, as is possible with the ODF, due to the limitations of microscopy of opaque materials. Recovery of the Q_l^m coefficients requires the solution of an inverse problem, as will be described in Chapter 12.

5.3.2 Representation of ODFs using GSHFs

As a second example of representation using the classic basis functions, we return to the ODF, previously represented using the primitive basis. The basis functions to be used in this case are the GSHFs, which were introduced in Chapter 3. It is known that for any square-integrable function (e.g., the ODF), defined on the domain of all possible rotations ($SO(3)$), we can express that function as an infinite series of GSHFs, weighted by appropriate coefficients. Thus,

$$f(g) = \sum_{l=0}^{\infty} \sum_{m=-l}^{+l} \sum_{n=-l}^{+l} C_l^{mn} T_l^{mn}(g) \tag{5.35}$$

where $T_l^{mn}(g)$ are the GSHFs introduced in Chapter 3, and C_l^{mn} are the weighting coefficients. Following the usual pattern, we can solve for the weighting coefficients by making use of the orthogonality relations given in relation (3.41):

$$C_l^{mn} = (2l+1) \iiint_{SO(3)} f(g) T_l^{*mn}(g) dg \tag{5.36}$$

Note that the invariant measure $dg = dR$ is that given by (3.42). If discrete sampling J orientations of the ODF is available, such that the ODF can be expressed by (5.14), then the C_l^{mn} coefficients may be estimated by the expression

$$C_l^{mn} \approx \frac{(2l+1)}{J} \sum_{j=1}^{J} \iiint_{SO(3)} \delta\left(g - g_j\right) T_l^{*mn}(g) dg = \frac{(2l+1)}{J} \sum_{j=1}^{J} T_l^{*mn}\left(g_j\right) \tag{5.37}$$

When symmetries are considered, as we will show in Chapter 6, the number of basis functions required for representation can be substantially reduced. This reduction can be important for two reasons: (1) the numerical requirements of representation are reduced, and (2) some important insights into processing and properties can be clarified by careful examination of the significant terms that remain. Readers are reminded, however, that whether or not symmetry is explicitly considered, expression (5.35) is completely general and will be exact for any well-behaved (i.e., square-integrable) function on the domain $SO(3)$.

5.4 PRIMITIVE BASIS REPRESENTATION OF THE MICROSTRUCTURE FUNCTION

Recall the definition of the microstructure function given in relation (4.25). $M(\mathbf{x}, h)$ represents the microstructure of a sample that occupies region Ω of 3D space. It is a real-valued function on the product space $\Omega \times H$, where H denotes the comprehensive local state space, as described in Section 4.1. Consider rectangular samples of size $D_1 \times D_2 \times D_3$ within Ω. Thus, the real intervals $[0, D_k)$ represent the range of possible values for spatial variables x_k ($k = 1, 2, 3$) within a given rectangular region. For simplicity, the rectangular model is dimensioned such that it accepts partitioning into a discrete set of cubical voxels of equivalent size, called *subcells*. Furthermore, with an eye toward compatibility with representations by Haar wavelet functions, D_k are selected such that they accept partitioning into 2^{P_k} subintervals of fixed size $\delta = D_k/2^{P_k}$. (P_k denotes the *generation* of the Haar wavelet basis in the x_k variable and will be defined subsequently). Let the positive integers s_k, $1 \leq s_k \leq 2^{P_k}$ enumerate the cubical subcells of the sample, each occupying region $\omega_{s_1 s_2 s_3} \subset \Omega$. In a local coordinate frame the subcells contain the points (Figure 5.4)

$$\omega_{s_1 s_2 s_3} = \{(x_1, x_2, x_3) | (s_1 - 1)\delta \leq x_1 < s_1\delta, (s_2 - 1)\delta \leq x_2 < s_2\delta, (s_3 - 1)\delta \leq x_3 < s_3\delta\} \tag{5.38}$$

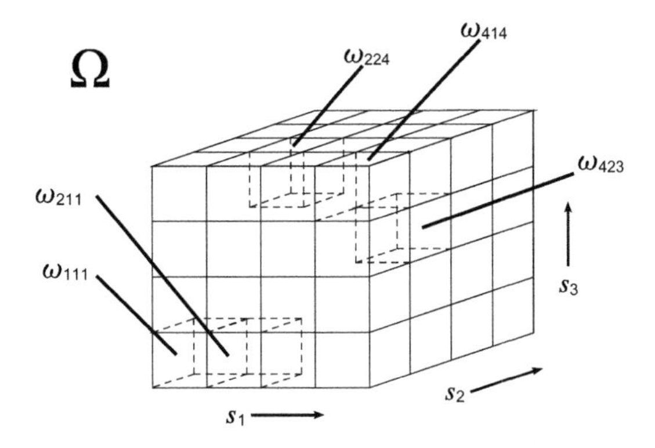

FIGURE 5.4

Sample partitioned into cubic subcells, with several of the cells labeled.

Invoking the index contraction convention, the subcells will be identified by a single index s: $\omega_{s_1 s_2 s_3} \leftrightarrow \omega_s$. The number of indices involved is $S = 2^{(P_1+P_2+P_3)}$. The subcells satisfy the following relations:

$$\omega_s \cap \omega_{s'} = \varnothing \left(\text{for } s \neq s'\right), \quad \bigcup_{s=1}^{S} \omega_s = \Omega \tag{5.39}$$

Spatial indicator functions can be defined for each subcell (Figure 5.5). These are

$$\chi_s(\mathbf{x}) = \begin{cases} 1 \text{ if } \mathbf{x} \in \omega_s \\ 0 \text{ otherwise} \end{cases} \tag{5.40}$$

Compact support for the indicator functions is evident, each in its specified region. Note that the spatial resolution of the indicator basis is just $(\delta^3/D_1 D_2 D_3) = 1/S$. Also, note that the spatial indicator functions satisfy the following orthogonality conditions:

$$\frac{1}{\delta^3} \iiint_{\Omega} \chi_s(\mathbf{x}) \chi_{s'}(\mathbf{x}) dx = \delta_{ss'} \tag{5.41}$$

where $\delta_{ss'}$ is the Kronecker delta, which equals 1 when $s = s'$, and 0 otherwise.

Similarly, the local state space H can be partitioned into N subregions, labeled γ_n, each containing an arbitrary local state h_n. These subregions, assumed to comprise continuous variables for the development shown here, are selected to have the following properties:

$$\gamma_n \cap \gamma_m = \varnothing \ (n \neq m), \ \bigcup_{n=1}^{N} \gamma_n = H, \ \iiint_{\gamma_n} dh = 1/N\left(\text{for all n}\right) \tag{5.42}$$

The last relation in (5.42) is taken for convenience; it means that the local state subregions have been defined such that each has the same (invariant) measure. Note that 1/N specifies the resolution of the continuous component of the local state variables.

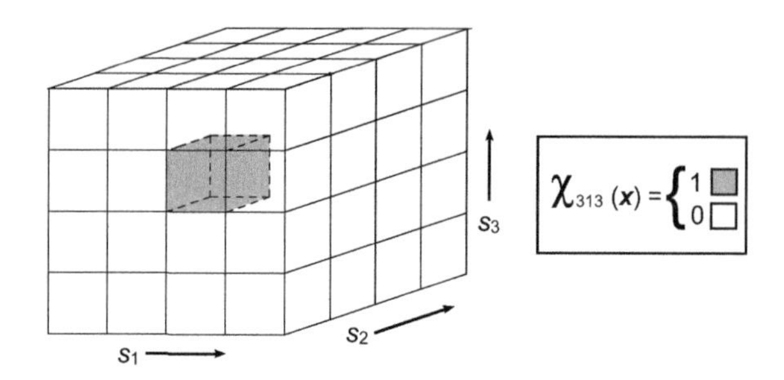

FIGURE 5.5

The shaded region gives the volume where the indicator function χ_{313} is nonzero. The index 313 refers to the values of s_1, s_2, s_3. In a 4 × 4 × 4 sample the index 313 would correspond to $s = 35$.

Associated with each subregion of the local state space, define a *local state indicator function* $\chi^n(h)$ such that

$$\chi^n(h) = \begin{cases} 1 \text{ if } h \in \gamma_n \\ 0 \text{ otherwise} \end{cases} \tag{5.43}$$

These must also satisfy orthogonality conditions

$$N \iiint_H \chi^n(h)\chi^{n'}(h)dh = \delta^{nn'} \tag{5.44}$$

The microstructure function accepts a primitive Fourier approximation in the product space of spatial and local state indicator functions:

$$M(\mathbf{x}, h) \approx D_s^n \chi_s(\mathbf{x})\chi^n(h) \tag{5.45}$$

Notice the application of the summation convention on the right side of (5.45), implying summation over the s and n variables. Equality in (5.45) with the complete (physical) microstructure function is normally achieved only in the limit that S and N both go to infinity.

When considering the normalization relation for the microstructure function, Eq. (4.26), one finds the following normalizations among the Fourier coefficients of the primitive basis:

$$\sum_{s=1}^{S} \sum_{n=1}^{N} D_s^n = SN \tag{5.46}$$

and

$$\sum_{n=1}^{N} D_s^n = N(\text{for all } s) \tag{5.47}$$

Also, given the definition of the $M(\mathbf{x}, h)$, Eq. (4.25), and normalization condition (5.47), it is readily seen that the texture coefficients are bounded:

$$0 \leq D_s^n \leq N \tag{5.48}$$

If subcell s^* contains only one orientation, say $h^* \in \gamma_{n^*}$, then $D_{s^*}^{n*} = N$ and all other $D_{s^*}^{n \neq n^*} = 0$. If more than one orientation subregion is represented in subcell s^*, the Fourier coefficients will be less than N but greater than 0, and they will sum to N according to (5.47). Relations (5.46) through (5.48) help establish the hull of microstructure functions that is described in a later section.

5.4.1 Revisiting the primitive basis for other distribution functions

Having introduced the representation of the microstructure function in the primitive basis, we are prepared to express other distribution functions of interest in the same representation. Considering relation (8.4.1) for the individual cubical subcells, one can describe the *local state distribution* for a single subcell of the sample as ${}_sf(h)$:

$$\frac{1}{\delta^3}\iiint\limits_{x\in\Omega} M(x,h)\chi_s(x)dx = \frac{1}{\delta^3}\iiint\limits_{x\in\omega_s} M(x,h)dx = {}_sf(h) = D_s^n\chi^n(h) \tag{5.49}$$

where, for simplicity, we omit reference to the particular sample of the ensemble. The local state distribution for a particular element of the ensemble, based on relations (4.29) and (5.49), is

$$\begin{aligned} f(h) &= \frac{1}{vol(\Omega)}\iiint\limits_{\Omega} M(x,h)dx = \frac{1}{\delta^3 S}\iiint\limits_{\Omega} M(x,h)dx \\ &= \frac{1}{\delta^3 S}\sum_{s=1}^{S}\iiint\limits_{\omega_s} M(x,h)dx = \frac{1}{S}\sum_{s=1}^{S} {}_sf(h) = \frac{1}{S}\sum_{s=1}^{S} D_s^n\chi^n(h) \\ &= \overline{D}_s^n\chi^n(h) \end{aligned} \tag{5.50}$$

Thus, the Fourier coefficients of the sample-level local state distribution, $\overline{D}_s^n$, are just the numerical averages of the coefficients of the cell-level local state distributions.

5.5 REPRESENTATIVE VOLUME ELEMENT

One significant question arising during the characterization and measurement of microstructures involves the appropriate size of the sample for it to be statistically representative of the overall material. This question relates to the concept of ensembles, discussed in Chapter 4. One element of an ensemble may be thought of as a single region scanned during a microscopy campaign. Consider the microstructure function ${}^{(k)}M(x,h)$ in the kth element of the ensemble, and its Fourier representation in the primitive basis:

$${}^{(k)}M(x,h) = {}^{(k)}D_s^n\chi_s(x)\chi^n(h) \tag{5.51}$$

This function is defined over the region Ω, and over all of its subcells, $\omega_s \in \Omega$. The question often arises in mechanics as to whether a region Ω of specified size and shape would be a *representative volume element* (RVE) in which the considerations of continuum mechanics could be applied. (Readers should note that the RVE differs from the concept of the statistical volume element, which was introduced in Chapter 4). When continuum mechanics is applied without consideration of microstructure, the question is normally framed in terms of the variation of effective properties of the material that occupies region Ω in the ensemble. Thus, if P is a property of interest, taken here as a scalar, and ${}^{(k)}P$ is the property of the kth element of the ensemble, then one can define the ensemble average and the variance of property P using the following relations:

$$\langle P \rangle = \frac{1}{K}\sum_{k=1}^{K} {}^{(k)}P, \quad \sigma_{P|\Omega}^2 = \frac{1}{K}\sum_{k=1}^{K} [{}^{(k)}P - \langle P \rangle]^2 \tag{5.52}$$

If the property measurements are normally distributed, then 68% of the measurements lie in the range $\langle P \rangle - \sigma_{P|\Omega} \le P \le \langle P \rangle + \sigma_{P|\Omega}$, and it can be stated that a sample of size Ω comprises an RVE of character $\sigma_{P|\Omega}$. Presumably, increasing the size of Ω decreases $\sigma_{P|\Omega}$, but modeling is generally limited in the size of the region that can be handled numerically.

Can the quantitative description of microstructure itself also be used to define the character of an RVE? The answer to this question is that it can; and just as the character of the RVE would vary depending on the choice of property selected in the previous treatment, so the character of the RVE would vary with choice of quantitative representation of microstructure. Consider as an example the microstructure function given in Eq. (5.51). A natural measure of *distance between microstructure functions* can be expressed as

$$d_k^{k'} = d({}^{(k)}M, {}^{(k')}M) = \frac{1}{vol(\Omega)} \iiint_{\Omega} \iiint_{H} \left| {}^{(k)}M(x,h) - {}^{(k')}M(x,h) \right| dh dx$$

$$= \frac{1}{vol(\Omega)} \sum_{s=1}^{S} \sum_{n=1}^{N} \left| {}^{(k)}D_s^n - {}^{(k')}D_s^n \right| \iiint_{\Omega} \chi_s(x) dx \iiint_{H} \chi^n(h) dh$$

$$d_k^{k'} = \frac{1}{SN} \sum_{s=1}^{S} \sum_{n=1}^{N} \left| {}^{(k)}D_s^n - {}^{(k')}D_s^n \right| \tag{5.53}$$

The distance $d_k^{k'}$ is also presumed to be normally distributed, with ensemble average and variance defined by

$$\langle d \rangle = \frac{1}{2K^2} \sum_{k=1}^{K} \sum_{k'=1}^{K} d_k^{k'}, \; \sigma_{d|\Omega}^2 = \frac{1}{2K^2} \sum_{k=1}^{K} \sum_{k'=1}^{K} \left[d_k^{k'} - \langle d \rangle \right]^2 \tag{5.54}$$

If the distance measurements are normally distributed, then 68% of the measurements lie within the range $\langle d \rangle - \sigma_{d|\Omega} \le d_k^{k'} \le \langle d \rangle + \sigma_{d|\Omega}$, and it can be stated that a sample of size Ω comprises an RVE of character $\sigma_{d|\Omega}$. Presumably, increasing the size of Ω decreases $\sigma_{d|\Omega}$, but modeling is generally limited in the size of the region that can be handled numerically.

Since the microstructure function is rather inaccessible to experimental observation, it may often be preferable to use another representation of microstructure in defining the size of the RVE. Several choices are available. The local state distribution function, $f(h)$, would often be a logical choice. Expressing $f(h)$ in terms of the microstructure function,

$$
\begin{aligned}
{}^{(k)}f(h) &= \frac{1}{vol(\Omega)} \iiint\limits_{\Omega} {}^{(k)}M(x,h)dx \\
&= \frac{1}{vol(\Omega)} \sum_{s=1}^{S} \sum_{n=1}^{N} {}^{(k)}D_s^n \chi^n(h) \iiint\limits_{\Omega} \chi_s(x)dx \\
&= \frac{1}{S} \sum_{s=1}^{S} \sum_{n=1}^{N} {}^{(k)}D_s^n \chi^n(h)
\end{aligned}
$$

$$
{}^{(k)}f(h) = \sum_{n=1}^{N} {}^{(k)}D^n \chi^n(h) \left(\text{where} \, {}^{(k)}D^n = \frac{1}{S} \sum_{s=1}^{S} {}^{(k)}D_s^n \right) \tag{5.55}
$$

Therefore, relation (5.53) can be rewritten as

$$
\begin{aligned}
d_k^{k'} &= d({}^{(k)}f, {}^{(k')}f) = \iiint\limits_{H} \left| {}^{(k)}f(h) - {}^{(k')}f(h) \right| dh \\
&= \sum_{n=1}^{N} \left| {}^{(k)}D^n - {}^{(k')}D^n \right| \iiint\limits_{H} \chi^n(h)dh
\end{aligned}
$$

$$
d_k^{k'} = \frac{1}{N} \sum_{n=1}^{N} \left| {}^{(k)}D^n - {}^{(k')}D^n \right| \tag{5.56}
$$

Relation (5.54) and its interpretation remain the same.

It is instructive to consider whether the characters of the microstructure-based and property-based RVEs are related. An elementary approach can be taken based on the elementary first-order homogenization relationship. Assume the property ${}^{(k)}P$ in the kth element of the ensemble to be estimated by

$$
{}^{(k)}P \approx \iiint\limits_{H} {}^{(k)}f(h)p(h)dh \approx \frac{1}{N} \sum_{n=1}^{N} {}^{(k)}D^n Q^n \tag{5.57}
$$

where $p(h)$ is the local property associated with local state h, which has been represented in the Fourier series as

$$
p(h) = \sum_{n=1}^{N} Q^n \chi^n(h) \tag{5.58}
$$

Relation (5.52) can thus be expressed as

$$
\langle P \rangle = \frac{1}{KN} \sum_{k=1}^{K} \sum_{n=1}^{N} {}^{(k)}D^n Q^n, \quad \sigma_{P|\Omega}^2 = \frac{1}{K} \sum_{k=1}^{K} \left[\frac{1}{N} \sum_{n=1}^{N} {}^{(k)}D^n Q^n - \langle P \rangle \right]^2 \tag{5.59}
$$

Notice that the form of (5.59) is very similar to that of (5.52), with ${}^{(k)}P$, which is a scalar, to be replaced with the scalar product of two vectors, one representing the local state distribution function with coefficients ${}^{(k)}D^n$ and the other representing the local properties with coefficients Q^n. The latter are independent of k.

5.5.1 Statistical homogeneity and ergodicity

Consider the local state distribution in an individual subcell s^*. This can be defined from

$$
\begin{aligned}
{}^{(k)}f_{s^*}(h) &= \frac{1}{vol(\omega_{s*})}\iiint_{\omega_{s*}} {}^{(k)}M(x,h)dx \\
&= \frac{1}{\delta^3}\sum_{s=1}^{S}\sum_{n=1}^{N} {}^{(k)}D_s^n \chi^n(h)\iiint_{\omega_{s*}} \chi_s(x)dx \\
&= \sum_{n=1}^{N} {}^{(k)}D_{s*}^n \chi^n(h)
\end{aligned}
\tag{5.60}
$$

If the ensemble average of ${}^{(k)}D_s^n$, $\langle D_s^n\rangle$ is independent of subcell s, then we say that the ensemble of microstructure possesses *statistical homogeneity*. Recall from relation (5.55) that sample coefficients of the local state distribution function, ${}^{(k)}D^n$, are related to the subcell coefficients by the relation

$$
{}^{(k)}D^n = \frac{1}{S}\sum_{s=1}^{S} {}^{(k)}D_s^n \tag{5.61}
$$

If the difference between this ensemble average $\langle D_s^n\rangle$ and ${}^{(k)}D^n$ itself satisfies the relation

$$
\left|{}^{(k)}D^n - \langle D_s^n\rangle\right| \le \varepsilon(\text{for all } k,n) \tag{5.62}
$$

for sufficiently small ε, then we say that the microstructure is *ergodic*. In loose terms, statistical homogeneity suggests that all locations in Ω are equivalent as far as statistical properties of the stochastic processes by which the ensemble is created are concerned. This equivalence suggests the ergodic hypothesis embodied in relation (5.62)—namely, that averaging over all realizations of the ensemble is equivalent to taking a volume average over any element of the ensemble, provided it occupies a sufficiently large region, Ω. Presumably, ergodicity is fully established in the limit as $vol(\Omega)\to\infty$ and $\varepsilon\to 0$.

SUMMARY

In this chapter all five spectral representations introduced in Chapter 3 were applied to distribution functions that capture the concept of microstructure. In subsequent chapters the same spectral ideas are applied to structure–property relations (Chapter 7), and specifically to homogenization relations (Chapter 8), to arrive at a full spectral framework for microstructure design. However, before turning to property relations, Chapter 6 incorporates the important concept of symmetry into both the description of microstructure and its spectral representation.

Exercises

Section 5.1

1. Partition the domain of the grain size distribution, shown in Figure 4.9, into nine subintervals. Then find the coefficients of the primitive representation for this distribution. What factors contribute to the choice of tessellation scheme in the primitive representations?

2. Verify that the primitive representation of discrete orientation data given by (5.12) satisfies the normalization relationship for the ODF given by Eq. (4.14).

Section 5.2

3. **(small project).** Again, using Figure 4.9 and the binning determined in Exercise 4.4, obtain a discrete representation of the grain size distribution representing the measured distribution. Augment the distribution in the way described in the text, and obtain the Fourier-transformed coefficients of this distribution. Evaluate the number of terms that should be kept among these transformed coefficients. Use this smaller set to determine an estimate of the discrete representation in the direct space. Evaluate the accuracy of your estimation, and comment on your results.

CHAPTER

Symmetry in Microstructure Representation

6

CHAPTER OUTLINE

Symmetry exists in many physical arrangements of matter. Fiber-reinforced composites often have several associated symmetries, that is, transformations that leave the properties of the composite unchanged. Similarly, polycrystalline materials may have symmetries both at the crystal level and at the sample level. This chapter introduces various common symmetries and includes them in the description of microstructure that was developed in Chapters 4 and 5. Symmetry relations associated with the local state space give rise to symmetries in microstructure distribution functions, and these must be given careful consideration, not only to effect the greatest economy of representation but also to guide the investigator in physical understanding of microstructure–properties relationships.

6.1 POINT SYMMETRY SUBGROUPS OF THE CRYSTAL LATTICE

Consider transformations of the reference lattice L^o by orthogonal tensors $\mathbf{Q} \in O(3)$ to obtain a new lattice $\mathbf{Q}L^o$:

$$\mathbf{Q}L^o = \{\mathbf{x} | \mathbf{x} = m_1\mathbf{Q}\mathbf{a}_1^o + m_2\mathbf{Q}\mathbf{a}_2^o + m_3\mathbf{Q}\mathbf{a}_3^o,\ m_i \in Z\} \tag{6.1}$$

This is just a special case of the general affine transformation. The *symmetry subgroup* of the lattice consists of the set of all $\mathbf{Q}$ for which $\mathbf{Q}L^o = L^o$. Specifically, let Γ be the symmetry subgroup of the reference lattice. Then,

$$\Gamma = \{\mathbf{Q}|\mathbf{Q}L^o = L^o,\ \mathbf{Q} \in O(3)\} \quad (6.2)$$

It is known that 14 distinct lattice types are possible (e.g., the Bravais lattices; see Appendix 1), each of which corresponds to a distinct finite subgroup Γ. Readers are reminded that when one considers the superposition of a microscopic basis (motif) to each lattice point, the symmetry may be altered from that of the lattice. Appendix 1 lists the 32 classes of symmetry subgroups that occur in natural crystals. Associated with each class is a symmetry subgroup called the *point group* of the crystal. Further discussion of the 32 point groups is beyond the scope of this book, but see Nye (1953) and Hamermesh (1962) for a complete description.

Since only the pure rotational part of $\mathbf{Q}$ is readily measurable in most diffraction experiments, those elements of the point group that are pure rotations, $\mathbf{R} \in SO(3)$, are of special interest here. These form a subgroup $G \subset \Gamma$ defined by

$$G = \{\mathbf{R}|\mathbf{R}L^o = L^o,\ \mathbf{R} \in SO(3)\} \quad (6.3)$$

Hereafter our interest is in the rotational subgroup G rather than the complete subgroup Γ.

6.1.1 Equivalence classes and crystal orientation

In the absence of material symmetry (e.g., the triclinic phases), all of the orientations enumerated in (4.14) must be considered physically distinctive. When the crystalline phases of the material exhibit symmetry, more than one rotation of the reference crystal can be used to describe the orientation of the local crystal. Specifically, all rotations $\mathbf{R} \in G$ of the reference crystal possessing symmetry subgroup G leave the reference crystal in a new orientation that is indistinguishable from the original reference crystal, as described in (6.3).

For any rotation $\mathbf{R}$ we can define an *equivalence class*, which is the left coset of G, according to

$$\mathbf{R}G = \left\{\mathbf{R}\tilde{\mathbf{R}}^{(1)}, \mathbf{R}\tilde{\mathbf{R}}^{(2)}, \ldots, \mathbf{R}\tilde{\mathbf{R}}^{(M)}\right\} \quad (6.4)$$

where $G = \{\tilde{\mathbf{R}}^{(1)}, \tilde{\mathbf{R}}^{(2)}, \ldots, \tilde{\mathbf{R}}^{(M)}\}$. Every element of $\mathbf{R}G$ is indistinguishable from orientation $\mathbf{R}$, and hence $\mathbf{R}G$ is an equivalence class in $SO(3)$. Each element of $\mathbf{R}G$ is G-equivalent to every other element of $\mathbf{R}G$.

A second type of equivalence class is important to the efficient description of crystalline microstructures. Typical processing procedures can impart *statistical symmetry* (also known as *processing* or *sample symmetry*) to the distribution of local states. For example, a material that has been heavily rolled develops symmetry in the sample such that if the sample is rotated by π about its rolling direction, or about the direction normal to the sheet, or about a direction perpendicular to the rolling and normal directions, the statistical distribution of orientations is unchanged. When combined with the identity element, this set of four rotations defines the dihedral symmetry subgroup D_2:

$$D_2 = \{\mathbf{I}, \mathbf{R}(\pi\mathbf{e}_1), \mathbf{R}(\pi\mathbf{e}_2), \mathbf{R}(\pi\mathbf{e}_3)\} \tag{6.5}$$

Thus, if orientation $\mathbf{R}$ occurs in the rolled sample, then orientations $\mathbf{R}$, $\mathbf{R}(\pi\mathbf{e}_1)\mathbf{R}$, $\mathbf{R}(\pi\mathbf{e}_2)\mathbf{R}$, $\mathbf{R}(\pi\mathbf{e}_3)\mathbf{R}$ all occur with the same statistical frequency.

Following these notions, for any rotation $\mathbf{R}$ we can define another kind of equivalence class, which is the right coset of P, according to

$$P\mathbf{R} = \left\{\tilde{\tilde{\mathbf{R}}}^{(1)}\mathbf{R},\ \tilde{\tilde{\mathbf{R}}}^{(2)}\mathbf{R}, \ldots,\ \tilde{\tilde{\mathbf{R}}}^{(N)}\mathbf{R}\right\} \tag{6.6}$$

where $P = \{\tilde{\tilde{\mathbf{R}}}^{(1)},\ \tilde{\tilde{\mathbf{R}}}^{(2)}, \ldots,\ \tilde{\tilde{\mathbf{R}}}^{(N)}\}$ is the processing symmetry subgroup. Every element of $P\mathbf{R}$ is statistically indistinguishable from orientation $\mathbf{R}$ in the sample, and hence $P\mathbf{R}$ is an equivalence class in $SO(3)$. Each element of $P\mathbf{R}$ is P-equivalent to every other element of $P\mathbf{R}$. Readers are cautioned not to impute property equivalence to the elements of $P\mathbf{R}$; such is not the case. The right coset of P identifies orientations that occur with approximate statistical equivalence in the microstructure. But each element of $P\mathbf{R}$ will generally have different local properties.

Note that the two kinds of symmetry elements—crystallographic and statistical—can be combined. If G and P both occur in the microstructure then a combined equivalence class in $SO(3)$ can be defined:

$$P\mathbf{R}G = \left\{ \begin{array}{l} \tilde{\tilde{\mathbf{R}}}^{(1)}\mathbf{R}\tilde{\mathbf{R}}^{(1)},\ \tilde{\tilde{\mathbf{R}}}^{(1)}\mathbf{R}\tilde{\mathbf{R}}^{(2)}, \ldots,\ \tilde{\tilde{\mathbf{R}}}^{(1)}\mathbf{R}\tilde{\mathbf{R}}^{(M)}, \\ \tilde{\tilde{\mathbf{R}}}^{(2)}\mathbf{R}\tilde{\mathbf{R}}^{(1)},\ \tilde{\tilde{\mathbf{R}}}^{(2)}\mathbf{R}\tilde{\mathbf{R}}^{(2)}, \ldots,\ \tilde{\tilde{\mathbf{R}}}^{(2)}\mathbf{R}, \ldots,\ \tilde{\tilde{\mathbf{R}}}^{(2)}\mathbf{R}\tilde{\mathbf{R}}^{(M)}, \\ \ldots, \tilde{\tilde{\mathbf{R}}}^{(N)}\mathbf{R}\tilde{\mathbf{R}}^{(M)} \end{array} \right\} \tag{6.7}$$

The same caution is repeated concerning elements of $P\mathbf{R}G$. Only elements of the subclass $\tilde{\tilde{\mathbf{R}}}^{(n)}\mathbf{R}G = \{\tilde{\tilde{\mathbf{R}}}^{(n)}\mathbf{R}\tilde{\mathbf{R}}^{(1)},\ \tilde{\tilde{\mathbf{R}}}^{(n)}\mathbf{R}\tilde{\mathbf{R}}^{(2)}, \ldots,\ \tilde{\tilde{\mathbf{R}}}^{(n)}\mathbf{R}\tilde{\mathbf{R}}^{(M)}\} \subset P\mathbf{R}G$ can be expected to exhibit the same local properties.

6.1.2 Fundamental zones of crystal orientation

The terms *asymmetrical domain* and *fundamental zone* refer to the set of all physically distinct orientations of the local crystal that can occur in nature. We prefer the term *fundamental zone* and label it as *FZ*. The set of all (distinct) equivalence classes $\mathbf{R}G$ in $SO(3)$ is the fundamental zone of crystal orientations; its mathematical symbol is $SO(3)/G = FZ$ (or *FZ*(G) when we wish to specify the particular symmetry subgroup involved). Note that every element $\mathbf{R} \in SO(3)$ belongs to only one equivalence class $\mathbf{R}G$. *FZ* consists of the complete set of all disjoint equivalence classes that are G-related on $SO(3)$. Formally,

$$FZ = SO(3)/G = \{\mathbf{R}G | \mathbf{R} \in SO(3)\} \tag{6.8}$$

Geometrical representations of *FZ* are available for the Euler-angle parameterizations for all crystal types (Bunge, 1993; Morawiec, 2004). These geometrical representations are not unique. Selecting among the possibilities requires an additional constraint on the particular elements that belong in the chosen fundamental zone. This constraint distinguishes between different (equivalent) fundamental zones. Typically, the natural constraint is that the particular rotation of the minimum rotation angle is selected from each **R***G*, and the others are excluded; however, readers should be aware that other choices can and have been made.

Fundamental zone for cubic lattices

As mentioned in Chapter 4, two fundamental zones are of further interest in this book. The first is for centro-symmetric cubic crystals possessing the rotational symmetry subgroup O, also known as the octahedral group. O contains the identity element $\mathbf{I}$, 3 rotations of the type $\mathbf{R}(\pi\langle 100\rangle)$, 6 rotations of the type $\mathbf{R}(\pi/2\langle 100\rangle)$, 6 rotations of the type $\mathbf{R}(\pi\langle 110\rangle)$, and 8 rotations of the type $\mathbf{R}(2\pi/3\langle 111\rangle)$, for a total of 24. A common choice for the fundamental zone of cubic crystals is defined by the "loaf" in orientation space (Figure 6.1):

$$FZ(O) = \left\{(\varphi_1, \Phi, \varphi_2)|0 \le \varphi_1 < 2\pi,\ \cos^{-1}\left(\frac{\cos\varphi_2}{\sqrt{1+\cos^2\varphi_2}}\right) \le \Phi \le \pi/2,\ 0 \le \varphi_2 \le \pi/4\right\} \quad (6.9)$$

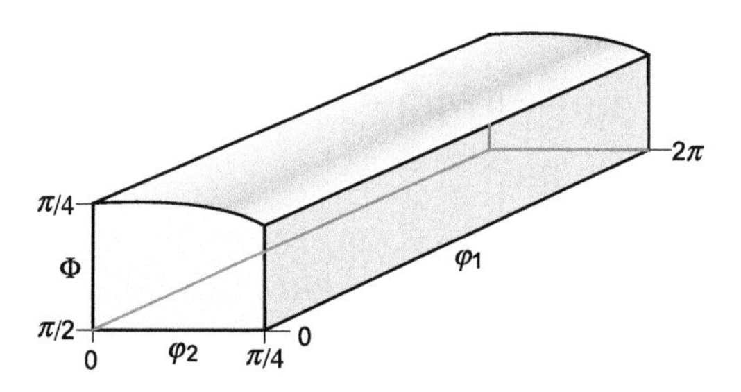

FIGURE 6.1

Fundamental zone for cubic material, *FZ*(*O*).

Fundamental zone for hexagonal lattices

For hexagonal crystals with lattice symmetry subgroup D_6, the subgroup G contains 6 rotations of the form $\mathbf{R}(n\pi/3\langle 0001\rangle)$ ($n = 1, 2, \ldots, 6$), 3 rotations of the form $\mathbf{R}(\pi\langle 11\bar{2}0\rangle)$, and 3 rotations of the form $\mathbf{R}(\pi\langle 10\bar{1}0\rangle)$, for a total of 12. A common choice for the fundamental zone of hexagonal crystals is (Figure 6.2)

$$FZ(D_6) = \{(\varphi_1, \Phi, \varphi_2)|0 \le \varphi_1 < 2\pi,\ 0 \le \Phi \le \pi/2,\ 0 \le \varphi_2 \le \pi/3\} \quad (6.10)$$

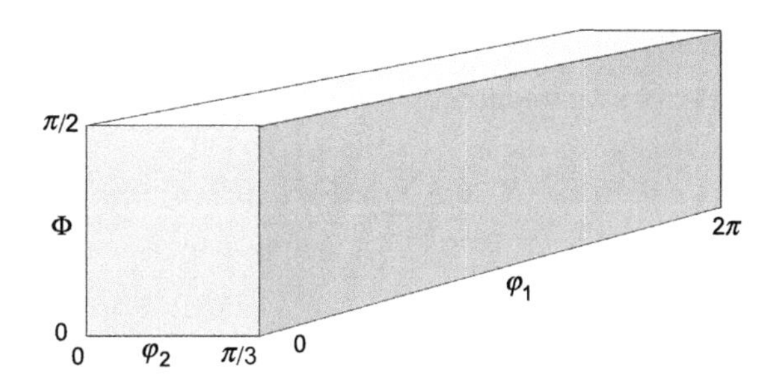

FIGURE 6.2

Fundamental zone for hexagonal material, $FZ(D_6)$.

6.2 SYMMETRY CONSIDERATIONS IN $SO(3)$ AND S^2

Many microstructure representations pertain to the fundamental zones of equivalence classes of $SO(3)$ (and S^2) as described in Section 6.1. The question that arises is how to economically represent functions on the pertinent fundamental zones via Fourier series, given the presence of symmetries that reduce the relevant spaces. For the primitive basis, this is accomplished by simply restricting the basis functions to the appropriate fundamental zone; but for the classic Fourier bases, which have global support over the full domain of rotations, we must understand how symmetries can be implemented. Two methods have been used in the past. In one of these the symmetries are imposed on the coefficients of the Fourier series; in the other the symmetries are embedded in the basis functions themselves. In the sequel we follow the second of these, which was pioneered by Bunge (1993).

Two types of symmetry are of interest in describing microstructures: crystallographic symmetry, pertaining to the crystal lattice, and statistical symmetry among the crystallite orientations, induced by symmetries in processing the material. Statistical symmetry is often imperfectly formed in polycrystals, and, consequently, care must be taken in any decision to impose statistical symmetry. Forcing imperfectly formed statistical symmetry can lead to substantial error, affecting the subsequent microstructure–properties relations in nonphysical ways. As computer code is readily available for computing the symmetric GSHFs and SSHFs, for all pertinent crystal symmetry subgroups, the description provided here is mainly aimed at providing some understanding of how the symmetry conditions are introduced, and how the available codes may be tested for fidelity to the pertinent symmetry conditions. Many further details on this subject can be found in the work of Bunge (1993). The shorthand notation for these functions will not be employed in this section since considerations of symmetry require the full complement of indices in their explicit form.

6.2.1 Crystallographic symmetry

Consider real-valued functions $f(R) \in L^2(SO(3)/G)$ that possess crystallographic symmetry:

$$f(R\tilde{R}^{(j)}) = f(R) \quad (\forall \tilde{R}^{(j)} \in G) \tag{6.11}$$

This relationship is not generally fulfilled by the GSHFs, $T_l^{mn}(R)$. It can be shown, however, that new, symmetric functions $\dot{T}_l^{mn}(R)$ can be introduced,

$$\dot{T}_l^{\mu n}(R) = \sum_{m=-l}^{+l} \dot{A}_l^{m\mu} T_l^{mn}(R) \tag{6.12}$$

that fulfill the condition

$$\dot{T}_l^{\mu n}(R) = \dot{T}_l^{\mu n}(R\tilde{R}^{(j)}) \tag{6.13}$$

Readers should note that crystallographic symmetry is *right-handed symmetry* in terms of active rotations R. In other words, the symmetry element $\tilde{R}^{(j)}$ appears to the right of rotation R. This amounts to a prerotation of the reference crystal lattice by$\tilde{R}^{(j)}$, which leaves the crystal in a physical orientation that is indistinguishable from the original reference crystal. (In other works on texture representation, such as Bunge (1993), the passive description of orientation is used, as was described in Section 2.4. In this case, the symmetry element falls on the left side of the rotation and is called a left-handed symmetry element. Since the passive and active frames are related by transposition, it is apparent that right-handed must become left-handed and vice versa when comparing the two notations.) Readers should also notice that the symmetry coefficients $\dot{A}_l^{m\mu}$ in relation (6.12) are independent of index n (see actual values in Appendix 2).

The coefficients $\dot{A}_l^{m\mu}$ must be selected to satisfy the condition

$$\sum_{m=-l}^{+l} \dot{A}_l^{m\mu}[T_l^{ms}(\tilde{R}^{(j)}) - \delta_{ms}] = 0 \ \ (\text{for } \tilde{R}^{(j)} \in G; \ -l \le s \le +l) \tag{6.14}$$

The index μ enumerates over the linearly independent solutions

$$1 \le \mu \le M(l) \tag{6.15}$$

$M(l)$ is a function of the crystal symmetry subgroup G, and is described for triclinic (C_1), monoclinic (C_2), orthorhombic (D_2), tetragonal (D_4), hexagonal (D_6), and cubic (O) lattice symmetries for $l \le 50$ by Bunge (1993). These are shown in Figures 6.3 and 6.4.

From Eq. (3.55) we extend the foregoing to obtain

$$\dot{T}_l^{\mu 0}(\Phi, \varphi_2) = \sqrt{\frac{4\pi}{2l+1}}\dot{k}_l^{\mu}(\Phi, \varphi_2 - \pi/2) = \sqrt{\frac{4\pi}{2l+1}}\dot{k}_l^{*\mu}(\Phi, \beta) \tag{6.16}$$

where $\beta = \dfrac{\pi}{2} - \varphi_2$, and $\dot{k}_l^{\mu}(\Phi, \beta)$ are the surface spherical harmonic functions satisfying the symmetry conditions

$$\dot{k}_l^{\mu}(\hat{\mathbf{n}}) = \dot{k}_l^{\mu}(\hat{\mathbf{n}}\tilde{R}^{(j)}) \ \ (\forall \ \tilde{R}^{(j)} \in G) \tag{6.17}$$

The symmetry coefficients, in this case, must satisfy the relations

$$\dot{k}_l^{\mu}(\hat{\mathbf{n}}) = \sum_{m=-l}^{+l} \dot{A}_l^{*m\mu} k_l^{m}(\hat{\mathbf{n}}) \tag{6.18}$$

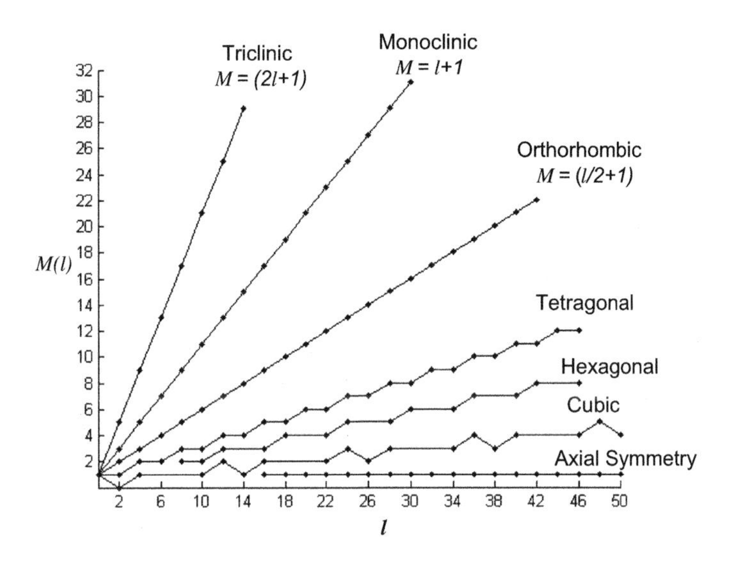

FIGURE 6.3

The number of linearly independent spherical harmonics, $M(l)$ versus the degree, l, for various symmetry groups (even values of l). The top line is triclinic; the bottom is for axial symmetry.

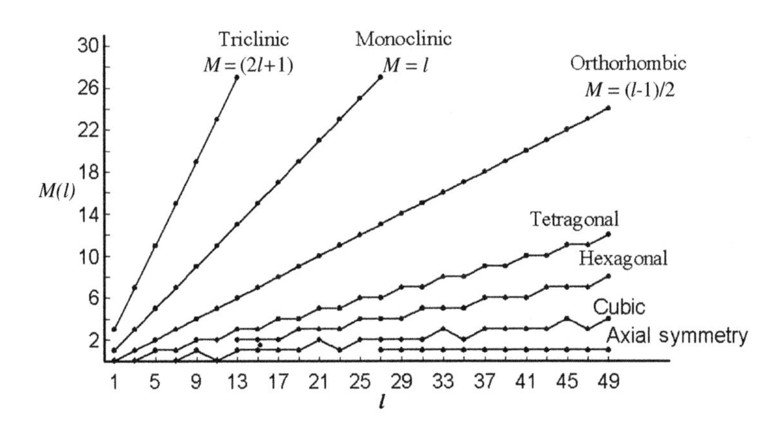

FIGURE 6.4

The number of linearly independent spherical harmonics, $M(l)$ versus the degree, l, for various symmetry groups (odd values of l). The top line is triclinic; the bottom is for axial symmetry.

(the asterisk indicates a complex conjugate). The $\dot{\dot{A}}_l^{m\mu}$ coefficients are the same as those defined for the generalized spherical harmonics in Eq. (6.12).

Calculation of the surface spherical harmonic functions, $\dot{\dot{k}}_l^{\mu}(\Phi, \beta)$, is undertaken using the following equation; values of coefficients $B_l^{m\mu}$ are given in Table A2.1 of Appendix 2. See also Eq. (3.52).

$$\dot{k}_l^{\mu}(\Phi,\beta) = \sum_{m=0}^{l} B_l^{m\mu}\overline{P}_l^{m}(\Phi)\cos m\beta \tag{6.19}$$

It can be shown, further, that the symmetry coefficients can be chosen to be real (Bunge, 1993):

$$\dot{A}_l^{m\mu} = \dot{A}_l^{*m\mu} \tag{6.20}$$

and, after adding the further requirement of orthonormality, the following relationship among the symmetry coefficients is obtained:

$$\sum_{m=-l}^{+l} \dot{A}_l^{m\mu}\dot{A}_l^{m\mu'} = \delta_{\mu\mu'} \tag{6.21}$$

Consistent with (6.21) the orthonormality relationships are

$$\iiint_{SO(3)/G} \dot{T}_l^{\mu n}(R)\dot{T}_{l'}^{*\mu' n'}(R)dR = \frac{1}{2l+1}\delta_{ll'}\delta_{\mu\mu'}\delta_{nn'} \tag{6.22}$$

and

$$\iiint_{S^2/G} \dot{k}_l^{\mu}(\hat{\mathbf{n}})\dot{k}_{l'}^{*\mu'}(\hat{\mathbf{n}})d\hat{n} = \delta_{ll'}\delta_{\mu\mu'} \tag{6.23}$$

6.2.2 Statistical symmetry

Statistical symmetry is now considered in the same way, but readers are cautioned that these relations are rarely perfectly formed in real polycrystals. Consider real-valued functions $f(R) \in L^2(S\ SO(3))$ that possess statistical symmetry belonging to *statistical symmetry subgroup S*:

$$f(\tilde{R}^{(j)}R) = f(R) \ \ (\forall \tilde{R}^{(j)} \in S) \tag{6.24}$$

This relationship is not generally fulfilled by the GSHF, $T_l^{mn}(R)$. It can be shown, however, that new symmetric functions $\dot{T}_l^{mn}(R)$ can be introduced,

$$\dot{T}_l^{m\nu}(R) = \sum_{n=-l}^{+l} \dot{A}_l^{n\nu} T_l^{mn}(R) \tag{6.25}$$

that fulfill the condition

$$\dot{T}_l^{m\nu}(R) = \dot{T}_l^{m\nu}(\tilde{\tilde{\mathrm{R}}}^{(j)}R) \ \ (\forall \tilde{\tilde{\mathrm{R}}}^{(j)} \in S) \tag{6.26}$$

Readers should note that crystallographic symmetry is *left-handed symmetry* in terms of active rotations R. In other words, the symmetry element $\tilde{\tilde{\mathrm{R}}}^{(j)}$ appears to the left of rotation R. This amounts to a postrotation of the local crystal lattice by $\tilde{\tilde{\mathrm{R}}}^{(j)}$, which leaves the crystal in a new physical orientation that is statistically indistinguishable from the local crystal orientation. Note that the symmetry coefficients $\dot{A}_l^{n\nu}$ in relation (6.25) are independent of index m.

The coefficients $\dot{A}_l^{n\nu}$ must be selected to satisfy the condition

$$\sum_{n=-l}^{+l} \dot{A}_l^{n\nu}[T_l^{sn}(\tilde{\tilde{\mathrm{R}}}^{(j)}) - \delta_{sn}] = 0 \quad (\text{for } \tilde{\tilde{\mathrm{R}}}^{(j)} \in S;\ -l \leq s \leq +l) \tag{6.27}$$

The index ν enumerates over the linearly independent solutions

$$1 \leq \nu \leq N(l) \tag{6.28}$$

$N(l)$ is a function of the statistical symmetry subgroup S, and is described for triclinic (C_1), monoclinic (C_2), orthorhombic (D_2), and axial (D_∞) symmetries for $l \leq 50$ by Bunge (1993), by simply substituting $N(l)$ for $M(l)$.

From Eq. (3.55) we extend the foregoing to obtain

$$\dot{T}_l^{0n}(\varphi_1, \Phi) = \sqrt{\frac{4\pi}{2l+1}}\dot{k}_l^{\nu}(\Phi, \varphi_1 - \pi/2) = \sqrt{\frac{4\pi}{2l+1}}\dot{k}_l^{\nu}(\Phi, \gamma) \tag{6.29}$$

where $\gamma = \varphi_1 - \frac{\pi}{2}$, and $\dot{k}_l^{\mu}(\Phi, \gamma)$ are the surface spherical harmonic functions satisfying the symmetry conditions

$$\dot{k}_l^{\nu}(\hat{n}) = \dot{k}_l^{\nu}(\tilde{\tilde{\mathrm{R}}}^{(j)}\hat{n}) \quad (\forall\ \tilde{\tilde{\mathrm{R}}}^{(j)} \in S) \tag{6.30}$$

The symmetry coefficients, in this case, must satisfy the relations

$$\dot{k}_l^{\mu}(\hat{\mathbf{n}}) = \sum_{n=-l}^{+l} \dot{A}_l^{n\nu} k_l^{n}(\hat{\mathbf{n}}) \tag{6.31}$$

As with the lattice symmetry coefficients, the statistical coefficients can be chosen to be real:

$$\dot{A}_l^{n\nu} = \dot{A}_l^{*n\nu} \tag{6.32}$$

and after adding the further requirement of orthonormality the following relationship among the symmetry coefficients is obtained:

$$\sum_{n=-l}^{+l} \dot{A}_l^{n\nu} \dot{A}_l^{n\nu'} = \delta_{\nu\nu'} \tag{6.33}$$

Consistent with (6.33), the orthonormality relationships are

$$\iiint\limits_{S\ SO(3)} \dot{T}_l^{m\nu}(R)\dot{T}_{l'}^{m'\nu'}(R)dR = \frac{1}{2l+1}\delta_{ll'}\delta_{mm'}\delta_{\nu\nu'} \tag{6.34}$$

and

$$\iiint\limits_{S\ S^2} \dot{k}_l^{\nu}(\hat{\mathbf{n}})\dot{k}_{l'}^{*\nu'}(\hat{n})d\hat{n} = \delta_{ll'}\delta_{\nu\nu'} \tag{6.35}$$

6.2.3 Combined crystallographic and statistical symmetry

When both crystallographic (right-handed) and statistical (left-handed) symmetry are present in the representation, relations (6.32) and (6.35) can be combined to form

$$\dddot{T}_l^{\mu\nu}(R) = \sum_{m=-l}^{+l}\sum_{n=-l}^{+l} \dot{A}_l^{m\mu}\dot{A}_l^{n\nu}T_l^{mn}(R) \tag{6.36}$$

Readers should note the presence of three symmetry dots above the GSHFs, indicating the presence of both types of symmetry. These functions satisfy the relations

$$\dddot{T}_l^{\mu\nu}(R) = \dddot{T}_l^{\mu\nu}(\tilde{\tilde{R}}^{(j)}R\tilde{R}^{(k)}) \ (\forall \tilde{\tilde{R}}^{(j)} \in S, \ \tilde{R}^{(k)} \in G) \tag{6.37}$$

The doubly symmetric functions must satisfy the orthogonality relations

$$\iiint_{S\,SO(3)/G} \dddot{T}_l^{\mu\nu}(R)\dddot{T}_{l'}^{\mu'\nu'}(R)dR = \frac{1}{2l+1}\delta_{ll'}\delta_{\mu\mu'}\delta_{\nu\nu'} \tag{6.38}$$

For calculation purposes we use the following equations, with the coefficients a_l^{mns} and $B_l^{m\mu}$ available in tables or data files:

$$\varepsilon_\nu = \begin{cases} 1 \text{ for } \nu = 0 \\ \sqrt{2} \text{ for } \nu \neq 0 \end{cases}$$

$$\dddot{T}_l^{\mu\nu}(\varphi_1\Phi\varphi_2) = p_l^{\mu\nu}(\Phi\varphi_2)\cos\nu\varphi_1 + q_l^{\mu\nu}(\Phi\varphi_2)\sin\nu\varphi_1 \tag{6.39}$$

$$p_l^{\mu\nu}(\Phi\varphi_2) = \varepsilon_\nu\sqrt{2\pi}\sum_{m=0}^{l/4} B_l^{(4m)\nu}\sum_{s=0,2,..}^{l} a_l^{(4m)ns}\cos s\Phi\cos 4m\varphi_2$$

$$q_l^{\mu\nu}(\Phi\varphi_2) = -\varepsilon_\nu\sqrt{2\pi}\sum_{m=0}^{l/4} B_l^{(4m)\nu}\sum_{s=1,3,..}^{l} a_l^{(4m)ns}\cos s\Phi\sin 4m\varphi_2$$

6.2.4 Explicit treatment of symmetry elements

Next, we consider how the symmetry coefficients $\dot{A}_l^{m\mu}$ and $\dot{A}_l^{n\nu}$ might be constructed from the foregoing relationships. Note that the specific coefficients obtained are not unique; the particular ones described here are those given by Bunge (1993). The main results are obtained by detailed consideration of relations (6.14) and (6.27), which treat independently the crystallographic and statistical symmetries. Given that the form of these relationships is essentially identical, results obtained for elements of crystallographic symmetry are basically interchangeable with the equivalent elements of statistical symmetry. In the simplest cases we consider the occurrence of *selection rules* in the treatment of low-order symmetry groups. These mainly affect the presence of particular coefficients or their combinations in the Fourier series.

Rotation axes of order n

Begin with the presence of an nth-order rotation axis about $\Phi = 0$. Relations (3.52) and (6.18) can be combined to obtain

$$\dot{\dot{k}}_l^{\mu}(\Phi,\beta) = \sum_{m=-l}^{+l} \dot{\dot{A}}_l^{m\mu}\left[\frac{1}{\sqrt{2\pi}}e^{im\beta}\overline{P}_l^m(\cos\Phi)\right] = \sum_{m=-l}^{+l} \dot{\dot{A}}_l^{m\mu}\left[\frac{1}{\sqrt{2\pi}}e^{im(\beta+2\pi/n)}\overline{P}_l^m(\cos\Phi)\right] \tag{6.40}$$

from which it follows that m must take on the values

$$m = n\mu \tag{6.41}$$

where μ is a positive or negative integer. Other functions with indices m not satisfying (6.41) must not belong to the set of symmetric functions. Thus, we see the meaning of the label *selection rules.* The symmetry coefficients themselves can be readily expressed as

$$\dot{\dot{A}}_l^{m\mu} = \delta_{m(n\mu)} \tag{6.42}$$

Clearly, if relation (6.40) must be true for all β, then we must have $m = 0$ and the *rotationally symmetric* SSHFs are $\dot{\dot{k}}_l^0(\Phi,\beta) = \overline{P}_l^0(\Phi)/\sqrt{2\pi}$. Of course, the axial symmetry implied in this case does not occur among the crystallographic symmetry groups. Among the statistical symmetries, however, axial symmetry is a technologically important case.

Mirror plane perpendicular to $\Phi = 0$

Consider the presence of a mirror plane perpendicular to the direction $\Phi = 0$, which implies that the function must be unchanged if Φ is replaced with $\pi - \Phi$. In this case, given that $\overline{P}_l^m(\pi-\Phi) = (-1)^{l+m}\,\overline{P}_l^m(\Phi)$, it must follow that

$$\dot{\dot{A}}_l^{m\mu} = \begin{cases} +1 \text{ if } l+m \text{ is an even integer} \\ 0 \text{ otherwise} \end{cases} \tag{6.43}$$

Mirror plane in the plane $\beta = 0$

If $\beta = 0$ contains a mirror plane then it is easily shown that $\dot{\dot{A}}_l^{-m\mu} = (-1)^m\,\dot{\dot{A}}_l^{+m\mu}$, and hence only half of the symmetry coefficients are independent. One convenient choice of symmetry coefficients is

$$\dot{\dot{A}}_l^{+m\mu} = \frac{1}{\sqrt{2}}\delta_{m\mu},\quad \dot{\dot{A}}_l^{-m\mu} = \frac{(-1)^m}{\sqrt{2}}\delta_{m\mu},\quad \dot{\dot{A}}_l^{m0} = \delta_{m0} \tag{6.44}$$

Readers should note that in (6.44) μ takes on values $0 \le \mu \le l$.

2n-fold rotation axes in $\Phi = 0$

If a $2n$-fold rotation reflection axis in $\Phi = 0$ transforms $\Phi,\beta \to (\pi-\Phi),(\pi/n+\beta)$, one must have

$$k_l^m\left(\pi-\Phi,\frac{\pi}{n}+\beta\right) = (-1)^{l+m+m/n}k_l^m(\Phi,\beta) \tag{6.45}$$

from which we obtain the selection rule

$$m = m'n, \; l + m'(n+1) = 2l' \quad (l', m' \in Z) \tag{6.46}$$

This set of selection rules can be expressed in terms of the symmetry coefficients as

$$\dot{A}_l^{m\mu} = \delta_{m(m'n)} \, \delta_{(l+m'(n+1))(2l')} \tag{6.47}$$

Twofold rotation axis in the direction $\Phi = 90°$, $\beta = 0°$

If we consider a twofold rotation axis in the direction $\Phi = 90°$, $\beta = 0°$, this transforms Φ, $\beta \to \pi - \Phi, -\beta$, which leads to the condition

$$\dot{A}_l^{-m\mu} = (-1)^l \dot{A}_l^{+m\mu} \tag{6.48}$$

which must be fulfilled. This leads to linearly independent functions of the form

$$\dot{k}_l^{\mu}(\Phi,\beta) = \frac{1}{\sqrt{4\pi}} \overline{P}_l^{\mu}(\Phi)\left[e^{i\mu\beta} + (-1)^{l+\mu} e^{-i\mu\beta}\right], \quad \dot{k}_l^{0}(\Phi,\beta) = \frac{1}{\sqrt{2\pi}} \overline{P}_l^{0}(\Phi) \text{ (for even } l\text{)} \tag{6.49}$$

6.2.5 Combining selection rules for low-order symmetry subgroups

Typical symmetry subgroups for crystal lattices or statistical considerations comprise several of the symmetry elements described in the previous section, and thus several selection rules can be important in establishing the effects of the full set of symmetry elements. For example, the hexagonal symmetry subgroup D_{6h} combines a sixfold rotation axis in the direction $\Phi = 0°$ with a twofold rotation axis in the $\Phi = 90°$, $\beta = 0°$, and a mirror plane perpendicular to $\Phi = 0°$. Thus, (6.42), (6.43), and (6.48) must simultaneously be fulfilled. In this case it can be shown that the coefficients shown in Table 6.1 are applicable.

If the mirror plane is missing (as with D_6), relation (6.43) is not fulfilled, and additional coefficients appear in the Fourier series. Other low-symmetry subgroups, such as orthorhombic (D_2 or D_{2h}), tetragonal (D_4 or D_{4h}), or axial symmetry (C_∞), are readily obtained by the same methods.

Table 6.1 Symmetry Coefficients for the Hexagonal Symmetry Subgroup D_{6h}

μ/m	0	±6	±12	±18 ...
1	1	0	0	0
2	0	$1/\sqrt{2}$	0	0
3	0	0	$1/\sqrt{2}$	0
4	0	0	0	$1/\sqrt{2}$
⋮				
$M(l)$				

Source: Adapted from Bunge (1993), Table 14.4.

6.2.6 Higher-symmetry subgroups (cubic)

So-called *higher-symmetry subgroups* contain additional elements of symmetry that cannot be satisfied by selection rules on the indices only. These higher-symmetry cases are generally distinguished from the lower-symmetry cases by the presence of the diagonal threefold rotation axes. Examples include the tetrahedral, octahedral, and icosahedral subgroups. The latter occur neither as crystal nor statistical symmetry in known systems. Consideration of the threefold rotations about diagonal axes can only be treated with linear combinations of the original basis functions. Thus, in terms of the symmetries of the SSHFs, one can write

$$\dot{k}_l^{\mu}(\Phi,\beta) = \sum_{m=-l}^{+l} \dot{A}_l^{m\mu}\left(\frac{1}{\sqrt{2\pi}}\right) e^{im\beta}\overline{P}_l^{m}(\Phi) \tag{6.50}$$

Following the approach taken by Bunge (1993), the $\overline{P}_l^m(\Phi)$ functions are themselves expressed as a Fourier series in terms of the exponential functions:

$$\overline{P}_l^{m}(\Phi) = \sum_{s=-l}^{+l} a_l^{ms} e^{is\Phi} \tag{6.51}$$

Introducing (6.51) into (6.50) we obtain

$$\dot{k}_l^{\mu}(\Phi,\beta) = \left(\frac{1}{\sqrt{2\pi}}\right) \sum_{s=-l}^{+l}\sum_{m=-l}^{+l} \dot{A}_l^{m\mu} a_l^{ms} e^{im\beta} e^{is\Phi} \tag{6.52}$$

The threefold axis can be shown to require the following equality:

$$\dot{k}_l^{\mu}(\Phi = \alpha, \beta = 0) = \dot{k}_l^{\mu}\left(\Phi = \frac{\pi}{2}, \beta = \alpha\right) \tag{6.53}$$

But introducing (6.53) into (6.52) leads to the condition

$$\sum_{m=-l}^{+l}\sum_{s=-l}^{+l} \dot{A}_l^{m\mu} a_l^{ms} e^{is\alpha} = \sum_{n=-l}^{+l}\sum_{s=-l}^{+l} \dot{A}_l^{s\mu} a_l^{sn} e^{in\pi/2} e^{is\alpha} \tag{6.54}$$

that must be true for all α. This implies that the coefficients of $e^{is\alpha}$ must agree, and therefore

$$\sum_{m=-l}^{+l} \dot{A}_l^{m\mu} a_l^{ms} = \dot{A}_l^{s\mu} \sum_{n=-l}^{+l} a_l^{sn} e^{in\pi/2} \tag{6.55}$$

Setting

$$b_l^s = \sum_{n=-l}^{+l} a_l^{sn} e^{in\pi/2} \tag{6.56}$$

the following equation is found for determination of the coefficients $\dot{A}_l^{m\mu}$:

$$\sum_{m=-l}^{+l} \dot{A}_l^{m\mu}[a_l^{ms} - \delta_{ms} b_l^s] = 0 \tag{6.57}$$

Various selection rules may now be added to this basic relationship to obtain the symmetry coefficients for the various cubic symmetry subgroups. Consider, for example, the octahedral group (O) that must be subject to the fourfold selection rule described in subsection "Rotation Axes of Order n" in Section 6.14. Thus, the index m must be an integer multiple of 4. The octahedral group also requires symmetry generated by an additional fourfold axis about $\Phi = \frac{\pi}{2}$, $\beta = \frac{\pi}{2}$. This symmetry axis requires that the function be fourfold along $\beta = 0$. Thus, the coefficients on the left of (6.54) must vanish for non-fourfold s. This leads to the condition

$$\sum_{m=-l}^{+l} \dot{A}_l^{m\mu} a_l^{ms} = 0;\ m = 4m'\ (m' \in Z) \tag{6.58}$$

whereby either

$$\begin{aligned} & l = 2l';\ s = 4s' + 2 \\ \text{or} \quad & l = 2l' + 1;\ s = 2s' + 1 \end{aligned} \quad (s', l' \in Z) \tag{6.59}$$

If the interest is in O_h, which also contains a center of symmetry, then the additional condition

$$l = 2l' \tag{6.60}$$

must hold.

The foregoing is intended to be an introduction to the question of symmetry in the classic GSHFs and SSHFs. Other methods for introducing symmetry into these functions or, as an alternative, into the Fourier coefficients, can be found in the literature noted in the Bibliography. Since commercial code is available for calculating the symmetric GSHFs and SSHFs for all symmetry subgroups of interest, the prequel is intended primarily as background for understanding and testing the available codes.

6.2.7 Combined higher symmetry: Cubic orthorhombic

For cubic-orthorhombic materials, the fundamental zone of distinct crystal orientations can be defined as

$$SO(3)/O = \left\{ (\varphi_1, \Phi, \varphi_2) | 0 \le \varphi_1 < \frac{\pi}{2}, \cos^{-1}\left(\frac{\cos\varphi_2}{\sqrt{1 + (\cos\varphi_2)^2}} \right) \le \Phi \le \frac{\pi}{2}, 0 \le \varphi_2 \le \frac{\pi}{4} \right\} \tag{6.61}$$

where (φ_1, Φ, φ_2) represent the set of Bunge–Euler angles commonly used to describe a crystal orientation.

Following the pattern introduced in Chapter 4, the orientation distribution function (ODF) can be represented as a Fourier series of symmetrized generalized spherical harmonics as

$$f(g) = \sum_{\ell=0}^{\infty} \sum_{\mu=1}^{M(\ell)} \sum_{\eta=1}^{N(\ell)} C_\ell^{\mu\eta} \dot{\dot{T}}_\ell^{\mu\eta}(g) \tag{6.62}$$

where $C_\ell^{\mu\eta}$ represent the Fourier coefficients, and $\dot{\dot{T}}_l^{\mu\eta}(g)$ constitute a complete set of orthonormal basis functions for cubic materials of which the ODFs exhibit orthorhombic sample symmetry. Each distinct

ODF can be represented by a unique set of Fourier coefficients in the infinite-dimensional Fourier space.

SUMMARY

This chapter added the concept of symmetry to the description of microstructure and to the spectral representations of that structure developed in Chapters 4 and 5. It is now apparent that the generalized spherical harmonic and surface spherical harmonic representations of distributions on the state space of crystal orientations are particularly efficient at representing distribution functions, including any implicit symmetry. This makes them ideal spectral bases for ODFs. The primitive and discrete Fourier representations do not capture symmetry, at least in their form as presented in this book. However, there are other benefits to the representations and we will continue to consider all the presented spectral methods as we begin to look at structure–property relations, homogenization relations, and subsequently at microstructure design.

Exercises

Section 6.1

6.1. Using the symmetry elements of D_6, demonstrate that the hexagonal fundamental zone is that given by Eq. (6.10) and illustrated in Figure 6.2.

Section 6.2

6.2. What is the rotation matrix associated with point operator C_{4z}^{-} ? Find this in the following two ways.

(a) Form the three columns of the matrix using the transformed positions of the vectors $\mathbf{e}_1$, $\mathbf{e}_2$, $\mathbf{e}_3$, respectively. (*Note:* The "Transformation" column in Appendix 1 gives the passive description, that is, the vectors that transform to the new basis vectors.)

(b) Use the usual 3D rotation matrix definition for a rotation of θ about an axis n (use θ and n as given in Appendix 1).

6.3. Prove, from the symmetry conditions, relations (6.41), (6.42), (6.43), (6.44), (6.47), and (6.48).

CHAPTER

Structure–Property Relations

Continuum Mechanics

7

CHAPTER OUTLINE

In Chapter 2, we discussed the concept of a tensor. In this chapter, we will employ the same concepts to describe several physical quantities of interest in establishing the effective properties of composite materials based on the morphology and response of their local constituents.

7.1 POTENTIALS AND GRADIENTS

Most physical phenomena are formulated and studied using the concepts of *scalar potentials* and their fields. Examples include electrical conduction, diffusion, dielectrics, magnetism, thermal conduction, flow in porous media, and solid mechanics. A fundamental hypothesis in all of these phenomena is the existence of a scalar potential that can be expressed as an inner product of a pair of conjugate tensorial physical quantities such as current and field in electrical conduction, heat flux and temperature gradient in thermal conduction, and stress and strain in solid mechanics.

In conduction-type problems, the field equations take on fairly simple forms. Here, the conjugate tensorial quantities can be identified as a *flux* denoted by **J** (e.g., current, magnetic induction, heat flux, mass flux) and a *gradient* denoted by **E** (e.g., electric field, magnetic field, concentration gradient, pressure gradient, temperature gradient). The spatial variation of the flux variable in a given volumetric domain is controlled by a *conservation principle* (e.g., mass conservation, charge conservation, heat conservation), and in the absence of sources is expressed as

$$\nabla \cdot \mathbf{J} = 0 \tag{7.1}$$

where the ∇ operator has been defined in Eq. (2.65). The spatial variation of the conjugate variable $\mathbf{E}$ in a given volumetric domain is constrained to follow a *continuity equation*:

$$\nabla \times \mathbf{E} = 0 \tag{7.2}$$

Equation (7.2) is generally required of all smooth fields of gradients. (Note that $\mathbf{E}$ is often defined as a gradient of the potential in these problems.)

7.2 STRESS

The underlying equations for mechanics of materials are often written in terms of stress. In this section, we develop the necessary tensorial description of stress that will be required in the formation of structure-property relations later on.

7.2.1 Stress tensor

The elastic–plastic theories of solid mechanics are in general much more complicated than the simple description of the conduction-type phenomena described later in Section 7.4. First, the field quantities, namely stress and strain, are second-rank tensors. Second, in large plastic strains on polycrystalline metals, the phenomena are associated with highly nonlinear constitutive behavior as well as geometric nonlinearities arising from large changes in shape and size of the volumetric domain being studied. The various physical quantities encountered in theories of solid mechanics are described next.

When a body is subjected to force(s), it experiences internal stress. Stress at a material point is fundamentally defined as force per unit area in the neighborhood of that material point. However, this definition of stress requires specification of the orientation of the infinitesimal area element chosen at the material point. Since it is possible to choose an infinite number of area elements with different orientations at the material point, it follows that one can define an infinite number of stresslike quantities at the material point. The concept of a stress tensor helps us resolve this dilemma.

Figure 7.1 shows stresses on the surfaces of an infinitesimal cuboidal volume element at a material point of interest in the current loaded configuration. The stress vectors on each surface have been resolved into the three orthonormal basis directions, giving a total of nine different stress components. Note that the stress vectors in this figure are depicted only on the three positive surfaces (normals on these surfaces are along the positive directions of the orthonormal basis). It can be shown that the stress vectors on the negative surfaces are exactly equal and opposite to the corresponding ones on the positive surfaces by virtue of force equilibrium. The nine stress components are labeled using two indices, with each index ranging from 1 to 3. The first index identifies the plane on which the stress component is described and the second index refers to the direction of the stress component. In the indicial notation, the stress components are simply referred to as σ_{ij}. Satisfying equilibrium of moments on the cuboid shown in Figure 7.1 yields the important result that the *stress tensor* defined here is symmetric:

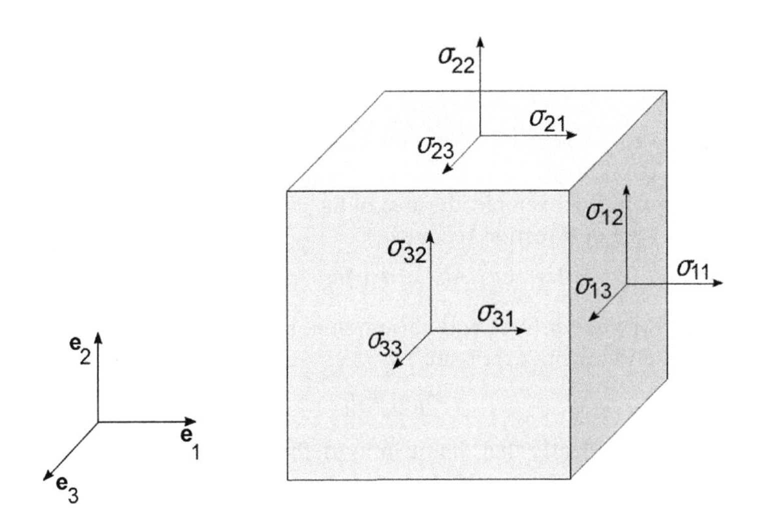

FIGURE 7.1

Description of the stress tensor at a material point.

$$\sigma_{ji} = \sigma_{ij} \tag{7.3}$$

As previously noted, it is possible to define stress vectors on any plane passing through a material point. To avoid confusion between the definition of the stress tensor at a material point and the stress vectors on any plane passing through the material point, the latter will henceforth be referred to as the *traction vectors*. The stress components shown in Figure 7.1 describe only the traction vectors on three specific (orthonormal) planes passing through the material point. However, once these nine components are specified, it is possible to uniquely determine the traction vector on any plane passing through the material point. To accomplish this, one can section the cuboidal volume element shown in Figure 7.1 and apply equilibrium on the sectioned part (see Figure 7.2).

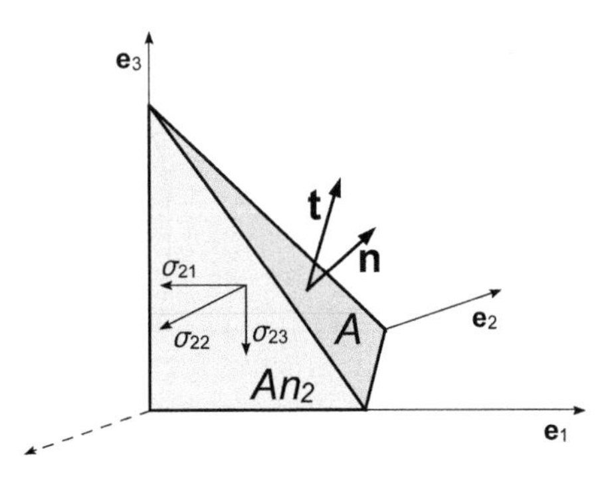

FIGURE 7.2

Sectioning of the volume element in Figure 7.1 to establish the relationship between the stress tensor and the traction vector on a plane with a normal of **n**.

In Figure 7.2, the sectioned volume of the cuboid is depicted with the cut section shaded darker than the faces of the cuboid. Let **n** denote the unit normal of the cut section, and **t** the traction vector (force per unit area) on this section. For clarity, the stress components are shown only on one face of the cuboid in this figure (i.e., the negative face normal to $\mathbf{e}_2$). Let A denote the area of the cut section. Note that the areas of the cut faces of the cuboid can be expressed simply as the projections of A in the directions of the face normals. For example, the area of the cuboid face normal to $\mathbf{e}_2$ is simply given by An_2. Force equilibrium in the $\mathbf{e}_1$ direction requires

$$At_1 = \sigma_{11}An_1 + \sigma_{21}An_2 + \sigma_{31}An_3 \tag{7.4}$$

Similar equations can be written for force equilibrium in $\mathbf{e}_2$ and $\mathbf{e}_3$ directions. The resulting equations can be compacted into the following expression:

$$\boldsymbol{\sigma}_{ji}n_j = t_i \tag{7.5}$$

Equation (7.5) establishes clearly that the quantity defined by $\boldsymbol{\sigma}$ is indeed a second-rank tensor because it describes a linear transformation mapping of all vectors in Euclidean vector space into itself. The physical quantity stress tensor is essentially defined through Eq. (7.5), although it was introduced here in a different manner. In tensor notation, Eq. (7.5) can be expressed as

$$\boldsymbol{\sigma}^T\mathbf{n} = \mathbf{t} \tag{7.6}$$

Noting that the stress tensor is symmetric (Eq. 7.3), Eq. (7.6) can also be expressed as

$$\boldsymbol{\sigma}\,\mathbf{n} = \mathbf{t} \tag{7.7}$$

Example 7.1

A cylindrical rod is pulled in tension by a force of 1000 N (Figure 7.3). What is the traction vector on a plane at an angle of 45° to the tensile direction, given a rod diameter of 1 cm?

Solution

The only nonzero stress component is σ_{11} and is computed as

$$\sigma_{11} = \frac{\mathrm{F}}{\pi r^2} = \frac{1000\ \mathrm{N}}{\pi(0.5\ \mathrm{cm})^2} = 1273\frac{\mathrm{N}}{\mathrm{cm}^2} = 12.73\ \mathrm{MPa}$$

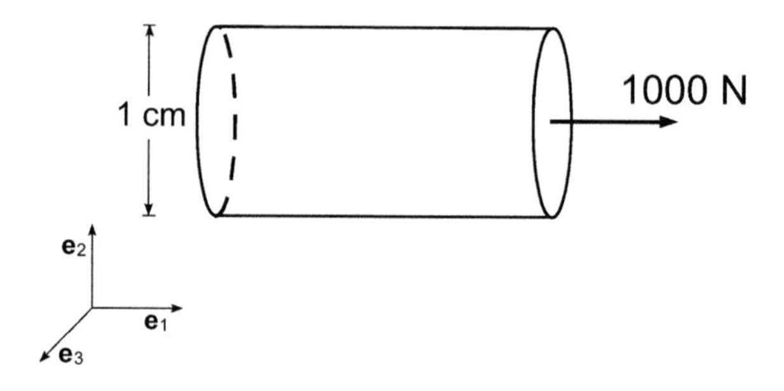

FIGURE 7.3

Schematic of cylindrical rod under tension.

The stress tensor can be expressed as

$$[\sigma] = \begin{pmatrix} 12.73 & 0 & 0 \\ 0 & 0 & 0 \\ 0 & 0 & 0 \end{pmatrix} \text{MPa}$$

The normal of the plane of interest can be expressed as

$$\{\mathbf{n}\} = \begin{Bmatrix} 1/\sqrt{2} \\ 1/\sqrt{2} \\ 0 \end{Bmatrix}$$

The traction vector on the plane of interest is computed using Eq. (7.6):

$$\{\mathbf{t}\} = \begin{pmatrix} 12.73 & 0 & 0 \\ 0 & 0 & 0 \\ 0 & 0 & 0 \end{pmatrix} \begin{Bmatrix} 0.707 \\ 0.707 \\ 0 \end{Bmatrix} = \begin{Bmatrix} 9 \\ 0 \\ 0 \end{Bmatrix} \text{MPa}$$

Alternatively, this problem can also be solved completely using tensor notation (instead of matrix algebra) as follows:

$$\sigma = 12.73(\mathbf{e}_1 \otimes \mathbf{e}_1)\ \text{MPa}$$

$$\mathbf{t} = \sigma\mathbf{n} = 12.73(\mathbf{e}_1 \otimes \mathbf{e}_1)\left(\frac{1}{\sqrt{2}}\mathbf{e}_1 + \frac{1}{\sqrt{2}}\mathbf{e}_2\right) = \frac{12.73}{\sqrt{2}}\mathbf{e}_1 = 9\mathbf{e}_1\ \text{MPa}$$

It is sometimes convenient to decompose a stress tensor into *hydrostatic* and *deviatoric* components as

$$\boldsymbol{\sigma} = \frac{1}{3}tr\,(\boldsymbol{\sigma})\,\mathbf{I} + \boldsymbol{\sigma}' \tag{7.8}$$

where $\boldsymbol{\sigma}'$ denotes the deviatoric stress tensor and $\frac{1}{3}tr\,(\boldsymbol{\sigma})$ denotes the hydrostatic component of the stress tensor. We will find this decomposition particularly useful in studying plastic deformation in metals. Note also that a similar decomposition can be applied to the symmetric infinitesimal strain tensor $\boldsymbol{\varepsilon}$ that will be defined later.

Example 7.2

Evaluate the hydrostatic and deviatoric components of the stress tensor corresponding to a uniaxial stress state.

Solution

A uniaxial stress state in the $\mathbf{e_1}$ direction can be expressed as

$$[\sigma] = \begin{pmatrix} \sigma_{11} & 0 & 0 \\ 0 & 0 & 0 \\ 0 & 0 & 0 \end{pmatrix}$$

The hydrostatic and deviatoric components of this stress tensor are computed using Eq. (7.8) as

$$[\sigma] = \begin{pmatrix} \frac{\sigma_{11}}{3} & 0 & 0 \\ 0 & \frac{\sigma_{11}}{3} & 0 \\ 0 & 0 & \frac{\sigma_{11}}{3} \end{pmatrix} + \begin{pmatrix} \frac{2\sigma_{11}}{3} & 0 & 0 \\ 0 & -\frac{\sigma_{11}}{3} & 0 \\ 0 & 0 & -\frac{\sigma_{11}}{3} \end{pmatrix}$$

Since the stress tensor has been shown to be a symmetric second-rank tensor, it displays all of the properties of a symmetric second-rank tensor described in Chapter 2. For example, the coordinate transformation laws and the procedures for finding eigenvalues and eigenvectors are directly applicable to the stress tensor (as well as other numerous second-rank tensors that will be introduced later in this chapter). More specifically, the symmetry of the stress tensor allows the following representation:

$$\boldsymbol{\sigma} = \sum_{i=1}^{3} \sigma_i \, \mathbf{p}_i \otimes \mathbf{p}_i \tag{7.9}$$

where the eigenvalues σ_i are guaranteed to be real. In particular, if the eigenvalues are ordered such that $\sigma_1 \geq \sigma_2 \geq \sigma_3$, it is easy to show that the normal stress on any plane passing through the point of interest (where the stress tensor is defined) is indeed σ_1 and the maximum shear stress on any plane through the same point is $1/2(\sigma_1 - \sigma_3)$.

7.2.2 Field equations for stress analysis

Consider a subdomain D in a body. Let ∂D denote the bounding surface of this subdomain (Figure 7.4). Let $\mathbf{t}$ represent the traction vector on the bounding surface and $\mathbf{b}$ the body force per unit volume in D. Applying Newton's laws of motion ($\mathbf{F} = m\mathbf{a}$ for a single particle under a single force) on D yields

$$\int_D \rho \, \mathbf{a} \, dV = \int_{\partial D} \mathbf{t} \, dA + \int_D \mathbf{b} \, dV \tag{7.10}$$

where ρ denotes the density and $\mathbf{a}$ denotes the acceleration vector. Substituting Eqs. (7.6) and (2.72) into Eq. (7.10) yields

$$\int_D \rho \, \mathbf{a} \; dV = \int_D \operatorname{div} \boldsymbol{\sigma}^T dV + \int_D \mathbf{b} \, dV \tag{7.11}$$

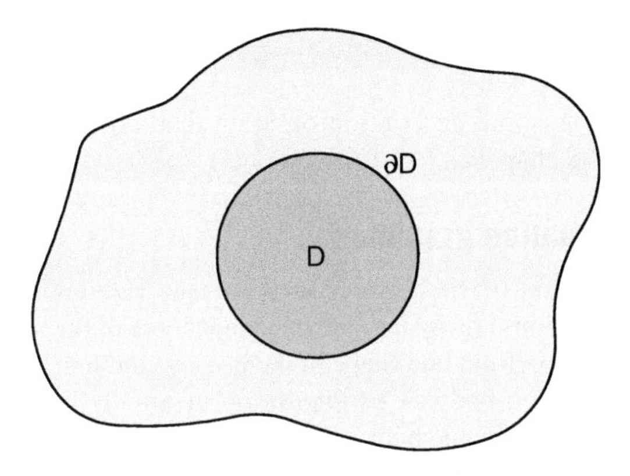

FIGURE 7.4

A volumetric subdomain, D, of interest in a body.

Since Eq. (7.11) is expected to be satisfied for any subdomain of the body, we can infer that the following field equation should be applicable at all material points in the body:

$$\text{div}\,\boldsymbol{\sigma}^T + \mathbf{b} = \rho\,\mathbf{a} \tag{7.12}$$

Assuming static or quasistatic conditions (i.e., ignoring the inertia term) yields the following simple expression for satisfying equilibrium conditions in the body:

$$\text{div}\,\boldsymbol{\sigma}^T + \mathbf{b} = \text{div}\,\boldsymbol{\sigma} + \mathbf{b} = 0 \tag{7.13}$$

Using indicial notation, this equation can be expressed as

$$\sigma_{ji,j} + b_i = \sigma_{ij,j} + b_i = 0 \tag{7.14}$$

Note that in the absence of body forces, Eq. (7.14) takes the simple form $\nabla\cdot\boldsymbol{\sigma} = 0$, which is very similar to the conservation equation (Eq. 7.1) for conduction-type problems. Eq. (7.14) is in effect an embodiment of the *conservation of linear momentum* principle.

In addition to the conservation principle, the solid mechanics field theories require satisfaction of the continuity condition. The continuity condition arises from the need for smooth fields of the displacement gradients in the volumetric domain under consideration. This requires that $\nabla \times \nabla u_i = \mathbf{0}$, which expressed in terms of the strain components reduces to

$$\varepsilon_{ij,kl} + \varepsilon_{kl,ij} = \varepsilon_{ik,jl} + \varepsilon_{jl,ik} \tag{7.15}$$

The concept of work is frequently used in mechanics. The *rate of work* done on subdomain D, called *power* and denoted ω, is expressed as

$$\omega = \int_{\partial D} (\boldsymbol{t}\cdot\boldsymbol{v})dA + \int_{D} (\boldsymbol{b}\cdot\boldsymbol{v})dV \tag{7.16}$$

7.3 STRAIN AND MOTION

In this section, we present a tensorial description of strain that will be required in both elastic and plastic frameworks later in the chapter.

7.3.1 Motion and deformation gradients

Kinematics refers to a description of the changes in shape and size undergone by a given body as a function of time. Such descriptions are, by definition, independent of the material behavior since they simply describe the transformation from one shape of the body to another. In this context, it is useful to define a "configuration" of a given body. A *configuration* of a body is simply the collection of all positions (points in space) occupied by the body at any given time. For example, Figure 7.5 shows two configurations of a body **B** at two different instants, 0 and t, denoted as **B**(0) and **B**(t), respectively. In this figure, **B**(0) simply represents the collection of all spatial points occupied by the body at time 0, and is denoted as the "initial configuration." **B**(t) represents a similar collection of all points occupied by the body at time t, and is denoted as the *deformed configuration*.

In Figure 7.5, **x** represents the position occupied by a specific material point in the initial configuration **B**(0). The position occupied by the same material point in the deformed configuration **B**(t) is shown as **y** in this figure. Although the sets of points occupied by the body at these two instants are shown here as mutually exclusive sets—that is, there are no common elements in **B**(0) and **B**(t)—this need not be the case. The only restriction in defining different configurations of the body is that we allow one material point to occupy only one spatial position in a given configuration. This means that we assume that there exists a unique invertible mapping function that maps material points in one configuration to those in another configuration. Such a mapping function is called the *motion* of body **B**, and can be mathematically expressed as

$$\mathbf{y} = \mathbf{y}(\mathbf{x}, t) \tag{7.17}$$

The condition that this function is invertible implies the existence of the following function:

$$\mathbf{x} = \mathbf{x}(\mathbf{y}, t) \tag{7.18}$$

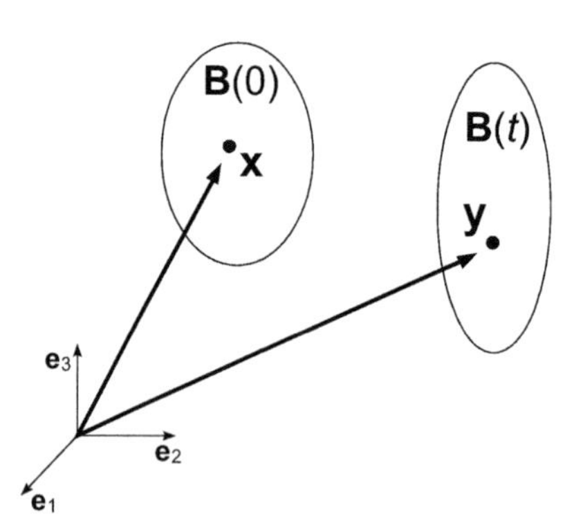

FIGURE 7.5

Schematic description of motion undergone by a body. **B**(0) represents the initial configuration and **B**(t) represents the deformed configuration at time t.

Furthermore, we assume that the body is comprised of a set of continuous material points that remain continuous in the different deformed configurations. Physically, this implies that a specific material point sees the same material points in its neighborhood during the deformation. This is a reasonable assumption except when there are cracks developing in the material or phase transformations occurring in the material or there is material sliding such as in grain boundary sliding; these processes are excluded from consideration in the present discussion. Mathematically, the *continuity* assumption implies that the function described in Eq. (7.17) can be differentiated to yield

$$\mathbf{F}(\mathbf{x}, t) = \frac{\partial \mathbf{y}}{\partial \mathbf{x}} \tag{7.19}$$

$\mathbf{F}(\mathbf{x}, t)$ is a second-rank tensor, as defined through Eq. (7.19), and is called the *deformation gradient tensor.* The deformation gradient tensor contains important information regarding the strain experienced by the body, the details of which shall be addressed later. Invertibility of the mapping function in Eq. (7.17) implies that $\mathbf{F}^{-1}(\mathbf{x}, t)$ exists and therefore requires that

$$\det(\mathbf{F}(\mathbf{x}, t)) \neq 0 \tag{7.20}$$

The difference between the position of a material point in the deformed configuration and its position in the initial configuration is defined as the *displacement vector* denoted $\mathbf{u}$:

$$\mathbf{u} = \mathbf{y} - \mathbf{x} \tag{7.21}$$

The spatial gradient of the displacement vector is denoted $\mathbf{H}$ and is called the *displacement gradient tensor*:

$$\mathbf{H}(\mathbf{x}, t) = \frac{\partial \mathbf{u}}{\partial \mathbf{x}} \tag{7.22}$$

Substituting Eqs. (7.21) and (7.19) in Eq. (7.22), we obtain

$$\mathbf{H} = \mathbf{F} - \mathbf{I} \tag{7.23}$$

Deformation fields where $\mathbf{F}$ is uniform (i.e., not dependent on spatial location) are referred to as *homogeneous deformations*. We consider the following few examples of homogeneous deformation fields and compute the corresponding deformation gradient and displacement gradient tensors.

Example 7.3

Figure 7.6 shows the initial and deformed configurations and coordinates of some of the relevant corners of the body in these configurations. Evaluate the **F** and **H** tensors for this deformation.

Solution

The motion associated with this deformation can be expressed as

$$y_1 = \frac{L_1}{L_o} x_1 + a$$

$$y_2 = \frac{W_1}{W_o} x_2 + b$$

$$y_3 = x_3$$

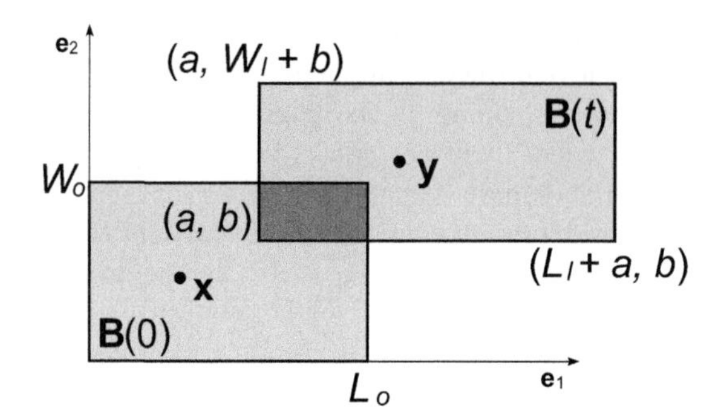

FIGURE 7.6

Schematic of a deformed body at times 0 and t.

The corresponding deformation and displacement gradient tensors are

$$[\mathbf{F}] = \begin{bmatrix} L_1/L_o & 0 & 0 \\ 0 & W_1/W_o & 0 \\ 0 & 0 & 1 \end{bmatrix} \quad [\mathbf{H}] = \begin{bmatrix} L_1/L_o - 1 & 0 & 0 \\ 0 & W_1/W_o - 1 & 0 \\ 0 & 0 & 0 \end{bmatrix}$$

Note that the rigid-body translation component of the motion has been effectively filtered out in the definitions of the gradient tensors.

Example 7.4

Figure 7.7 shows the initial and deformed configurations in a pure rotation about $\mathbf{e}_3$. Evaluate the **F** and **H** tensors for this deformation.

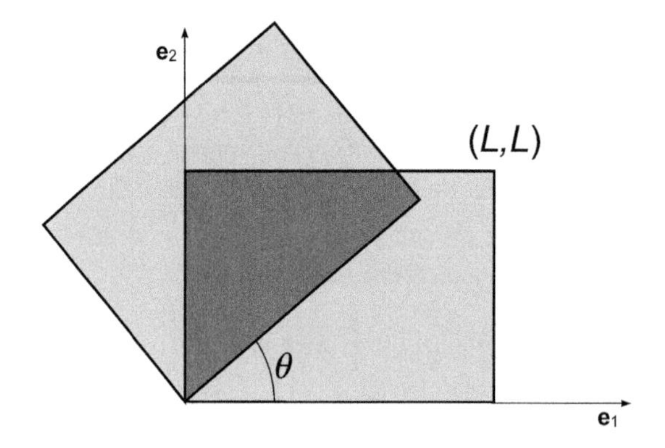

FIGURE 7.7

Schematic of a rotated body before and after the rotation.

Solution

The motion associated with this deformation can be expressed as

$$y_1 = (\cos\theta)x_1 - (\sin\theta)x_2$$
$$y_2 = (\sin\theta)x_1 + (\cos\theta)x_2$$
$$y_3 = x_3$$

The corresponding deformation and displacement gradient tensors are

$$[\mathbf{F}] = \begin{bmatrix} \cos\theta & -\sin\theta & 0 \\ \sin\theta & \cos\theta & 0 \\ 0 & 0 & 1 \end{bmatrix} \quad [\mathbf{H}] = \begin{bmatrix} \cos\theta - 1 & -\sin\theta & 0 \\ \sin\theta & \cos\theta - 1 & 0 \\ 0 & 0 & 0 \end{bmatrix}$$

Assuming small rotation angles, $\theta << 1$, **H** can be approximated as

$$[\mathbf{H}] = \begin{bmatrix} 0 & -\theta & 0 \\ \theta & 0 & 0 \\ 0 & 0 & 0 \end{bmatrix}$$

Note that for pure rotation, **F** is orthogonal (i.e., in polar decomposition of **F**, we would obtain **F** = **R** and **U** = **I**). During small deformations, **H** is skew-symmetric for a pure rotation.

Example 7.5

Figure 7.8 shows the initial and deformed configurations in simple shear deformation. Evaluate the **F** and **H** tensors for this deformation.

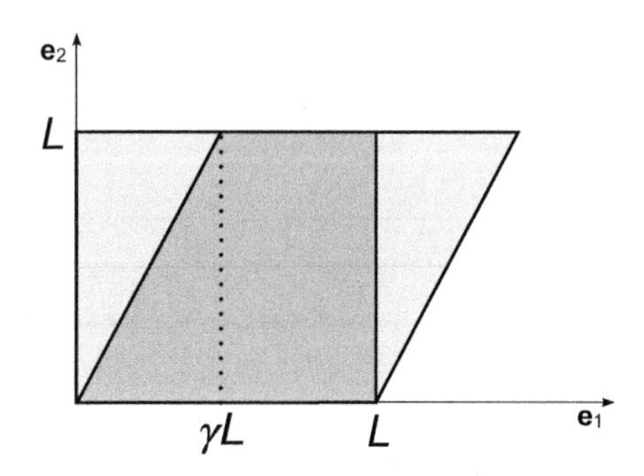

FIGURE 7.8

Schematic of a deformed body under simple shear, before and after deformation.

Solution

The motion associated with this deformation can be expressed as

$$y_1 = x_1 + \gamma x_2$$
$$y_2 = x_2$$
$$y_3 = x_3$$

The corresponding deformation and displacement gradient tensors are

$$[\mathbf{F}] = \begin{bmatrix} 1 & \gamma & 0 \\ 0 & 1 & 0 \\ 0 & 0 & 1 \end{bmatrix}$$

$$[\mathbf{H}] = \begin{bmatrix} 0 & \gamma & 0 \\ 0 & 0 & 0 \\ 0 & 0 & 0 \end{bmatrix} = \begin{bmatrix} 0 & \gamma/2 & 0 \\ \gamma/2 & 0 & 0 \\ 0 & 0 & 0 \end{bmatrix} + \begin{bmatrix} 0 & \gamma/2 & 0 \\ -\gamma/2 & 0 & 0 \\ 0 & 0 & 0 \end{bmatrix}$$

The symmetric and skew-symmetric components of the displacement gradient tensor have special meanings in the case of small strains, that is, $\gamma << 1$. The first term, corresponding to the symmetric part, represents a pure shear deformation as shown in Figure 7.9.

The skew-symmetric component of **H** for the simple shear deformation represents a rotation of $\gamma/2$ about $\mathbf{e}_3$. The superposition of the two deformations corresponding to the symmetric and the skew-symmetric parts of **H** produces the net deformation described in simple shear.

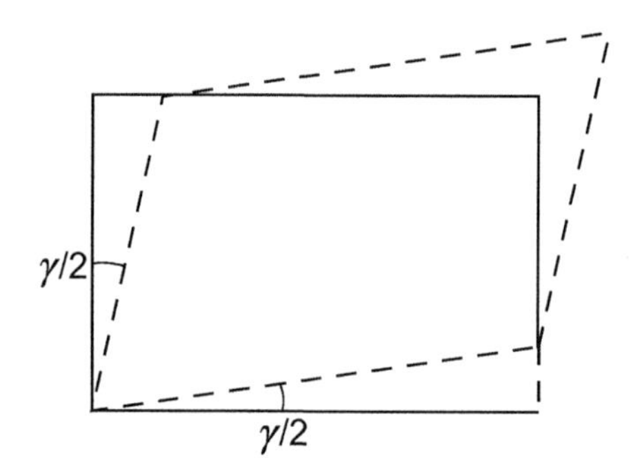

FIGURE 7.9

Schematic of a body under pure shear.

It is instructive to take a closer look at how the neighborhood of a point transforms during a prescribed motion. The deformation gradient tensor can be interpreted as a linear transformation mapping of infinitesimal vectors in the neighborhood of $\mathbf{x}$ to their new locations in the neighborhood of $\mathbf{y}$. As an example, select three mutually orthogonal infinitesimal vectors in the neighborhood of

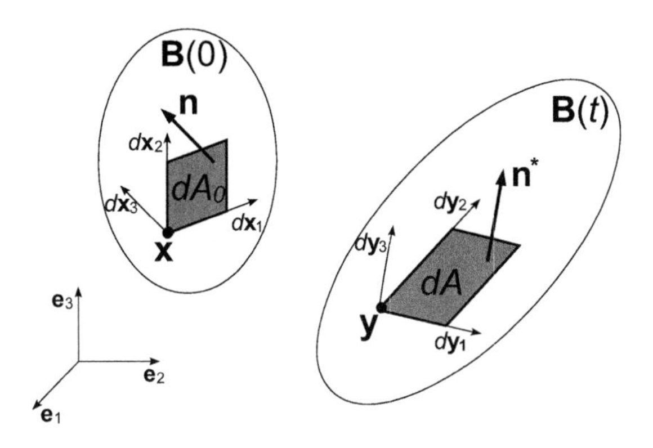

FIGURE 7.10

Transformation of the infinitesimal area elements in the neighborhood of a material point during motion.

$\mathbf{x}$ and label them $d\mathbf{x}_1$, $d\mathbf{x}_2$, and $d\mathbf{x}_3$. Denote their corresponding locations in the deformed configuration as $d\mathbf{y}_1$, $d\mathbf{y}_2$, and $d\mathbf{y}_3$, respectively (see Figure 7.10).

The deformation gradient tensor provides the following linear transformation mapping (see Eq. 7.19):

$$d\mathbf{y}_i = \mathbf{F} d\mathbf{x}_i \tag{7.24}$$

Since $\mathbf{F}$ describes a linear transformation mapping, Eq. (7.24) can be extended to map all infinitesimal line elements in the neighborhood of $\mathbf{x}$ in $\mathbf{B}(0)$ to their new configurations in the neighborhood of $\mathbf{y}$ in $\mathbf{B}(t)$.

An infinitesimal area element (a parallelogram defined by two nonparallel vectors) can be specified as a vector with a magnitude that reflects the area and a direction that reflects the unit normal of the area element. For example, the area element contained by the infinitesimal vectors $d\mathbf{x}_1$ and $d\mathbf{x}_2$ in Figure 7.10 can be expressed as

$$dA_o \mathbf{n} = d\mathbf{x}_1 \times d\mathbf{x}_2 \tag{7.25}$$

where $\mathbf{n}$ represents the unit normal for the area element and dA_o represents the area. Similarly, the same area element in the deformed configuration can be represented by

$$dA\, \mathbf{n}^* = d\mathbf{y}_1 \times d\mathbf{y}_2 \tag{7.26}$$

where $\mathbf{n}^*$ and dA represent its unit normal and the area, respectively. Substituting Eq. (7.24) in Eqs. (7.25) and (7.26) produces the following results:

$$\mathbf{n}^* = \frac{\mathbf{F}^{-T}\mathbf{n}}{|\mathbf{F}^{-T}\mathbf{n}|} \tag{7.27}$$

$$dA = dA_o(\det \mathbf{F})\ |\mathbf{F}^{-T}\mathbf{n}| \tag{7.28}$$

Equations (7.27) and (7.28) are applicable to all area elements in the neighborhood of $\mathbf{x}$ and are referred to as the *Nansen's formulae*. Note also that $\mathbf{n}$ and $\mathbf{n}^*$ are not necessarily parallel to each other.

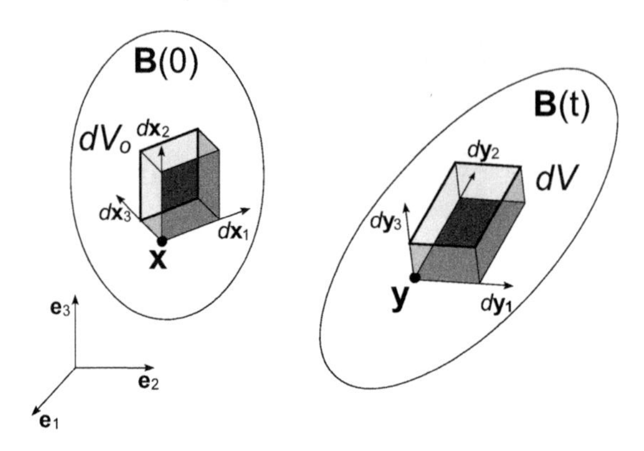

FIGURE 7.11

The infinitesimal volume elements in the neighborhood of a material point during motion.

An infinitesimal volume element can be associated with any three mutually nonparallel vectors in Euclidean 3-space. For example, Figure 7.11 shows a volume element defined by $d\mathbf{x}_1$, $d\mathbf{x}_2$, and $d\mathbf{x}_3$ in the initial configuration. The corresponding volume element in the deformed configuration is defined by $d\mathbf{y}_1$, $d\mathbf{y}_2$, and $d\mathbf{y}_3$:

$$dV_o = (d\mathbf{x}_1 \times d\mathbf{x}_2) \cdot d\mathbf{x}_3 \tag{7.29}$$

$$dV = (d\mathbf{y}_1 \times d\mathbf{y}_2) \cdot d\mathbf{y}_3 \tag{7.30}$$

Note that although $d\mathbf{x}_1$, $d\mathbf{x}_2$, and $d\mathbf{x}_3$ may be taken to be orthonormal to each other, $d\mathbf{y}_1$, $d\mathbf{y}_2$, and $d\mathbf{y}_3$ are not necessarily orthonormal.

Substituting Eq. (7.24) in Eqs. (7.27) and (7.28) yields the following result:

$$\frac{dV}{dV_o} = \frac{(\mathbf{F}d\mathbf{x}_1 \times \mathbf{F}d\mathbf{x}_2) \cdot \mathbf{F}d\mathbf{x}_3}{(d\mathbf{x}_1 \times d\mathbf{x}_2) \cdot d\mathbf{x}_3} = \det(\mathbf{F}) \tag{7.31}$$

It follows from Eq. (7.31) that for any physically realizable deformation, the following condition should be met:

$$\det(\mathbf{F}) > 0 \tag{7.32}$$

As a corollary, we note that for *isochoric* (volume-conserving) deformations, $\det(\mathbf{F}) = 1$.

By virtue of the polar decomposition theorem, the nonsingular $\mathbf{F}$ can be decomposed into a positive definite symmetric tensor $\mathbf{U}$ (or $\mathbf{V}$) and an orthogonal tensor $\mathbf{R}$ (see Eqs. 2.61–2.64). $\mathbf{R}$ is called the *rotation tensor*, $\mathbf{U}$ the *right stretch tensor*, and $\mathbf{V}$ the *left stretch tensor*. The eigenvalues, λ_i, of $\mathbf{U}$ and $\mathbf{V}$ are real and are referred to as the *stretches* or *stretch ratios*. This is because, implicit in the polar decomposition is the idea that application of $\mathbf{U}$ causes a stretch of λ_1 along $\mathbf{r}_1$, λ_2 along $\mathbf{r}_2$, and λ_3 along $\mathbf{r}_3$, without any change in the respective angles between the line segments along these three directions. That is, after application of $\mathbf{U}$ the line elements along $\mathbf{r}_i$ remain orthonormal. The application

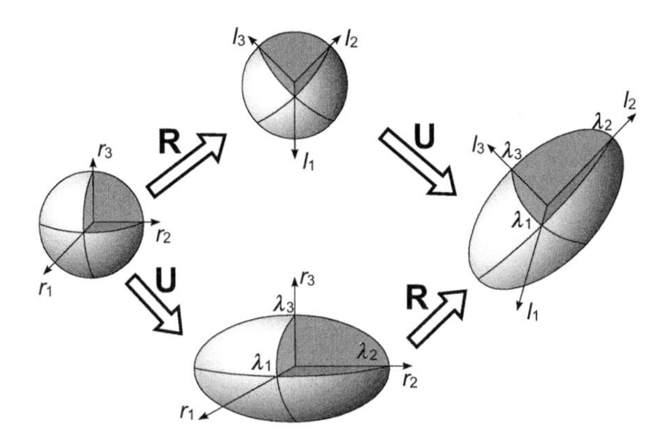

FIGURE 7.12

Physical interpretation of the polar decomposition of **F** described through changes in configuration of an initial unit sphere. The unit sphere can be stretched first (**U**) and then rotated (**R**), or rotated first (**R**) and then stretched (**V**) to reach the same final combination.

of **R**, however, causes a rigid-body rotation without any change in length of any line element in the body.

Physically, polar decomposition of **F** implies that the overall change in geometry between the initial and the deformed configurations can be decomposed conveniently into two idealized components in two equivalent ways, as shown in Figure 7.12. These are described here using an orthonormal reference frame that is aligned with the eigenvectors of **U**. In this specific reference frame, it is possible to visualize the overall deformation imposed on a unit sphere as either of the following:

1. Applying first the stretches, λ_i, along the $\mathbf{r}_i$, and then subjecting the stretched sphere to a rotation **R**.
2. Rotating the sphere by **R** first and then applying the stretches, λ_i, along the new rotated directions of $\mathbf{r}_i$ (these are denoted as $\mathbf{l}_i$ and are indeed the eigenvectors of **V**).

7.3.2 Strain and strain rate tensors

It should be clear from the earlier discussion that deformation gradients contain the essential information about the strain experienced by a body. Rigid-body translations, for example, do not involve any deformation gradients (see Example 7.3) and therefore do not produce any strain in the body. The polar decomposition of the deformation gradient tensor, **F**, reveals that one of its components is actually a pure rotation, **R**, which does not produce any strain in the solid. It therefore follows that an appropriate definition of strain should be based on the stretch tensor **U** (or equivalently on **V**). A *generalized definition of strain* (for finite deformations) can be expressed as

$$\mathbf{E} = \sum_{i=1}^{3} f(\lambda_i)\, \mathbf{r}_i \otimes \mathbf{r}_i \tag{7.33}$$

Note that Eq. (7.33) implies that the eigenvectors (also called principal directions) of the generalized strain measure **E** are the same as the eigenvectors of **U**, while the eigenvalues of strain are defined through $f(\lambda_i)$. In addition, we impose the following restrictions on $f(\lambda_i)$:

$$f(1) = 0 \tag{7.34}$$

$$\frac{df}{d\lambda_i} > 0 \tag{7.35}$$

$$\left(\frac{df}{d\lambda_i}\right)_{\lambda_i = 1} = 1 \tag{7.36}$$

Equation (7.34) requires that a stretch ratio of 1 (i.e., no change in length) corresponds to zero eigenstrains (also called principal strains). Equation (7.35) requires that an increase in stretch ratio should produce an increase in the corresponding eigenstrain. Equation (7.36) ensures that in the limit of small strains (i.e., $\lambda_i \approx 1$) an increment of stretch produces an equal increment of strain.

Multiple definitions of strain are possible with the restrictions described in Eqs. (7.34) through (7.36). One frequently used strain measure is defined using the following function:

$$f(\lambda_i) = \frac{1}{2}\left(\lambda_i^2 - 1\right) \tag{7.37}$$

Equation (7.37) results in *Green's definition of strain* and can be expressed as

$$\mathbf{E}^G = \frac{1}{2}\left(\mathbf{U}^2 - \mathbf{I}\right) = \frac{1}{2}\left(\mathbf{F}^T\mathbf{F} - \mathbf{I}\right) \tag{7.38}$$

Substitution of Eq. (7.23) in Eq. (7.38) yields

$$\mathbf{E}^G = \frac{1}{2}\left(\mathbf{H} + \mathbf{H}^\mathbf{T} + \mathbf{H}^\mathbf{T}\mathbf{H}\right) \tag{7.39}$$

For deformations involving small strains (i.e., $\lambda_i \approx 1$), it can be shown that the contribution from the $\mathbf{H}^\mathbf{T}\mathbf{H}$ term in Eq. (7.39) is negligible. This produces the familiar definition of small strain as the symmetric component of the displacement gradient tensor.

$$\boldsymbol{\varepsilon} = \frac{1}{2}\left(\mathbf{H} + \mathbf{H}^\mathbf{T}\right) \tag{7.40}$$

As noted earlier in Examples 7.4 and 7.5, the anti-symmetric component of **H** carries important information on rotations in the case of deformations involving small strains. Formally, this is expressed as

$$\boldsymbol{\omega} = \frac{1}{2}\left(\mathbf{H} - \mathbf{H}^T\right) \tag{7.41}$$

It should be noted that Eq. (7.40) is used extensively in the description of elastic deformations in crystalline solids (e.g., metals and ceramics) where the elastic stretches are very close to 1 and the resulting elastic strains are quite small.

Linear elastic behavior of most crystalline materials is a good example of path-independent constitutive response of a material. In the path-independent material behavior, the current stress in the body can be directly related to the current strain in the body without any regard to the deformation history experienced by the body. However, the constitutive response of several materials becomes strongly path-dependent when we explore their plastic response; that is, it is not enough to describe the initial and final configurations, but it is actually necessary to incorporate the entire deformation path in the specification of their constitutive behavior. In describing path-dependent behavior, it is essential to formulate the constitutive response in a rate form (e.g., current stress is related to plastic strain rate in metal plasticity). Therefore, there is a need to develop appropriate descriptions of strain rate tensors that describe the instantaneous strain rates in the body.

Realizing that strain rates are nothing but spatial gradients of the velocity (analogous to strain being a spatial gradient of the displacement), we first define a velocity in the current configuration as an instantaneous time derivative of the position vector. It should be noted that the velocity can be expressed either as a function of position in the current configuration or the position in the initial configuration. In other words, if we follow a particular material point with a position in the initial configuration that was denoted by $\mathbf{x}$, we can define a material description of *velocity field* as

$$\dot{\mathbf{y}} = \frac{d}{dt}(\mathbf{y}(\mathbf{x}, t)) \tag{7.42}$$

Alternatively, we can express the same velocity field in terms of the positions in the current configuration as

$$\mathbf{v} = \mathbf{v}(\mathbf{y}, \mathbf{t}) = \dot{\mathbf{y}}(\mathbf{x}(\mathbf{y}, t), t) \tag{7.43}$$

A *velocity gradient tensor* can now be defined in the current configuration to reflect the instantaneous strain rate in the body as

$$\mathbf{L} = \frac{\partial \mathbf{v}}{\partial \mathbf{y}} \tag{7.44}$$

Note that the spatial gradient in $\mathbf{L}$ is taken with respect to $\mathbf{y}$ unlike in the case of $\mathbf{F}$ where the spatial gradient was taken with respect to $\mathbf{x}$. Consequently, $\mathbf{L}$ is not the same as $\dot{\mathbf{F}}$ (henceforth a dot over a variable denotes its time derivative). In fact, a relationship between $\mathbf{L}$ and $\dot{\mathbf{F}}$ can be derived as follows:

$$\mathbf{L} = \frac{\partial \mathbf{v}}{\partial \mathbf{y}} = \frac{\partial \mathbf{v}}{\partial \mathbf{x}}\frac{\partial \mathbf{x}}{\partial \mathbf{y}} = \frac{\partial \dot{\mathbf{y}}}{\partial \mathbf{x}}\frac{\partial \mathbf{x}}{\partial \mathbf{y}} = \dot{\mathbf{F}}\mathbf{F}^{-1} \tag{7.45}$$

Further, $\mathbf{L}$ can be uniquely decomposed into its symmetric and anti-symmetric components as

$$\mathbf{D} = \frac{1}{2}(\mathbf{L} + \mathbf{L}^T) \tag{7.46}$$

$$\mathbf{W} = \frac{1}{2}(\mathbf{L} - \mathbf{L}^T) \tag{7.47}$$

D is called the *stretching tensor*, while **W** is called the *spin tensor*. It will become clear from the following examples that the stretching tensor **D** reflects the instantaneous strain rates in the body, while the spin tensor **W** reflects the instantaneous rotation rates in the body.

Example 7.6

Assuming that the rigid-body rotation described in Example 7.4 is performed at a constant rotation rate, evaluate **L**, **D**, and **W** tensors for this deformation.

Solution

Let $\theta = \omega t$, where ω represents the rotation rate experienced by the body. The motion of the body is described by the following equations:

$$y_1 = (\cos \omega t)x_1 - (\sin \omega t)x_2$$
$$y_2 = (\sin \omega t)x_1 + (\cos \omega t)x_2$$
$$y_3 = x_3$$

The material description of the velocity field can be expressed as

$$\dot{y}_1 = -\omega(\sin \omega t)x_1 - \omega(\cos \omega t)x_2$$
$$\dot{y}_2 = \omega(\cos \omega t)x_1 - \omega(\sin \omega t)x_2$$
$$\dot{y}_3 = 0$$

The corresponding spatial description of the velocity field can be expressed as

$$v_1 = -\omega y_2$$
$$v_2 = \omega y_1$$
$$v_3 = 0$$

The velocity gradient tensor can be computed as

$$[\mathbf{L}] = \begin{bmatrix} 0 & -\omega & 0 \\ \omega & 0 & 0 \\ 0 & 0 & 0 \end{bmatrix}$$

It is evident that for this special motion $\mathbf{L} = \mathbf{W}$, $\mathbf{D} = \mathbf{0}$, indicating the body is undergoing a pure rotation without any strain.

Example 7.7

Figure 7.13 shows the motion experienced by a body subjected to isochoric plane strain compression at a constant true strain. Evaluate **L**, **D**, and **W** tensors for this deformation.

Solution

Constant true strain rate deformation is characterized by the requirement

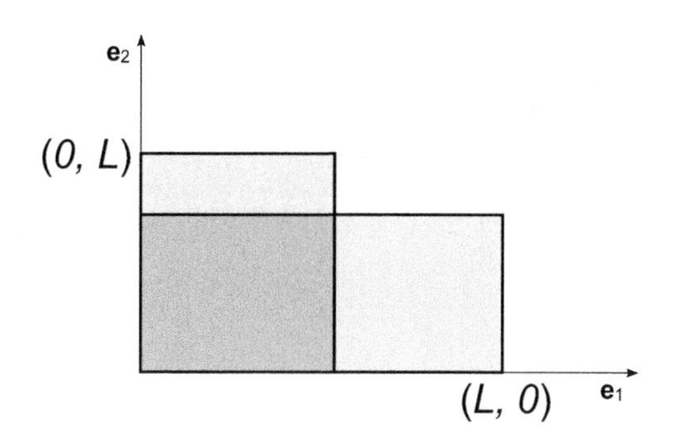

FIGURE 7.13

Schematic of a body subjected to isochoric plane strain.

$$\frac{\dot{l}}{l} = \dot{\varepsilon} = \text{constant}$$

where $\mathbf{l}$ represents the instantaneous length of a line segment and $\dot{\varepsilon}$ the strain rate. For a constant imposed strain rate, the previous condition reduces to

$$l = L \exp(\dot{\varepsilon} t)$$

where $\mathbf{L}$ denotes the initial length of the line segment. Using these notions, the body motion during isochoric plane strain compression can be described by

$$\begin{aligned} y_1 &= \exp(\dot{\varepsilon}\, t) x_1 \\ y_2 &= \exp(-\dot{\varepsilon}\, t) x_2 \\ y_3 &= x_3 \end{aligned}$$

The corresponding deformation gradient tensor can be computed as

$$[\mathbf{F}] = \begin{bmatrix} \exp(\dot{\varepsilon}\, t) & 0 & 0 \\ 0 & \exp(-\dot{\varepsilon}\, t) & 0 \\ 0 & 0 & 1 \end{bmatrix}$$

In the previous example we looked at the spatial description of the velocity field to compute the velocity gradient tensor. Although the exact same approach may be taken for this problem, an alternative approach to compute **L** makes use of Eq. (7.45). These two methods will yield identical results:

$$[\mathbf{L}] = \begin{bmatrix} \dot{\varepsilon} & 0 & 0 \\ 0 & -\dot{\varepsilon} & 0 \\ 0 & 0 & 0 \end{bmatrix}$$

It is also evident that for this deformation, $\mathbf{L} = \mathbf{D}$, $\mathbf{W} = \mathbf{0}$, indicating that there is no redundant rotation.

Example 7.8

Show that power on the subdomain D can be expressed as

$$\frac{d}{dt}\int_D \frac{1}{2}\rho v_i v_i \, dV + \int_D \sigma_{ij}\dot{\varepsilon}_{ij} \, dV$$

Solution

$$\omega = \int_{\partial D} (\mathbf{t}\cdot\mathbf{v})dA + \int_D (\mathbf{b}\cdot\mathbf{v})dV = \int_{\partial D} (\sigma\mathbf{n}\cdot\mathbf{v})dA + \int_D (\mathbf{b}\cdot\mathbf{v})dV$$

$$= \int_{\partial D} \sigma_{ij}n_j v_i \, dA + \int_D b_i v_i \, dV = \int_D (\sigma_{ij}v_i)_{,j} \, dV + \int_D b_i v_i \, dV$$

$$= \int_D \left(\sigma_{ij,j}v_i + b_i v_i + \sigma_{ij}v_{i,j}\right)dV$$

$$= \int_D \left(\rho\dot{v}_i v_i + \sigma_{ij}v_{i,j}\right)dV = \int_D \left(\frac{1}{2}\rho(v_i v_i)^{\cdot} + \sigma_{ij}L_{ij}\right)dV$$

$$= \frac{d}{dt}\int_D \frac{1}{2}\rho v_i v_i \, dV + \int_D \sigma_{ij}\dot{\varepsilon}_{ij} \, dV$$

7.4 CONDUCTIVITY

A large number of physical phenomena are successfully described mathematically by invoking the existence of a *scalar potential field*, $\phi(\mathbf{x})$. Conductivity (electrical and thermal) is a prime example of such physical phenomena. In these problems, the expansion of the potential in a binomial series, ignoring the terms beyond the second order, results in the definition of a linear constitutive relationship between the flux vector and the applied gradient field using a symmetric, positive-definite, second-order tensor. For example, the linear constitutive relationship for conductivity can be expressed as

$$\mathbf{J} = \frac{\partial \phi}{\partial \mathbf{E}} = \mathbf{KE}, \qquad K_{ij} = \left.\frac{\partial^2 \phi}{\partial E_i \partial E_j}\right|_{\mathbf{E}=0} \tag{7.48}$$

where $\mathbf{J}$ denotes the appropriate vector flux (defining the rate of energy or charge flow through a surface of unit area) and $\mathbf{E}$ denotes the applied vector field (defined as the negative of the thermal, charge, or chemical potential gradients; the negative sign reflects the fact that the flux occurs in the direction opposite to the applied field gradient). **K** is a material property that is independent of the geometry and is symmetric and positive-definite:

$$K_{ij} = K_{ji}, \quad \mathbf{KE}\cdot\mathbf{E} > 0 \quad \forall \ |\mathbf{E}| \neq 0 \tag{7.49}$$

The material property tensors of the type just defined often exhibit additional symmetries that are derived from the symmetry inherent to the internal structure of the material. In crystalline solids, the symmetries exhibited by the crystal lattice have a profound influence on the number of independent material constants needed in the complete description of these material property tensors, when expressed in a principal material reference frame. Next, we will illustrate these concepts through some special cases.

Example 7.9

Express the conductivity tensor in the principal material reference frame with the minimum number of constants for the following crystal symmetries: (1) monoclinic, (2) orthorhombic, (3) hexagonal, and (4) cubic.

Solution

A monoclinic crystal is one with a lattice that exhibits one plane of mirror symmetry and/or a two-fold rotation axis. Here we assume that the crystal has a two-fold rotation axis, but note that the final result is the same for either symmetry element. Let $\{\mathbf{e}_i\}$ denote the orthonormal basis associated with the crystal. Further, let the e_2 plane be the axis of two-fold symmetry. Let $\{\mathbf{e}_i'\}$ be a new frame defined by an 180° rotation of $\{\mathbf{e}_i\}$ about e_2. The transformation matrix from $\{\mathbf{e}_i\}$ to $\{\mathbf{e}_i'\}$ is given by

$$[Q] = \begin{bmatrix} -1 & 0 & 0 \\ 0 & 1 & 0 \\ 0 & 0 & -1 \end{bmatrix}$$

Consider the value of K_{21} and K_{21}'. The coordinate transformation law requires that $K_{21}' = Q_{2p}Q_{1q} K_{pq} = -K_{21}$. However, symmetry of the crystal lattice requires $K_{ij}' = K_{ij}$ for the selected coordinate frames (specifically selected to reflect the crystal symmetry being investigated). This leads to the conclusion that $K_{21} = 0$, as it is the only way to reconcile these contradicting requirements. Similarly, one can conclude that $K_{23} = 0$ for this crystal symmetry. In summary, the symmetric conductivity tensor for monoclinic crystals can be expressed as

$$[K]_{monoclinic} = \begin{bmatrix} K_{11} & 0 & K_{13} \\ 0 & K_{22} & 0 \\ K_{13} & 0 & K_{33} \end{bmatrix}$$

An orthorhombic crystal exhibits two additional two-fold rotation axes of symmetry. The additional transformation matrices that need to be considered include

$$[Q] = \begin{bmatrix} 1 & 0 & 0 \\ 0 & -1 & 0 \\ 0 & 0 & -1 \end{bmatrix}, \quad \begin{bmatrix} -1 & 0 & 0 \\ 0 & -1 & 0 \\ 0 & 0 & 1 \end{bmatrix}$$

Using the same approach that was described earlier allows us to conclude that $K_{13} = 0$ for the orthorhombic crystals. Therefore, the corresponding conductivity tensor can be expressed as

$$[K]_{orthorhombic} = \begin{bmatrix} K_{11} & 0 & 0 \\ 0 & K_{22} & 0 \\ 0 & 0 & K_{33} \end{bmatrix}$$

A hexagonal crystal exhibits a six-fold rotation axis (assumed to be aligned with $\mathbf{e}_3$ in the following discussion). A transformation matrix for this symmetry element can be expressed as

$$[Q] = \begin{bmatrix} cos\,60 & sin\,60 & 0 \\ -sin\,60 & cos\,60 & 0 \\ 0 & 0 & 1 \end{bmatrix}$$

Requiring that the components of the conductivity tensor not change value with this transformation results in the following simplified description:

$$[K]_{hexagonal} = \begin{bmatrix} K_{11} & 0 & 0 \\ 0 & K_{11} & 0 \\ 0 & 0 & K_{33} \end{bmatrix}$$

It is worth noting that a hexagonal crystal exhibits additional symmetry elements, other than the one used before. However, consideration of these additional elements does not lead to any further simplification in the description of the conductivity matrix. As an aside, it is also worth noting that changing the previous symmetry element to an arbitrary rotation about $\mathbf{e}_3$ does not reduce the conductivity matrix any further, indicating that the previous form of the conductivity matrix is also valid for all solids that exhibit transverse planar isotropy.

In addition to the symmetry elements of the orthorhombic lattice structure, the cubic crystal exhibits symmetry with 90° rotations about the $\mathbf{e}_1$, $\mathbf{e}_2$, and $\mathbf{e}_3$ axes. The additional transformation matrices that need to be considered include

$$[Q] = \begin{bmatrix} 0 & 1 & 0 \\ -1 & 0 & 0 \\ 0 & 0 & 1 \end{bmatrix}, \quad \begin{bmatrix} 0 & 0 & 1 \\ 0 & 1 & 0 \\ -1 & 0 & 0 \end{bmatrix}, \quad \begin{bmatrix} 1 & 0 & 0 \\ 0 & 0 & -1 \\ 0 & 1 & 0 \end{bmatrix}$$

Following the same approach as before, we can now show that the conductivity tensor for cubic crystals can be expressed simply as

$$[K]_{cubic} = k \begin{bmatrix} 1 & 0 & 0 \\ 0 & 1 & 0 \\ 0 & 0 & 1 \end{bmatrix}, \text{ or } K_{ij} = k\,\delta_{ij}$$

Note that this result implies that the conductivity tensor for cubic crystals is actually an isotropic property. In other words, additional symmetries will not further reduce the representation of the $[\mathbf{K}]$ matrix.

Clearly, if an isotropic phase is present, then an infinite set of additional symmetry elements (rotations and rotation-inversions) must be present, but these cannot reduce the number of independent constants of the conductivity tensor below 1, which is already required by the symmetry of cubic crystals. This is an example of Neumann's principle, which states that *every physical property of a crystal must possess at least the symmetry of the point group of the crystal.*

7.5 ELASTICITY

Elastic materials are generally defined as those materials that do not exhibit a permanent change of geometry in any closed-loop loading/unloading cycle. In the following treatment, we will assume that the elastic materials are also *hyperelastic*, that is, they do not dissipate energy in any closed-loop loading/unloading cycle. Implicitly, this means that elastic response of materials is assumed to be path-independent. Let $W(\mathbf{x},t)$ represent the strain energy density in a body of interest. The path independence of elastic response, together with the expectation that the elastic stored energy is unaffected by rigid-body translations and rotations, permits the *strain energy density function* to be expressed as $W(\mathbf{x},t) = W(\boldsymbol{\varepsilon}(\mathbf{x},t))$. Thus, the elastic strain energy density depends only on the local elastic strain. Assuming that the density changes in the solid are negligible, the strain (potential) energy component of power (see Example 7.8) can be expressed as

$$\frac{d}{dt}\int_D W dV = \int_D \sigma_{ij}\dot{\varepsilon}_{ij}\, dV \Rightarrow \int_D \left(\sigma_{ij} - \frac{\partial W}{\partial \varepsilon_{ij}}\right)\dot{\varepsilon}_{ij}\, dV = 0$$
$$\Rightarrow \sigma_{ij} = \frac{\partial W(\boldsymbol{\varepsilon})}{\partial \varepsilon_{ij}} \tag{7.50}$$

Next, we expand the strain energy density function as a series in the neighborhood of $\boldsymbol{\varepsilon} = \mathbf{0}$ (with the expectation that the elastic strains in crystalline materials will be small).

$$W(\boldsymbol{\varepsilon}) = W(\mathbf{0}) + \left.\frac{\partial W}{\partial \varepsilon_{ij}}\right|_{\boldsymbol{\varepsilon}=\mathbf{0}} \varepsilon_{ij} + \frac{1}{2}\left.\frac{\partial^2 W}{\partial \varepsilon_{ij}\partial \varepsilon_{kl}}\right|_{\boldsymbol{\varepsilon}=\mathbf{0}} \varepsilon_{ij}\varepsilon_{kl} + \cdots \tag{7.51}$$

Setting $W(\mathbf{0}) = 0$ and $\boldsymbol{\sigma}|_{\varepsilon=0} = 0$, one can approximate Eq. (7.51) as

$$W \approx \frac{1}{2}\left.\frac{\partial^2 W}{\partial \varepsilon_{ij}\partial \varepsilon_{kl}}\right|_{\boldsymbol{\varepsilon}=\mathbf{0}} \varepsilon_{ij}\varepsilon_{kl} = \frac{1}{2}C_{ijkl}\varepsilon_{ij}\varepsilon_{kl}$$
$$\Rightarrow \sigma_{ij} = C_{ijkl}\varepsilon_{kl}, \quad W = \frac{1}{2}\sigma_{ij}\varepsilon_{ij} \tag{7.52}$$

where $\mathbf{C}$ represents a positive-definite, fourth-rank *elastic stiffness tensor.* The linear elastic relationship between the stress and strain components is fully invertible and can also be expressed as

$$W = \frac{1}{2}C_{ijkl}\varepsilon_{ij}\varepsilon_{kl} = \frac{1}{2}S_{ijkl}\sigma_{ij}\sigma_{kl}, \quad \varepsilon_{ij} = S_{ijkl}\sigma_{kl}, \quad \mathbf{S}^{-1} = \mathbf{C} \tag{7.53}$$

where $\mathbf{S}$ represents a positive-definite, fourth-rank *elastic compliance tensor.* The fourth-rank tensors previously described obey the coordinate transformation law described in Eq. (2.40).

The elastic stiffness and compliance tensors previously described possess certain symmetries that are guaranteed by the symmetric nature of the stress and the strain tensors, and the fact that $\partial^2 W/\partial \varepsilon_{ij}\partial \varepsilon_{kl} = \partial^2 W/\partial \varepsilon_{kl}\partial \varepsilon_{ij}$. These symmetries can be expressed as

$$C_{ijkl} = C_{jikl} = C_{ijlk} = C_{klij}, \quad S_{ijkl} = S_{jikl} = S_{ijlk} = S_{klij} \tag{7.54}$$

As a consequence of these symmetries, the total number of *independent elastic constants* needed to describe either the $\mathbf{S}$ or the $\mathbf{C}$ tensors, in the most general case, is 21.

It is often convenient to express the linear elastic stress–strain relations in the following matrix form:

$$\begin{Bmatrix} \sigma_{11} \\ \sigma_{22} \\ \sigma_{33} \\ \sigma_{23} \\ \sigma_{13} \\ \sigma_{12} \end{Bmatrix} = \begin{bmatrix} C_{1111} & C_{1122} & C_{1133} & C_{1123} & C_{1113} & C_{1112} \\ C_{1122} & C_{2222} & C_{2233} & C_{2223} & C_{2213} & C_{2212} \\ C_{1133} & C_{2233} & C_{3333} & C_{3323} & C_{3313} & C_{3312} \\ C_{1123} & C_{2223} & C_{3323} & C_{2323} & C_{2313} & C_{2312} \\ C_{1113} & C_{2213} & C_{3313} & C_{2313} & C_{1313} & C_{1312} \\ C_{1112} & C_{2212} & C_{3312} & C_{2312} & C_{1312} & C_{1212} \end{bmatrix} \begin{Bmatrix} \varepsilon_{11} \\ \varepsilon_{22} \\ \varepsilon_{33} \\ 2\varepsilon_{23} \\ 2\varepsilon_{13} \\ 2\varepsilon_{12} \end{Bmatrix} \tag{7.55}$$

The notation in Eq. (7.55) can be further simplified using the conversion $C_{ijkl} = C_{mn}$ (note that this simple conversion rule is only applied to stiffness components, the corresponding rule for the compliance components will have a more complex form due to the requirement that the contracted stiffness and compliance matrices are inverses of each other; see Eqs. 7.56 and 7.57), with

ij or kl	11	22	33	23	31	12
m or n	1	2	3	4	5	6

For example, $C_{1233} = C_{63}$. Note that symmetry of the tensor means that component 31 is the same as component 13, and so forth.

$$\begin{Bmatrix} \sigma_{11} \\ \sigma_{22} \\ \sigma_{33} \\ \sigma_{23} \\ \sigma_{13} \\ \sigma_{12} \end{Bmatrix} = \begin{bmatrix} C_{11} & C_{12} & C_{13} & C_{14} & C_{15} & C_{16} \\ C_{12} & C_{22} & C_{23} & C_{24} & C_{25} & C_{26} \\ C_{13} & C_{23} & C_{33} & C_{34} & C_{35} & C_{36} \\ C_{14} & C_{24} & C_{34} & C_{44} & C_{45} & C_{46} \\ C_{15} & C_{25} & C_{35} & C_{45} & C_{55} & C_{56} \\ C_{16} & C_{26} & C_{36} & C_{46} & C_{56} & C_{66} \end{bmatrix} \begin{Bmatrix} \varepsilon_{11} \\ \varepsilon_{22} \\ \varepsilon_{33} \\ 2\varepsilon_{23} \\ 2\varepsilon_{13} \\ 2\varepsilon_{12} \end{Bmatrix} \tag{7.56}$$

Note that there are 21 independent elastic constants in the stiffness matrix just described. The compliance tensor has the following analogous description:

$$\begin{Bmatrix} \varepsilon_{11} \\ \varepsilon_{22} \\ \varepsilon_{33} \\ 2\varepsilon_{23} \\ 2\varepsilon_{13} \\ 2\varepsilon_{12} \end{Bmatrix} = \begin{bmatrix} S_{11} & S_{12} & S_{13} & S_{14} & S_{15} & S_{16} \\ S_{12} & S_{22} & S_{23} & S_{24} & S_{25} & S_{26} \\ S_{13} & S_{23} & S_{33} & S_{34} & S_{35} & S_{36} \\ S_{14} & S_{24} & S_{34} & S_{44} & S_{45} & S_{46} \\ S_{15} & S_{25} & S_{35} & S_{45} & S_{55} & S_{56} \\ S_{16} & S_{26} & S_{36} & S_{46} & S_{56} & S_{66} \end{bmatrix} \begin{Bmatrix} \sigma_{11} \\ \sigma_{22} \\ \sigma_{33} \\ \sigma_{23} \\ \sigma_{13} \\ \sigma_{12} \end{Bmatrix} \tag{7.57}$$

As noted earlier, most engineering materials exhibit additional symmetries that are attributable to the symmetry in their internal structure. The symmetries exhibited by the crystal lattice have a profound influence on the number of independent terms in the stiffness and compliance tensors previously represented. In the following examples, we will consider some special cases of symmetry in the crystal lattice. As with the description of the conductivity tensor, we will start with the low-symmetry lattice structures and progressively increase the number of symmetry elements in the lattice structure.

Example 7.10

Describe the stiffness matrix for a monoclinic crystal in its most compact form using the minimum number of elastic constants.

Solution

The transformation matrix for the symmetry element of the monoclinic single crystal was described earlier in Example 7.9. Consider the value of C_{2221} and C'_{2221}. The coordinate transformation law requires that $C'_{2221} = Q_{2p}Q_{2q}Q_{2r}Q_{1s}C_{pqrs} = -C_{2221}$. However, symmetry of the crystal lattice requires $C'_{ijkl} = C_{ijkl}$ for the selected coordinate frames. This leads to the conclusion that $C_{2221} = 0$, as it is the only way to reconcile these contradicting requirements. Similar investigation of other components provides these additional results: $C_{3332} = C_{1112} = C_{2223} = C_{3211} = C_{3213} = C_{2113} = C_{3321} = 0$. Similar conclusions can also be drawn for the components of the compliance tensor. In summary, the linear elastic stress–strain relations for monoclinic crystals can be expressed as

$$\begin{Bmatrix} \sigma_{11} \\ \sigma_{22} \\ \sigma_{33} \\ \sigma_{23} \\ \sigma_{13} \\ \sigma_{12} \end{Bmatrix} = \begin{bmatrix} C_{11} & C_{12} & C_{13} & 0 & C_{15} & 0 \\ C_{12} & C_{22} & C_{23} & 0 & C_{25} & 0 \\ C_{13} & C_{23} & C_{33} & 0 & C_{35} & 0 \\ 0 & 0 & 0 & C_{44} & 0 & C_{46} \\ C_{15} & C_{25} & C_{35} & 0 & C_{55} & 0 \\ 0 & 0 & 0 & C_{46} & 0 & C_{66} \end{bmatrix} \begin{Bmatrix} \varepsilon_{11} \\ \varepsilon_{22} \\ \varepsilon_{33} \\ 2\varepsilon_{23} \\ 2\varepsilon_{13} \\ 2\varepsilon_{12} \end{Bmatrix}$$

Example 7.11

Derive the stiffness matrix for an orthorhombic crystal in its most compact form using the minimum number of elastic constants.

Solution

The relevant transformation matrices for this symmetry subgroup were described in Example 7.9. Following the same logic that was described in Example 7.10, we arrive at the following simplified description of the linear elastic stress–strain relations for the orthotropic crystal lattice:

$$\begin{Bmatrix} \sigma_{11} \\ \sigma_{22} \\ \sigma_{33} \\ \sigma_{23} \\ \sigma_{13} \\ \sigma_{12} \end{Bmatrix} = \begin{bmatrix} C_{11} & C_{12} & C_{13} & 0 & 0 & 0 \\ C_{12} & C_{22} & C_{23} & 0 & 0 & 0 \\ C_{13} & C_{23} & C_{33} & 0 & 0 & 0 \\ 0 & 0 & 0 & C_{44} & 0 & 0 \\ 0 & 0 & 0 & 0 & C_{55} & 0 \\ 0 & 0 & 0 & 0 & 0 & C_{66} \end{bmatrix} \begin{Bmatrix} \varepsilon_{11} \\ \varepsilon_{22} \\ \varepsilon_{33} \\ 2\varepsilon_{23} \\ 2\varepsilon_{13} \\ 2\varepsilon_{12} \end{Bmatrix}$$

Example 7.12

Derive the stiffness matrix for a hexagonal crystal in its most compact form with the minimum number of elastic constants.

Solution

Following the same approach as before, and the symmetry transformation of the hexagonal crystal described in Example 7.9, we arrive at the following simplified description of the linear elastic stress–strain relations for a hexagonal crystal:

$$\begin{Bmatrix} \sigma_{11} \\ \sigma_{22} \\ \sigma_{33} \\ \sigma_{23} \\ \sigma_{13} \\ \sigma_{12} \end{Bmatrix} = \begin{bmatrix} C_{11} & C_{12} & C_{13} & 0 & 0 & 0 \\ C_{12} & C_{11} & C_{13} & 0 & 0 & 0 \\ C_{13} & C_{13} & C_{33} & 0 & 0 & 0 \\ 0 & 0 & 0 & C_{44} & 0 & 0 \\ 0 & 0 & 0 & 0 & C_{44} & 0 \\ 0 & 0 & 0 & 0 & 0 & (C_{11}-C_{12})/2 \end{bmatrix} \begin{Bmatrix} \varepsilon_{11} \\ \varepsilon_{22} \\ \varepsilon_{33} \\ 2\varepsilon_{23} \\ 2\varepsilon_{13} \\ 2\varepsilon_{12} \end{Bmatrix}$$

Changing this symmetry element to be an arbitrary rotation about $\mathbf{e}_3$ does not reduce the stiffness matrix any further, indicating that the previous description of the stress–strain relation is also valid for all solids that exhibit transverse planar isotropy.

Example 7.13

Derive the stiffness matrix for a cubic crystal in its most compact form with the minimum number of elastic constants.

Solution

Using the symmetry elements of the cubic crystal described in Example 7.9, we arrive at the following simplified description of the linear elastic stress–strain relations for cubic crystals:

$$\begin{Bmatrix} \sigma_{11} \\ \sigma_{22} \\ \sigma_{33} \\ \sigma_{23} \\ \sigma_{13} \\ \sigma_{12} \end{Bmatrix} = \begin{bmatrix} C_{11} & C_{12} & C_{12} & 0 & 0 & 0 \\ C_{12} & C_{11} & C_{12} & 0 & 0 & 0 \\ C_{12} & C_{12} & C_{11} & 0 & 0 & 0 \\ 0 & 0 & 0 & C_{44} & 0 & 0 \\ 0 & 0 & 0 & 0 & C_{44} & 0 \\ 0 & 0 & 0 & 0 & 0 & C_{44} \end{bmatrix} \begin{Bmatrix} \varepsilon_{11} \\ \varepsilon_{22} \\ \varepsilon_{33} \\ 2\varepsilon_{23} \\ 2\varepsilon_{13} \\ 2\varepsilon_{12} \end{Bmatrix}$$

The elastic stiffness and compliance tensors for the cubic crystal can be expressed as

$$\begin{aligned} C_{ijkl} &= C_{12}\delta_{ij}\delta_{kl} + C_{44}\left(\delta_{ik}\delta_{jl} + \delta_{il}\delta_{jk}\right) + (C_{11} - C_{12} - 2C_{44})\sum_{r=1}^{3} \delta_{ir}\delta_{jr}\delta_{kr}\delta_{lr} \\ S_{ijkl} &= S_{12}\delta_{ij}\delta_{kl} + \frac{S_{44}}{4}\left(\delta_{ik}\delta_{jl} + \delta_{il}\delta_{jk}\right) + \left(S_{11} - S_{12} - \frac{S_{44}}{2}\right)\sum_{r=1}^{3} \delta_{ir}\delta_{jr}\delta_{kr}\delta_{lr} \end{aligned} \tag{7.58}$$

It is also worth noting that for materials that exhibit *isotropic* properties, such as amorphous phases, the elastic stiffness and compliance tensors can be expressed using only two independent constants. Since multiple choices are possible in the selection of the two independent elastic constants, the stiffness and compliance tensors for isotropic materials can be expressed in several different forms, all of which are equivalent to each other. Here, we represent them as

$$\begin{aligned} C_{ijkl} &= C_{12}\delta_{ij}\delta_{kl} + C_{44}\left(\delta_{ik}\delta_{jl} + \delta_{il}\delta_{jk}\right) \\ S_{ijkl} &= S_{12}\delta_{ij}\delta_{kl} + \frac{S_{44}}{4}\left(\delta_{ik}\delta_{jl} + \delta_{il}\delta_{jk}\right) \end{aligned} \tag{7.59}$$

Comparison of Eqs. (7.58) and (7.59) allows us to define an *anisotropy ratio* for cubic crystals as

$$A = \frac{2C_{44}}{C_{11} - C_{12}} \tag{7.60}$$

Note that for the isotropic case, $A = 1$. The deviation of A from a value of 1 serves as a measure of the degree of elastic anisotropy in the cubic crystal.

7.5.1 Crystal property relations in the sample frame

Next, we turn our attention to properties of a single crystal in the sample frame. The coordinate transformation law for the second-rank conductivity tensor follows relation (2.39):

$$K'_{rs} = Q_{ri}Q_{sj}K_{ij}$$

Great care must be taken in exercising this relationship, to set up the sense of the transformation in the correct way. The coefficients K_{ij} on the right side of this expression are the crystal properties given in a convenient crystal frame attached to the crystal lattice. These are the known property coefficients, and they possess the symmetries of the crystal lattice. The primed coefficients on the right side, K'_{rs}, are these same properties, but expressed in the sample or global frame. These coefficients no longer exhibit the simplicity of form that is evident in the crystal frame.

It is typically in the sample frame that macroscopic properties are to be expressed. So the key to successful coordinate transformation of the conductivity coefficients from the crystal frame to the sample frame is in correctly defining the direction cosines Q_{ij}. If the orthonormal set $\{\mathbf{e}_i^c\}$ is attached to the crystal lattice, and $\{\mathbf{e}_i\}$ to the sample, the transformation we seek is

$$Q_{il} = \mathbf{e_i} \cdot \mathbf{e_l^c}$$

In terms of Bunge–Euler angles, as defined by relation (2.3.4), we can easily show that

$$Q_{il} = g_{il}^{c \to s} = g_{li}^{s \to c} = Q_{lk}(\varphi_1)Q_{kj}(\Phi)Q_{ji}(\varphi_2)$$

where the direction cosines $Q_{lk}(\varphi_1)$, $Q_{kj}(\Phi)$, $Q_{ji}(\varphi_2)$ are those defined in Section 2.3.

If the coordinate transformation relation is applied to orthorhombic crystalline material, we must have

$$\begin{aligned} K'_{rs} &= Q_{r1}Q_{s1}K_{11} + Q_{r2}Q_{s2}K_{22} + Q_{r3}Q_{s3}K_{33} \\ &= g_{1r}^{s \to c}g_{1s}^{s \to c}K_{11} + g_{2r}^{s \to c}g_{2s}^{s \to c}K_{22} + g_{3r}^{s \to c}g_{3s}^{s \to c}K_{33} \end{aligned}$$

When the material is transversely isotropic or the crystal lattice possesses hexagonal symmetry, this transformation relation simplifies to

$$K'_{rs} = (Q_{r1}Q_{s1} + Q_{r2}Q_{s2})K_{11} + Q_{r3}Q_{s3}K_{33}$$
$$= \left(g_{1r}^{s\to c}g_{1s}^{s\to c} + g_{2r}^{s\to c}g_{2s}^{s\to c}\right)K_{11} + g_{3r}^{s\to c}g_{3s}^{s\to c}K_{33}$$

And for the highly symmetrical case of the cubic crystal, we must have

$$K'_{rs} = (Q_{r1}Q_{s1} + Q_{r2}Q_{s2} + Q_{r3}Q_{s3})k = k\delta_{rs}$$

where we have used relation (2.36). The additional symmetry possessed by an amorphous phase gives no further reduction, and therefore the conductivity of amorphous materials has the same form as cubic crystals.

The components of the elastic stiffness tensor for a cubic material in the sample frame can be derived using the same procedure. Consider the expressions for elastic stiffness and compliance for cubic crystals in the sample frame. These are given in terms of the three fundamental elastic constants C_{11}, C_{12}, and C_{44}:

$$\begin{aligned} C'_{\mathrm{abcd}}(g) &= Q_{ai}Q_{bj}Q_{ck}Q_{dl}C_{ijkl} \\ &= C_{12}\delta_{\mathrm{ab}}\delta_{\mathrm{cd}} + C_{44}(\delta_{\mathrm{ac}}\delta_{\mathrm{bd}} + \delta_{\mathrm{ad}}\delta_{\mathrm{bc}}) \\ &\quad + (C_{11} - C_{12} - 2C_{44})\sum_{r=1}^{3} Q_{ar}Q_{br}Q_{cr}Q_{dr} \\ &= C_{12}\delta_{\mathrm{ab}}\delta_{\mathrm{cd}} + C_{44}(\delta_{\mathrm{ac}}\delta_{\mathrm{bd}} + \delta_{\mathrm{ad}}\delta_{\mathrm{bc}}) \\ &\quad + (C_{11} - C_{12} - 2C_{44})\sum_{r=1}^{3} g_{ra}^{s\to c}g_{rb}^{s\to c}g_{rc}^{s\to c}g_{rd}^{s\to c} \end{aligned} \tag{7.61}$$

A corresponding equation for the components of the elastic compliance tensor in the sample frame is expressed as

$$S'_{\mathrm{abcd}}(g) = S_{12}\delta_{\mathrm{ab}}\delta_{\mathrm{cd}} + \frac{S_{44}}{4}(\delta_{\mathrm{ac}}\delta_{\mathrm{bd}} + \delta_{\mathrm{ad}}\delta_{\mathrm{bc}}) + \left(S_{11} - S_{12} - \frac{S_{44}}{2}\right)\sum_{r=1}^{3} g_{ra}^{s\to c}g_{rb}^{s\to c}g_{rc}^{s\to c}g_{rd}^{s\to c} \tag{7.62}$$

where S_{11}, S_{12}, and S_{44} are the fundamental elastic compliances that can be related to C_{11}, C_{12}, and C_{44} (e.g., Herakovich, 1998). Only the last term in Eqs. (7.61) and (7.62), $\sum_{r=1}^{3} g_{ra}^{s\to c}g_{rb}^{s\to c}g_{rc}^{s\to c}g_{rd}^{s\to c}$, is actually dependent on the crystal orientation, and therefore only this term needs to be expressed in a Fourier series representation of the functions $C'_{\mathrm{abcd}}(g)$ and $S'_{\mathrm{abcd}}(g)$.

7.5.2 Fourier representations of the crystal property relations: The discrete Fourier transform

The coordinate transformation relationships developed in the previous section constitute functions of the Euler-angle parameters. For example, the right sides of relations (7.61) and (7.62) contain two constant terms and a term with four products of the direction cosines. Referring to relation (2.34) for the definition of the components of the direction cosine matrix, we can see that the Euler-angle dependence includes polynomial terms of the products of sines and cosines of the Euler angles up to fourth order. The same general picture holds for second-order tensors like the conductivity, except that the order of polynomial terms is two, corresponding to the order of the tensor. These crystal property functions do not depend on grain size at ordinary length scales, and thus can be expressed in discrete form at a finite number of points associated with the binning of the augmented fundamental zone of orientations: FZ_3 for cubic or FZ_H for hexagonal phases.

Following the constructs described in Section 5.2.2, the discrete Fourier representation of any property function, say $p(\varphi_1, \Phi, \varphi_2)$, can be expressed as a set of B numbers, p_b, representing the values of the function in each bin. These values of the function can be described with the discrete Fourier series

$$p_b = \frac{1}{B} \sum_{k=0}^{B-1} P_k e^{i2\pi kb/B} \tag{7.63}$$

where the P_k coefficients are the discrete Fourier transforms for the property function, defined by

$$P_k = \sum_{b=0}^{B-1} p_b e^{-i2\pi kb/B} \tag{7.64}$$

Homogenization relations to be developed in Chapter 8 will show that first-order estimates of macroscopic elastic properties depend only on simple volume averages of the corresponding local elastic properties. These averaging procedures for conductivities and elastic properties are readily constructed with the discrete Fourier coefficients obtained for the crystal properties relations and the ODF. We will return to the problem of averaging in Chapter 8.

Generalized spherical harmonic representation

Using GSHFs, the Fourier representations needed for the rotation-dependent components of Eqs. (7.63) and (7.64) can be summarized as (assuming both crystal and sample symmetry are present; see Eq. (3.39) and Section 6.2.3)

$$\sum_{r=1}^{3} g_{ra} g_{ra} g_{ra} g_{ra} = \sum_{\ell=0}^{4} \sum_{\mu=1}^{M(\ell)} \sum_{\eta=1}^{N(\ell)} {}_{aaaa}A_\ell^{\mu\eta} \dot{\dot{T}}_\ell^{\mu\eta}(g),$$

$$\sum_{r=1}^{3} g_{ra} g_{rb} g_{ra} g_{rb} = \sum_{r=1}^{3} g_{ra} g_{ra} g_{rb} g_{rb} = \sum_{\ell=0}^{4} \sum_{\mu=1}^{M(\ell)} \sum_{\eta=1}^{N(\ell)} {}_{abab}A_\ell^{\mu\eta} \dot{\dot{T}}_\ell^{\mu\eta}(g), \tag{7.65}$$

The Fourier coefficients in Eq. (7.65) are computed using the standard Fourier methods, exploiting the orthonormality of the Fourier basis. For example,

$${}_{abab}A_\ell^{\mu\eta} = (2l+1) \iiint_{FZ(O)} \sum_{r=1}^{3} g_{ra} g_{ra} g_{rb} g_{rb} \dot{\dot{T}}_\ell^{\mu\eta}(g) dg \tag{7.66}$$

and the Fourier coefficients may be evaluated using numerical integration techniques (e.g., the Simpson rule). Note that, due to the inherent symmetry of the description in Eq. (7.65), only six of these sets of coefficients are found to be distinct (${}_{abab}A_\ell^{\mu\eta} = {}_{aabb}A_\ell^{\mu\eta}$). Furthermore, as the elastic stiffness tensor is a fourth-rank linear tensor, Eq. (7.66) reveals that only the terms of the Fourier series

Table 7.1 Values of the $_{aaaa}A$ and $_{abab}A$ Coefficients (Eq. 7.66)

	Fourier Coefficients $\ell\ \mu\ \nu$			
	011	**411**	**412**	**413**
$_{1111}A$	0.6	0.196438	−0.29277	0.3873
$_{2222}A$	0.6	0.196438	0.29277	0.3873
$_{3333}A$	0.6	0.523788	0	0
$_{1212}A = {}_{1122}A$	0.2	0.0654792	0	−0.3873
$_{1313}A = {}_{1133}A$	0.2	−0.261871	0.29277	0
$_{2323}A = {}_{2233}A$	0.2	−0.261871	−0.29277	0

up to ℓ equal to 4 are nonzero. Therefore, representation of the orthorhombic crystal stiffness coefficients in the sample reference frame, for any cubic crystal, needs only 19 nonzero Fourier coefficients. These are presented in Table 7.1.

7.5.3 Thermo-elastic constitutive behavior

The purely elastic stress–strain relationships described in the previous section need to be suitably modified to incorporate coupling effects of thermal, electric, and magnetic fields, when they exist in the problem being studied. In a first-order approach, coupled thermo-elastic, piezo-electric, and magnetostrictive constitutive behaviors are all described by adding together the strains produced by the different fields. For example, the *thermo-elastic constitutive behavior* is expressed as

$$\varepsilon_{ij} = S_{ijkl}\sigma_{kl} + \varepsilon_{ij}^{th} \tag{7.67}$$

where $\boldsymbol{\varepsilon}^{th}$ is the strain induced by the thermal field. In a linear description of the thermal response of the material, the thermal strains are expressed as

$$\boldsymbol{\varepsilon}^{th} = \boldsymbol{\alpha}\,\theta \tag{7.68}$$

where θ denotes the change in temperature (from a chosen reference value), and $\boldsymbol{\alpha}$ is a symmetric second-rank tensor describing the thermal expansion coefficient of the material. $\boldsymbol{\alpha}$ is expected to be a material property that is independent of the geometry of the sample being studied. The treatment described here is easily extended to piezo-electric and magnetostrictive constitutive behaviors, while recognizing that the new materials properties that arise in those relationships are third-rank tensors.

7.6 CRYSTAL PLASTICITY

Crystal plasticity refers to study of plastic deformation in single-crystal and polycrystalline materials while attempting to take into account explicitly the details of physics and geometry of deformation at the crystal (also called grain) level. At low homologous temperatures, the dominant mode of plastic deformation in crystalline materials is slip on specific crystallographic planes in specific directions. For metals with low-stacking fault energy and/or low-symmetry crystal structures, deformation twinning (also on specific crystallographic planes along specific directions) is an additional mode of plastic deformation and plays an influential role in both the stress–strain response of the material and the evolution of the underlying microstructure in the material. Crystal plasticity attempts to incorporate these details in the description and formulation of elastic–plastic constitutive models for crystalline materials.

Although the crystal plasticity theory can be applied to a wide range of crystalline materials including metals, rocks, and crystalline and semi-crystalline polymers, a large fraction of the modeling studies reported in literature have focused on single-phase, high-stacking fault energy metals such as aluminum and copper. One of the major successes of the crystal plasticity theory has been the ability to predict both the anisotropic stress–strain response of these metals as well as the evolution of the averaged crystallographic texture in the material in a variety of large deformation processes. At the present time, serious efforts are being made to extend these successes to other classes of materials. In this section, we collate and summarize the major accomplishments to date in the development and use of the crystal plasticity theory and models.

7.6.1 Small strains

It has been long observed that plastic deformation occurs in metals by simple shear caused by gliding of specific crystallographic planes in specific crystallographic directions lying on those planes; this basic process is hereafter referred to as *crystallographic slip*. Line defects in the crystal structure, called *dislocations*, play an important role in facilitating crystallographic slip. Dislocation theory not only explains why most solids yield at stress levels that are much smaller than the theoretical shear strength, but also suggests that slip (glide of dislocations) occurs along the close-packed directions (shortest Burgers vector, $\vec{\mathbf{b}}$) and on close-packed planes. Furthermore, the shear stress on the slip plane resolved along the slip direction was identified as the primary driving force for slip. This concept is referred to as the *Schmid law*.

Let us now examine, in detail, how a block of metal deforms. For pedagogy, we will consider first purely plastic behavior (i.e., no elastic deformation, also referred as rigid-plastic deformation). Consider a single crystal cylinder of aluminum subjected to tension, as shown in Figure 7.14. Since

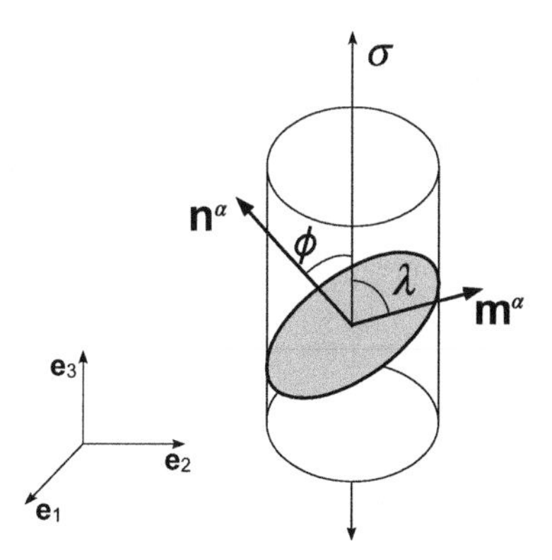

FIGURE 7.14

Schematic description of a slip plane in a sample subjected to tension.

aluminum has an fcc crystal structure, it has 12 potential slip systems of the type $\{111\}\langle 1\bar{1}0\rangle$. Let the shaded plane in the cylinder represent one of the potential slip systems in the crystal with slip plane normal $\mathbf{n}^{\alpha}$ and a slip direction $\mathbf{m}^{\alpha}$. The *resolved shear stress* (τ^{α}) on this slip system (labeled α) can be expressed for the most general stress state as (see also Exercise 7.3)

$$\tau^{\alpha} = \boldsymbol{\sigma} \cdot \mathbf{m}^{\alpha} \otimes \mathbf{n}^{\alpha} \tag{7.69}$$

For the uniaxial loading condition shown in Figure 7.14, Eq. (7.69) reduces to

$$\tau^{\alpha} = \sigma_{33} m_3 n_3 = \sigma \cos\phi \cos\lambda \tag{7.70}$$

The term $\cos\phi \cos\lambda$ is often referred to as the *Schmid factor*. It can be shown that the maximum value of the Schmid factor is 0.5.

The resolved shear stress on a slip system produces slip. The precise relationship between the resolved shear stress and the amount of slip will be discussed later. For now, let $\dot{\gamma}^{\alpha}$ denote the effective shear rate on this slip system. The corresponding *plastic velocity gradient tensor*, $\mathbf{L}^{p}$, is expressed as

$$\mathbf{L}^{p} = \dot{\gamma}^{\alpha} \mathbf{m}^{\alpha} \otimes \mathbf{n}^{\alpha} \tag{7.71}$$

When slip occurs on multiple slip systems, the corresponding plastic velocity gradient tensor is expressed as the sum of all contributions from all slip systems:

$$\mathbf{L}^{p} = \sum_{\alpha} \dot{\gamma}^{\alpha} \mathbf{m}^{\alpha} \otimes \mathbf{n}^{\alpha} \tag{7.72}$$

In the absence of elastic deformations, the difference between the *applied velocity gradient tensor*, $\mathbf{L}^{app}$, and that accommodated by the slip processes, $\mathbf{L}^{p}$, needs to be accommodated by a *rigid-body spin*, $\mathbf{W}^{*}$. This rigid-body spin causes a lattice rotation, $\mathbf{R}^{*}$. Therefore, the relationship between the lattice rotation tensor and the plastic velocity gradient tensor can be expressed as (apply relation (7.45) to the case where $\mathbf{F} = \mathbf{R}$)

$$\mathbf{W}^{*} = \dot{\mathbf{R}}^{*} \mathbf{R}^{*T} = \mathbf{L}^{app} - \mathbf{L}^{p} \tag{7.73}$$

The additive decomposition of the velocity gradient tensors into symmetric and anti-symmetric components allows us to reorganize (7.73) as

$$\mathbf{W}^{*} = \frac{1}{2}\left(\mathbf{L}^{app} - \mathbf{L}^{appT}\right) - \mathbf{W}^{p} = \frac{1}{2}\left(\mathbf{L}^{app} - \mathbf{L}^{appT}\right) - \sum_{\alpha} \frac{\dot{\gamma}^{\alpha}}{2}\left(\mathbf{m}^{\alpha} \otimes \mathbf{n}^{\alpha} - \mathbf{n}^{\alpha} \otimes \mathbf{m}^{\alpha}\right) \tag{7.74}$$

Also,

$$\mathbf{D}^{app} \equiv \frac{1}{2}\left(\mathbf{L}^{app} + \mathbf{L}^{appT}\right) = \mathbf{D}^{p} = \sum_{\alpha} \frac{\dot{\gamma}^{\alpha}}{2}\left(\mathbf{m}^{\alpha} \otimes \mathbf{n}^{\alpha} + \mathbf{n}^{\alpha} \otimes \mathbf{m}^{\alpha}\right) \tag{7.75}$$

where $\mathbf{D}^{p}$ and $\mathbf{W}^{p}$ are the symmetric and anti-symmetric components of $\mathbf{L}^{p}$. It is convenient to introduce here the following definitions of a slip tensor, $\mathbf{S}^{\alpha}$, and its symmetric components (the Schmid tensor, $\mathbf{P}$) and skew-symmetric components:

$$\mathbf{S}^{\alpha} = \mathbf{m}^{\alpha} \otimes \mathbf{n}^{\alpha}, \ \mathbf{P}^{\alpha} \equiv 1/2(\mathbf{m}^{\alpha} \otimes \mathbf{n}^{\alpha} + \mathbf{n}^{\alpha} \otimes \mathbf{m}^{\alpha}), \ \mathbf{Q}^{\alpha} \equiv 1/2(\mathbf{m}^{\alpha} \otimes \mathbf{n}^{\alpha} - \mathbf{n}^{\alpha} \otimes \mathbf{m}^{\alpha}) \tag{7.76}$$

Table 7.2 $\mathbf{P}^\alpha$ for 12 fcc Crystal Slip Systems

	Slip System	$\mathbf{P}^\alpha$		Slip System	$\mathbf{P}^\alpha$
1	$(111)\,[\bar{1}01]$	$\frac{1}{\sqrt{6}}\begin{bmatrix} -1 & -\frac{1}{2} & 0 \\ -\frac{1}{2} & 0 & \frac{1}{2} \\ 0 & \frac{1}{2} & 1 \end{bmatrix}$	7	$(1\bar{1}1)\,[10\bar{1}]$	$\frac{1}{\sqrt{6}}\begin{bmatrix} 1 & -\frac{1}{2} & 0 \\ -\frac{1}{2} & 0 & \frac{1}{2} \\ 0 & \frac{1}{2} & -1 \end{bmatrix}$
2	$(111)\,[0\bar{1}1]$	$\frac{1}{\sqrt{6}}\begin{bmatrix} 0 & -\frac{1}{2} & \frac{1}{2} \\ -\frac{1}{2} & -1 & 0 \\ \frac{1}{2} & 0 & 1 \end{bmatrix}$	8	$(1\bar{1}1)\,[011]$	$\frac{1}{\sqrt{6}}\begin{bmatrix} 0 & \frac{1}{2} & \frac{1}{2} \\ \frac{1}{2} & -1 & 0 \\ \frac{1}{2} & 0 & 1 \end{bmatrix}$
3	$(111)\,[1\bar{1}0]$	$\frac{1}{\sqrt{6}}\begin{bmatrix} 1 & 0 & \frac{1}{2} \\ 0 & -1 & -\frac{1}{2} \\ \frac{1}{2} & -\frac{1}{2} & 0 \end{bmatrix}$	9	$(1\bar{1}1)\,[110]$	$\frac{1}{\sqrt{6}}\begin{bmatrix} 1 & 0 & \frac{1}{2} \\ 0 & -1 & \frac{1}{2} \\ \frac{1}{2} & \frac{1}{2} & 0 \end{bmatrix}$
4	$(\bar{1}11)\,[101]$	$\frac{1}{\sqrt{6}}\begin{bmatrix} -1 & \frac{1}{2} & 0 \\ \frac{1}{2} & 0 & \frac{1}{2} \\ 0 & \frac{1}{2} & 1 \end{bmatrix}$	10	$(11\bar{1})\,[101]$	$\frac{1}{\sqrt{6}}\begin{bmatrix} 1 & \frac{1}{2} & 0 \\ \frac{1}{2} & 0 & \frac{1}{2} \\ 0 & \frac{1}{2} & -1 \end{bmatrix}$
5	$(\bar{1}11)\,[01\bar{1}]$	$\frac{1}{\sqrt{6}}\begin{bmatrix} 0 & -\frac{1}{2} & \frac{1}{2} \\ -\frac{1}{2} & 1 & 0 \\ \frac{1}{2} & 0 & -1 \end{bmatrix}$	11	$(11\bar{1})\,[\bar{1}10]$	$\frac{1}{\sqrt{6}}\begin{bmatrix} -1 & 0 & \frac{1}{2} \\ 0 & 1 & -\frac{1}{2} \\ \frac{1}{2} & -\frac{1}{2} & 0 \end{bmatrix}$
6	$(\bar{1}11)\,[110]$	$\frac{1}{\sqrt{6}}\begin{bmatrix} -1 & 0 & \frac{1}{2} \\ 0 & 1 & \frac{1}{2} \\ \frac{1}{2} & \frac{1}{2} & 0 \end{bmatrix}$	12	$(11\bar{1})\,[011]$	$\frac{1}{\sqrt{6}}\begin{bmatrix} 0 & \frac{1}{2} & \frac{1}{2} \\ \frac{1}{2} & 1 & 0 \\ \frac{1}{2} & 0 & -1 \end{bmatrix}$

Table 7.2 presents the twelve slip systems for an fcc crystal along with their corresponding $\mathbf{P}^\alpha$ tensors expressed in the crystal reference frame.

Now we turn our attention to the precise relationship between resolved shear stress on a slip system and the amount of slip on the slip system. In the rate-independent approach, it is assumed that once the resolved shear stress (τ^α defined in Eq. 7.69) reaches the value of the *critical resolved shear stress*, then slip can occur on that slip system essentially at any desired rate. Figure 7.15 shows a graphical representation of the rate-independent slip law as a solid line. In this figure, we have introduced a new variable called the *slip resistance*, denoted s^α.

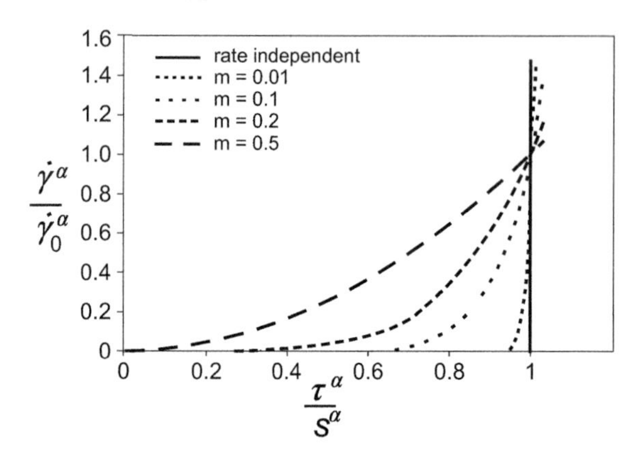

FIGURE 7.15

Schematic representation of rate dependence of slip.

For rate-independent plasticity, the slip resistance has exactly the same role as the critical resolved shear stress. The rate-independent slip law often leads to nonunique solutions for slip activity in selected combinations of loading conditions and crystal orientations. In an effort to obtain unique solutions, rate-dependent approaches have been adopted in the literature that attribute the rate dependence to the "viscosity" of dislocation motion. In this book, we will emphasize the rate-dependent approach, thereby avoiding problems with uniqueness.

The dashed lines in Figure 7.15 describe the rate dependence of crystallographic slip, invoking the following power law to describe phenomenologically the dependence of dislocation velocity on the imposed resolved shear stress on the dislocation (Asaro and Needleman, 1985):

$$\dot{\gamma}^{\alpha} = \dot{\gamma}^{0}\left(\frac{|\tau^{\alpha}|}{s^{\alpha}}\right)^{\frac{1}{m}}\operatorname{sgn}(\tau^{\alpha}) \tag{7.77}$$

Note that the rate-independent response described earlier can be thought of as a limiting case of the rate-dependent response described in Eq. (7.77), by letting m approach 0. More important, the substitution of Eqs. (7.77) and (7.69) into Eq. (7.75) produces a set of five independent nonlinear equations in five unknowns that are guaranteed to produce unique solutions. To formulate such a well-posed system of equations, we also need to recognize the pressure independence of crystal plasticity and use the deviatoric stress tensor, σ' in Eq. (7.69), that is, $\tau^{\alpha} = \boldsymbol{\sigma}\cdot\mathbf{P}^{\alpha} = \boldsymbol{\sigma}'\cdot\mathbf{P}^{\alpha}$. Incorporating these changes, Eq. (7.75) can now be cast as

$$\mathbf{D}^{p} = \sum_{\alpha}\left(\dot{\gamma}^{0}\left(\frac{|\boldsymbol{\sigma}'\cdot\mathbf{P}^{\alpha}|}{s^{\alpha}}\right)^{\frac{1}{m}}\operatorname{sgn}(\boldsymbol{\sigma}'\cdot\mathbf{P}^{\alpha})\right)\mathbf{P}^{\alpha} \tag{7.78}$$

For later use in homogenization theories, we recast Eq. (7.78) as

$$\mathbf{D}^{p} = \sum_{\alpha}\left(\dot{\gamma}^{0}\frac{|\boldsymbol{\sigma}'\cdot\mathbf{P}^{\alpha}|^{\frac{1}{m}-1}}{(s^{\alpha})^{\frac{1}{m}}}(\mathbf{P}^{\alpha}\otimes\mathbf{P}^{\alpha})\right)\boldsymbol{\sigma}' = \mathbf{M}(\boldsymbol{\sigma}')\boldsymbol{\sigma}' \tag{7.79}$$

Equation (7.79) is invertible (assuming $tr(\mathbf{D}^p) = 0$) and can be expressed as (useful for homogenization theories discussed later)

$$\boldsymbol{\sigma}' = \mathbf{M}^{-1}\mathbf{D}^p = \mathbf{N}\mathbf{D}^p \tag{7.80}$$

Note that Eq. (7.78) can be solved for either a prescribed stress tensor or a prescribed velocity gradient tensor. Solving for $\mathbf{D}^p$ having prescribed the stress tensor is relatively straightforward. In this regard, it would be beneficial to note that $\boldsymbol{\sigma}' \cdot \mathbf{P}^\alpha$ can be replaced with $\boldsymbol{\sigma} \cdot \mathbf{P}^\alpha$ whenever needed. However, if $\mathbf{D}^p$ is prescribed, we can only solve for the deviatoric stress tensor, $\boldsymbol{\sigma}'$. We would need additional information (on the hydrostatic component of the stress tensor) to establish the complete stress tensor, $\boldsymbol{\sigma}$.

Note also that the structure of Eq. (7.78) allows only for the imposition of a traceless $\mathbf{D}^p$. The numerical effort associated with solving Eq. (7.78) for $\boldsymbol{\sigma}'$ for an imposed $\mathbf{D}^p$ scales dramatically with decreasing values of the slip rate sensitivity parameter m. For most metals, at low homologous temperatures, experimental measurements indicate a value for m in the range of 0.01 to 0.02. Equation (7.78) becomes numerically stiff with these values because small changes in the values of $\boldsymbol{\sigma}'$, especially when $\boldsymbol{\sigma}' \cdot \mathbf{P}^\alpha \approx s^\alpha$, produce very large changes in the values of $\mathbf{D}^p$. In any implementation of a numerical solution of Eq. (7.78), this aspect needs careful detailed attention.

Example 7.14

Solve for the plastic stretching and lattice spin tensors in an fcc single crystal loaded in tension along the <100> direction (Figure 7.16).

Solution

The imposed stress tensor is expressed as $\begin{bmatrix} \sigma & 0 & 0 \\ 0 & 0 & 0 \\ 0 & 0 & 0 \end{bmatrix}$. The resolved shear stresses on the 12 slip systems described in Table 7.2 are given as

$$\tau^1 = \tau^4 = \tau^6 = \tau^{11} = -\frac{\sigma}{\sqrt{6}}, \quad \tau^3 = \tau^7 = \tau^9 = \tau^{10} = \frac{\sigma}{\sqrt{6}}, \quad \tau^2 = \tau^5 = \tau^8 = \tau^{12} = 0$$

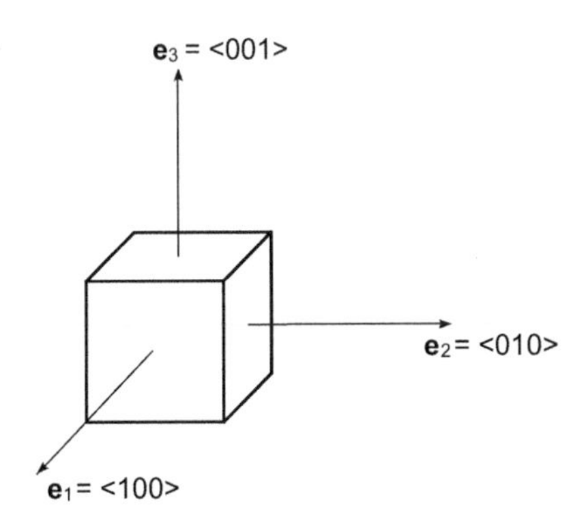

FIGURE 7.16

fcc crystal loaded along <100> direction.

The corresponding slip rates are given by (assuming $s^\alpha = s$)

$$\dot{\gamma}^1 = \dot{\gamma}^4 = \dot{\gamma}^6 = \dot{\gamma}^{11} = -\dot{\gamma}^0\left(\frac{\sigma}{s\sqrt{6}}\right)^{1/m}, \quad \dot{\gamma}^3 = \dot{\gamma}^7 = \dot{\gamma}^9 = \dot{\gamma}^{10} = \dot{\gamma}^0\left(\frac{\sigma}{s\sqrt{6}}\right)^{1/m},$$
$$\dot{\gamma}^2 = \dot{\gamma}^5 = \dot{\gamma}^8 = \dot{\gamma}^{12} = 0$$

The corresponding plastic stretching tensor is computed to be

$$\mathbf{D}^p = \frac{\dot{\gamma}^0}{\sqrt{6}}\left(\frac{\sigma}{s\sqrt{6}}\right)^{1/m}\begin{bmatrix} 8 & 0 & 0 \\ 0 & -4 & 0 \\ 0 & 0 & -4 \end{bmatrix}$$

Note that, as expected, the solution indicates that the crystal will contract equally in the lateral plane and that there are no shears produced when loaded in this highly symmetric orientation. Assuming that there is no superimposed applied spin—that is, $\frac{1}{2}(\mathbf{L}^{app} - \mathbf{L}^{appT}) = \mathbf{0}$—the lattice spin is computed from Eq. (7.74) as $\mathbf{W}^* = \mathbf{0}$, indicating that there would be no lattice rotation in this highly symmetric loading situation.

Example 7.15

Solve for the plastic stretching and lattice spin tensors in an fcc single crystal loaded in compression along the <110> direction (Figure 7.17).

Solution

It is convenient to assign two reference frames as depicted before. The imposed compressive stress tensor is expressed in the global frame as

$$\begin{bmatrix} -\sigma & 0 & 0 \\ 0 & 0 & 0 \\ 0 & 0 & 0 \end{bmatrix}$$

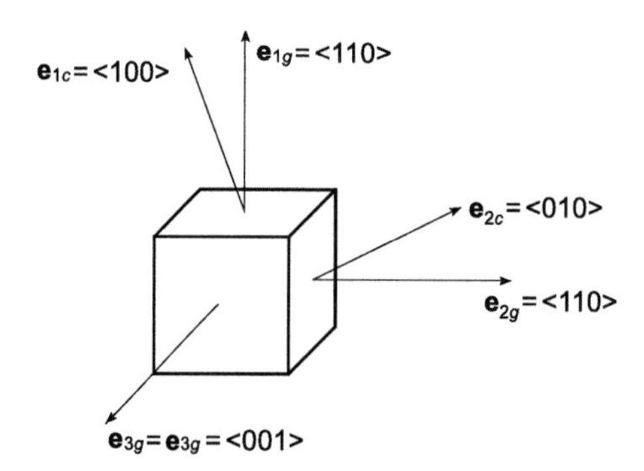

FIGURE 7.17

fcc crystal loaded in compression along <110> direction.

The same stress tensor in the crystal reference frame is expressed as

$$\sigma^c = \begin{bmatrix} -\frac{\sigma}{2} & -\frac{\sigma}{2} & 0 \\ -\frac{\sigma}{2} & -\frac{\sigma}{2} & 0 \\ 0 & 0 & 0 \end{bmatrix}$$

The resolved shear stresses on the 12 slip systems described in Table 7.2 are given as

$$\tau^1 = \tau^2 = \frac{\sigma}{\sqrt{6}}, \quad \tau^{10} = \tau^{12} = -\frac{\sigma}{\sqrt{6}}, \quad \tau^3 = \tau^4 = \tau^5 = \tau^6 = \tau^7 = \tau^8 = \tau^9 = \tau^{11} = 0$$

The corresponding slip rates are given by (assuming $s^\alpha = s$)

$$\dot{\gamma}^1 = \dot{\gamma}^2 = -\dot{\gamma}^{10} = -\dot{\gamma}^{12} = \dot{\gamma}^0 \left(\frac{\sigma}{s\sqrt{6}}\right)^{1/m}, \quad \dot{\gamma}^3 = \dot{\gamma}^4 = \dot{\gamma}^5 = \dot{\gamma}^6 = \dot{\gamma}^7 = \dot{\gamma}^8 = \dot{\gamma}^9 = \dot{\gamma}^{11} = 0$$

The corresponding plastic stretching tensor is computed in the crystal frame to be

$$[\mathbf{D}^p]^c = \frac{\dot{\gamma}^0}{\sqrt{6}} \left(\frac{\sigma}{s\sqrt{6}}\right)^{1/m} \begin{bmatrix} -2 & -2 & 0 \\ -2 & -2 & 0 \\ 0 & 0 & 4 \end{bmatrix}$$

Transformed to the global reference frame, the plastic stretching tensor is expressed as

$$[\mathbf{D}^p]^g = \frac{\dot{\gamma}^0}{\sqrt{6}} \left(\frac{\sigma}{s\sqrt{6}}\right)^{1/m} \begin{bmatrix} -4 & 0 & 0 \\ 0 & 0 & 0 \\ 0 & 0 & 4 \end{bmatrix}$$

The solution here indicates that the crystal will expand only in the $\mathbf{e}_{3g}$ direction, with zero expansion in the $\mathbf{e}_{2g}$ direction. Effectively, the crystal undergoes plane strain compression, even though no constraints have been applied to stop the expansion of the crystal in the $\mathbf{e}_{2g}$ direction. This example brings out clearly the inherent plastic anisotropy exhibited by fcc single crystals. This prediction has been verified experimentally in single crystals of copper. Figure 7.18 shows the cross-sections of a cylinder of copper single crystal loaded in compression in the <110> direction, without the application of any lateral constraints to the expansion of the

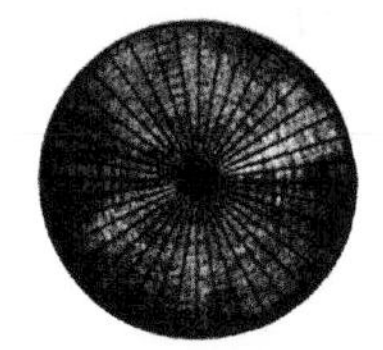
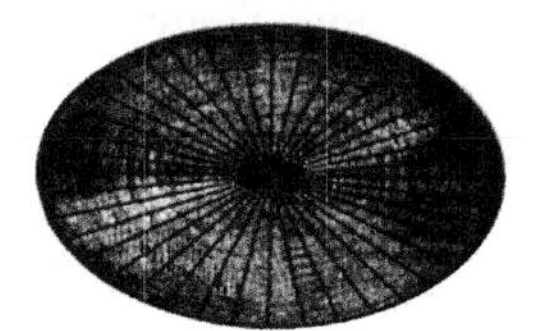

FIGURE 7.18

Demonstration of anisotropy of copper (*left*) before and (*right*) after deformation. Source: *Kalidindi and Anand (1993).*

cylinders. The initial cross-section was chosen to be circular. The deformed cross-section is found to be elliptical with the short dimension essentially unchanged from the original dimension.

Assuming that there is no superimposed applied spin—that is, $\frac{1}{2}(\mathbf{L}^{app} - \mathbf{L}^{appT}) = \mathbf{0}$—the lattice spin is computed from Eq. (7.74) as $\mathbf{W}^* = \mathbf{0}$, indicating that there would be no lattice rotation in this loading condition.

In describing fcc crystal plasticity it has been convenient to express the pertinent tensors, $\boldsymbol{\sigma}$, $\mathbf{L}$, $\mathbf{D}$, $\mathbf{W}$, $\mathbf{m}^a$, $\mathbf{n}^a$, S^a, P^a, Q^a, in the local crystal frame where their form is as simple as possible. Readers are reminded, however, that these and all tensors can be expressed in any orthonormal coordinate frame by the standard transformation laws described in Chapter 2. And whereas the geometrical form of the tensors m^a, n^a, S^a, P^a, Q^a of crystal plasticity will often be simplest in the local crystal frame, the field variables $\boldsymbol{\sigma}$, $\mathbf{L}$, $\mathbf{D}$, $\mathbf{W}$ are generally required in the sample or global frame. Thus, the ability to transform the components of these tensors from one coordinate frame to another is fundamental to the application of crystal plasticity. An example of this need was illustrated in Example 7.15. Readers are reminded that the coordinate transformation laws are described in Section 2.3.

Although the examples used so far have been restricted to fcc crystals, the framework described here can be extended easily to other crystal lattice structures. In bcc crystals, the slip directions are the $\langle 111 \rangle$ directions (closed-packed directions), but the slip planes are not well defined. A number of crystallographic planes containing the $\langle 111 \rangle$ directions can potentially serve as slip planes; this is generally referred to as *pencil-glide*. In many numerical simulations reported in the literature, it is generally assumed that the slip planes in bcc crystals are restricted to $\{110\}$ and $\{121\}$ planes.

The slip process in hcp crystals is significantly more complicated. The easy glide (slip) systems in hcp crystals are usually either the basal slip systems or the prismatic slip systems. Generally, the c/a ratio controls which of these two slip families constitutes the easy glide system in a given hexagonal close packed (hcp) crystal. When the c/a ratio is less than the ideal c/a ratio, the basal planes are closed-packed planes and constitute the easy glide planes, and the **a** directions become the slip directions (closed-packed directions). On the other hand, if the c/a ratio is larger than the ideal c/a ratio, the prismatic planes are the preferred slip planes, with the **a** directions being the preferred slip directions.

Although the preceding correlations are generally expected, one should note that there are significant exceptions to these rules. Table 7.3 lists the possible slip systems in hcp crystals. Note that

Table 7.3 List of Potential Slip Systems in hcp Crystals

Slip Family	(Slip Plane) [Slip Direction]
Basal slip	$(0001)\ [1\bar{2}10]$, $(0001)\ [2\bar{1}\bar{1}0]$, $(0001)\ [11\bar{2}0]$
Prism slip	$(10\bar{1}0)\ [1\bar{2}10]$, $(01\bar{1}0)\ [2\bar{1}\bar{1}0]$, $(\bar{1}100)\ [1120]$
Pyramidal $\langle \mathbf{a} \rangle$	$(10\bar{1}1)\ [1\bar{2}10]$, $(01\bar{1}1)\ [2\bar{1}\bar{1}0]$, $(\bar{1}101)\ [11\bar{2}0]$, $(\bar{1}011)\ [1\bar{2}10]$, $(0\bar{1}11)\ [2\bar{1}\bar{1}0]$, $(1\bar{1}01)\ [11\bar{2}0]$
Pyramidal $\langle \mathbf{c} + \mathbf{a} \rangle$	$(10\bar{1}1)\ [\bar{2}113]$, $(10\bar{1}1)\ [1\bar{2}13]$, $(01\bar{1}1)\ [\bar{1}\bar{1}23]$, $(01\bar{1}1)\ [1\bar{2}13]$, $(\bar{1}101)\ [1\bar{2}13]$, $(\bar{1}101)\ [2\bar{1}\bar{1}3]$, $(\bar{1}011)\ [2\bar{1}\bar{1}3]$, $(\bar{1}011)\ [11\bar{2}3]$, $(0\bar{1}11)\ [11\bar{2}3]$, $(0\bar{1}11)\ [\bar{1}2\bar{1}3]$, $(1\bar{1}01)\ [\vec{1}\ 2\bar{1}3]$, $(1\bar{1}01)\ [\bar{2}113]$

the slip planes and slip directions are identified here using the conventional four-index labeling system. To perform the type of computations described earlier, one has to first transform these into descriptions in a 3D orthonormal crystal reference frame. Note also that the consideration of only the basal, prism, and pyramidal <**a**> slip systems does not provide the necessary five independent slip systems to accommodate an arbitrary traceless velocity gradient tensor. This is because none of these slip systems is capable of producing a strain (length change) along the crystal **c**-axis. Ductile hcp metals are known to either slip on the pyramidal <**c**+**a**> slip systems or produce deformation twins to address this deficiency.

7.6.2 Crystal plasticity: Finite strains

The theory presented in the last section considers only small plastic strains in the crystal. However, in dealing with uniform ductility of single and polycrystalline metals, we will encounter finite plastic strains. Here, we present briefly a rigorous formulation of elastic–plastic constitutive relations for crystalline materials subjected to finite deformations at low homologous temperatures where plastic deformation is assumed to occur primarily by crystallographic slip. It is based largely on the concepts described earlier, but has been generalized to a finite-deformation, frame-indifferent form. The development of this modern formulation can be traced through the papers of Hill (1966), Lee (1969), Rice (1971), Hill and Rice (1972), Asaro and Rice (1977), Asaro (1983a,b), Peirce, et al. (1983), Asaro and Needleman (1985), and Kalidindi et al. (1992). The central tenet of the crystal plasticity framework is that plastic deformation occurs in metals by flow of material through the crystal lattice without distorting the lattice itself, and then the lattice with the embedded deformed material undergoes elastic deformation and rigid rotation (see Figure 7.19).

The total deformation gradient ($\mathbf{F}$) is decomposed into the elastic ($\mathbf{F}^*$) and plastic components ($\mathbf{F}^p$) as

$$\mathbf{F} = \mathbf{F}^* \mathbf{F}^p \tag{7.81}$$

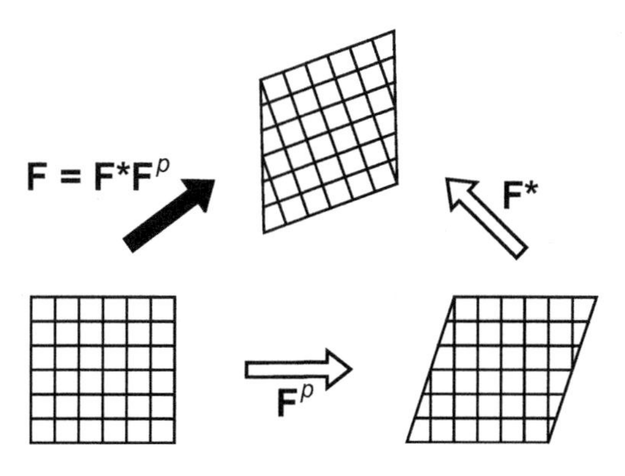

FIGURE 7.19

A schematic description of the multiplicative decomposition of total deformation in single crystals into elastic and plastic components.

$\mathbf{F}^*$ contains deformation gradients due to elastic stretching as well as lattice rotation. It is important to note that the deformations shown in Figure 7.19 represent the total deformation imposed on a crystalline region starting from an initial stress-free condition, and not the incremental (or relative) deformations. The intermediate relaxed configuration shown in Figure 7.19 is a hypothetical configuration, in that the material will not recover to this configuration on unloading from the loaded configuration.

The constitutive equation for elastic deformations in the crystal can be expressed as

$$\mathbf{T}^* = C[\mathbf{E}^*] \tag{7.82}$$

where $\mathbf{C}$ is the fourth-order elasticity tensor, and $\mathbf{T}^*$ and $\mathbf{E}^*$ are a pair of work conjugate stress–strain measures defined using the elastic deformation gradient tensor as

$$\mathbf{T}^* = \mathbf{F}^{*-1}\{(\det \mathbf{F}^*)\mathbf{T}\}\mathbf{F}^{*-T}, \qquad \mathbf{E}^* = \frac{1}{2}\{\mathbf{F}^{*T}\mathbf{F}^* - I\} \tag{7.83}$$

The stress in the crystal for a particular applied deformation can be computed using Eqs. (7.81) through (7.83), provided $\mathbf{F}^p$ is known. The evolution of $\mathbf{F}^p$ (called the *flow-rule*) is path-dependent and is therefore expressed in a rate form:

$$\dot{\mathbf{F}}^p = \mathbf{L}^p\,\mathbf{F}^p, \quad \mathbf{L}^p = \sum_\alpha \dot{\gamma}^\alpha \mathbf{S}_o^\alpha, \qquad \mathbf{S}_o^\alpha = \mathbf{m}_o^\alpha \otimes \mathbf{n}_o^\alpha \tag{7.84}$$

where the subscripts o have been used on $\mathbf{m}$ and $\mathbf{n}$ to indicate that these directions are taken from the stress-free initial configuration and remain constant during deformation. The same power law slip that was described earlier in Eq. (7.77) can be used here with the resolved shear stress (τ^α) defined as

$$\tau^\alpha = \left(\mathbf{F}^{*T}\mathbf{F}\right)\mathbf{T}^* \cdot \mathbf{S}_o^\alpha \approx \mathbf{T}^* \cdot \mathbf{S}_o^\alpha \tag{7.85}$$

Further details of this model and approaches for efficient numerical integration of these equations to simulate large deformation processes (along with predictions of the anisotropic stress–strain curves and texture evolution) can be found in Kalidindi et al. (1992). These details are beyond the scope of this book, and the interested reader is encouraged to follow the papers to gain a better understanding of this modeling framework.

7.6.3 Slip hardening

We now shift our attention to another important aspect of metal plasticity (i.e., *strain hardening*). Cold-working is an important step in all wrought alloy manufacturing processes. In fact, the ability to easily strengthen metals by cold-working is one of the main reasons for their widespread use in a majority of applications requiring load-bearing capacity. The main mechanism of strengthening during cold-work is dislocation production through various sources (e.g., the Frank–Reed source). Dislocation theory predicts that, when a dislocation is pinned by obstacles, the stress required to make the dislocation loop traverse the plane (thereby producing plastic strain) is inversely proportional to the pinning distance between the obstacles. In a cold-worked material, dislocations themselves serve as pinning obstacles to dislocations moving on other planes. The average pinning distance of the dislocations is inversely proportional to the square root of the dislocation density. Therefore, as the metal is cold-worked, the dislocation density increases dramatically, which in turn increases the stress required to operate a Frank–Reed source and produce further plastic deformation.

In the framework of the crystal plasticity model described in the previous section, an important consequence of plastic deformation in metals at low homologous temperatures is that the slip resistance (s^α) evolves with slip activity in the crystal. In cubic metals, slip hardening can be captured phenomenologically using the following simple expression:

$$\dot{s}^\alpha = \sum_\beta h^{\alpha\beta} |\dot{\gamma}^\beta| \tag{7.86}$$

where $h^{\alpha\beta}$ describes the hardening of slip system α due to slip activity on slip system β. In general, one could decompose $h^{\alpha\beta}$ into two components: (1) a self-hardening component, h^α, and (2) a latent hardening component, $q^{\alpha\beta}$.

The self-hardening component may be conveniently described phenomenologically by a saturation-type hardening law, while the latent hardening component may be described by a set of coefficients that describe the influence of slip activity on one system on the hardening of the other slip systems. While a number of descriptions have been proposed in the literature for latent hardening in cubic crystals (Kocks, 1975; Bassani and Wu, 1991), there continue to be several gaps in our understanding of the slip hardening phenomena. In this book, we employ the following highly simplified description:

$$\dot{s}^\alpha = h^o \left(1 - \frac{s^\alpha}{s^s}\right)^a \sum_\beta |\dot{\gamma}^\beta| \tag{7.87}$$

where h^o, s^s, and a denote hardening parameters. Note also that Eq. (7.87) implicitly assumes that slip activity on one slip system contributes equally to hardening of all slip systems in the crystal.

7.6.4 Fourier representation of crystal plasticity

Various crystal plasticity relations can be expressed in the Fourier domain by selecting an appropriate domain for the local state variables on which these relations depend. Two complications typically arise, however. The first is that many of these relations are nonlinear. For example, relations (7.79) and (7.80) involve tensors **M** and **N**, which are functions not only of the lattice orientation parameters but also of either the stress tensor or the strain tensor. The second complication is that in some interesting cases the input slip resistance, s^α, could be considered to be a function of grain size, l. For example, in materials that exhibit a strong grain size dependence on the yield strength, it may be appropriate to include local grain size in the list of pertinent local state variables. And it must be anticipated that other models for plastic behavior may include still other local state variables—variables that are not discussed in the present work. For all of these cases, however, the procedure for discrete Fourier representation of crystal properties follows the pattern illustrated in Section 7.5.2, making appropriate adjustments to the pertinent local state space and covering that space with an appropriate discrete set of bins.

The typical approach to arriving at a discrete Fourier representation of crystal plasticity relations involves assuming a standard form for the representation and then fitting Fourier coefficients to match results arising from an appropriate phenomenological model. As an example, the Fourier representation that describes the functional dependence of the local stresses in the constituent crystals on their orientations for a hexagonal material is presented in Wu (2007) (no summation implied on repeated index i):

$$\sigma'_{ii}(g,q) = \sum_{l=0}^{l^*} \sum_{\mu=1}^{N(l)} \sum_{\nu=1}^{M(l)} {}^{ii}_{y1}S_l^{\mu\nu}(q) T_l^{\mu\nu}(g) \tag{7.88}$$

where the Fourier coefficients ${}^{ii}_{y1}S_l^{\mu\nu}(q)$ for hcp crystals experiencing slip on prism, basal, and pyramidal slip systems are calculated via a curve-fitting method to detailed crystal plasticity calculations over a range of possible strain paths (Wu, 2007).

In describing the plastic properties we needed a lot more Fourier coefficients than those needed to describe elastic properties. For practical reasons, the series is truncated after a desirable accuracy has been attained, and this is reflected in Eq. (7.88) by restricting the series to terms corresponding to $l \leq l^*$. For example, for hexagonal–orthorhombic textures, it was observed that 57 terms were required for reasonable accuracy (corresponding to $l^* = 12$). The desired accuracy in property predictions therefore dictates the number of Fourier dimensions that need to be explored by the MSD methodology for a selected design application.

7.7 MACROSCALE PLASTICITY

The crystal plasticity theory described in the previous section, while incorporating several of the details of the physics and the geometry of the deformation at the crystal scale, is substantially difficult to implement in the design of engineering components. The design and simulation tools currently used by most designers typically employ grossly simplified phenomenological descriptions of plasticity at the macroscale (i.e., polycrystal) level. This formalism is briefly introduced here and is used later in linking the polycrystal constitutive response to the popular formalisms for plastic behavior of metals used in design codes.

In design, considerable effort is often focused on avoiding plastic deformation as it leads to inelastic deformation and failure over a period of time. Design and simulation tools often incorporate a yield criterion that indicates the onset of plastic deformation in the component. The yield criterion is most generally described as a function of the local stress components as

$$f(\sigma_{ij}) = f(\sigma_{11}, \sigma_{22}, \sigma_{33}, \sigma_{23}, \sigma_{13}, \sigma_{12}) = k \tag{7.89}$$

where $f(\sigma_{ij})$ denotes the yield function and k denotes a material property.

If the yield function is assumed to be isotropic, then one can compact the description of the yield function as

$$\text{Isotropy} \Rightarrow f(\sigma_1, \sigma_2, \sigma_3) = k \quad \text{or} \quad f(I_1, I_2, I_3) = k \tag{7.90}$$

where σ_i are the principal stresses and I_i are the principal invariants of the stress tensor. Furthermore, if plastic yield is assumed to be pressure-independent, the description of the yield function can be further simplified as

$$\text{Pressure} - \text{independence} \Rightarrow f(J_2, J_3) = k \tag{7.91}$$

where J_i are the principal invariants of the deviatoric stress tensor. It is a common practice in engineering design to ignore the dependence of plastic yield on J_3 in fully dense metals. The von Mises yield function is the most common example of yield descriptions used in the design of metallic structural components, and is available as a standard option in virtually all simulation packages used by the designers. The von Mises yield function can be expressed as

$$\bar{\sigma} = \sqrt{3J_2} = \sigma_y = \left[\frac{1}{2}\left\{(\sigma_{11} - \sigma_{22})^2 + (\sigma_{11} - \sigma_{33})^2 + (\sigma_{22} - \sigma_{33})^2\right\} + 3\left(\sigma_{12}^2 + \sigma_{13}^2 + \sigma_{23}^2\right)\right]^{1/2} \tag{7.92}$$

where $\bar{\sigma}$ is called the von Mises stress or the equivalent stress, and σ_y is the isotropic yield strength of the material in simple tension (and in simple compression).

Development of robust and reliable anisotropic yield functions for polycrystalline metals is currently an active area of research. The most commonly available formalism in currently used design codes is the anisotropic Hill's yield criterion. For a material that exhibits orthotropic symmetry in its macroscale yield response, the anisotropic Hill's yield criterion can be expressed as

$$\begin{aligned} &\frac{1}{2}\left(\frac{1}{\sigma_{y2}^2} + \frac{1}{\sigma_{y3}^2} - \frac{1}{\sigma_{y1}^2}\right)(\sigma_{22} - \sigma_{33})^2 + \frac{1}{2}\left(\frac{1}{\sigma_{y3}^2} + \frac{1}{\sigma_{y1}^2} - \frac{1}{\sigma_{y2}^2}\right)(\sigma_{33} - \sigma_{11})^2 \\ &+ \frac{1}{2}\left(\frac{1}{\sigma_{y1}^2} + \frac{1}{\sigma_{y2}^2} - \frac{1}{\sigma_{y3}^2}\right)(\sigma_{11} - \sigma_{22})^2 + \frac{1}{\tau_{y12}^2}\tau_{12}^2 + \frac{1}{\tau_{y23}^2}\tau_{23}^2 + \frac{1}{\tau_{y13}^2}\tau_{13}^2 = 1 \end{aligned} \tag{7.93}$$

where σ_{y1}, σ_{y2}, σ_{y3} describe the tensile (or compressive) yield strengths along the 1-, 2-, and 3-axes, respectively, and τ_{y12}, τ_{y13}, τ_{y23} describe the shear yield strengths in the 1-2, 1-3, and 2-3 planes, respectively. Note that Eq. (7.93) embodies pressure independence.

SUMMARY

In this chapter readers were presented with the basic continuum mechanics descriptions for the constitutive behavior of materials. In particular, we reviewed all of the important concepts related to the descriptions of conductivity, thermo-elasticity, and plasticity in materials. We also explored the important consequences of symmetry in the description of these constitutive responses. Most important, we introduced the idea of spectral representations for describing the structure–property relationships of interest in MSDPO. It is hoped that readers appreciate the tremendous value of the spectral representations in providing computationally efficient and compact representations of the desired relationships. It will be seen in later chapters that these spectral relationships are central to MSDPO.

Exercises

Section 7.2.1

1. Evaluate the hydrostatic and deviatoric components for the stress state corresponding to simple shear. Find the principal stress components and principal stress directions corresponding to the simple shear stress state.

2. Prove that any pure deviatoric stress state can be expressed as a sum of two simple shear stress states. Specify the planes on which these simple shears are to be prescribed.

3. Consider a body subjected to uniaxial stress in the $\mathbf{e}_1$ direction. Compute the resolved shear stress in the direction **m** on a plane with a normal that is given by **n.** The ratio of the resolved shear stress to the applied uniaxial stress is called the Schmid factor. Prove that the highest possible value for the Schmid factor for the uniaxial stress state is 0.5.

Section 7.3.1

4. Show that for simple shear γ on a plane of which the normal is **n** with a shear direction **m**, the deformation gradient tensor can be expressed as $\mathbf{F} = \mathbf{I} + \gamma\, \mathbf{m} \otimes \mathbf{n}$.
5. Prove Eqs. (7.27) and (7.28).
6. How could Nansen's formulae be used in conjunction with Bragg's law of diffraction to determine the elastic strain present in the crystal lattice? Describe the equations you would use and any limitations that are evident.
7. A traction-free surface of normal **n** must satisfy the following condition in the near vicinity of the surface: $\sigma n = 0$. Return to Exercise 7.6, and reconsider the determination of the near-surface elastic strain, this time including the effects of a traction-free surface.

Section 7.3.2

8. Hencky strain measure is defined using the function $f(\lambda_i) = \ln(\lambda_i)$. Prove that this function satisfies the conditions stipulated in Eqs. (7.34) through (7.36). Note that this definition corresponds to the familiar definition of true or logarithmic strain.
9. Prove that for isochoric (constant-volume) small strains, $tr(\varepsilon) = 0$.
10. Prove that for general isochoric motions, $tr(\mathbf{L}) = 0$.
11. Evaluate **L**, **D**, and **W** for simple shear deformation defined in Exercise 7.1.
12. Prove the following identities:

$$\mathbf{W} = \dot{\mathbf{R}}\mathbf{R}^T + \frac{1}{2}\mathbf{R}\left(\dot{\mathbf{U}}\mathbf{U}^{-1} - \left(\dot{\mathbf{U}}\mathbf{U}^{-1}\right)^T\right)\mathbf{R}^T$$

$$\mathbf{D} = \frac{1}{2}\mathbf{R}\left(\dot{\mathbf{U}}\mathbf{U}^{-1} + \left(\dot{\mathbf{U}}\mathbf{U}^{-1}\right)^T\right)\mathbf{R}^T$$

13. Show that a pure spin of $\dot{\theta}$ about an axis **n** can be represented as

$$\mathbf{R} = \mathbf{n}\otimes\mathbf{n} + (\mathbf{I} - \mathbf{n}\otimes\mathbf{n})\cos\theta + \mathbf{N}\sin\theta$$
$$\mathbf{W} = \mathbf{N}\dot{\theta}$$

where **n** is the dual vector of the anti-symmetric tensor **N**.

14. Show that a pure spin of $\dot{\theta}$ about an axis **n** can be represented as

$$\mathbf{R} = \mathbf{n}\otimes\mathbf{n} + (\mathbf{I} - \mathbf{n}\otimes\mathbf{n})\cos\theta + \mathbf{N}\sin\theta$$
$$\mathbf{W} = \mathbf{N}\dot{\theta}$$

where **n** is the dual vector of the anti-symmetric tensor **N**.

CHAPTER

Homogenization Theories

8

CHAPTER OUTLINE

In Chapter 7, we focused on the inherently anisotropic constitutive behavior of a single crystal, the basic building block of a polycrystalline material. In this chapter, we turn our attention to the effective properties exhibited by polycrystals, which are treated here essentially as composites of single crystals. The word "effective" is referring to the behavior exhibited by the polycrystal at the length scale of the representative volume element (RVE) of the polycrystal. At this higher-length scale, it is assumed that the polycrystal can be replaced by a homogenized material that exhibits the effective behavior identified for the polycrystal. The theories that predict the effective properties of a composite material are generically known as composite theories or homogenization theories.

8.1 INTRODUCTION

Let us first establish some of the underlying assumptions in the class of *homogenization theories* to be discussed here. It has been assumed that the composite microstructures of interest have at least two different length scales that differ from each other by several orders of magnitude. At the lower-length scale, a material point in the microstructure is assumed to be significantly larger (by several orders of magnitude) than the atomic length scales, and it is assumed that the continuum descriptions of the constitutive behavior and the field equations presented in Chapter 7 can be applied locally at this lower-length scale. In a polycrystal, this length scale will be of the order of the grain size, and is often referred to as the *mesoscale*. As one moves from one grain to another the local state (or structure) of the material fluctuates, as do the resultant properties. However, at the higher-length scale of the representative volume element (which may be orders of magnitude above the mesoscale) these fluctuations in properties generally average out, resulting in smoothly varying (or constant) effective properties.

The nature of the variations in these effective properties depends on the assumptions that are made about the statistical nature of the structure. The most common assumptions include periodicity, quasiperiodicity, and statistical homogeneity. In this book, we focus only on statistically homogeneous materials. For the first-order statistical measures employed in the first part of this book, this means that any RVE chosen from a material sample will have the same volume fraction of all of the different mesoscale constituents to some tolerance. The effective properties of the material are therefore constant throughout the material.

Averaging methods underlie a variety of homogenization methods. Using continuum mechanics and variational principles, the response of the material at the RVE level may be estimated, or bounded, in terms of global boundary assumptions, thus leading to computation of macroscopic properties. The relationships between local and global responses often rely on the fact that for many properties of composites the macroscopic, or effective, values of the property obey the same field equation as the local (microscale) values. There is a vast literature in this general area based on pioneering work by Hill (1952), Kröner (1977), Hashin and Shtrikman (1963), Willis (1981), Suquet (1987), and others. These homogenization methods might include the rule of mixtures (Voigt, 1910), the inverse rule of mixtures (Reuss, 1929), the Mori–Takana method (Mori and Tanaka, 1973), Kerner's method (Kerner, 1956), the dilute distribution model (Eshelby, 1957), the self-consistent model (Hill, 1965), and others. Excellent treatment of the overall subject may be found in Nemat-Nasser and Hori (1999), Milton (2002), and various other books and articles. The general framework resulting from this approach may be applied to a range of material responses with the same basic equations; these include fluid flow, diffusion, electromagnetism, and elasticity (Milton, 2002).

Examples of homogenization processes are replete in the engineering world. If an engineer wishes to choose a steel bolt that will provide desired rigidity while supporting a given load, she or he will generally be presented with a single material property (such as Young's modulus) for all bolts arising from a certain manufacturing route. This is in spite of the fact that the engineer is aware that at the mesoscale the stiffness tensor is more complex and varies significantly more than the assumed effective isotropic properties. Not only would it be impossible for the manufacturer to give mesoscale data for each bolt, but it would be extremely burdensome for the engineer to interpret the data with respect to the given engineering problem.

To define the concept of effective properties more precisely, we hypothesize that the continuum theories developed at the mesoscale are also correct at the RVE scale. For example, in the theory of elasticity we require that an effective stiffness tensor satisfy

$$\overline{\boldsymbol{\sigma}} = \mathbf{C}^{eff}\,\overline{\boldsymbol{\varepsilon}} \tag{8.1}$$

at the RVE scale, where the volume average values of stress and strain are used as the homogenized quantities:

$$\overline{\boldsymbol{\sigma}} = \frac{1}{V(\Omega)} \int_{\Omega} \boldsymbol{\sigma}(\mathbf{x})\, d\mathbf{x}, \quad \overline{\boldsymbol{\varepsilon}} = \frac{1}{V(\Omega)} \int_{\Omega} \boldsymbol{\varepsilon}(\mathbf{x})\, d\mathbf{x} \tag{8.2}$$

Although this form of the homogenization theory is hypothesized here, we will later show that it is indeed rigorous for elasticity. Note that the volume-averaged quantities are independent of location of the RVE in the microstructure. This arises as a consequence of the assumption of statistical homogeneity in the microstructure.

Readers should note that, in general, $\mathbf{C}^{eff} \neq \overline{\mathbf{C}}$. In fact, taking the average stiffness as the effective stiffness is equivalent to assuming the commonly used rule of mixtures. An alternative homogenization method that is often used is the inverse rule of mixtures. For most material structures neither of these approaches yields an accurate effective property tensor, but in certain cases they are good estimates.

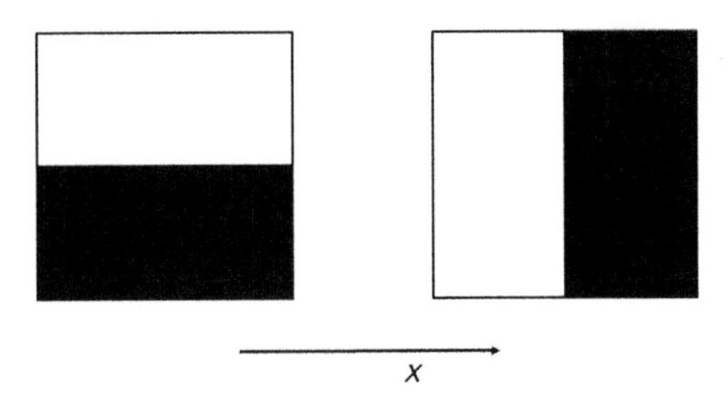

FIGURE 8.1

Schematic of a material sample of two materials aligned in different directions, resulting in an effective modulus in the x-direction equal to the arithmetic and harmonic averages of the individual material modulae, respectively.

A simple example of such cases may be demonstrated by forming a hypothetical material sample from two different isotropic materials. We will calculate the effective Young's modulus in the x-direction for samples made up of 50% each of the two isotropic materials with individual Young's modulae of E_1 and E_2, respectively. The two cases are shown in Figure 8.1. In both cases we are looking for $E_x^{eff} = \overline{\sigma}_x / \overline{\varepsilon}_x$.

If the left edges of the samples are fixed, and a displacement is applied uniformly to the right edges, in the first example the strain is equal in both materials, $\varepsilon_1 = \varepsilon_2 = \overline{\varepsilon}$. Hence, using Hook's law, $\sigma_1 = E_1 \varepsilon_1$, and so forth:

$$E_x^{eff} = \frac{\overline{\sigma}_x}{\overline{\varepsilon}_x} = \frac{0.5(\sigma_1 + \sigma_2)}{0.5(\varepsilon_1 + \varepsilon_2)} = \frac{E_1 \overline{\varepsilon} + E_2 \overline{\varepsilon}}{\overline{\varepsilon} + \overline{\varepsilon}} = \frac{E_1 + E_2}{2} = \overline{E} \tag{8.3}$$

In the second case, the stress is equal in both materials; hence

$$E_x^{eff} = \frac{\overline{\sigma}_x}{\overline{\varepsilon}_x} = \frac{0.5(\sigma_1 + \sigma_2)}{0.5(\varepsilon_1 + \varepsilon_2)} = \frac{\overline{\sigma} + \overline{\sigma}}{\overline{\sigma}/E_1 + \overline{\sigma}/E_2} = \frac{1}{1/(2E_1) + 1/(2E_2)} \tag{8.4}$$

In fact, as we shall see in the next section, the rule of mixtures and the inverse rule of mixtures sometimes determine the maximum and minimum values an effective property can take for a fixed volume fraction of the constituents. This ability to place bounds on an effective property is extremely useful when it may not be possible to determine an accurate value, and it will be discussed in more detail in the next section.

8.2 FIRST-ORDER BOUNDS FOR ELASTICITY

As discussed in Chapter 4, the most basic, first-order, statistical information that can be extracted from a material is its one-point statistics (i.e., the volume fractions of the distinct local states in the microstructure); bounds derived using only this information on the microstructure are referred to as the *first-order bounds*. We start with a derivation of the rigorous first-order bounds for elasticity.

A representative volume element of a composite material system is shown in Figure 8.2. The partial differential equations that control the various field quantities in this volume can be expressed for elasticity as

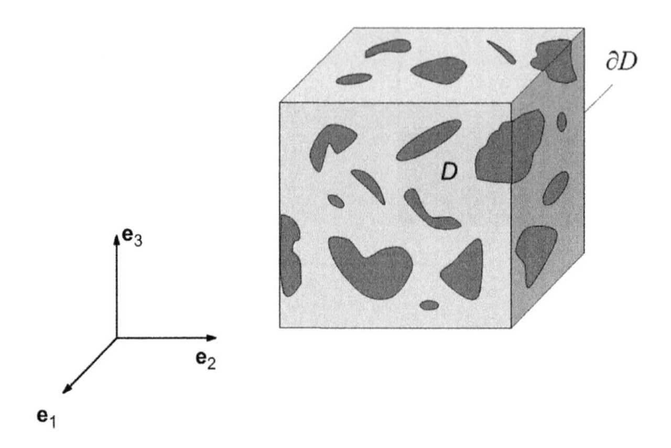

FIGURE 8.2

Schematic description of the internal structure of a composite material. The dark and bright regions represent two different phases.

$$\begin{aligned} \boldsymbol{\sigma}(\mathbf{x}) &= \mathbf{C}(\mathbf{x})\,\boldsymbol{\varepsilon}(\mathbf{x}), \quad \mathbf{C} = \mathbf{C}^T \\ \mathbf{C}\boldsymbol{\varepsilon}\cdot\boldsymbol{\varepsilon} &> 0 \quad \forall\ |\boldsymbol{\varepsilon}| \neq 0 \\ \nabla\cdot\boldsymbol{\sigma} &= 0, \quad \nabla\times\boldsymbol{\varepsilon} = 0 \end{aligned} \tag{8.5}$$

We wish to homogenize these differential equations. The homogenization problem is stated as finding $\mathbf{C}^{eff}$ such that the effective constitutive relation at the higher-length scale can be expressed as (see Exercise 8.2)

$$\overline{\boldsymbol{\sigma}} = \mathbf{C}^{eff}\overline{\boldsymbol{\varepsilon}}, \quad \overline{\boldsymbol{\sigma}\cdot\boldsymbol{\varepsilon}} = \overline{\mathbf{C}\boldsymbol{\varepsilon}\cdot\boldsymbol{\varepsilon}} = \overline{\boldsymbol{\sigma}}\cdot\overline{\boldsymbol{\varepsilon}} = \mathbf{C}^{eff}\overline{\boldsymbol{\varepsilon}}\cdot\overline{\boldsymbol{\varepsilon}} > 0 \quad \forall|\overline{\boldsymbol{\varepsilon}}| \neq 0 \tag{8.6}$$

where the bar over a field quantity denotes its volume-averaged value. To establish the equivalence $\overline{\boldsymbol{\sigma}\cdot\boldsymbol{\varepsilon}} = \overline{\boldsymbol{\sigma}}\cdot\overline{\boldsymbol{\varepsilon}}$, we must first establish a result for the average elastic energy in the volume.

We wish to prove that the volume-averaged elastic stored energy in a composite material can be expressed as $\overline{\boldsymbol{\varepsilon}}\cdot\mathbf{C}^{eff}\overline{\boldsymbol{\varepsilon}}$. We begin by expressing the local stress and strain fields as a sum of the volume-averaged quantities and local fluctuations:

$$\boldsymbol{\sigma}(\mathbf{x}) = \overline{\boldsymbol{\sigma}} + \boldsymbol{\sigma}'(\mathbf{x}), \qquad \boldsymbol{\varepsilon}(\mathbf{x}) = \overline{\boldsymbol{\varepsilon}} + \boldsymbol{\varepsilon}'(\mathbf{x}) \tag{8.7}$$

The volume-averaged elastic stored energy is given as (see also Eq. 7.52)

$$\overline{W} = \frac{1}{2}\overline{C_{ijkl}\varepsilon_{ij}\varepsilon_{kl}} = \frac{1}{2}\overline{S_{ijkl}\sigma_{ij}\sigma_{kl}} = \frac{1}{2}\overline{\sigma_{ij}\varepsilon_{ij}} \tag{8.8}$$

Substituting Eq. (8.7) yields

$$\overline{W} = \frac{1}{2}\overline{(\overline{\sigma}_{ij} + \sigma'_{ij})(\overline{\varepsilon}_{ij} + \varepsilon'_{ij})} = \frac{1}{2}\overline{(\overline{\sigma}_{ij}\overline{\varepsilon}_{ij} + \sigma'_{ij}\overline{\varepsilon}_{ij} + \overline{\sigma}_{ij}\varepsilon'_{ij} + \sigma'_{ij}\varepsilon'_{ij})} \tag{8.9}$$

Noting that $\overline{\sigma'_{ij}} = \overline{\varepsilon'_{ij}} = 0$ and that $\overline{\sigma}_{ij}$ and $\overline{\varepsilon}_{ij}$ can be treated as constants in the volume averaging, we obtain

$$\overline{W} = \frac{1}{2}\left(\overline{\sigma}_{ij}\overline{\varepsilon}_{ij} + \overline{\sigma'_{ij}\varepsilon'_{ij}}\right) \tag{8.10}$$

Focusing on the second term in expression (8.10) and using the definition of strain tensor, the symmetry of the stress tensor, the equilibrium equation, and the Gauss theorem (Eq. 2.72) allows us to obtain the following expression:

$$\overline{\sigma'_{ij}\varepsilon'_{ij}} = \overline{\sigma'_{ij}u'_{i,j}} = \overline{(\sigma'_{ij}u'_i)_{,j} - \sigma'_{ij,j}u'_i} = \overline{(\sigma'_{ij}u'_i)_{,j}} = \frac{1}{V}\int_{\partial D} \sigma'_{ij}u'_i n_j dS = \frac{1}{V}\int_{\partial D} t'_i u'_i dS \tag{8.11}$$

where ∂D denotes the boundary of the RVE of the composite microstructure being studied. In homogenization theories, it is customary to choose boundary conditions on the representative volume element such that either $\mathbf{t}' = 0$ or $\mathbf{u}' = 0$ (depending on whether you are applying a stress tensor or a strain tensor to the RVE). Therefore, $\overline{\sigma'_{ij}\varepsilon'_{ij}} = 0$ and

$$\overline{W} = \frac{1}{2}\overline{\sigma}_{ij}\overline{\varepsilon}_{ij} = \frac{1}{2}\overline{\sigma}_{ij}S^{eff}_{ijkl}\overline{\sigma}_{kl} = \frac{1}{2}\overline{\varepsilon}_{ij}C^{eff}_{ijkl}\overline{\varepsilon}_{kl} \tag{8.12}$$

The final result for use is obtained by restating the goal and the result:

$$\overline{W} = \frac{1}{2}\overline{\varepsilon_{ij}C_{ijkl}\varepsilon_{kl}} = \frac{1}{2}\overline{\varepsilon}_{ij}C^{eff}_{ijkl}\overline{\varepsilon}_{kl} \tag{8.13}$$

Now consider the application of Eq. (8.5) to the perturbation strain field $\overline{\boldsymbol{\varepsilon}}$. The symmetric positive definiteness of $\mathbf{C}$ requires that

$$\overline{\mathbf{C}\boldsymbol{\varepsilon}\cdot\boldsymbol{\varepsilon}} = \overline{\mathbf{C}(\boldsymbol{\varepsilon}-\overline{\boldsymbol{\varepsilon}})\cdot(\boldsymbol{\varepsilon}-\overline{\boldsymbol{\varepsilon}})} \geq 0 \tag{8.14}$$

Multiplying out the brackets, taking the averages represented by the overbars, and noting that constants such as $\overline{\boldsymbol{\varepsilon}}$ can be taken outside of the averaging process, one arrives at

$$\begin{aligned}\overline{\mathbf{C}(\boldsymbol{\varepsilon}-\overline{\boldsymbol{\varepsilon}})\cdot(\boldsymbol{\varepsilon}-\overline{\boldsymbol{\varepsilon}})} &= \overline{\mathbf{C}\boldsymbol{\varepsilon}\cdot\boldsymbol{\varepsilon}} - \overline{\mathbf{C}\boldsymbol{\varepsilon}\cdot\overline{\boldsymbol{\varepsilon}}} - \overline{\mathbf{C}\overline{\boldsymbol{\varepsilon}}\cdot\boldsymbol{\varepsilon}} + \overline{\mathbf{C}\overline{\boldsymbol{\varepsilon}}\cdot\overline{\boldsymbol{\varepsilon}}} \\ &= \overline{\mathbf{C}\boldsymbol{\varepsilon}}\cdot\overline{\boldsymbol{\varepsilon}} - \overline{\mathbf{C}\boldsymbol{\varepsilon}}\cdot\overline{\boldsymbol{\varepsilon}} - \overline{\mathbf{C}\boldsymbol{\varepsilon}}\cdot\overline{\boldsymbol{\varepsilon}} + \overline{\mathbf{C}}\overline{\boldsymbol{\varepsilon}}\cdot\overline{\boldsymbol{\varepsilon}} = \overline{\mathbf{C}}\overline{\boldsymbol{\varepsilon}}\cdot\overline{\boldsymbol{\varepsilon}} - \overline{\mathbf{C}\boldsymbol{\varepsilon}}\cdot\overline{\boldsymbol{\varepsilon}} \geq 0\end{aligned} \tag{8.15}$$

where the first term in the expansion is obtained from the result in Eq. (8.6) ($\overline{\boldsymbol{\sigma}\cdot\boldsymbol{\varepsilon}} = \overline{\boldsymbol{\sigma}}\cdot\overline{\boldsymbol{\varepsilon}}$), and the third term is obtained by noting that $\overline{\mathbf{C}\overline{\boldsymbol{\varepsilon}}\cdot\boldsymbol{\varepsilon}} = \overline{\mathbf{C}^T\boldsymbol{\varepsilon}}\cdot\overline{\boldsymbol{\varepsilon}} = \overline{\mathbf{C}\boldsymbol{\varepsilon}}\cdot\overline{\boldsymbol{\varepsilon}}$ since $\mathbf{C}^T = \mathbf{C}$; this is possibly easier to see if it is expanded in indicial form. By employing the result of Eqs. (8.6) and (8.13), we may arrive at the following rigorous bound:

$$\begin{aligned}&\overline{\mathbf{C}}\overline{\boldsymbol{\varepsilon}}\cdot\overline{\boldsymbol{\varepsilon}} - \overline{\mathbf{C}\boldsymbol{\varepsilon}}\cdot\overline{\boldsymbol{\varepsilon}} \geq 0 \Rightarrow \overline{\mathbf{C}}\overline{\boldsymbol{\varepsilon}}\cdot\overline{\boldsymbol{\varepsilon}} - \mathbf{C}^{eff}\overline{\boldsymbol{\varepsilon}}\cdot\overline{\boldsymbol{\varepsilon}} \geq 0 \\ &\Rightarrow \overline{\mathbf{C}}\overline{\boldsymbol{\varepsilon}}\cdot\overline{\boldsymbol{\varepsilon}} \geq \mathbf{C}^{eff}\overline{\boldsymbol{\varepsilon}}\cdot\overline{\boldsymbol{\varepsilon}} \quad \forall\overline{\boldsymbol{\varepsilon}}\end{aligned} \tag{8.16}$$

A similar treatment starting with $\boldsymbol{\varepsilon}(\mathbf{x}) = \mathbf{C}^{-1}(\mathbf{x})\boldsymbol{\sigma}(\mathbf{x}) = \mathbf{S}(\mathbf{x})\boldsymbol{\sigma}(\mathbf{x})$ at Eq. (8.5), where $\mathbf{S}$ denotes the inverse of the stiffness (i.e., the compliance tensor), results in another bound:

$$\mathbf{S}^{eff}\overline{\boldsymbol{\sigma}}\cdot\overline{\boldsymbol{\sigma}} \leq \overline{\mathbf{S}}\overline{\boldsymbol{\sigma}}\cdot\overline{\boldsymbol{\sigma}} \ \ \forall\overline{\boldsymbol{\sigma}} \tag{8.17}$$

Recalling that $\overline{\mathbf{S}}$ and $\mathbf{S}^{eff}$ tensors are both symmetric and positive definite, we can assume that the square-roots of these tensors exist and that they are also symmetric. Therefore, inequality (8.17) can be transformed as

$$\mathbf{S}^{eff}\boldsymbol{\sigma}\cdot\boldsymbol{\sigma} \leq \overline{\mathbf{S}}\boldsymbol{\sigma}\cdot\boldsymbol{\sigma} = \overline{\mathbf{S}}^{1/2}\overline{\mathbf{S}}^{1/2}\boldsymbol{\sigma}\cdot\boldsymbol{\sigma} \Rightarrow \overline{\mathbf{S}}^{-1/2}\mathbf{S}^{eff}\overline{\mathbf{S}}^{-1/2}\boldsymbol{\sigma}\cdot\boldsymbol{\sigma} \leq \mathbf{I}\boldsymbol{\sigma}\cdot\boldsymbol{\sigma} \tag{8.18}$$

Since this is true for any $\boldsymbol{\sigma}$, it is clear from inequality (8.18) that the eigenvalues of $\overline{\mathbf{S}}^{-1/2}\mathbf{S}^{eff}\overline{\mathbf{S}}^{-1/2}$ are numerically less than or equal to 1 (conversely, interpreting the equation in matrix terms, which will be more familiar to the student, if $\overline{\mathbf{S}}^{-1/2}\mathbf{S}^{eff}\overline{\mathbf{S}}^{-1/2}$ were a matrix with an eigenvalue greater than 1, then one could simply choose $\boldsymbol{\sigma}$ to be the eigenvector of that eigenvalue for inequality (8.18) to be false). Therefore, the following inequality can be inferred:

$$\begin{aligned}(\overline{\mathbf{S}}^{-1/2}\mathbf{S}^{eff}\overline{\mathbf{S}}^{-1/2})^{-1}\boldsymbol{\sigma}\cdot\boldsymbol{\sigma} \geq \mathbf{I}\boldsymbol{\sigma}\cdot\boldsymbol{\sigma} &\Rightarrow \overline{\mathbf{S}}^{1/2}\mathbf{S}^{eff^{-1}}\overline{\mathbf{S}}^{1/2}\boldsymbol{\sigma}\cdot\boldsymbol{\sigma} \geq \mathbf{I}\boldsymbol{\sigma}\cdot\boldsymbol{\sigma} \\ &\Rightarrow \mathbf{S}^{eff^{-1}}\boldsymbol{\sigma}\cdot\boldsymbol{\sigma} \geq \overline{\mathbf{S}}^{-1}\boldsymbol{\sigma}\cdot\boldsymbol{\sigma}\end{aligned} \tag{8.19}$$

Recalling that $\mathbf{S}^{eff^{-1}} = \mathbf{C}^{eff}$ (Eq. 7.53), inequality (8.19) can be recast as

$$\overline{\boldsymbol{\varepsilon}}\cdot\mathbf{C}^{eff}\overline{\boldsymbol{\varepsilon}} \geq \overline{\boldsymbol{\varepsilon}}\cdot\overline{\mathbf{S}}^{-1}\overline{\boldsymbol{\varepsilon}} \tag{8.20}$$

Finally, adding the other bound for elastic stiffness from Eq. (8.16), we can express rigorous bounds for elastic stiffness (Hill, 1952; Paul, 1960; Milton, 2002) as

$$\overline{\varepsilon}_{ij}\overline{S}_{ijkl}^{-1}\overline{\varepsilon}_{kl} \leq \overline{\varepsilon}_{ij}C_{ijkl}^{eff}\overline{\varepsilon}_{kl} \leq \overline{\varepsilon}_{ij}\overline{C}_{ijkl}\overline{\varepsilon}_{kl} \tag{8.21}$$

where $\overline{S}_{ijkl}^{-1}$ denote the components of the inverse of the averaged elastic compliance tensor (not the reciprocal of the components of the averaged elastic compliance tensor). The bounds in Eq. (8.21) can also be expressed using the compliance tensor as

$$\overline{\sigma}_{ij}\overline{C}_{ijkl}^{-1}\overline{\sigma}_{kl} \leq \overline{\sigma}_{ij}S_{ijkl}^{eff}\overline{\sigma}_{kl} \leq \overline{\sigma}_{ij}\overline{S}_{ijkl}\overline{\sigma}_{kl} \tag{8.22}$$

Note that the bounds in Eqs. (8.21) and (8.22) are expressed in terms of the stored elastic strain energy, and not directly on the components of the stiffness and compliance tensors. Since the bounds are to be valid for any choice of the averaged macroscale strain or strain tensors, certain smart choices of these tensors will yield simple bounds for the components of the effective stiffness tensor. For example, if the averaged strain tensor is selected to have only one nonzero component (either normal or shear component), the following bounds result for the diagonal terms of the elastic stiffness tensor (no summation on i, j):

$$\overline{S}_{ijij}^{-1} \leq C_{ijij}^{eff} \leq \overline{C}_{ijij} \tag{8.23}$$

8.2.1 Degraded bounds for off-diagonal components

The derivation of the bounds for the off-diagonal terms of the stiffness tensor needs a more sophisticated approach (Proust and Kalidindi, 2006) than that required to derive Eq. (8.23). We illustrate this by deriving the bounds for the off-diagonal components of the effective stiffness tensors, assuming that the composite material of interest exhibits orthorhombic symmetry in the properties at the macroscale.

For materials with orthorhombic symmetry at the macroscale, the effective stiffness tensor can be expressed in the contracted notation with nine independent parameters as shown in Example 7.11. Of these nine parameters, the six diagonal terms can be bounded by relatively simple choices of the imposed averaged strain tensors.

We take as an example the derivation for the bounds on C^{eff}_{1122}. We choose the average strain tensor to be applied to have the following form:

$$\overline{\varepsilon}_{ij} = \begin{pmatrix} \overline{\varepsilon}_{11} & 0 & 0 \\ 0 & \overline{\varepsilon}_{22} & 0 \\ 0 & 0 & 0 \end{pmatrix} \tag{8.24}$$

Substituting this strain tensor in the upper bound for stiffness produces the following inequality:

$$\overline{\varepsilon}_{11} C^{eff}_{1111} \overline{\varepsilon}_{11} + 2\overline{\varepsilon}_{11} C^{eff}_{1122} \overline{\varepsilon}_{22} + \overline{\varepsilon}_{22} C^{eff}_{2222} \overline{\varepsilon}_{22} \le \overline{\varepsilon}_{11} \overline{C}_{1111} \overline{\varepsilon}_{11} + 2\overline{\varepsilon}_{11} \overline{C}_{1122} \overline{\varepsilon}_{22} + \overline{\varepsilon}_{22} \overline{C}_{2222} \overline{\varepsilon}_{22} \tag{8.25}$$

Let $\overline{\varepsilon}_{22} = \alpha\, \overline{\varepsilon}_{11}$. We have to treat two conditions separately, with α being positive or negative. For the case where α is positive, inequality (8.25) yields

$$C^{eff}_{1122} \le \overline{C}_{1122} + \frac{1}{2\alpha} \left(\overline{C}_{1111} - C^{eff}_{1111} \right) + \frac{\alpha}{2} \left(\overline{C}_{2222} - C^{eff}_{2222} \right) \tag{8.26}$$

A *degraded bound* can be obtained by using the minimum possible values for C^{eff}_{1111} and C^{eff}_{2222}. To this end, we use $\overline{S}^{-1}_{1111}$ and $\overline{S}^{-1}_{2222}$ for values of C^{eff}_{1111} and C^{eff}_{2222}, respectively. C^{eff}_{1122} then becomes inferior to a function Φ dependent on α:

$$C^{eff}_{1122} \le \Phi(\alpha) = \overline{C}_{1122} + \frac{1}{2\alpha} \left(\overline{C}_{1111} - \overline{S}^{-1}_{1111} \right) + \frac{\alpha}{2} \left(\overline{C}_{2222} - \overline{S}^{-1}_{2222} \right) \tag{8.27}$$

The minimum value of Φ can be found by equating the derivative of Φ to zero. This yields the corresponding value of α:

$$\alpha = \sqrt{\frac{\overline{C}_{1111} - \overline{S}^{-1}_{1111}}{\overline{C}_{2222} - \overline{S}^{-1}_{2222}}} \tag{8.28}$$

Computing the corresponding value of C^{eff}_{1122} yields

$$C^{eff}_{1122} \le \overline{C}_{1122} + \sqrt{\Delta_{11}\Delta_{22}}, \quad \Delta_{11} = \overline{C}_{1111} - \overline{S}^{1}_{1111}, \quad \Delta_{22} = \overline{C}_{2222} - \overline{S}^{-1}_{2222} \tag{8.29}$$

For the case where α is negative, the upper-bound inequality yields

$$C^{eff}_{1122} \ge \overline{C}_{1122} + \frac{1}{2\alpha} \left(\overline{C}_{1111} - C^{eff}_{1111} \right) + \frac{\alpha}{2} \left(\overline{C}_{2222} - C^{eff}_{2222} \right) \tag{8.30}$$

As before, with the goal of deriving a degraded bound, we substitute C^{eff}_{1111} and C^{eff}_{2222} with $\overline{S}^{-1}_{1111}$ and $\overline{S}^{-1}_{2222}$, respectively. Following the same approach as before, we obtain this time a lower bound for C^{eff}_{1122}, which when combined with the upper bound obtained earlier allows us to formulate the following bounds:

$$\overline{C}_{1122} - \sqrt{\Delta_{11}\Delta_{22}} \le C^{eff}_{1122} \le \overline{C}_{1122} + \sqrt{\Delta_{11}\Delta_{22}} \tag{8.31}$$

Repeating this derivation with the lower bound on the elastic stiffness for the same average strain tensor yields

$$2\bar{\varepsilon}_{11}C_{1122}^{eff}\bar{\varepsilon}_{22} \geq 2\bar{\varepsilon}_{11}\overline{S}_{1122}^{-1}\bar{\varepsilon}_{22} + \bar{\varepsilon}_{11}(\overline{S}_{1111}^{-1} - C_{1111}^{eff})\bar{\varepsilon}_{11} + \bar{\varepsilon}_{22}(\overline{S}_{2222}^{-1} - C_{2222}^{eff})\bar{\varepsilon}_{22} \tag{8.32}$$

Once again, working the cases where α is positive and where α is negative separately, we obtain the following bounds:

$$\overline{S}_{1122}^{-1} - \sqrt{\Delta_{11}\Delta_{22}} \leq C_{1122}^{eff} \leq \overline{S}_{1122}^{-1} + \sqrt{\Delta_{11}\Delta_{22}} \tag{8.33}$$

We therefore obtain two sets of lower and upper bounds for C_{1122}^{eff}, and they can be combined as

$$\max(\overline{C}_{1122}, \overline{S}_{1122}^{-1}) - \sqrt{\Delta_{11}\Delta_{22}} \leq C_{1122}^{eff} \leq \min(\overline{C}_{1122}, \overline{S}_{1122}^{-1}) + \sqrt{\Delta_{11}\Delta_{22}} \tag{8.34}$$

As a check on derivation (8.34), we need to verify that the upper-bound value is actually higher than the lower bound. Considering the two cases where $\overline{C}_{1122} \geq \overline{S}_{1122}^{-1}$ and $\overline{C}_{1122} \leq \overline{S}_{1122}^{-1}$ separately results in the requirement

$$-2\sqrt{\Delta_{11}\Delta_{22}} \leq \overline{C}_{1122} - \overline{S}_{1122}^{-1} \leq 2\sqrt{\Delta_{11}\Delta_{22}} \tag{8.35}$$

It can be shown that this requirement is automatically satisfied by the requirement on the overall bounds, that is, $\bar{\varepsilon}_{ij}\overline{S}_{ijkl}^{-1}\,\bar{\varepsilon}_{kl} \leq \bar{\varepsilon}_{ij}\overline{C}_{ijkl}\,\bar{\varepsilon}_{kl}$ (this can be seen using again the special strain tensor with $\bar{\varepsilon}_{22} = \alpha\,\bar{\varepsilon}_{11}$).

8.3 HOMOGENIZATION OF OTHER PHYSICAL PROPERTIES

The derivations given in Section 8.2 for elasticity may be extended to other physical phenomena with similar underlying theories. Table 8.1 summarizes the constitutive laws and field equations employed at the lower-length scale and the corresponding effective macroscale constitutive laws sought in the homogenization theories for a few example problems. The bars on top of various field quantities indicate the volume-averaged quantities, while *eff* as a superscript indicates the effective (homogenized) material property at the higher-length scale.

8.4 FIRST-ORDER BOUNDS FOR THERMAL EXPANSION

As an example of a problem that does not include only elasticity, in this section the first-order bounds for the thermal expansion tensor are coupled with the bounds for elastic stiffness. We present a brief derivation of these bounds following the methods described by Milton (2002) and Rosen and Hashin (1970).

Consider a multiphase composite material system subjected to a uniform increment in temperature of θ. The thermo-elastic constitutive relationship that needs to be satisfied at each material point in the composite can be expressed as

$$\varepsilon_{ij}(\mathbf{x}) = \varepsilon_{ij}^{e}(\mathbf{x}) + \varepsilon_{ij}^{th}(\mathbf{x}) = S_{ijkl}(\mathbf{x})\sigma_{kl}(\mathbf{x}) + \alpha_{ij}(\mathbf{x})\theta \tag{8.36}$$

where ε_{ij}^{e} and ε_{ij}^{th} represent the elastic and thermal strain fields, and α_{ij} represents the thermal expansion coefficient tensor. It has been implied in Eq. (8.36) that the elastic and thermal strain

Table 8.1 Summary of the Homogenization Problem Statement for Example Systems

Class of Problems	Conductivity	Elasticity	Rigid-Plasticity	Thermo-Elasticity
Lower-Length Scale				
Field variable	(ϕ, j_i, e_i)	$(\phi^e, \sigma_{ij}, \varepsilon_{ij})$	$(\phi^p, \sigma_{ij}, \dot{\varepsilon}_{ij}{}^p)$	$(\phi^e, \sigma_{ij}, \varepsilon_{ij})$
Constitutive laws	$j_i = \frac{\partial \phi}{\partial e_i} = K_{ij} e_j$ $K_{ij} = \frac{\partial^2 \phi}{\partial e_i \partial e_j} = K_{ji}$ $K_{ij} e_i e_j \geq 0$	$\sigma_{ij} = \frac{\partial \phi^e}{\partial \varepsilon_{ij}} = C_{ijkl} \varepsilon_{kl}$ $C_{ijkl} = \frac{\partial^2 \phi^e}{\partial \varepsilon_{ij} \partial \varepsilon_{kl}} = C_{klij}$ $C_{ijkl} \varepsilon_{ij} \varepsilon_{kl} \geq 0$	$\sigma_{ij} = \frac{\partial \phi^p}{\partial \dot{\varepsilon}_{ij}{}^p}$ $\sigma_{ij} = N_{ijkl} \dot{\varepsilon}_{kl}{}^p$ $N_{ijkl} = N_{klij}$ $N_{ijkl} \dot{\varepsilon}_{ij}{}^p \dot{\varepsilon}_{kl}{}^p \geq 0$	$\sigma_{ij} = C_{ijkl}(\varepsilon_{kl} - \alpha_{kl} T)$ $\alpha_{kl} = \alpha_{lk}$
Conservation principles	$\nabla \cdot \mathbf{j} = 0$	$\nabla \cdot \boldsymbol{\sigma} = 0$	$\nabla \cdot \boldsymbol{\sigma} = 0$	$\nabla \cdot \boldsymbol{\sigma} = 0$
Continuity equations	$\mathbf{e} = -\nabla \phi$ $\nabla \times \nabla \phi = 0$	$\boldsymbol{\varepsilon} = \frac{1}{2}(\nabla \mathbf{u} + (\nabla \mathbf{u})^T)$ $\nabla \times \nabla u_i = 0$	$\dot{\boldsymbol{\varepsilon}}^p = \frac{1}{2}(\nabla \dot{\mathbf{u}} + (\nabla \dot{\mathbf{u}})^T)$ $\nabla \times \nabla \dot{u}_i = 0$	$\boldsymbol{\varepsilon} = \frac{1}{2}(\nabla \mathbf{u} + (\nabla \mathbf{u})^T)$ $\nabla \times \nabla u_i = 0$
Higher-Length Scale				
Effective constitutive laws	$\bar{j}_i = K_{ij}^{eff} \bar{e}_j$ $K_{ij}^{eff} = K_{ji}^{eff}$ $K_{ij}^{eff} \bar{e}_i \bar{e}_j \geq 0$	$\bar{\sigma}_{ij} = C_{ijkl}^{eff} \bar{\varepsilon}_{kl}$ $C_{ijkl}^{eff} = C_{klij}^{eff}$ $C_{ijkl}^{eff} \bar{\varepsilon}_{ij} \bar{\varepsilon}_{kl} \geq 0$	$\bar{\sigma}_{ij} = N_{ijkl}^{eff} \dot{\bar{\varepsilon}}_{kl}{}^p$ $N_{ijkl}^{eff} \dot{\bar{\varepsilon}}_{ij}{}^p \dot{\bar{\varepsilon}}_{kl}{}^p \geq 0$	$\bar{\sigma}_{ij} = C_{ijkl}^{eff}(\bar{\varepsilon}_{kl} - \alpha_{kl}^{eff} T)$ $\alpha_{kl}^{eff} = \alpha_{lk}^{eff}$

components are additive; in most crystalline materials these strains are infinitesimal, and therefore this is a reasonable assumption. The effective thermo-elastic constitutive response of the composite is expressed as

$$\bar{\varepsilon}_{ij} = S^{eff}_{ijkl}\bar{\sigma}_{kl} + \alpha^{eff}_{ij}\theta \text{ or } \bar{\sigma}_{ij} = C^{eff}_{ijkl}(\bar{\varepsilon}_{kl} - \alpha^{eff}_{kl}\theta) \tag{8.37}$$

where C^{eff}_{ijkl} and S^{eff}_{ijkl} are the effective stiffness and compliance tensors (see Section 8.2), and α^{eff}_{ij} is the effective thermal expansion coefficient's tensor. To obtain an expression for α^{eff}_{ij}, one needs to apply the constitutive equations of (8.5) at each point in the composite material and solve the stress equilibrium equations (see Table 8.1) with the appropriate boundary conditions. Note also that the formulation of the effective thermo-elastic constitutive response in Eq. (8.37) implicitly assumes that α^{eff}_{ij} is independent of the applied loading conditions (i.e., independent of the applied stress field on the composite). Therefore, one can choose any convenient loading condition and solve for α^{eff}_{ij}. Next, we will select a specific loading condition that results in uniform stress and strain fields in the composite and yields an exact solution for α^{eff}_{ij}. This would of course be possible only for specific selected material systems; only these are considered in the following discussion.

Consider a two-phase composite where the constituent phases exhibit anisotropic thermo-elastic properties. Let this composite be subjected to a uniform stress field (i.e., $\sigma_{ij} = \bar{\sigma}_{ij}$ everywhere in the composite). Application of Eq. (8.36) to each constituent phase results in

$$\varepsilon^{(1)}_{ij} = S^{(1)}_{ijkl}\bar{\sigma}_{kl} + \alpha^{(1)}_{ij}\theta, \qquad \varepsilon^{(2)}_{ij} = S^{(2)}_{ijkl}\bar{\sigma}_{kl} + \alpha^{(2)}_{ij}\theta \tag{8.38}$$

where the superscripts (1) and (2) identify the phases present in the composite. The stress state to be imposed on the composite will be selected such that $\varepsilon^{(1)}_{ij} = \varepsilon^{(2)}_{ij} = \bar{\varepsilon}_{ij}$, which then automatically satisfies compatibility. This choice requires (equating the two strains in Eq. 8.38):

$$\bar{\sigma}_{ij} = P_{ijkl}\left(\alpha^{(2)}_{kl} - \alpha^{(1)}_{kl}\right)\theta, \qquad \mathbf{P} = (\mathbf{S}^{(1)} - \mathbf{S}^{(2)})^{-1} \tag{8.39}$$

Substitution of Eq. (8.40) in Eq. (8.39) and Eq. (8.38) yields

$$\alpha^{eff}_{ij} = \alpha^{(p)}_{ij} + \left(S^{eff}_{ijkl} - S^{(p)}_{ijkl}\right)P_{klmn}(\alpha^{(1)}_{mn} - \alpha^{(2)}_{mn}) \tag{8.40}$$

where the superscript (p) refers to either of the constituent phases. For the special case, where the constituent phases are isotropic, Eq. (8.41) reduces to

$$\alpha^{eff}_{ij} = \alpha^{(p)}\delta_{ij} + \frac{\alpha^{(1)} - \alpha^{(2)}}{\left(\frac{1}{K^{(1)}} - \frac{1}{K^{(2)}}\right)}\left(3S^{eff}_{ijkk} - \frac{1}{K^{(p)}}\delta_{ij}\right) \tag{8.41}$$

where $K^{(p)}$ denotes the bulk modulus of phase p. First-order bounds on α^{eff}_{ij} can be derived using the first-order bounds on $\mathbf{S}^{eff}$ described earlier in Section 8.2.

It is also possible to obtain closed-form expressions for effective thermal expansion of polycrystals for certain classes of crystal symmetry. In particular, when the crystals exhibit an axis of transverse isotropy in the description of S_{ijkk} and α_{ij} in a principal crystal reference frame (e.g., hexagonal, tetragonal, and trigonal crystals exhibit this feature), it is possible to solve for α^{eff}_{ij} by considering a purely hydrostatic uniform stress field in the composite, that is, $\sigma_{ij} = \bar{\sigma}_{ij} = \sigma\delta_{ij}$. The corresponding strain field in any of the constituent crystals, obtained through the use of Eq. (8.5), can be expressed in the crystal reference frame as

$$[\varepsilon_{ij}] = \begin{bmatrix} \sigma S_{11kk} + \theta\alpha_1 & 0 & 0 \\ 0 & \sigma S_{22kk} + \theta\alpha_2 & 0 \\ 0 & 0 & \sigma S_{22kk} + \theta\alpha_2 \end{bmatrix} \tag{8.42}$$

Note that the symmetry in the properties arising from the symmetry of the crystal lattice (assuming that the crystal 1-axis is the axis of transverse isotropy) has been taken into account in formulating Eq. (8.42). A closer look at Eq. (8.42) reveals that a specific value of the hydrostatic stress component can be selected such that the strain field in the polycrystal is uniform and purely dilatational. More specifically, if we select

$$\sigma = -\frac{\alpha_1 - \alpha_2}{S_{11kk} - S_{22kk}}\theta \tag{8.43}$$

the corresponding strain field is given by

$$\varepsilon_{ij} = \overline{\varepsilon}_{ij} = \frac{\alpha_2 S_{11kk} - \alpha_1 S_{22kk}}{S_{11kk} - S_{22kk}}\theta\,\delta_{ij} \tag{8.44}$$

Since the resulting strain field is purely dilatational and uniform (the same in all of the constituent crystals), it satisfies the required compatibility conditions. Substitution of these special uniform strain and stress fields (Eqs. 8.43 and 8.44) in the effective thermo-elastic relation of Eq. (8.37) yields the desired relationship for α_{ij}^{eff} as

$$\alpha_{ij}^{eff} = \frac{\alpha_1 - \alpha_2}{S_{11kk} - S_{22kk}} S_{ijkk}^{eff} + \frac{\alpha_2 S_{11kk} - \alpha_1 S_{22kk}}{S_{11kk} - S_{22kk}}\delta_{ij} \tag{8.45}$$

SUMMARY

This chapter provided a concise refresher in the composite theories available in current literature for the rigorous first-order bounds for effective mechanical properties of composites. It was also shown that in some instances only degraded bounds are available. It is hoped that readers appreciate the value of these first-order bounds in that they clearly identify the property combinations that are not feasible in the given composite system. Often, these first-order theories produce widely separated bounds and identification of the specific combinations that are feasible within the bounds requires specification of higher-order details of the microstructure.

Exercises

8.1 (short project). Research the mechanics literature to learn about the theorems of minimum potential energy and complementary energy. Then use these relationships to obtain an alternative derivation (differing from the derivation in this chapter) for the upper and lower bounds on elastic stiffness and compliance. You need only be concerned about the diagonal components of the 6 × 6 elasticity matrix for this exercise.

8.2. Prove that $\varepsilon_{ij}C_{ijkl}\varepsilon_{kl}$ may be written as $\mathbf{C}\boldsymbol{\varepsilon}\cdot\boldsymbol{\varepsilon}$.

8.3. By writing in indicial form, prove that $\overline{\mathbf{C}\overline{\boldsymbol{\varepsilon}}\cdot\boldsymbol{\varepsilon}} = \overline{\mathbf{C}^T\boldsymbol{\varepsilon}\cdot\overline{\boldsymbol{\varepsilon}}} = \overline{\mathbf{C}\boldsymbol{\varepsilon}}\cdot\overline{\boldsymbol{\varepsilon}}$.

CHAPTER

Microstructure Hull and Closures

CHAPTER OUTLINE

Previous chapters have built a framework for both representing the structure of materials and for relating that structure to effective properties. The concepts have also been embedded in spectral relations to arrive at an efficient mathematical basis for microstructure design. In this chapter we pull together these ideas to form the two underlying design spaces of microstructure-sensitive design: the microstructure hull and the property closure. The first encompasses (in spectral space) the envelope of all physically realizable microstructures (based on a given definition of structure, such as orientation distribution), and the second defines the space of possible property combinations (for a subset of properties of interest and for a given homogenization relation that provides values of those properties). These constructs then provide the foundation for the design approach introduced and demonstrated in Chapter 10.

9.1 MICROSTRUCTURE HULL

In this section we introduce the *microstructure hull*. This is the set of all possible microstructures within the modeling framework chosen for homogenization. The reason it is called a *hull* is that it is found to be a convex hull in the Fourier space of microstructure function representations. Thus, taking any two points lying in the microstructure hull (representing any two microstructures), the line segment of points connecting these two also represent microstructures that exist, and these belong to the same microstructure hull. Points in the Fourier space of microstructures, which do not lie within the microstructure hull, must be considered to be unphysical. Microstructure design demands that we select microstructures that belong to the microstructure hull. In the following we carefully describe the microstructure hull for the orientation distribution function $f(g)$, approximated in its primitive basis as a piecewise continuous function. We will see that the microstructure hull is a convex polytope of a particular structure.

9.1.1 Hull for the primitive basis

Recall the Fourier representation and approximation of the microstructure function $M(\mathbf{x},h)$ in its primitive Fourier basis (see Eqs. 5.45–5.48):

$$M(\mathbf{x},h) \approx D_s^n \chi_s(\mathbf{x}) \chi^n(h) \tag{9.1}$$

where summation over subcells $1 \leq s \leq S$ and subdomains $1 \leq n \leq N$ is implied by the repeated indices on the right side. It is understood that the nature of the primitive basis functions, $\chi_s(\mathbf{x})$, $\chi^n(h)$, leads to piecewise-continuous approximations to the true microstructure function. Conservation of mass requires that the Fourier coefficients of the microstructure satisfy relations (5.47) and (5.48) for each subcell s.

Relation (5.48) suggests an elementary geometrical interpretation. For each subcell s, consider relation (5.48) for $D_s^1, D_s^2, ..., D_s^{N-1}$. This relation describes an $N-1$ dimensional hypercube, bounded by the set of $N-1$ half-spaces embodied in expressions of the form $D_s^n \geq 0$, and a second set of $N-1$ half-spaces described by expressions of the form $D_s^n \leq N$. The intersection of these $2(N-1)$ half-spaces is a hypercube. Readers should note that it is not necessary to describe D_s^N since by relation (5.4.10) we must have

$$D_s^N = N - \sum_{n=1}^{N-1} D_s^n \tag{9.2}$$

Let $d_s = (D_s^1, D_s^2, ..., D_s^{N-1})$ represent a generic point in the Fourier space, representing subcell s. Notice that relation (9.2) defines a family of hyperplanes that cut through the hypercube. All physically real d_s must satisfy (9.2). If $D_s^N = 0$ (i.e., subdomain γ_N is not represented in the local state distribution of subcell s), then the sum over $D_s^1 + D_s^2 +, ..., + D_s^{N-1}$ must equal N. This is essentially a matter of conservation of mass. Note that this is the equation of a hyperplane that crosses each axis $D_s^n (1 \leq n \leq N-1)$ at N. If D_s^N takes some value greater than 0, but less than N, then relation (9.2) describes the equation of a hyperplane that crosses each axis $D_s^n (1 \leq n \leq N-1)$ at $N - D_s^N$. In the limit that $D_s^N = N$, the hyperplane just touches the point $D_s^1 = D_s^2 =, ..., = D_s^{N-1} = 0$ of the hypercube. Since all physically real d_s must satisfy relations (5.47), (5.48), and (9.2), it follows that the intersection of the hyperplanes defined by (9.2) with the hypercube described by (5.47) and (5.48) defines the set of all real d_s. Thus, the set of all possible $d_s = (D_s^1, D_s^2, ..., D_s^{N-1})$ lies in an N-pyramid consisting of the convex combination of the apex point $D_s^1 = D_s^2 =, ..., = D_s^{N-1} = 0$, and the $N-1$ vertices which have the form

$$\hat{d}_s^{(1)} = (N, 0, ..., 0), \ \hat{d}_s^{(2)} = (0, N, ..., 0), ..., \ \hat{d}_s^{(N-1)} = (0, 0, ..., N) \tag{9.3}$$

It is convenient to take the apex point to be $\hat{d}_s^{(N)}$, where

$$\hat{d}_s^{(N)} = (0, 0, ..., 0) \tag{9.4}$$

All physically allowed local state distributions d_s must belong to the convex set $\mathbf{M}_s$, defined by exercising the convex combinations of the vertices (including the apex):

$$\mathbf{M}_s = \left\{ d_s \middle| d_s = \sum_{n=1}^{N} \alpha_n \hat{d}_s^{(n)}, \ \sum_{n=1}^{N} \alpha_n = 1, \ \alpha_n \geq 0 \right\} \tag{9.5}$$

The next task is to describe the complete microstructure hull **M** for all possible microstructure functions. Let d represent the ordered set of local state distributions in the complete set of subcells:

$$d = (d_1, d_2, \ldots, d_S)\ (d_s \in \mathbf{M}_s) \tag{9.6}$$

Notice that d belongs to the product space

$$d \in \mathbf{M}_1 \otimes \mathbf{M}_2 \otimes \ldots \otimes \mathbf{M}_S \tag{9.7}$$

In other words, d belongs to a large set of all possible ordered sets of local state distribution. Each element of this ordered set is a local state distribution in one of the subcells of the rectangular model. This product space is the microstructure hull, **M**, for all piecewise continuous rectangular models of the microstructure function:

$$\mathbf{M} = \mathbf{M}_1 \otimes \mathbf{M}_2 \otimes \ldots \otimes \mathbf{M}_S \tag{9.8}$$

Applying the notion of *hull* to the set **M** suggests that we must prove that **M** is a convex hull. Consider two points in **M**, $d = (d_1, d_2, \ldots, d_S)$ and $d' = (d'_1, d'_2, \ldots, d'_S)$. By definition, **M** is convex if all $\lambda d + (1-\lambda)d'$ $(0 \le \lambda \le 1)$, for any choice of d and d', also belong to **M**. Notice that since all $d_s, d'_s \in \mathbf{M}_s$ it must hold that $\lambda d_s + (1-\lambda)d'_s \in \mathbf{M}_s$ $(0 \le \lambda \le 1)$. This follows from the convexity of $\mathbf{M}_s$ as expressed in (9.5). Also notice that

$$\lambda d + (1-\lambda)d' = ((\lambda d_1 + (1-\lambda)d'_1), \ldots, (\lambda d_S + (1-\lambda)d'_S))\quad (0 \le \lambda \le 1) \tag{9.9}$$

But since each component of the ordered set $\lambda d + (1-\lambda)d'$ $(0 \le \lambda \le 1)$ must lie within its appropriate $\mathbf{M}_s$, it follows that $\lambda d + (1-\lambda)d'$ must lie in **M**, and therefore the set **M** is convex. Hereafter, we refer to **M** simply as the *microstructure hull.*

Consideration of the *vertices* of the microstructure hull follows from the vertices in each of the component subspaces, $\mathbf{M}_s$. Including the apex, each subspace $\mathbf{M}_s$ contains N vertices; these vertices are labeled $\hat{d}_s^{(n)}$ $(n = 1, 2, \ldots, N)$. The vertices in the microstructure hull **M** are the microstructures

$$\hat{d}^{(k)} \in \left\{ \begin{array}{l} \left(\hat{d}_1^{(1)}, \hat{d}_2^{(1)}, \hat{d}_3^{(1)}, \ldots, \hat{d}_S^{(1)}\right), \left(\hat{d}_1^{(2)}, \hat{d}_2^{(1)}, \hat{d}_3^{(1)}, \ldots, \hat{d}_S^{(1)}\right), \left(\hat{d}_1^{(1)}, \hat{d}_2^{(2)}, \hat{d}_3^{(1)}, \ldots, \hat{d}_S^{(1)}\right), \\ \left(\hat{d}_1^{(1)}, \hat{d}_2^{(1)}, \hat{d}_3^{(2)}, \ldots, \hat{d}_S^{(1)}\right), \ldots, \left(\hat{d}_1^{(1)}, \hat{d}_2^{(1)}, \hat{d}_3^{(1)}, \ldots, \hat{d}_S^{(2)}\right), \left(\hat{d}_1^{(2)}, \hat{d}_2^{(2)}, \hat{d}_3^{(1)}, \ldots, \hat{d}_S^{(1)}\right), \\ \ldots, \left(\hat{d}_1^{(N)}, \hat{d}_2^{(N)}, \hat{d}_3^{(N)}, \ldots, \hat{d}_S^{(N-1)}\right), \left(\hat{d}_1^{(N)}, \hat{d}_2^{(N)}, \hat{d}_3^{(N)}, \ldots, \hat{d}_S^{(N)}\right) \end{array} \right\} \tag{9.10}$$

The number of vertices is equal to N^S, and hence the range of k in (9.10) is $1 \le k \le N^S$. *Notice that the number of vertices grows exponentially with the number of subcells S of the rectangular model.* This is called the *large numbers problem* of microstructure design. All possible microstructures belonging to the microstructure hull can be described as convex combinations of the vertices $\hat{d}^{(k)}$:

$$d \in \mathbf{M} = \left\{ d \,\middle|\, d = \sum_{k=1}^{N^S} \alpha^{(k)} \hat{d}^{(k)}, \sum_{k=1}^{N^S} \alpha^{(k)} = 1, \alpha^{(k)} \ge 0 \right\} \tag{9.11}$$

Relation (9.11) is known as Caratheodory's theorem. It is one of the basic results in convexity (Caratheodory, 1911).

From a practical standpoint, expanding the ordered sets that belong to the microstructure hull, any microstructure $d \in \mathbf{M}$ can be written in terms of its complete set of Fourier components:

$$d = \left(D_1^1, D_1^2, \ldots, D_1^N, D_2^1, D_2^2, \ldots, D_2^N, \ldots, D_S^1, D_S^2, \ldots, D_S^N\right) \tag{9.12}$$

We are reminded that these coefficients are subject to the constraints expressed in (5.47) and (5.48). Given relation (5.47) it is not necessary to describe all N-terms of the local state distribution in each subcell, but only $N - 1$. Thus, it is sufficient to describe the $(N - 1)S$ components:

$$d = \left(D_1^1, D_1^2, \ldots, D_1^{N-1}, D_2^1, D_2^2, \ldots, D_2^{N-1}, \ldots, D_S^1, D_S^2, \ldots, D_S^{N-1}\right) \tag{9.13}$$

Some microstructures of special character

Here we briefly mention some very special cases of microstructure that occur within the microstructure hull $\mathbf{M}$. *Single-state microstructures*, or materials comprising a single local state, say $h \in \gamma_{n*}$, have Fourier components of the form

$$D_s^n = \begin{cases} N & \text{if } n = n^* \\ 0 & \text{otherwise} \end{cases} \quad \text{(for all } s\text{)} \tag{9.14}$$

A second class of important microstructures is the class of *perfectly disordered microstructures*. For these each subcell has the same local state distribution (if the cell size is large enough compared to any local heterogeneity in the material). Having specified the distribution in one subcell, say $s = s*$, then all other subcells in this class have the same distribution of local state. Put succinctly,

$$D_s^n = D_{s^*}^n \ (\text{for } 1 \le s \le S,\ 1 \le n \le N) \tag{9.15}$$

Finally, we mention the class of *random, perfectly disordered microstructures.* In this case each Fourier coefficient has the same value:

$$D_s^n = 1 \ (\text{for } 1 \le n \le N,\ 1 \le s \le S) \tag{9.16}$$

We visualize elements of the microstructure hull for a simple example where $N = 2$ and $S = 4$ in Figure 9.1. The microstructures labeled at various corners are given in that figure.

At the corners of various polytopes in Figure 9.1 are microstructures where each cell only contains a single state. Thus, D_s^n is equal to 0 or N for each n, s. Such microstructures will be termed *eigen-microstructures*. Figure 9.2 illustrates eigen-microstructures for the case $S = 4$ and $N = 2$.

9.1.2 Hull for the discrete Fourier basis

Recall the definition of the ODF for a given bin, ω_n:

$$\int_{\omega_n} f(\varphi_1, \Phi, \varphi_2) \sin \Phi d\varphi_1 d\Phi d\varphi_2 = p(g_s \in \{\omega_n\}) \tag{9.17}$$

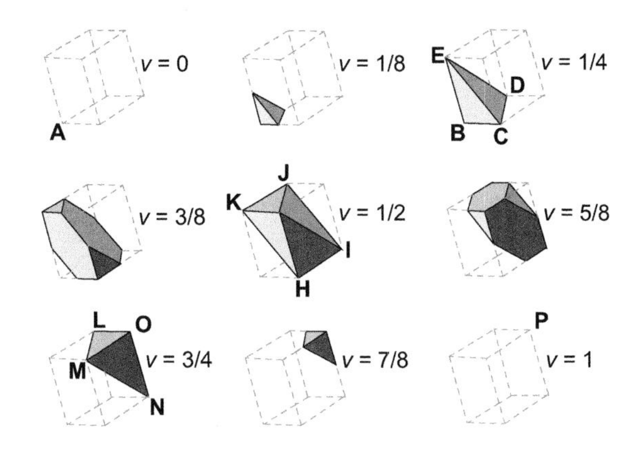

FIGURE 9.1

Polytopes representing the microstructure hull for $S = 4$ and $N = 2$, for selected volume fractions, v. Axes are D_1^1, D_2^1, D_3^1.

where g_s is the orientation of an arbitrary material point taken from a sample of the material, and p indicates probability. The usual normalization of Euler space means that (assuming cubic symmetry as an example and the corresponding fundamental zone, *FZ*3, as given in Section 5.2.2):

$$\frac{1}{8\pi^2}\int_E f(\varphi_1, \Phi, \varphi_2)\sin\Phi d\varphi_1 d\Phi d\varphi_2 = \frac{1}{\pi^2}\int_{FZ3} f(\varphi_1, \Phi, \varphi_2)\sin\Phi d\varphi_1 d\Phi d\varphi_2 = 1 \tag{9.18}$$

where E indicates integration over the full Euler space. Let us discretize *FZ*3 such that the value of the ODF in each bin, ω_n, is given by a single value f^n (as in Eq. 5.32), and capture the volume of the individual bins in the parameter λ_n:

$$\lambda_n = \frac{1}{\pi^2}\int_{\omega_n} \sin\Phi d\varphi_1 d\Phi d\varphi_2 \tag{9.19}$$

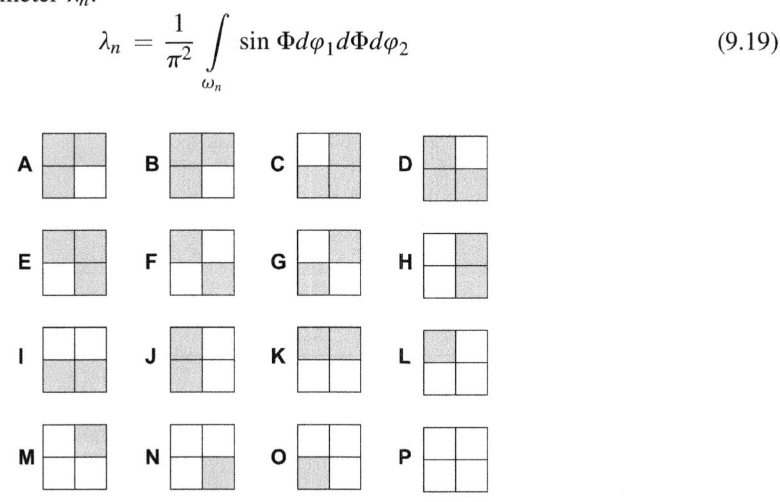

FIGURE 9.2

Eigen-microstructures for $S = 4$ and $N = 2$. Labels correspond to points in Figure 9.1.

Then it will be convenient to define a normalized, discrete ODF by (no summation on n):

$$\tilde{f}^{n} = f^{n}\lambda_{n} \tag{9.20}$$

Hence,

$$\sum_{n}\tilde{f}^{n} = \sum_{n} f^{n}\lambda_{n} = 1 \tag{9.21}$$

Then, from Eqs. (5.32), (9.19), and (9.20), we have that for a single crystal sample with orientation in bin c, the ODF is given by

$$\tilde{f}_{c}^{n} = \begin{cases} 1/3 & \text{if } \omega_{n} \cong \omega_{c} \\ 0 & \text{otherwise} \end{cases} \tag{9.22}$$

where the notation $\omega_n \cong \omega_c$ indicates that bin number n in the Euler space is equivalent to bin number c under the assumed symmetry conditions. In this discrete space any ODF may written as a convex combination of single-crystal ODFs:

$$\begin{aligned} \tilde{f}^{n} &= \sum_{c}\alpha_{c}\tilde{f}_{c}^{n} \\ \alpha_{c} &\geq 0, \quad \sum_{c}\alpha_{c} = 1 \end{aligned} \tag{9.23}$$

The Fourier representation of this normalized ODF is given by

$$\tilde{F}^{k} = \Im(\tilde{f}^{n}) = \sum_{n=0}^{N-1}\tilde{f}^{n}e^{-2\pi ikn/N} \tag{9.24}$$

providing a discrete Fourier representation of the ODF. Then the hull of Fourier coefficients (representing the space of all possible microstructures) is defined by the constraints on $\tilde{F}^{k}$. Since the ODF must be real, Eqs. (3.18) and (3.19) must be satisfied; that is, only half of the Fourier coefficients are independent. Hence, only those independent coefficients should be included in the description of the hull. Furthermore (by definition of the transform and Eq. 9.21),

$$\left|\tilde{F}^{k}\right| \leq \sum_{n=0}^{N-1}\tilde{f}^{k} = 1 \tag{9.25}$$

and (from Eq. 3.30),

$$\sum_{k=0}^{N-1}\tilde{F}^{k} = N\tilde{f}^{0} \leq N \tag{9.26}$$

The result is a convex hull of microstructures in Fourier space that represents all possible ODFs. We do not discuss this representation of the hull in more detail at this point, although there is a more detailed presentation of the two-point hull in Section 12.3.

However, as for the case of the original ODF, an alternative form of this representation (in terms of single crystals) is sometimes desirable. In particular, it makes the hull easier to search (as shown in Section 10.2.4). By linearity of the Fourier transform, taking the Fourier transform of Eq. (9.23):

$$\begin{aligned} \tilde{F}^k &= \sum_c \alpha_c \tilde{F}_c^k \\ \alpha_c \geq 0, \quad \sum_c \alpha_c &= 1 \end{aligned} \tag{9.27}$$

The space of all possible textures (at the resolution of the current discretization) may then be defined in α-space by a simple solid polytope representing the space bound by the inequalities and sum in Eq. (9.27).

At this point the reason for defining a hull in Fourier space may not be obvious—it appears simpler to define the space of all possible $\tilde{f}^n$ rather than $\tilde{F}^k$. However, the logic will become clear when we define the property closure and argue that the number of parameters required in Fourier space for an accurate determination of properties is much smaller than the number required in real space. Hence, the dimensionality of the hull is much smaller and therefore much easier to define and search.

9.1.3 Microstructure hull using spherical harmonics functions

As described in Section 5.3.2, the ODF can be represented as a Fourier series of generalized spherical harmonics as

$$f(g) = \sum_{l=0}^{\infty} \sum_{\mu=1}^{M(l)} \sum_{\nu=1}^{N(l)} \mathrm{F}_l^{\mu\nu} \dot{\dot{T}}_l^{\mu\nu}(g) \tag{9.28}$$

where $\mathrm{F}_l^{\mu\nu}$ represents the Fourier coefficients, and $\dot{\dot{T}}_l^{\mu\nu}(g)$ constitutes a complete set of orthonormal basis functions for cubic materials of which the ODFs exhibit orthorhombic sample symmetry. Each distinct ODF can be represented by a unique set of Fourier coefficients in the infinite-dimensional Fourier space. The complete set of feasible textures can be demonstrated to produce a convex hull in the infinite-dimensional Fourier space, which is also convex in any of its projections in the finite-dimensional subspaces. Such a texture hull for cubic-orthorhombic materials is shown in Figure 9.3 in the first three dimensions of the Fourier space. The most important attribute of these convex and compact hulls in the Fourier space is that all physically realizable textures lie inside them, allowing us to explore the *complete* set of textures in microstructure design.

Recognition of the fact that ODF provides information only about the volume fraction of the various orientations present in the polycrystalline sample permits an alternate mathematical description that results in an approximate hull and is relatively easy to map out. Suppose that a set of crystals, g^k, are chosen in the fundamental zone such that any orientation in the fundamental zone is within the neighborhood of some g^k to some tolerance. Then any ODF can be approximated by a distribution defined in terms of this chosen set of crystals. Mathematically, a continuous ODF function is defined in terms of the discrete set of crystals by using delta functions:

$$f(g) = \sum_k a_k \delta(g - g^k), \quad 0 \leq a_k \leq 1, \quad \sum_k a_k = 1 \tag{9.29}$$

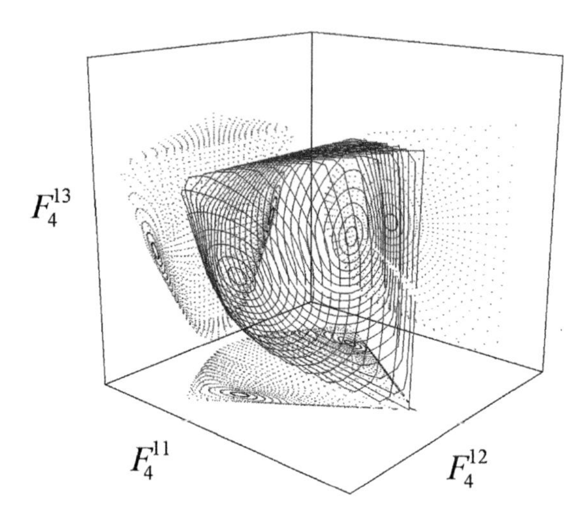

FIGURE 9.3

Representation of the texture hull for cubic-orthorhombic materials in the first three dimensions of the Fourier space.

where the Dirac-delta function, $\delta(g - g^k)$, represents the ODF of a single crystal of orientation g^k; a_k denotes its volume fraction in the polycrystal and hence must sum to 1. Let ${}^kF_l^{\mu\nu}$ define the Fourier coefficients of the single-crystal (g^k) ODFs. It is then possible to define a convex and compact texture hull (Adams et al., 2001), M, that approximates any physically possible ODF for the given symmetry:

$$M = \left\{ F_l^{\mu\upsilon} \middle| F_l^{\mu\upsilon} = \sum_k a_k {}^kF_l^{\mu\upsilon}, {}^kF_l^{\mu\upsilon} \in M^k, a_k \geq 0, \sum_k a_k = 1 \right\} \tag{9.30}$$

where

$$M^k = \left\{ {}^kF_l^{\mu\upsilon} \middle| {}^kF_l^{\mu\upsilon} = \frac{1}{2l+1} \dot{T}_l^{:*\mu\upsilon}(g^k) \right\} \tag{9.31}$$

The asterisk denotes a complex conjugate. It should be recognized that M represents the complete set of all theoretically feasible ODFs, several of which have not yet been realized in practice or even targeted for manufacture by materials specialists. It is also important to recognize that all textures that can be physically conceived will have representations inside these hulls. Just as important is the notion that points outside these hulls do not correspond to any physically conceivable textures.

9.2 PROPERTY CLOSURES

We now introduce the concept of a *property closure*, which, simply stated, is the set of all values that a property (or set of properties) can take for physically realizable microstructures. In this context we are generally referring to effective (i.e., global) properties, since the properties of individual crystals are assumed to be well defined.

For example, assume that we are interested in the property C^{eff}_{1111}. From Section 8.2 it is clear that, using first-order homogenization, the set of values that C^{eff}_{1111} can take is given by $\overline{S}^{-1}_{1111} \leq C^{eff}_{1111} \leq \overline{C}_{1111}$; hence, this defines the 1D property closure for this property. If we were to extend the example to a pair of properties, say C^{eff}_{1111} and C^{eff}_{1212}, then we might assume that the combined property closure for the ordered pair $(C^{eff}_{1111}, C^{eff}_{1212})$ would be given by $\{C^{eff}_{1111} \in [\overline{S}^{-1}_{1111}, \overline{C}_{1111}], C^{eff}_{1212} \in [\overline{S}^{-1}_{1212}, \overline{C}_{1212}]\}$. However, while this set may encompass the combined property closure, it is clear that since C^{eff}_{1111} and C^{eff}_{1212} are not independent, there will be points in this combined space that are not physically realizable. For example, it is unlikely that C^{eff}_{1111} would take its maximum value while C^{eff}_{1212} took its minimum value for the same microstructure.

9.2.1 Spectral representation

The delineation of a property closure is simplified using a spectral representation. Consider first the discrete representation of distribution functions and related properties. For simplicity, we will consider a state space, H, with a single variable, such as crystal orientation, or material phase. Assume for the purpose of calculating properties that we have discretized the state space into N partitions. Hence, the state distribution function is described by the variables $\tilde{f}^n$ as defined earlier (Eq. 5.55). The constraint on the distribution function is that $\sum_{n=1}^{N} \tilde{f}^n = 1$.

Now take a pair of properties associated with this state space, p and q; assume that p takes the local property p^n when the local state takes the value n, and similarly for q. Then the averaged (or upper-bound) material property is given by $\overline{p} = \tilde{f}^n p^n$, where summation is assumed over the n's. Clearly, equality is only approximate for a discretized space, but we will allow abuse of notation in this manner where the approximation is obvious.

Thus, repeating the previous example, we may write

$$\left(\tilde{f}^n S^n_{1111}\right)^{-1} \leq C^{eff}_{1111} \leq \tilde{f}^n C^n_{1111} \tag{9.32}$$

It may be argued that the discrete representation of the state space (the ODF) and the property space is a spectral representation of sorts, using the primitive basis (see Sections 3.1 and 5.1). However, much greater efficiency is gained if a de facto Fourier representation is used. Just as the Fourier transform of the ODF was introduced in Section 5.2.2, we may introduce the Fourier transform of a property p:$P^k = \Im(p^n)$. Then, using Plancherel's theorem, Eq. (3.28),

$$\overline{p} = \tilde{f}^n p^n = \frac{1}{N}(\tilde{F}^k)^* P^k \tag{9.33}$$

It is generally the case that p^n is a smooth function on the local state space (represented by n). Hence, the Fourier transform is generally dominated by a few important terms; the other terms may often be ignored without significant loss of accuracy. This leads to an efficient method of determining the macroscopic properties and the bounds represented by Eq. (9.32).

We present two examples for producing full closures, with and without the need for spectral efficiency.

Example 9.1

Determine the first-order property closure for C^*_{1111} and C^*_{1212} for two isotropic phases with the properties $\lambda_1 = 70$ GPa, $\mu_1 = 35$ GPa, $\lambda_2 = 100$ GPa, $\mu_2 = 55$ GPa.

Solution

For an isotropic medium, the stiffness matrix is given by (see Section 7.5 for much of the following)

$$C_{ijkl} = \mu\left(\delta_{ik}\delta_{jl} + \delta_{il}\delta_{jk}\right) + \lambda\delta_{ij}\delta_{kl}$$

where the delta function is defined as

$$\delta_{ij} = \begin{cases} 1 & \text{if} \quad i = j \\ 0 & \text{if} \quad i \neq j \end{cases}$$

Hence, $C^1_{1111} = 140$ GPa, $C^2_{1111} = 210$ GPa, $C^1_{1212} = 35$ GPa, $C^2_{1212} = 55$ GPa, and so forth.

To find the compliance tensor, **S**, we change the stiffness tensor to a matrix form using the conversion $C_{ijkl} = C_{mn}$, with:

ij or *kl*	11	22	33	23	31	12
m or *n*	1	2	3	4	5	6

(*Note:* The symmetry of the tensor means that component 31 is the same as component 13, etc.)
Define

$$M = \begin{bmatrix} 1 & 0 & 0 & 0 & 0 & 0 \\ 0 & 1 & 0 & 0 & 0 & 0 \\ 0 & 0 & 1 & 0 & 0 & 0 \\ 0 & 0 & 0 & 1/2 & 0 & 0 \\ 0 & 0 & 0 & 0 & 1/2 & 0 \\ 0 & 0 & 0 & 0 & 0 & 1/2 \end{bmatrix}$$

Then **S** is given by $\mathbf{S} = M\mathbf{C}^{-1}M$ (the factor M is required due to reducing the size of the matrix from 9×9 to 6×6).

The discrete state space has only two cells with variables f^n ($n = 1/2$) giving the volume average of each phase. The state space variables take values determined by $f^1 \in [0, 1]$; $f^2 = 1 - f^1$ (f^2 is determined from the constraint that $f^1 + f^2 = 1$). Hence, we may find points on the boundary of the property closure by varying f^1 and calculating the maximum and minimum bounds for each value according to $\overline{S}_{11}^{-1} \leq C^{eff}_{11} \leq \overline{C}_{11}$, that is, $(f^n\mathbf{S}^n)^{-1}_{11} \leq C^{eff}_{11} \leq f^n C^n_{11}$ (and similarly for C^{eff}_{66}); see Figure 9.4.

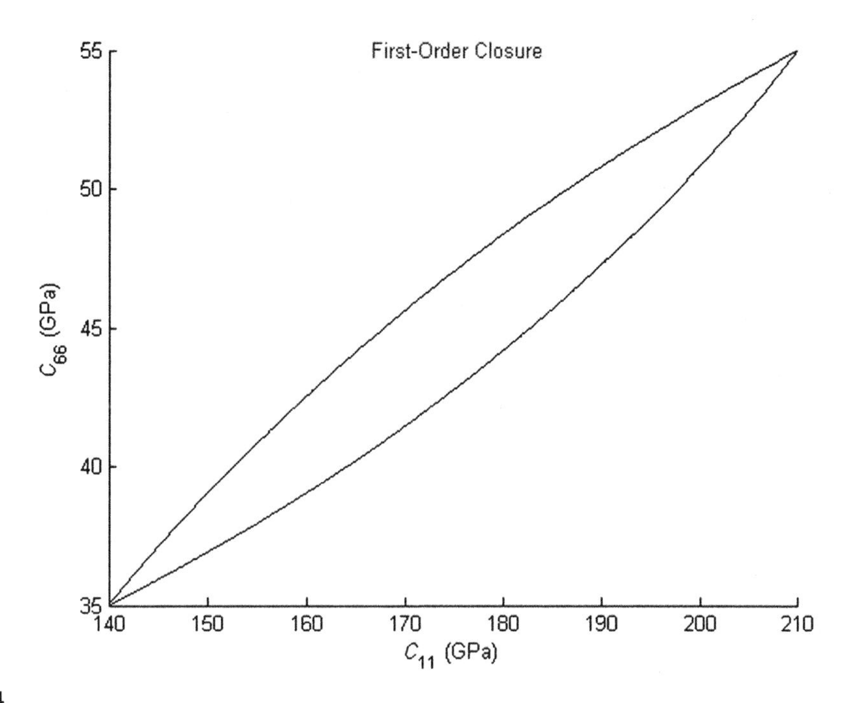

FIGURE 9.4

Property first-order closure for C_{11} and C_{66} for the composite of two isotropic phases.

Since there are only two points in the discrete state space there is no advantage to using the spectral form of the bounds. However, this will not be the case for a much larger state space, such as the ODF for a polycrystal.

Example 9.2

We demonstrate the polycrystal case with a simple example first by repeating the previous example for an anisotropic copper, with the fundamental zone divided into only two cells. The properties of the (copper) crystal are given by $C_{1111} = 168.4$ GPa, $C_{1122} = 121.4$ GPa, and $C_{2323} = 75.4$ GPa. The two orientations represented in the fundamental zone have Euler angles $(\varphi_1, \varphi, \varphi_2) = (3.51, 0.95, 0.58)$ and $(1.38, 1.07, 0.38)$.

Solution

For each orientation, the associated rotation matrix, g, is given by Eq. (2.38). The values of the local stiffness matrix can be calculated from

$$C_{ijkl} = C_{1122}\delta_{ij}\delta_{kl} + C_{2323}\left(\delta_{ik}\delta_{jl} + \delta_{il}\delta_{jk}\right) + (C_{1111} - C_{1122} - 2C_{2323})\sum_{m=1}^{3} g_{im}g_{jm}g_{km}g_{lm}$$

Hence, $C^1_{1111} = 228$ GPa, $C^2_{1111} = 208.2$ GPa, $C^1_{1212} = 50.6$ GPa, $C^2_{1212} = 71.2$ GPa, and so on.

We now follow the same process as in Example 9.1, resulting in a closure as shown in Figure 9.5.

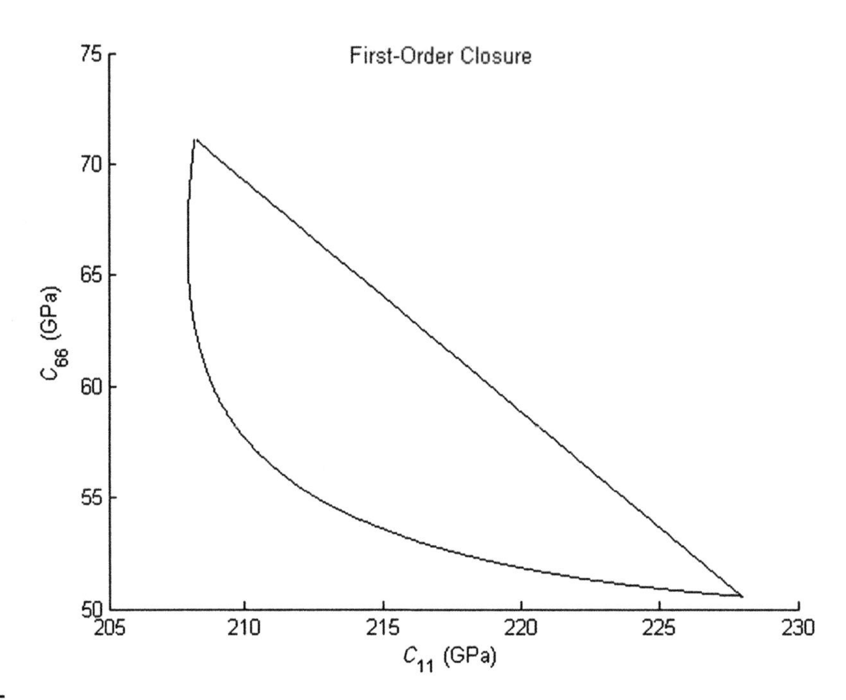

FIGURE 9.5

Property first-order closure for C_{11} and C_{66} for copper.

If the fundamental zone were now divided into, say, five-degree bins (5832 bins in total), the spectral form of the equations would require that we use less than 5% of the terms in the nonspectral formulation.

SUMMARY

This chapter combined the various concepts of the previous chapters to form the two underlying design spaces for microstructure-sensitive design: the microstructure hull and the property closure. A designer will focus on a particular closure relevant to the design problem of interest, in a similar way in which engineers have used Ashby charts to select materials that suit particular applications. However, in this new approach, not only does each point in the closure represent a realizable property combination (based on the assumed homogenization relation), but a particular microstructure (or set of microstructures) that achieve the desired property combination may be found in the relevant hull. This process is demonstrated in Chapter 10.

CHAPTER

Design for Performance Optimization

10

CHAPTER OUTLINE

With the design spaces for structure and properties defined in Chapter 9, the microstructure design framework may now be applied to a range of engineering problems. In this chapter a process for first-order microstructure-sensitive design is briefly explained, and then demonstrated through a series of case studies. By "first-order" we mean that the structure metrics employed in the design only contain volume fraction information (in the form of distributions, such as the orientation distribution function) and no geometrical information.

The performance of an engineered component is generally determined by several key properties, such as stiffness, thermal expansion coefficient, and toughness. With knowledge of the property closure for these key properties, a designer may search for the material design that optimizes the properties required for performance given geometrical and other constraints.

For our current first-order analysis, each point in the property closure is associated with a point (or many points) in state space H, which gives the distribution function for each parameter in H, and thus defines the design of the microstructure to first order. For example, if we consider only the effects of crystal orientation on the properties of interest, then a point in the property closure will be associated with an orientation distribution function (or a set of ODFs), thus defining the texture(s) of the optimal material.

Chapter 9 discussed a method of finding points on the boundary of the property closure, albeit in a somewhat ad hoc manner. More sophisticated optimization techniques will be introduced in

Chapter 14, but any simple method of varying the parameters of H to find boundary points will suffice for now. It is also assumed for the present that the designer is interested in unconstrained optimization on the closure; that is, the required optimum will be a boundary point. This is clearly not generally the case but will suit a sufficient number of examples to be a useful short-term assumption.

The design process, then, involves the designer searching the property closure for the optimum property for a given component. Once this has been found, there must be a mapping from the property closure to state space H that determines the actual microstructure design. On the other hand, if the state space is kept small by discretization (as in Example 9.2, for example, where only two orientations are considered), then one may search directly in the microstructure hull, mapping each point to the property closure during the search. We leave consideration of this to the student via the exercises.

10.1 DESIGN PROCESS USING GSH

When the term microstructure refers to the distribution of crystal lattice orientations in a material, the most efficient representation of the structure usually involves generalized spherical harmonic functions. We demonstrate in this section how this efficiency can be used in the design process for materials with desired yielding properties.

10.1.1 Designing for yield in cubic-orthorhombic material

Here we introduce an example of how the bounding theory may be combined with the Fourier spectral analysis (in the form of harmonic functions) to solve a physical design problem. We repeat the first-order bounding relations obtained earlier for diagonal elements of the effective stiffness tensor (see Section 8.2):

$$(\overline{S})^{-1}_{ijij} \le C^{*}_{ijij} \le \overline{C}_{ijij} \tag{10.1}$$

and for off-diagonal terms:

$$\begin{aligned} &\max\left(\overline{C}_{iijj}, (\overline{S})^{-1}_{iijj}\right) - \sqrt{\Delta_i \Delta_j} \le C^{*}_{iijj} \le \min\left(\overline{C}_{iijj}, (\overline{S})^{-1}_{iijj}\right) + \sqrt{\Delta_i \Delta_j}, \quad \text{when } i \neq j \\ &\Delta_i = \overline{C}_{iiii} - (\overline{S})^{-1}_{iiii} \end{aligned} \tag{10.2}$$

For the cubic-orthorhombic microstructures considered here, Eqs. (10.1) and (10.2) provide all of the nonzero effective stiffness components needed by mechanical designers for their finite-element analyses software.

A common formalism for describing anisotropic yield surfaces, which is also available as a standard option in many commercial finite-element software used by mechanical designers, is the orthorhombic Hill's yield surface description (Hill, 1948):

$$\begin{aligned} &\frac{1}{2}\left(\frac{1}{\sigma_{y2}^2} + \frac{1}{\sigma_{y3}^2} - \frac{1}{\sigma_{y1}^2}\right)(\overline{\sigma}_{22} - \overline{\sigma}_{33})^2 + \frac{1}{2}\left(\frac{1}{\sigma_{y3}^2} + \frac{1}{\sigma_{y1}^2} - \frac{1}{\sigma_{y2}^2}\right)(\overline{\sigma}_{33} - \overline{\sigma}_{11})^2 \\ &+ \frac{1}{2}\left(\frac{1}{\sigma_{y1}^2} + \frac{1}{\sigma_{y2}^2} - \frac{1}{\sigma_{y3}^2}\right)(\overline{\sigma}_{11} - \overline{\sigma}_{22})^2 + \frac{1}{\tau_{y12}^2}\overline{\tau}_{12}^2 + \frac{1}{\tau_{y23}^2}\overline{\tau}_{23}^2 + \frac{1}{\tau_{y13}^2}\overline{\tau}_{13}^2 = 1 \end{aligned} \tag{10.3}$$

where σ_{y1}, σ_{y2}, σ_{y3} denote the tensile (or compressive) yield strengths and τ_{y12}, τ_{y13}, τ_{y23} denote the shear yield strengths. The six anisotropic yield parameters in Eq. (10.3) can be bounded by a modified version of Sachs' model (Sachs, 1928) and an extended version of Taylor's plasticity model (Taylor, 1938). Both these models consider plasticity at the single-crystal level and make assumptions regarding the stress and strain distributions in the constituent crystals in arriving at the effective response of the polycrystal.

The crystal plasticity framework used in this example is a visco-plastic model, in which a power law has been used to describe the rate dependence of the slip (Hutchinson, 1976):

$$\dot{\gamma}^{\alpha} = \dot{\gamma}^{o}\left(\frac{|\tau^{\alpha}|}{s}\right)^{\frac{1}{m}}\text{sign}(\tau^{\alpha}) \tag{10.4}$$

where $\dot{\gamma}^{\alpha}$ and τ^{α} are the slip rate and the resolved shear stress, respectively, on the slip system α, $\dot{\gamma}^{o}$ and m are material parameters representing the reference shear rate and the instantaneous rate sensitivity of the slip, respectively, and s is the slip system deformation resistance (assumed to be equal for all slip systems in a given metal, a reasonable approximation for the cubic polycrystalline metals considered in this study).

Assuming quasistatic loading conditions and low homologous temperatures, in the present example we have set $\dot{\gamma}^{o} = 0.001\ s^{-1}$ and $m = 0.01$. The resolved shear stress is related to the applied stress tensor, $\boldsymbol{\sigma}$, as

$$\tau^{\alpha} = \boldsymbol{\sigma}\cdot\frac{1}{2}(\mathbf{m}^{\alpha}\otimes\mathbf{n}^{\alpha} + \mathbf{n}^{\alpha}\otimes\mathbf{m}^{\alpha}) \tag{10.5}$$

where $\mathbf{m}^{\alpha}$ and $\mathbf{n}^{\alpha}$ represent the slip direction and the slip plane normal for slip system α, respectively. Eqs. (10.4) and (10.5) can be combined to yield the following expression for the plastic power, $\dot{\omega}^{p}$, in each crystal:

$$\dot{\omega}^{p} = \frac{\dot{\gamma}^{o}}{s^{\frac{1}{m}}}\sum_{\alpha}\left|\frac{\sigma_{ij}}{2}\left(m_i^{\alpha}n_j^{\alpha} + n_i^{\alpha}m_j^{\alpha}\right)\right|^{1+\frac{1}{m}} \tag{10.6}$$

Because plastic deformation occurs in the visco-plastic models at all nonzero values of stress, the definition of a yield point presents some difficulties. Here, we follow an approach used in prior MSD case studies, and define the macroscale yield by setting the volume-averaged plastic power in the polycrystal to a critical value, that is, $\bar{\dot{\omega}}^{p} = \dot{\omega}^{p}_{cr}$. In the present example, $\dot{\omega}^{p}_{cr}$ was taken as equal to the product of an average yield strength of the material (taken as three times the critical resolved shear strength) and 0.001 sec^{-1} (a typical value for a strain rate).

To determine the lower bound for the yield strength, we assume that in each grain the same stress tensor is applied and this stress tensor is equal to the average stress tensor, $\bar{\boldsymbol{\sigma}}$. For a selected choice of a unit stress tensor $\bar{\boldsymbol{\sigma}}/|\bar{\boldsymbol{\sigma}}|$, Eq. (10.6) can be used to find the effective yield strength. As an example, the desired expression for evaluating the lower bound on σ_{y1} is

$$\bar{\dot{\omega}}^{p} = \frac{\sigma_{y1}^{1+\frac{1}{m}}\dot{\gamma}^{o}}{s^{\frac{1}{m}}}\sum_{\alpha}\left|m_1^{\alpha}(g)n_1^{\alpha}(g)\right|^{1+\frac{1}{m}} = \dot{\omega}^{p}_{cr} \tag{10.7}$$

where the functional dependence of m_1^{α} and n_1^{α} (components of the slip directions and the slip plane normals) on the crystal orientation g is specifically noted. Expressions similar to Eq. (10.7) have been derived for all six parameters needed in the anisotropic yield surface adopted in this study (see Eq. 10.3).

For the upper bound, we use an extended version of Taylor's model (Taylor, 1938). The main assumption in this model is that the deformation gradient is uniform in all the grains, independent of the crystal orientation. Consequently, in this model, we impose known deformation histories and evaluate the resulting stress conditions (Kalidindi, 1992). In this study, the macroscopic velocity gradient tensor is provided as input to the Taylor-type crystal plasticity model, and the macroscopic stress history is obtained as an output from the model. As an example, in evaluating σ_{y1}, the macroscopic velocity gradient takes the form

$$\overline{\mathbf{L}} = \begin{pmatrix} \dot{\bar{\varepsilon}} & 0 & 0 \\ 0 & -\mathrm{q}\dot{\bar{\varepsilon}} & 0 \\ 0 & 0 & -(1-\mathrm{q})\dot{\bar{\varepsilon}} \end{pmatrix} \tag{10.8}$$

where q can take any value between 0 and 1. Because the deformation imposed by Eq. (10.8) is isochoric, the resulting stress field is purely deviatoric. Furthermore, the effective stress tensor is taken as the averaged quantity over the entire polycrystal. To calculate the tensile yield strength, the hydrostatic component is computed by establishing the value of q (denoted as q^*) for which the averaged lateral stresses over the polycrystal are equal to each other and added to the diagonal elements of the averaged deviatoric stress:

$$\overline{\sigma}'_{33}(\mathrm{q}^*) = \overline{\sigma}'_{22}(\mathrm{q}^*), \qquad \sigma_{y1} = \overline{\sigma}'_{11}(\mathrm{q}^*) - \overline{\sigma}'_{22}(\mathrm{q}^*) \tag{10.9}$$

where $\overline{\sigma}'$ denotes the volume-averaged deviatoric stress tensor.

Note that the first-order bounds shown in Eqs. (10.1) and (10.2) require evaluation of volume-averaged quantities. One major advantage of the spectral (Fourier) representation (see Section 6.2.7) is that such volume-averaged quantities can be computed most economically by exploiting the orthonormality of the Fourier basis. The following desired relationships can be derived by combining Eqs. (6.62), (7.61), and (7.65). For example, the volume-averaged $\overline{C}_{iiii}$ can be expressed as (no summation on repeated indices in the following equations)

$$\begin{aligned} \overline{C}_{iiii} &= \oint C_{iiii}(g) f(g)\, \mathrm{d}g \\ &= C_{12} + 2C_{44} + (C_{11} - C_{12} - 2C_{44}) \left[{}_{iiii}A_0^{11} + \frac{1}{9} \sum_{\nu=1}^{3} {}_{iiii}A_4^{1\nu} F_4^{1\nu} \right] \end{aligned} \tag{10.10}$$

In an analogous treatment, the following expressions have been derived:

$$\overline{C}_{ijij} = C_{44} + (C_{11} - C_{12} - 2C_{44}) \left[{}_{ijij}A_0^{11} + \frac{1}{9} \sum_{\nu=1}^{3} {}_{ijij}A_4^{1\nu} F_4^{1\nu} \right] \text{ when } i \neq j \tag{10.11}$$

$$\overline{C}_{iijj} = C_{12} + (C_{11} - C_{12} - 2C_{44}) \left[{}_{ijij}A_0^{11} + \frac{1}{9} \sum_{\nu=1}^{3} {}_{ijij}A_4^{1\nu} F_4^{1\nu} \right] \text{ when } i \neq j \tag{10.12}$$

$$\overline{S}_{iiii} = S_{12} + \frac{S_{44}}{2} + \left(S_{11} - S_{12} - \frac{S_{44}}{2}\right)\left[{}_{iiii}A_0^{11} + \frac{1}{9}\sum_{\nu=1}^{3} {}_{iiii}A_4^{1\nu} F_4^{1\nu}\right] \quad (10.13)$$

$$\overline{S}_{ijij} = \frac{S_{44}}{4} + \left(S_{11} - S_{12} - \frac{S_{44}}{2}\right)\left[{}_{ijij}A_0^{11} + \frac{1}{9}\sum_{\nu=1}^{3} {}_{ijij}A_4^{1\nu} F_4^{1\nu}\right] \text{ when } i \neq j \quad (10.14)$$

$$\overline{S}_{iijj} = S_{12} + \left(S_{11} - S_{12} - \frac{S_{44}}{2}\right)\left[{}_{ijji}A_0^{11} + \frac{1}{9}\sum_{\nu=1}^{3} {}_{ijij}A_4^{1\nu} F_4^{1\nu}\right] \text{ when } i \neq j \quad (10.15)$$

Note that Eqs. (10.10) through (10.15) describe iso-property planes for a prescribed value of the components of the volume-averaged stiffness and compliance tensors. The intersections of these planes with the microstructure hull depicted in Figure 9.3 identify the complete set of ODFs that possess a specified value of $\overline{C}_{ijkl}$ or $\overline{S}_{ijkl}$. In this work, the same equations are being used to obtain bounds for the effective properties. Indeed, some of the bounds are given directly by the volume-averaged quantities themselves, while others require additional computations as described in Eqs. (10.1) and (10.2). Consequently, some of the relationships between microstructure coefficients $F_\ell^{\mu\nu}$ and effective bounds are linear, while others exhibit significant nonlinearity.

Following an approach similar to the one described earlier for elastic properties, the relationship between microstructure and the effective plastic yield properties can also be expressed in the Fourier space (the texture dependence arises primarily from Eqs. (10.4) and (10.5)). For the effective plastic yield properties identified in Eq. (10.3), a significantly larger number of Fourier coefficients are required, compared to the representation of the elastic properties described earlier. Truncating the Fourier representation to terms up to ℓ equal to 6 was found to result in a truncation error less than 10%, which in the present example is deemed acceptable.

To compute the lower bounds using equations of the type that is shown in Eq. (10.7) using the MSD framework described here, we first need to obtain Fourier representations for terms such as $\sum_\alpha |m_1^\alpha(g) n_1^\alpha(g)|^{1+\frac{1}{m}}$. There are six such terms corresponding to the six different parameters in the anisotropic yield surface description of Eq. (10.3). As an example, the Fourier representation for one of these terms is expressed as

$$\sum_\alpha |m_1^\alpha(g) n_1^\alpha(g)|^{1+\frac{1}{m}} = \sum_{\ell=0}^{\infty}\sum_{\mu=1}^{M(\ell)}\sum_{\nu=1}^{N(\ell)} {}_{y1}S_\ell^{\mu\nu}\, \dot{\ddot{T}}_\ell^{\mu\nu}(g)$$

$${}_{y1}S_\ell^{\mu\nu} = (2l+1) \oint \sum_\alpha |m_1^\alpha(g) n_1^\alpha(g)|^{1+\frac{1}{m}} \ddot{T}_\ell^{\mu\nu}(g)\,dg \quad (10.16)$$

The volume-averaged quantity of interest is then given as

$$\overline{\sum_\alpha |m_1^\alpha(g) n_1^\alpha(g)|^{1+\frac{1}{m}}} = \sum_{\ell=0}^{\infty}\sum_{\mu=1}^{M(\ell)}\sum_{\nu=1}^{N(\ell)} \frac{{}_{y1}S_\ell^{\mu\nu} F_\ell^{\mu\nu}}{2l+1} \quad (10.17)$$

and the lower bound of the effective yield strength, σ_{y1}, is then obtained as (see Eq. 10.7)

$$\frac{\sigma_{y1}}{s} = \left(\frac{\dot{\omega}^{p}_{cr}}{s\,\dot{\gamma}^{o} \sum_{\ell=0}^{\infty} \sum_{\mu=1}^{M(\ell)} \sum_{\nu=1}^{N(\ell)} \frac{{}_{y1}S_{\ell}^{\mu\nu} F_{\ell}^{\mu\nu}}{2l+1}} \right)^{\frac{m}{m+1}} \tag{10.18}$$

Table 10.1 provides the values of $S_{\ell}^{\mu\nu}$ for the six different tensile and shear yield strengths.

As described earlier, the computation of the yield parameters from the upper-bound theory is driven by prescribed velocity gradient tensors as described in Eq. (10.8). To facilitate these computations, the local crystal stresses computed by the Taylor-type model are expressed in a Fourier series. As an example, the Fourier representations of the stresses needed for evaluating the upper bound for σ_{y1} are expressed as (no summation implied on repeated indices)

$$\begin{aligned} \sigma'_{ii}(g,\mathrm{q}) &= \sum_{\ell=0}^{\infty} \sum_{\mu=1}^{M(\ell)} \sum_{\nu=1}^{N(\ell)} {}^{ii}_{y1}S_{\ell}^{\mu\nu}(\mathrm{q})\, \ddot{\mathrm{T}}_{\ell}^{\mu\nu}(g) \\ {}^{ii}_{y1}S_{\ell}^{\mu\nu}(\mathrm{q}) &= (2l+1) \oint \sigma'_{ii}(g,\mathrm{q}) \ddot{\mathrm{T}}_{\ell}^{\mu\nu}(g)dg \\ \overline{\sigma}'_{ii}(\mathrm{q}) &= \sum_{\ell=0}^{\infty} \sum_{\mu=1}^{M(\ell)} \sum_{\nu=1}^{N(\ell)} \frac{{}^{ii}_{y1}S_{\ell}^{\mu\nu}(\mathrm{q})\, F_{\ell}^{\mu\nu}}{2l+1} \end{aligned} \tag{10.19}$$

These equations can then be used with Eq. (10.9) to find the upper bound for σ_{y1}. To facilitate these computations, it was found convenient to express each of the Fourier coefficients ${}^{ii}_{y1}S_{\ell}^{\mu\nu}(\mathrm{q})$ as a sixth-order polynomial function in the variable q. Table 10.2 provides a summary of the coefficients needed for the computation of the tensile yield strengths by the upper-bound theory. Each coefficient is expressed as a sixth-order polynomial of the form $a_0 + a_1 q + a_2 q^2 + a_3 q^3 + a_4 q^4 + a_5 q^5 + a_6 q^6$ in the case of the normal yield strengths. In this table the Fourier coefficients are given for ℓ equal up to 4, but the property closures were delineated using the Fourier coefficients for ℓ equal up to 6.

It is noted that the results presented in Table 10.2 provide a comprehensive list of microstructure–property linkages needed by the designers in employing the MSD framework. Furthermore, these are presented in a format that can be readily used by mechanical designers. The coefficients in Table 7.1

Table 10.1 $S_{\ell}^{\mu\nu}$ Coefficient Values for Lower-Bound Plastic Yield Strength Predictions

	Fourier Coefficients, $\ell\ \mu\ \nu$							
	011	411	412	413	611	612	613	614
σ_{y1}	2.233	−0.157	0.235	−0.311	0.172	−0.249	0.273	−0.370
σ_{y2}	2.233	−0.157	−0.235	−0.311	0.172	0.249	0.273	0.370
σ_{y3}	2.234	−0.423	0	0	−0.552	0	0	0
τ_{y12}	1.282	0.296	0	0.532	0.107	0	0.065	0
τ_{y13}	1.282	0.504	−0.186	0.286	−0.065	−0.059	0	−0.088
τ_{y23}	1.282	0.504	0.186	0.286	−0.065	0.059	0	0.088

Table 10.2 ${}_{ii}S_{\ell}^{\mu\nu}(q)$ Coefficient Values Needed to Compute Tensile and Shear Yield Strengths

	Fourier Coefficients, σ_{y1}											
	σ'_{11}				σ'_{22}				σ'_{33}			
$\ell\mu\nu$	011	411	412	413	011	411	412	413	011	411	412	413
a_0	1.6	−0.4	0.8	0	0	−0.1	−0.1	0.2	−1.6	0.6	−0.6	−0.2
a_1	1.3	−0.5	−1.5	−0.4	−2.8	1.4	3.4	1.5	1.5	−1	−1.9	−1
a_2	2.4	0.1	−1.1	−12.7	−2.4	−9.4	−5.4	16.6	0.2	9.1	6.2	−4.5
a_3	−19.2	19.7	17.5	50.3	20.5	6.4	−1.7	−63.2	−1.8	−24.9	−14.2	14.6
a_4	39.1	−61.1	−36.4	−80.8	−34.6	13.4	14.4	88.7	−3.7	44.8	19	−10.1
a_5	−35.4	67.9	27.4	59.3	23.6	−18.7	−18.2	−57.5	11.2	−46.3	−6.7	−0.3
a_6	11.8	−25.4	−6.6	−16.4	−5.8	6.7	7.9	14.5	−5.8	17.7	−1.9	1.6
	Fourier Coefficients, σ_{y2}											
	σ'_{11}				σ'_{22}				σ'_{33}			
$\ell\mu\nu$	011	411	412	413	011	411	412	413	011	411	412	413
a_0	−1.6	−0.3	−0.1	0.8	1.6	0.3	−0.1	−0.8	0	0	0.2	0
a_1	1.5	−2.2	0.8	0.4	1.3	−1.3	0.7	0.6	−2.9	3.5	−1.5	−1
a_2	0.2	3.5	−11.2	2.1	2.4	−10	−7.9	−0.8	−2.4	5.8	19	−1.2
a_3	−1.8	−6.5	30.5	−7.1	−19.2	54.5	13.6	9.2	20.5	−45.4	−44.5	−2.1
a_4	−3.8	19.9	−41	19	39.1	−103	−1.2	−31.1	−34.5	78.9	43.9	11.6
a_5	11.3	−21.3	28.9	−29.6	−35.4	84.8	−12.4	39.4	23.5	−59.9	−18.4	−9
a_6	−5.8	6.7	−7.9	14.5	11.8	−25.4	6.6	−16.4	−5.8	17.7	1.9	1.6
	Fourier Coefficients, σ_{y3}											
	σ'_{11}				σ'_{22}				σ'_{33}			
$\ell\mu\nu$	011	411	412	413	011	411	412	413	011	411	412	413
a_0	0	−1.6	1.6	0.1	−0.6	−0.2	0	0.2	0.1	0.1	0	0
a_1	−2.9	1.5	1.3	−0.2	0.8	3.5	−2.1	−1.4	1.9	0.9	0.1	−0.1
a_2	−2.4	0.2	2.4	11.9	9	−8.5	10.3	−2.2	−30.7	−13	2.5	−0.8
a_3	20.5	−1.8	−19.2	−43.4	−31.1	−4.3	−26	32.5	127.5	41.4	−12.6	4.7
a_4	−34.6	−3.7	39.1	62.5	37.7	30.6	44.4	−79.5	−246	−49.1	18.9	−3.6
a_5	23.5	11.2	−35.4	−39.2	−15.1	−33.2	−38.7	75.9	225.4	19.6	−11.3	−2.3
a_6	−5.8	−5.8	11.8	8.8	0	12	12	−25.3	−78.3	0	2.3	2.3
	Fourier Coefficients, τ_{y12}				**Fourier Coefficients, τ_{y13}**				**Fourier Coefficients, τ_{y23}**			
$\ell\mu\nu$	011	411	412	413	011	411	412	413	011	411	412	413
	2.28	2.21	2.1	1.97	1.59	0.7	−0.38	0.3	1.59	0.7	0.38	0.3

can be used with any cubic polycrystalline metal and provide all of the coefficients needed. The coefficients presented in Tables 10.1 and 10.2 can be used with any fcc polycrystalline metal with the $(111)\langle 1\bar{1}0 \rangle$ family of slip systems, and represent the dominant terms in the series (further terms are needed if higher accuracy is desired; higher-order terms in the series expansion have been evaluated and used in the results displayed later in Figure 10.2). The coefficients provided cover both bounds for all of the 15 elastic–plastic effective properties needed in typical mechanical design. This information should make the MSD framework much more accessible to mechanical designers.

Using the Fourier representations described before, we are now ready to compute various elastic–plastic property closures, which may in turn be searched for optimum designs. The overall strategy for obtaining the closures is described schematically in Figure 10.1. To delineate a property closure we fix the value of the first property P^* and identify the minimum and maximum values of the second property using standard optimization techniques. An important constraint in this process would be to ensure that the microstructure coefficients are restricted to lie inside the texture hull that we determined earlier (see Figure 9.3). Herein lies the main advantage of working in the Fourier space.

We describe the technique used to determine the maximum value of Q for a given value of P; a similar technique is used to determine the minimum value of Q. As explained earlier, some of the properties, such as the diagonal components of the elastic stiffness tensor, have linear relationships with the Fourier coefficients of the texture, and therefore are represented by an iso-property plane intersecting the texture hull. Other properties, such as the yield strengths, have nonlinear relationships with the Fourier coefficients of the texture; therefore, instead of planes, these properties are represented by curved iso-property surfaces. The intersection of two particular planes or surfaces inside the hull represents the set of all the textures possessing both properties simultaneously.

To determine the maximum of property Q for a given property $P = P^*$, we start by selecting a set of directions that intersect the iso-property surface for the property $P = P^*$. This way, we can determine the Fourier coefficients of the microstructures for which the property P takes the value of P^*. Once we have these coefficients, it is possible to calculate the value of the second property Q for those selected textures. In selecting these initial sets of microstructures on the hypersurface defined by P^*, we try to ensure that the microstructures selected correspond to high values for Q (among the directions we try) and that they are reasonably spread out from each other in the Fourier space (the idea is to cover as much of the iso-property surface as possible).

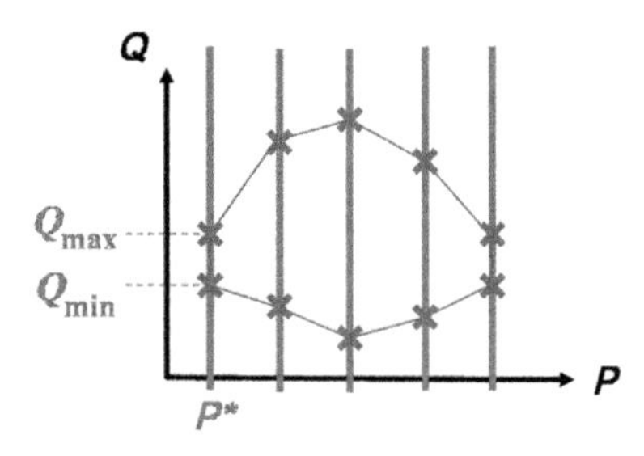

FIGURE 10.1

Schematic of the procedure to delineate the P-Q property closure. For a particular value of $P = P^*$, we are able to determine the maximum and minimum values of the second property Q while constraining the Fourier coefficients representing the texture to be inside the texture hull. The line joining all the points representing the different minimum and maximum values of Q for given values of P delineate the P-Q closure.

Once these textures are selected, we begin a search for the optimized value of Q using a gradient approach (Lyon and Adams, 2004). The gradient of a property defined as $\frac{\partial Q}{\partial F_{\ell}^{\mu\nu}}$ reveals the direction in which the function increases the most in the Fourier space. We determine the gradient at each selected point. We project this gradient onto the iso-property surface, which guarantees that the second property Q increases in that direction while remaining in the hypersurface $P = P^*$. For each preselected point on the hypersurface $P = P^*$, we determine the texture that has the highest value of Q in the direction of the projected gradient.

We continue this process either until the gradient at the optimized point is normal to the iso-property surface, in which case our search is done, or until the microstructure having the highest value for Q ends up on the surface of the texture hull. In this later case, we continue the search onto the surface of the hull until the value of Q does not increase anymore. This technique allows the determination of all the feasible combinations of properties for a given material, taking into consideration either of the bounding theories.

Some examples of property closures computed using the strategy just described are shown next. We first show examples of upper-bound closures for copper polycrystals in Figure 10.2. The dark gray regions in this figure represent the closures containing all of the feasible combinations (from the upper-bound relations) of values for the selected properties.

10.2 MICROSTRUCTURE DESIGN OF A COMPLIANT MECHANISM

Compliant mechanisms are mechanical devices that gain movement from parts that flex, bend, or have "springiness" to them (Midha et al., 1994). These mechanisms with built-in flexibility are practical since they are simple. They also eliminate the need for multiple rigid parts, pin joints, and add-on springs that can increase design complexity and cost. Compliant mechanisms are found in sensors, gearboxes, valves, bicycle derailleurs, and various other mechanical designs. In this case study, we seek the texture(s) that will maximize the deflection of the beam without initiating plastic deformation.

The compliant mechanism is idealized here as a long, slender cantilever beam of which the macroscale elastic–plastic properties exhibit orthorhombic symmetry (presumably the processing options have been restricted to accomplish this). The stress field in the cantilever beam, with one end fixed to a rigid surface and the other end subjected to a point load P (see Figure 10.3), is expressed as (Lekhnitskii, 1968)

$$\sigma_{11} = -\frac{12P}{hw^3} x_1 x_2 \tag{10.20}$$

where h and w are the beam height and width, respectively. Since the normal stresses are much higher than the shear stresses in a slender beam, we have ignored the shear stresses in this case study.

The application of Hill's anisotropic yield criterion (discussed in more detail in the following) requires

$$\frac{|\sigma_{11}|}{\sigma_{y1}} \leq 1 \tag{10.21}$$

The maximum deflection in the cantilever beam, at the time of the initiation of plastic strain, is expressed as

$$\delta = \frac{2}{3} \sigma_{y1} S_{1111} \frac{L^2}{w} \tag{10.22}$$

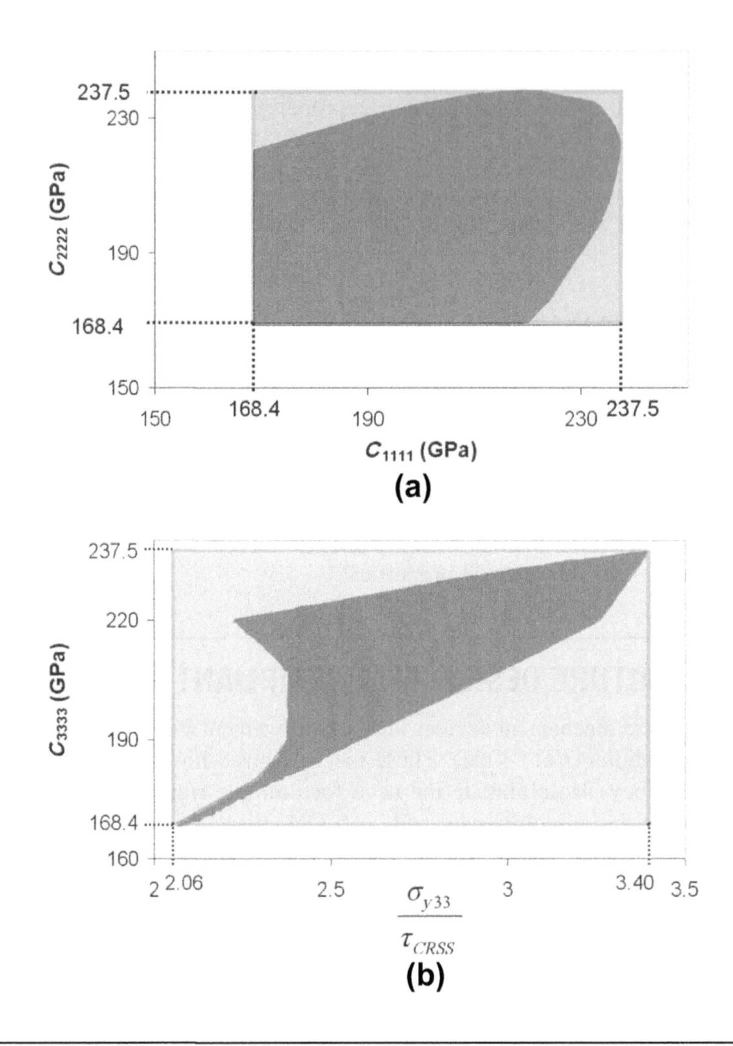

FIGURE 10.2

Examples of upper-bound closures for copper polycrystals: (a) C_{1111}/C_{2222} property closure, and (b) normalized yield strength/C_{3333} property closure.

where L is the length of the beam. For a fixed-beam geometry, the maximum deflection that can be attained without initiating plastic strain is therefore dependent only on the macroscale material properties S_{1111} (the first term of the compliance tensor) and σ_{y1}. In the case study presented here, the beam is assumed to have a square cross-section with $b = w = 18$ mm and $L = 180$ mm.

The microstructural design variable for this case study has been selected to be the ODF (or texture) in the beam. The material for the case study is assumed to be polycrystalline high-purity alpha-phase titanium (α-Ti). Since the sample is made from a hexagonal metal and the macroscale properties are expected to exhibit orthorhombic symmetry, the space of relevant ODFs for this case study is the class of hexagonal-orthorhombic ODFs. The application of the MSD framework to hcp polycrystals

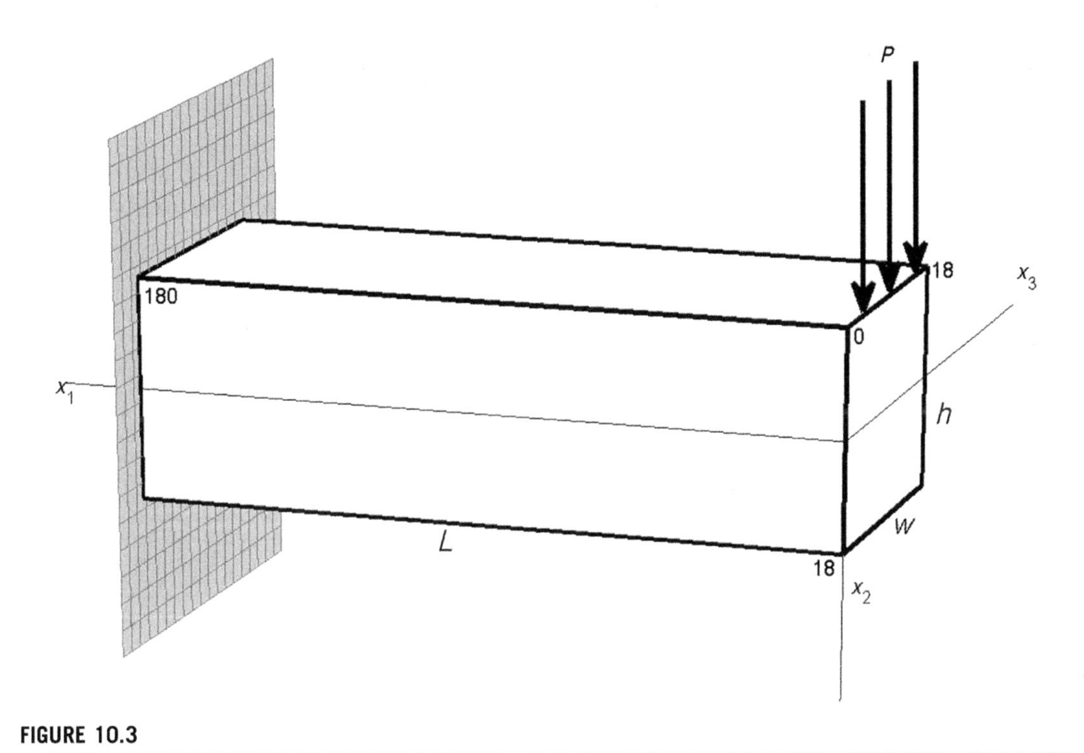

FIGURE 10.3

Schematic of the cantilever-compliant beam mechanism for the microstructure design case study.

required the consideration of a significantly larger number of dimensions of the texture hull when compared to the previous case studies that involved cubic polycrystals. Hence, some changes are required to deal with the mathematical description of the design space, that is, the microstructure (texture) hull. The approach is described and illustrated with two design case studies involving different assumptions of symmetry at the sample scale.

10.2.1 Microstructure representation for hexagonal material

Recall that the ODF, denoted as $f(g)$, reflects the normalized probability density associated with the occurrence of the crystallographic orientation g in the sample as

$$f(g)\,dg = \frac{V_g}{V}, \quad \int_{FZ} f(g)dg = 1 \tag{10.23}$$

where V denotes the total sample volume and V_g is the sum of all subvolume elements in the sample that are associated with a lattice orientation that lies within an incremental invariant measure, dg, of the orientation of interest, g. The lattice orientation, g, is defined by a set of three angles, $g = (\varphi_1, \Phi, \varphi_2)$. FZ denotes the fundamental zone of orientations and represents the local state space describing the set of distinct orientations relevant to a selected class of textures; for hexagonal-orthorhombic[1] textures it is described by

$$FZ = \left\{ g = (\varphi_1, \Phi, \varphi_2) \,\middle|\, 0 \le \varphi_1 \le \frac{\pi}{2}, 0 \le \Phi \le \frac{\pi}{2},\ 0 \le \varphi_2 \le \frac{\pi}{3} \right\} \tag{10.24}$$

The ODF can be expressed in a Fourier series using symmetrized generalized spherical harmonics (GSH), $T_l^{\mu\nu}(g)$, as

$$f(g) = \sum\nolimits_{l=0}^{\infty} \sum\nolimits_{\mu=1}^{M(l)} \sum\nolimits_{\nu=1}^{N(l)} F_l^{\mu\nu} T_l^{\mu\nu}(g) \tag{10.25}$$

This equation facilitates visualization of the ODF as a point in Fourier space the coordinates of which are given by$F_l^{\mu\nu}$. Recognition of the fact that ODF provides information only about the volume fraction of the various orientations present in the polycrystalline sample permits an alternate mathematical description as

$$f(g) = \sum_k \alpha_k\, \delta\left(g - g^k\right), \quad 0 \le \alpha_k \le 1\,, \ \sum_k \alpha_k = 1 \tag{10.26}$$

The Dirac-delta function $\delta(g - g^k)$ represents the ODF of a single crystal of orientation g^k, and α_k denotes its volume fraction in the polycrystal. Let ${}^kF_l^{\mu\nu}$ define the Fourier coefficients of the single-crystal ODFs. It is then possible to define a convex and compact microstructure hull, M, as

$$M = \left\{ F_l^{\mu\nu} \,\middle|\, F_l^{\mu\nu} = \sum_k \alpha_k {}^kF_l^{\mu\nu},\ {}^kF_l^{\mu\nu} \in M^k, \alpha_k \ge 0\,, \sum_k \alpha_k = 1 \right\} \tag{10.27}$$

where

$$M^k = \left\{ {}^kF_l^{\mu\nu} \,\middle|\, {}^kF_l^{\mu\nu} = \frac{1}{(2l+1)} T_l^{*\mu\nu}\left(g^k\right),\ g^k \in FZ \right\} \tag{10.28}$$

The asterisk after the GSH denotes a complex conjugate. It should be recognized that M represents the complete set of all theoretically feasible ODFs, many of which have not yet been realized in practice (or even been targeted for manufacture) by materials specialists.

10.2.2 Material property relations for hexagonal material

Next, we turn our attention to the homogenization theory to be adopted to arrive at the macroscale properties of the polycrystal for a given texture. The ODF described earlier constitutes a first-order description of the microstructure (also referred to as one-point statistics). Using this microstructure description, only the elementary bounds of the macroscale elastic–plastic properties can be evaluated. In this work, we have decided to employ the upper-bound theories for all macroscale elastic–plastic properties. The upper bounds on the diagonal components of the macroscale elastic compliance tensor, **S***, can be expressed as (see Section 8.2; no summation implied on repeated indices)

[1]The first symmetry in this standard notation used by the texture community refers to symmetry at the crystal level (resulting from the atomic arrangements in the crystal lattice), while the second refers to symmetry at the sample scale (resulting from processing history).

$$S^*_{ijij} = \langle S_{ijij} \rangle \tag{10.29}$$

where $\langle\rangle$ denotes an ensemble average (also equal to the volume average when the ergodic hypothesis is invoked). S is the local elastic compliance tensor in the sample reference frame that is defined using a coordinate transformation law for fourth-rank tensors as

$$S_{ijkl} = g_{ip} g_{jq} g_{kr} g_{ls} S^c_{pqrs} \tag{10.30}$$

where S^c is the local elastic compliance tensor in the local crystal reference frame, and g_{ij} are the components of the transformation matrix defined in terms of the Bunge–Euler angles.

The selected homogenization theories are then cast in the same Fourier space that has been used to represent the microstructure. The $S_{abcd}(g)$ functions can be represented in a Fourier series using GSH functions as

$$S_{abcd}(g) = \sum_{l=0}^{4} \sum_{\mu=1}^{N(l)} \sum_{\nu=1}^{M(l)} {}_{abcd}S_l^{\mu\nu} T_l^{\mu\nu}(g) \tag{10.31}$$

where ${}_{abcd}S_l^{\mu\nu}$ are referred to as the elastic Fourier coefficients and the details of their computation are given in Section 5.3.2. Note that the equation for the compliance tensor is a fourth-order polynomial of the transformation matrices. Hence, only the l terms up to fourth order are required to completely describe the function. The volume-averaged value for a particular material structure is then computed by exploiting the orthogonality of the Fourier basis as

$$S^*_{abcd} = \langle S_{abcd} \rangle = \oint S_{abcd}(g) f(g)\, dg = \sum_{l=0}^{4} \sum_{\mu=1}^{N(l)} \sum_{\nu=1}^{M(l)} \frac{1}{(2l+1)}\, {}_{abcd}S_l^{*\mu\nu} F_l^{\mu\nu} \tag{10.32}$$

The asterisk on the left side denotes an "effective" compliance tensor, and the asterisk on the right side indicates complex conjugation of the Fourier coefficients. This equation embodies one of the central features of MSD. It provides an efficient linkage between the complete set of feasible ODFs and the corresponding feasible combinations of macroscale elastic properties. Of particular significance is the fact that, unlike the Fourier representation of the ODF, the representation for properties often extends to only a finite number of terms in the Fourier expansion. As shown in Eqs. (10.31) and (10.32), in consideration of elastic properties, the only relevant Fourier coefficients are those that correspond to $l \leq 4$. The main advantage of the spectral methods described herein lies in the fact that they are able to formulate highly efficient structure–property relationships.

In many mechanical design problems, avoiding plastic deformation constitutes an important consideration. For this purpose, yield surfaces are often described in the 6D stress space.

A common formalism for describing anisotropic yield surfaces is the orthorhombic Hill's yield surface description (Hill, 1948):

$$\begin{aligned} &\frac{1}{2}\left(\frac{1}{\sigma_{y2}^2} + \frac{1}{\sigma_{y3}^2} - \frac{1}{\sigma_{y1}^2}\right)(\langle\sigma_{22}\rangle - \langle\sigma_{33}\rangle)^2 + \frac{1}{2}\left(\frac{1}{\sigma_{y3}^2} + \frac{1}{\sigma_{y1}^2} - \frac{1}{\sigma_{y2}^2}\right)(\langle\sigma_{33}\rangle - \langle\sigma_{11}\rangle)^2 \\ &+ \frac{1}{2}\left(\frac{1}{\sigma_{y1}^2} + \frac{1}{\sigma_{y2}^2} - \frac{1}{\sigma_{y3}^2}\right)(\langle\sigma_{11}\rangle - \langle\sigma_{22}\rangle)^2 + \frac{1}{\tau_{y12}^2}\langle\tau_{12}^2\rangle + \frac{1}{\tau_{y23}^2}\langle\tau_{23}^2\rangle + \frac{1}{\tau_{y13}^2}\langle\tau_{13}^2\rangle = 1 \end{aligned} \tag{10.33}$$

where σ_{y1}, σ_{y2}, σ_{y3} denote the macroscale tensile (or compressive) yield strengths and τ_{y12}, τ_{y13}, τ_{y23} denote the macroscale shear yield strengths; $\langle \sigma_{ij} \rangle$ and $\langle \tau_{ij} \rangle$ denote the macroscale normal and shear stress components, respectively, experienced by the material point under consideration.

The six anisotropic material yield parameters in this equation can be estimated for a polycrystalline material using an extended version of Taylor's model (Taylor, 1938), also commonly referred to as the Taylor-type crystal plasticity model. This model constitutes an upper bound and has been demonstrated (Wu et al., 2007) to provide reasonably accurate predictions of the macroscopic yield strengths for the polycrystalline α-Ti used in the case studies presented here. The Taylor-type crystal plasticity model assumes that each constituent single crystal experiences the same strain as the imposed strain at the macroscale, and computes the local stress in the crystal that allows the accommodation of the imposed plastic strain through slip on the multiple slip systems in the crystal. The volume-averaged stress (over the constituent single crystals) then provides an estimate of the macroscale yield strength for the polycrystal.

As an example, the following macroscopic velocity gradient is imposed on the polycrystal in evaluating σ_{y1}:

$$\langle \mathbf{L} \rangle = \begin{pmatrix} \langle \dot{\varepsilon} \rangle & 0 & 0 \\ 0 & -q\langle \dot{\varepsilon} \rangle & 0 \\ 0 & 0 & -(1-q)\langle \dot{\varepsilon} \rangle \end{pmatrix} \tag{10.34}$$

The parameter q is allowed to take any value between 0 and 1. Because the deformation imposed in this equation is isochoric, the resulting stress field is purely deviatoric. To calculate the tensile yield strength, the hydrostatic component is computed by establishing the value of q (denoted q^*) for which the averaged lateral stresses over the polycrystal are equal to each other and added to the diagonal elements of the averaged deviatoric stress:

$$\sigma_{y1} = \langle {\sigma'}_{11}(q^*) \rangle - \langle {\sigma'}_{22}(q^*) \rangle \tag{10.35}$$

where σ' denotes the volume-averaged deviatoric stress tensor for the polycrystal.

Fast computation of the yield strengths for polycrystals in the MSD framework entails the use of Fourier representations to describe the functional dependence of the local stresses in the constituent crystals on their orientations. The Fourier representations needed for evaluating the upper bound for σ_{y1} are expressed as (no summation implied on repeated index i) (Wu et al., 2007)

$$\sigma'_{ii}(g,q) = \sum_{l=0}^{l^*} \sum_{\mu=1}^{N(l)} \sum_{\nu=1}^{M(l)} {}^{ii}_{y1}S_l^{\mu\nu}(q)\, T_l^{\mu\nu}(g) \tag{10.36}$$

$$\langle \sigma'_{ii}(q) \rangle = \sum_{l=0}^{l^*} \sum_{\mu=1}^{N(l)} \sum_{\nu=1}^{M(l)} \frac{\overline{{}^{ii}_{y1}S_l^{\mu\nu}}(q) F_l^{\mu\nu}}{2l+1} \tag{10.37}$$

where the Fourier coefficients ${}^{ii}_{y1}S_l^{\mu\nu}(q)$ for hcp crystals experiencing slip on prism, basal, and pyramidal slip systems are calculated via a curve-fitting method to detailed crystal plasticity calculations over a range of possible strain paths (Wu et al., 2007).

In describing the plastic properties, we also observed that we needed a lot more Fourier coefficients than those needed to describe elastic properties. For practical reasons, we truncated the series after a desirable accuracy was attained, and this is reflected in Eqs. (10.36) and (10.37) by restricting the series to terms corresponding to $l \leq l^*$. For example, for hexagonal-orthorhombic textures, it was observed that the maximum error in Eq. (10.37) was less than 3% when the series was truncated to 57 terms (corresponding to $l^* = 12$). The desired accuracy in property predictions, therefore, dictates the number of Fourier dimensions that need to be explored by the MSD methodology for a selected design application. The Fourier coefficients relevant to both the elastic and plastic properties of this material were established in a recent study (Wu et al., 2007) and are used here.

10.2.3 Microstructure hull for hexagonal material

The main challenge encountered in the application of the MSD methodology to mechanical design case studies is that the microstructure–property linkages need to be explored in fairly large-dimensional Fourier spaces. In dealing with plastic properties of hexagonal polycrystals there is a need for consideration of a much higher-dimensional Fourier space compared to other work on the more symmetric cubic polycrystals. The increased demand in dimensionality is driven by two main factors: (1) The Fourier representation of the ODF for hcp polycrystals typically needs many more terms compared to the corresponding representations for cubic polycrystals; (2) the yield surfaces in hcp polycrystals are substantially more anisotropic compared to the yield surfaces for cubic polycrystals. For example, the design case studies reported by Houskamp et al. (Kalidindi et al., 2004; Houskamp et al., 2007) explored the cubic-orthorhombic texture hull in 12 dimensions, whereas consideration of similar problems here in hexagonal-orthorhombic texture hulls needs exploration of 57 dimensions of the microstructure hull (Wu et al., 2007).

Different methods can be used to search the space of all possible microstructures (the microstructure hull) for promising designs. These include binning the hull and searching the subsequent discrete space (Kalidindi et al., 2004; Sintay and Adams, 2005), using a generalized reduced gradient method to restrict the design space to the convex microstructure hull (Adams et al., 2001; Houskamp et al., 2007), or exploring the property closure directly as the design space (Saheli et al., 2004). The latter approach has a disadvantage when the property space is not convex, since defining the boundary of the property closure becomes extremely challenging, especially in design case studies where the overall performance is governed by a large number of macroscale properties. It should be noted that the property closures for a vast number of design problems are likely to be nonconvex.

The reduced gradient method helps to overcome one of the fundamental difficulties encountered when searching the microstructure hull—the issue of bounding the search space. An efficient Gram–Schmidt approach is used. However, it is difficult to apply this algorithm to the case studies described in this section because of the much larger number of Fourier dimensions involved. Hence, a simpler approach is taken to define the microstructure hull. Instead of using the Fourier coefficients $F_l^{\mu\nu}$ to define the design space, it is more convenient to define the design space in terms of the α_k (see Eq. 10.26).

Although the number of dimensions for the search space increases, the constraints on the space simplify. With this choice the design space is simply expressed as

$$\sum_k \alpha_k = 1, \quad \alpha_k \geq 0 \tag{10.38}$$

Note that the constraints described here can be implemented much more easily into any optimization search tool, compared to the constraints described in Eqs. (10.27) and (10.28). The next question is how many distinct orientations need to be considered in the fundamental zone (i.e., the range of k in Eq. (10.38)) for a given design problem.

The definition of the ODF (Eq. 10.23) and its Fourier representation (Eq. 10.25) guarantee that the complete set of all theoretically feasible ODFs forms a compact convex region in the Fourier space. As an example, the texture hull for hexagonal-orthorhombic textures is shown in Figure 10.4 in the first three dimensions of the Fourier space. The crystal orientations corresponding to the vertices of the polygonal approximation to the texture hull are called *principal orientations*. Also, as described earlier in Eqs. (10.32) and (10.37), different Fourier subspaces are relevant to different macroscale engineering properties of interest. The set of vertices of the Fourier subspace of interest is denoted M^p,

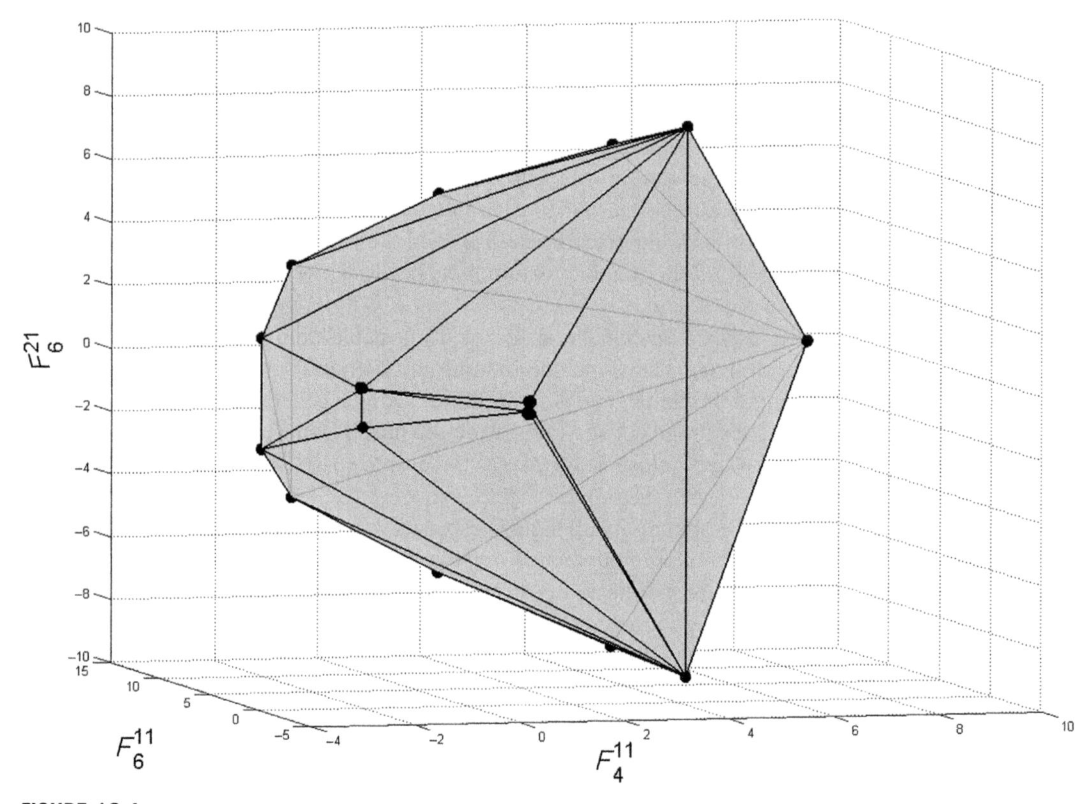

FIGURE 10.4

The hexagonal-orthorhombic texture hull projected in the three dimensions of the Fourier space that are common to the case studies presented in this section. The vertices of the hull are associated with the set of principal orientations for this subspace and are shown as *black circles*.

where p reminds us that this set corresponds to principal orientations. The texture hull in the selected subspace, $\widetilde{M}$, can then be described as

$$\widetilde{M} = \left\{ F_l^{\mu\nu} \middle| F_l^{\mu\nu} = \sum_p \alpha_p {}^p F_l^{\mu\nu},\ {}^p F_l^{\mu\nu} \in M^p, \alpha_p \geq 0\,, \sum_p \alpha_p = 1 \right\} \tag{10.39}$$

It should be noted that the main difference between Eq. (10.27) and Eq. (10.39) is that M describes the complete infinite-dimensional texture hull, whereas $\widetilde{M}$ defines only a selected subspace of M that is relevant to a given design problem. Consequently, the size of M^p should be significantly smaller than the size of M^k.

As an example, consider the fundamental zone of the hexagonal-orthorhombic textures described in Eq. (10.24). Discretizing this fundamental zone into 10-degree bins would result in 2590 lattice orientations (this would dictate the size of M^k). Our investigations have revealed that only the orientations on the fundamental zone's boundary can be principal orientations (i.e., only these orientations correspond to the vertices of the texture hulls). The same 10-degree binning of the fundamental zone's boundary produces only 339 distinct principal orientations. In other words, the size of M^p would be at least one order of magnitude lower than the corresponding size of M^k. If we further restrict our interest to only the three dimensions of the texture hull shown in Figure 10.4, we need to consider only 15 principal orientations. As can be seen from this example, the use of principal orientations and changing the microstructure design variable from $F_l^{\mu\nu}$ to α_p leads to major computational advantages in the MSD methodology.

It is important at this stage to reflect on some of the consequences of using α_p as the microstructure design variable. One might be tempted to conclude erroneously that a prescribed set of α_p corresponds to a unique ODF. Readers are cautioned that we are dealing with truncated Fourier spaces where the mapping between $F_l^{\mu\nu}$ and α_p is not one-to-one. Indeed, many different realizations of α_p can correspond to one set of truncated $F_l^{\mu\nu}$. A prescribed set of α_p, however, corresponds to a single set of truncated $F_l^{\mu\nu}$ coefficients, which in turn corresponds to a very large set of distinct ODFs. However, all of the distinct ODFs corresponding to the prescribed set of α_p are expected to exhibit the same (or approximately similar) values of macroscale properties of interest defined by the selected Fourier subspace (Adams et al., 2005; Proust and Kalidindi, 2006). Another important consequence of the approach presented here is that it naturally produces nonunique solutions (as expected and desired).

The approach of using α_p (henceforth denoted as $\boldsymbol{\alpha}$ for ease of notation) as the design variable can be conveniently applied to mechanical design problems as well as the delineation of property closures. The optimization problem for mechanical design involving components with statistically homogeneous microstructures can be formulated as

$$\text{Maximize } O(\rho), \text{where } \rho = (P_1(\boldsymbol{\alpha}), P_2(\boldsymbol{\alpha}), \ldots, P_N(\boldsymbol{\alpha})), \text{ subject to } \alpha_p \geq 0, \sum_p \alpha_p = 1 \tag{10.40}$$

where O denotes an objective function characterizing the performance of the mechanical component, and ρ is a set of relevant macroscale material properties (denoted as $P_i(\boldsymbol{\alpha})$) influencing the performance. In particular, it is noted that the functional dependence in $P_i(\boldsymbol{\alpha})$ can be highly nonlinear (especially with plastic properties; see Eqs. (10.33) through (10.37)).

10.2.4 Property closure and MSD for a compliant mechanism

In this section we demonstrate one method of obtaining the boundary points of property closures of interest to the case study. Briefly, one starts by identifying the potential range, R_i, of values for each property of interest, P_i, as

$$R_i = \left[\min\left\{P_i(\boldsymbol{\alpha})\middle|\alpha_p \geq 0, \sum_{p=1} \alpha_p = 1\right\}, \max\left\{P_i(\boldsymbol{\alpha})\middle|\alpha_p \geq 0, \sum_{p=1} \alpha_p = 1\right\}\right] \tag{10.41}$$

An N-property closure is then delineated by seeking the maximum and minimum values possible for one of the properties of interest while constraining all other properties to values preselected within their respective ranges. As an example, let $(P_2^*, \ldots, P_N^*)$ represent a selected combination of values for all properties of interest except P_1, where the values of each property are constrained to lie within their respective potential ranges, that is, $P_i^* \in R_i$. A restricted feasible range for P_1 can then be established as

$$R_1^* = \left\{P_1(\boldsymbol{\alpha})\middle|P_i(\boldsymbol{\alpha}) = P_i^* \ \forall\ i \neq 1, \alpha_p \geq 0, \sum_{p} \alpha_p = 1\right\} \tag{10.42}$$

If R_1^* is a nonempty set, it leads to the identification of points on the boundary of the property closure. Let P_1^{e*} denote the extrema of R_1^*. Then $(P_1^{e*}, P_2^*, \ldots, P_N^*)$ lie on the boundary of the property closure sought. It should be noted here that for several selections of $(P_2^*, \ldots, P_N^*)$, the corresponding R_1^* is indeed expected to be a null set.

In the present case study, all of the optimization problems formulated in Eqs. (10.40) through (10.42) were solved successfully using the `fmincon` function in MATLAB's Optimization Toolbox (Mathworks, 2007). The *fmincon* function uses sequential quadratic programming, chooses subsequent variables via the line-search method, and implements the Broyden–Fletcher–Goldfarb–Shanno quasi-Newton formula to define the Hessian (Mathworks, 2007). Convergence is defined using first-order necessary conditions.

A closure depicting the complete set of feasible combinations of S_{1111} and σ_{y1} for all theoretically feasible hexagonal-orthorhombic textures in the selected α-Ti metal was obtained using the procedures described earlier and plotted in Figure 10.5. Performance contours for maximum deflection—based on Eq. (10.22)—have been superimposed on this figure. The maximum deflection attainable in the compliant beam with isotropic properties (corresponding to a random texture) is 2.41 mm, while the expansion of the design space to the set of hexagonal-orthorhombic textures provides performances ranging from 1.01 to 3.31 mm. These results are summarized in Table 10.3. The best performance represents a 37% improvement over that of the isotropic solution. It is just as important to note that ignoring the inherent texture in the sample can result in extremely poor performance of the component (58% reduction in performance compared to the isotropic solution).

The best performance in this design case study corresponded to a yield strength of 322.6 MPa (close to the maximum yield strength possible in the selected material system, which was 329.23 MPa) and a compliance of 0.0086 GPa^{-1} (which is significantly lower than the maximum possible compliance of 0.0096 GPa^{-1}). This is because the combination of the highest yield point and the highest compliance is not feasible in any one texture. The best feasible performance resulted from a trade-off between yield strength and compliance. It is also worth noting that the worst performance corresponded to the lowest yield strength, in spite of the fact that it exhibited the highest compliance. Clearly, the value of the yield strength dominated this design.

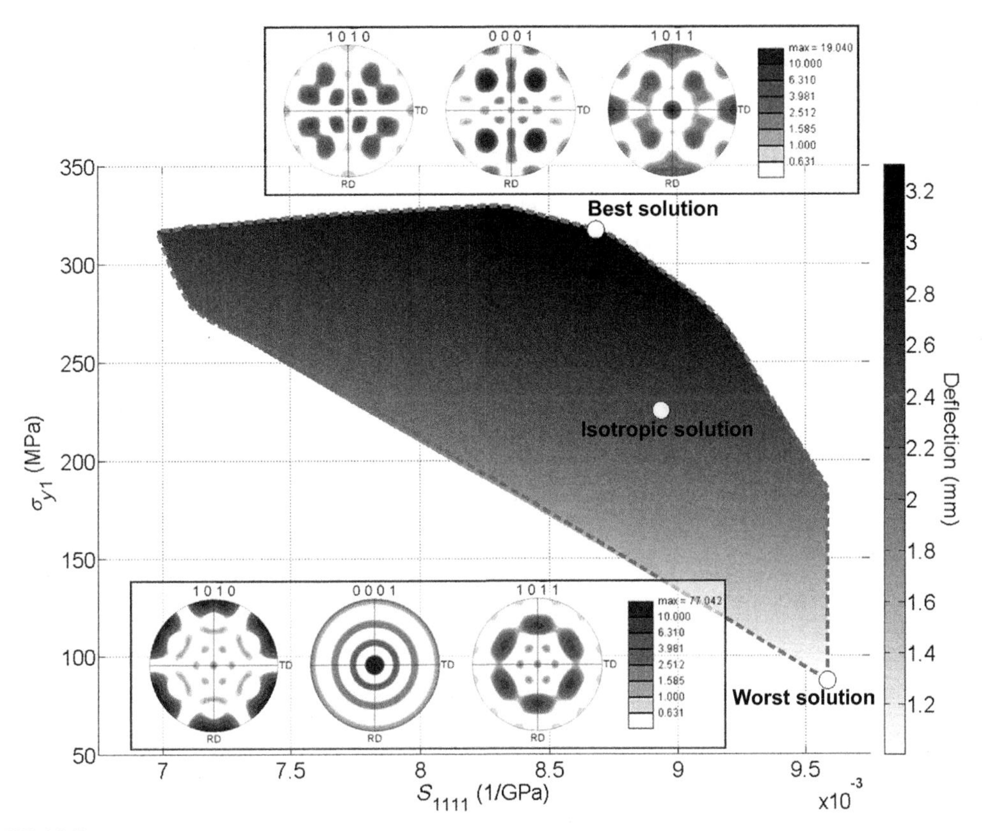

FIGURE 10.5

The relevant property closure for a cantilever-compliant beam made of high-purity polycrystalline α-Ti with hexagonal-orthorhombic textures. The textures shown are predicted to provide the best and the worst performances for this case study.

Table 10.3 MSD Results for the Compliant Beam Design

Case	Deflection (mm)	S_{1111} $\left(\frac{10^{-3}}{GPa}\right)$	σ_{y1} (MPa)
Best	3.31	8.6	322.6
Isotropic	2.41	8.9	225
Worst	1.01	9.58	87.6

The RD direction in the pole figures shown in Figure 10.5 corresponds to the beam axis (x_1-axis in Figure 10.3). The best performance was observed to correspond to a texture with the crystal (0001) planes inclined at a small angle to the RD axis, while the worst performance corresponded to a texture with the (0001) planes inclined at about 90 degrees to the RD axis. This is consistent with the results described in Table 10.3 because the yield strength of a titanium single crystal is expected to decrease significantly as the (0001) plane is tilted away from the loading direction.

10.3 MICROSTRUCTURE DESIGN OF A ROTATING DISK

In this section we apply the same approach developed in the previous section to the optimal design of a rotating titanium flywheel. Flywheel energy storage (FES) systems efficiently convert kinetic energy into useful power via an electric generator. The main functional component of an FES system is the rotating disc spinning around a shaft supported by magnetic bearings to reduce friction. High-strength materials are essential to enable the high rotational speeds needed to maximize the efficiency of the FES system. Typical FES systems comprise metallic flywheels operating around 4000 rpm (Arnold et al., 2001). The goal in this study is to identify the texture that will produce the maximum energy output in the flywheel without initiating plastic strain in the component.

The rotating disc in the FES system is treated as a transversely isotropic thin disc with the following radial and tangential stresses (in cylindrical coordinates r, θ, and z):

$$\sigma_{rr} = \rho\omega^2(3-\nu)\left(\frac{(a^2-b^2)}{4}+\frac{1}{8}\frac{(b^4-a^4)}{r^2}\right) \tag{10.43}$$

$$\sigma_{\theta\theta} = \rho\omega^2\left(r^2+(3-\nu)\left(\frac{(a^2-b^2)}{4}+\frac{1}{8}\frac{(b^4-a^4)}{r^2}\right)\right) \tag{10.44}$$

where a and b are the inner and outer radii (set equal to 50 mm and 120 mm, respectively), ρ is the density (taken as 4.51 g/cc for titanium), and ω is the angular velocity. The Poisson ratio, ν, is defined in the plane of the radial and tangential components as

$$\nu = 1-\frac{S_{rr\theta\theta}}{S_{rrrr}} \tag{10.45}$$

The cylindrical geometry and the loading conditions demand transverse isotropy in macroscale properties. Hill's anisotropic yield criterion (Eq. 10.32) for this symmetry and loading condition can be expressed as

$$\frac{(\sigma_{rr}-\sigma_{\theta\theta})^2}{\sigma_{yr}^2}+\frac{\sigma_{rr}\sigma_{\theta\theta}}{\sigma_{yz}^2}=1 \tag{10.46}$$

where σ_{yr} and σ_{yz} denote the yield stresses in the radial and thickness directions, respectively. By substituting the stresses in Eqs. (10.43) and (10.44) into the yield criterion (Eq. 10.46), the angular velocity (the variable governing kinetic energy) at the initiation of plastic yield can be solved from

$$\frac{1}{\rho\omega^2} = \max\left(\left.\left(\begin{array}{c}\left(\dfrac{(3+\nu)(b^4-a^4)-4r^4\sigma_{yr}^2}{4r^2\sigma_{yr}^2}\right)^2 + \\ \left(\dfrac{(3+\nu)}{\sigma_{yz}^2}\left(\left(\dfrac{a^2-b^2}{4}\right)+\dfrac{(b^4-a^4)}{8r^2}\right)\left(r^2+\dfrac{(3+\nu)}{\sigma_{yz}^2}\left(\left(\dfrac{a^2-b^2}{4}\right)-\dfrac{(b^4-a^4)}{8r^2}\right)\right)\right)\end{array}\right)^{\frac{1}{2}}\right|_{r\in[a,b]}\right) \tag{10.47}$$

Note that, in the isotropic solution to this problem, the plastic yielding always initiates along the inner surface of the disc. However, with the introduction of anisotropic constitutive behavior, the location of the initiation of plastic yielding can occur anywhere in the component. Note also that only three material properties—namely, ν, σ_{yr}, and σ_{yz}—influence the performance of the flywheel. The values of these properties are dependent on the texture in the sample.

The requirement of transverse isotropy in mechanical properties of the flywheel restricts attention to hexagonal-transversely isotropic ODFs. The corresponding fundamental zone can be expressed as (Bunge, 1993)

$$FZ = \left\{ (\Phi, \varphi_2) \,\middle|\, 0 \le \Phi < \frac{\pi}{2}, 0 \le \varphi_2 < \frac{\pi}{3} \right\} \tag{10.48}$$

Because of the higher symmetry compared to the orthorhombic case described earlier, the relevant Fourier subspace controlling the elastic–plastic yield properties is significantly smaller and involves only 12 dimensions. It was also observed that only 50 principal crystal orientations were needed to describe this reduced Fourier subspace.

Figure 10.6 depicts the relevant property closure for this case study involving three macroscale properties—namely, σ_{yr}, σ_{yz}, and ν. This property closure is highly nonconvex, and therefore a precise geometrical description of its boundary would be quite challenging.

Table 10.4 summarizes the best and worst possible performances predicted for the rotating disk using the MSD methodology (for the selected material and the selected homogenization theories). For comparison, the performance of the random texture (corresponding to a design with isotropic

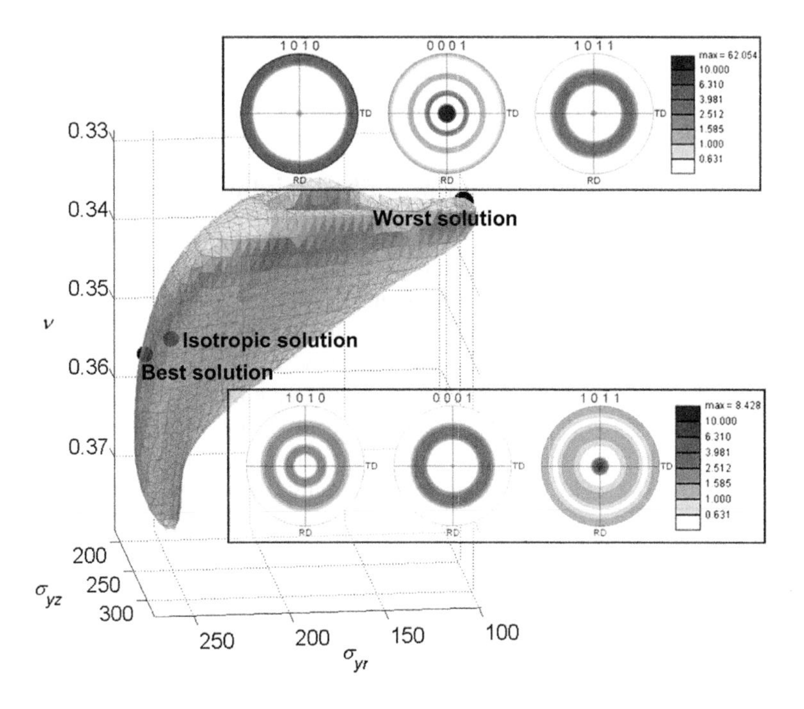

FIGURE 10.6

Property closure relevant to the performance of a high-purity polycrystalline α-Ti flywheel with hexagonal-transversely isotropic texture. Also shown are the pole figures corresponding to the best and worst performance.

Table 10.4 Predicted Angular Velocities and Corresponding Properties of a High-purity Polycrystalline α-Ti Flywheel with Hexagonal-transversely Isotropic Textures

Case	Specific Energy $\left(\frac{kJ}{kg}\right)$	ν	σ_{yr} (MPa)	σ_{yz} (MPa)
Best	25.6	0.35	266.5	275.5
Isotropic	24.1	0.35	250.7	250.7
Worst	9.99	0.33	105	306.4

properties) is also presented. An 8% improvement in performance was noted for the best texture compared to the random texture. On the other hand, it was noted that a poorly selected texture could degrade performance by as much as 58%. It is also noted that the best performance required high values of both the in-plane and the out-of-plane yield strengths (i.e., σ_{yr} and σ_{yz}, respectively), along with a higher value of the Poisson's ratio. The lowest predicted performance corresponded to the combination of the lowest Poisson's ratio, low radial yield strength (σ_{yr}), and high through-thickness yield strength (σ_{yz}).

10.4 MICROSTRUCTURE-SENSITIVE DESIGN OF A COMPOSITE PLATE

To demonstrate the versatility of the microstructure functions described in the previous chapters, we now describe its application to a totally different material system, namely, the continuous fiber-reinforced composite material systems. The main focus here is to elaborate on the advantages of using the concepts already presented in describing the internal structure of a continuous fiber-reinforced composite material system.

10.4.1 Microstructure representation of a fiber-reinforced composite

In the following discussion, for clarity of presentation, we present the microstructure functions for two specific classes of composites: 2D and 3D composites. Although the 2D composite microstructures are essentially a subset of the 3D composite microstructures, it will be seen that the Fourier basis required for the description of the complete set of 2D composite microstructures is substantially simpler. Presenting these two cases differently and simultaneously, as done here, should help readers in understanding better the mathematical concepts already presented.

The local state, h, at any location in the composite material is adequately described by an ordered pair of variables, the phase denoted ρ and the orientation of the material frame denoted g—that is, $h = (\rho, g)$ (see Figure 10.7). The local material frame is usually aligned along the principal axes of material anisotropy (in the present case, the local frame would be aligned with the fiber orientation). In the most general case, the local material frame is defined by a proper orthogonal rotation tensor with respect to the sample reference frame, that is, $g \in SO(3)$ just as in the case of the crystalline microstructures described earlier. Additionally, as discussed earlier, materials often exhibit certain symmetries in the description of the local properties that are derived from the detail of the internal structure at the local scale. To take these into account and represent economically the local state space, the fundamental zone of local material frames is identified as $FZ(\rho)$.

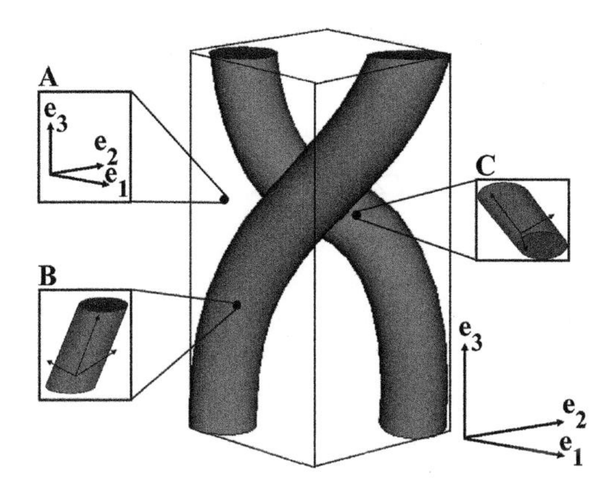

FIGURE 10.7

Schematic description of the different possible local states in the fiber-reinforced composite material system. Material point A lies in the isotropic matrix and is identified by h = (m). Material points B and C lie in the reinforcement phase and are identified by $h = (r, \Phi_B, \beta_B)$ and $h = (r, \Phi_C, \beta_C)$, respectively.

Here, we will further restrict our attention to composite materials with only two phases: an anisotropic-reinforcing fiber phase, *r*, and an isotropic matrix phase, *m*. Because of the isotropy of the matrix phase, it does not need a specification of material axes. However, the fiber-reinforcement phase needs specification of a local material frame at any fiber location in the material system. It will be further assumed here that the fiber phase has transversely isotropic mechanical properties with respect to the fiber axis. The local anisotropy in the fiber phase is then defined completely by the specification of a single material axis. In 2D composites where all fiber orientations are restricted to a single plane, the material axis can be specified by a single angle, $\theta \in [0, \pi)$. In 3D composites, the material axis can be specified by an ordered pair of angles, $\Phi \in [0, \pi)$, $\beta \in [0, \pi)$ (see Figure 10.8).

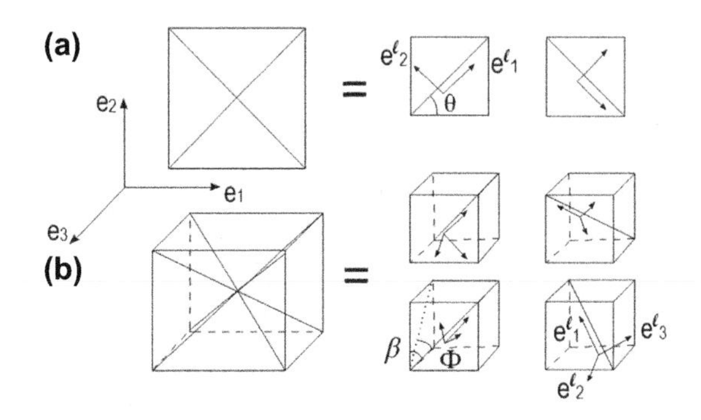

FIGURE 10.8

Idealization of a continuous fiber-reinforced composite into unidirectional laminate subcomponents for (a) 2D composites and (b) 3D composites.

In summary, the local state space representing the complete set of all possible distinct local states for the 2D and 3D composite systems can be expressed as

$$\begin{aligned} &\text{2D:} \quad H_2 = H^m \cup FZ^{r2}, \quad H^m = \{m\}, \quad FZ^{r2} = \{(r,\theta)|\theta \in [0,\pi)\} \\ &\text{3D:} \quad H_3 = H^m \cup FZ^{r3}, \quad FZ^{r3} = \{(r,\Phi,\beta)|\Phi \in [0,\pi), \beta \in [0,\pi)\} \end{aligned} \tag{10.49}$$

Since the local state space consists of some discrete local states (e.g., the separation of material into m and r) and some continuous state spaces (see Eq. 10.49), the local state distribution function for this microstructures class can be specified by a combination of the volume fraction of the reinforcement phase and a fiber orientation distribution function. For the idealized 2D composites, the local state distribution is specified by the set $\{V_r, f(\theta)\}$, where V_r is the total volume fraction of the reinforcing phase, and $f(\theta)$ is a distribution function of the type defined by Eqs. (4.13) and (4.14). Likewise, for the 3D composites, the local state distribution of the internal structure of the composites is described by $\{V_r, f(\Phi,\beta)\}$.

We have seen that Fourier representations of the local state distribution functions may include the primitive indicator functions or spherical harmonics, depending on the specific application. Another basis that is useful here for the 2D case is classic exponential functions. For this case study we adopt the following Fourier representations of the distributions for the 2D and 3D composite systems:

$$\begin{aligned} &\text{2D:} \quad f(\theta) = \sum_{l=-\infty}^{\infty} F_l e^{il\theta} \\ &\text{3D:} \quad f(\Phi,\beta) = \sum_{l=0}^{\infty} \sum_{m=-l}^{l} F_l^m k_l^m(\Phi,\beta) \end{aligned} \tag{10.50}$$

In the Fourier representations shown in Eq. (10.50), classic exponential functions are used for $f(\theta)$ and the surface spherical harmonic functions, $k_l^m(\Phi,\beta)$, were used for $f(\Phi,\beta)$. While other choices of Fourier basis are possible, it will be seen later that these choices of the Fourier basis produce the most economical representation of the microstructure-effective property relationships being sought. The coefficients $\{F_l\}$ and $\{F_l^m\}$ will be referred to as the Fourier coefficients of the material internal structure, and they represent uniquely the functions $f(\theta)$ and $f(\Phi,\beta)$ as distinct points in their respective infinite-dimensional Fourier spaces.

Material and processing symmetries arise in most problems, as we illustrated in previous examples. These symmetries not only reduce the local state space, they also reduce the number of terms needed in the Fourier representations shown in Eq. (10.50). In the problems under consideration here, the following symmetries reflect the fact that the assignment of local material axes at any fiber location can be accomplished in two equivalent ways (positive and negative directions along the fiber):

$$f(\theta+\pi) = f(\theta), \quad f(\pi-\Phi, \pi+\beta) = f(\Phi,\beta) \tag{10.51}$$

Imposition of the symmetries in Eq. (10.51) compact the Fourier representations in Eq. (10.50) into the following forms:

$$\begin{aligned} &\text{2D:} \quad f(\theta) = \sum_{l=-\infty,2}^{\infty} F_l e^{il\theta} \\ &\text{3D:} \quad f(\Phi,\beta) = \sum_{l=0,2}^{\infty} \sum_{m=-l}^{l} F_l^m k_l^m(\Phi,\beta) \end{aligned} \tag{10.52}$$

In addition to the implicit symmetries of the problem, such as those in Eq. (10.51), one can impose symmetries on the problem dictated by other factors. For example, one can impose orthorhombic sample symmetry; that is, only those internal structures that reflect orthorhombic sample symmetry (obtained presumably by choice of processing or manufacturing routes) are to be considered in the design. Imposition of orthorhombic sample symmetry requires the following:

$$\begin{aligned} &\text{2D:} \quad f(\pi - \theta) = f(\theta) \\ &\text{3D:} \quad f(\Phi, \beta) = f(\Phi, \pi - \beta) = f(\pi - \Phi, \pi - \beta) = f(\pi - \Phi, \beta) \end{aligned} \tag{10.53}$$

These requirements further restrict the associated local state spaces and further compact the Fourier representations of the local state distributions as follows:

$$\begin{aligned} &\text{2D:} \quad H_2 = H^m \cup FZ^{r2}, \quad H^m = \{m\}, \quad FZ^{r2} = \{(r, \theta) | \theta \in [0, \pi/2]\} \\ &\text{3D:} \quad H_3 = H^m \cup FZ^{r3}, \quad FZ^{r3} = \{(r, \Phi, \beta) | \Phi \in [0, \pi/2], \beta \in [0, \pi/2]\} \end{aligned} \tag{10.54}$$

$$\begin{aligned} &\text{2D:} \quad f(\theta) = F_0 + 2 \sum_{l=2,2}^{\infty} F_l \cos(l\theta) \\ &\text{3D:} \quad f(\Phi, \beta) = \sum_{l=0,2}^{\infty} \sum_{\mu=1}^{M(l)} F_l^{\mu} \dot{k}_l^{\mu}(\Phi, \beta) \end{aligned} \tag{10.55}$$

Note that the imposition of the orthorhombic sample symmetry precludes the need for the imaginary terms in the Fourier expansion of the distribution functions (i.e., $\dot{k}_l^{\mu}(\Phi, \beta)$ are real-valued functions, while $k_l^{m}(\Phi, \beta)$ are complex-valued functions), and reduces dramatically the number of independent terms in the series expansion. (*M(l)* indicates that the number of these functions depends on the value of *l* and these were tabulated in Figures 6.3 and 6.4.) Furthermore, the orthonormality of the Fourier basis, the normalization requirement, and the requirement for an invariant measure on the local state space result in the derivation of the following expressions for the Fourier coefficients in Eq. (10.55):

$$\begin{aligned} &F_l = \int_{H^{r2}} f(\theta) \cos(l\theta) dh, \qquad dh = \frac{2}{\pi} d\theta \\ &F_l^{\mu} = \iint_{H^{r3}} f(\Phi, \beta) \dot{k}_l^{\mu}(\Phi, \beta) dh, \qquad dh = \frac{2}{\pi} \sin \Phi \, d\Phi d\beta \end{aligned} \tag{10.56}$$

Note that the invariant measures of local state space used in Eq. (10.56) have been normalized to yield $f(\theta) = f(\Phi, \beta) = 1$ for material internal structures that contain a uniform distribution of all the local states from their respective fundamental sets of local state space. For these special microstructures, it also follows from Eq. (10.56) that only the first Fourier coefficient (corresponding to $l = 0$) is nonzero, and its value is equal to 1.

10.4.2 Material properties of a composite plate

For the present study, the following equations have been selected from the literature to estimate the effective properties of a unidirectional laminate subcomponent (Chamis, 1989):

$$\begin{aligned} S_{11} &= \left(E_m(1-V_r)+E_{f1}V_r\right)^{-1}, \qquad S_{22} = \left(1-\sqrt{V_r}\left(1-E_m/E_{f2}\right)\right)/E_m \\ S_{66} &= \left(1-\sqrt{V_r}\left(1-G_m/G_{f12}\right)\right)/G_m, \qquad S_{44} = \left(1-\sqrt{V_r}\left(1-G_m/G_{f23}\right)\right)/G_m \\ S_{12} &= \left(\nu_m(1-V_r)+\nu_{f12}V_r\right)/\left(E_m(1-V_r)+E_{f1}V_r\right) \end{aligned} \tag{10.57}$$

where S_{ij} are the five independent constants in the elastic compliance and stiffness tensors of a unidirectional laminate in its local frame assuming transverse isotropy (with the one-axis aligned with the fiber orientation); E_{f1}, E_{f2}, G_{f12}, G_{f23}, and ν_{f12} are the five independent elastic constants of the fiber phase (again assuming transverse isotropy with respect to fiber axis); and E_m and G_m are elastic properties of the matrix phase (assuming isotropy).

The local compliance and stiffness tensors of the laminate subsystems in a given 2D or 3D composite can be prescribed completely using the S_{ij} parameters described in Eq. (10.57). These local compliance and stiffness tensors can then be transformed into the global sample frame (see Figure 10.8) by the fourth-rank coordinate transformation law:

$$S_{abcd} = Q_{ap}Q_{bq}Q_{cr}Q_{ds}S^{l}_{pqrs}, \qquad C_{abcd} = Q_{ap}Q_{bq}Q_{cr}Q_{ds}C^{l}_{pqrs} \tag{10.58}$$

where l denotes that the tensor components are expressed in the local reference frame (see Figure 10.8), and Q_{ij} are the components of the transformation (rotation) matrix. The transformation matrices for the 2D and 3D composites of interest here can be expressed as

$$\begin{aligned} \text{2D:} \quad [Q] &= \begin{bmatrix} \cos\theta & -\sin\theta & 0 \\ \sin\theta & \cos\theta & 0 \\ 0 & 0 & 1 \end{bmatrix} \\ \text{3D:} \quad [Q] &= \begin{bmatrix} \cos\Phi & -\sin\Phi & 0 \\ \sin\Phi\cos\beta & \cos\Phi\cos\beta & -\sin\beta \\ \sin\Phi\sin\beta & \cos\Phi\sin\beta & \cos\beta \end{bmatrix} \end{aligned} \tag{10.59}$$

where the angles θ or Φ and β are defined individually for each 2D or 3D sublaminate, respectively, as shown in Figure 10.8.

10.4.3 First-order homogenization for a composite plate

As described earlier, the first-order bounds for orthotropic components of the fourth-rank elastic stiffness tensor can be expressed as (without summation over repeated indices)

$$\begin{aligned} \langle S\rangle^{-1}_{abab} &\le C^{*}_{abab} \le \langle C\rangle_{abab} \\ \max\left(\langle C\rangle_{aabb}, \langle S\rangle^{-1}_{aabb}\right) - \sqrt{\Delta_a\Delta_b} &\le C^{*}_{aabb} \le \min\left(\langle C\rangle_{aabb}, \langle S\rangle^{-1}_{aabb}\right) + \sqrt{\Delta_a\Delta_b} \\ \Delta_a &= \langle C\rangle_{aaaa} - \langle S\rangle^{-1}_{aaaa} \end{aligned} \tag{10.60}$$

where $\langle . \rangle$ denotes an ensemble or volume average, C_{abcd} and S_{abcd} denote the components of the laminate subsystem's elastic stiffness and compliance tensors in the sample reference frame (see Eq. 10.58), and C^*_{abcd} are the components of the effective elastic stiffness tensor in the sample reference frame.

In studies on selected classes of continuous fiber-reinforced composites, appropriately weighted averages of the first-order bounds have been found to yield reasonably accurate estimates of the effective properties. The weights used in these models were found to be dependent on the type of fibers and matrix materials used. In this study, it has been assumed that reasonably good estimates of the effective properties of the 2D and 3D composites being studied can be obtained using the following class of *weighted-average models* (WAMs) (Kregers and Teters, 1979):

$$C^*_{abcd} \approx \alpha\, C^{up}_{abcd} + (1-\alpha)\, C^{low}_{abcd} \tag{10.61}$$

where α is the weighting factor ($\alpha \in [0, 1]$), and the "up" and "low" denote the respective upper and lower bounds from Eq. (10.60).

10.4.4 Optimized design of a composite plate

A design study will invariably involve optimizing some aspect of the design with respect to a set of given constraints. We wish to optimize the design of a composite plate with a circular hole, subjected to in-plane compression. Hence, we assume that the constraint is given by the failure strength of a composite plate under this loading. The development of a comprehensive theoretical model for the effective failure properties of a composite material system is complicated. It is fully acknowledged that the development of robust linkages between the internal structure of the composite and the failure properties would require a sophisticated approach that takes into account the nature of the fiber-matrix interaction and a detailed description of the spatial distribution of the fiber reinforcement in the material's internal structure.

In the present study, however, we adopt the following simple model based on the Tsai–Hill description (Herakovich, 1998) and a consideration of only a limited number of failure modes or the failure strength of a unidirectional laminate in off-axis simple compression loading:

$$\begin{gathered}\sigma_f(\Phi) = \left(\frac{\cos^4\Phi}{X^2} + \left(\frac{1}{S^2} - \frac{1}{X^2}\right)\cos^2\Phi\sin^2\Phi + \frac{\sin^4\Phi}{Y^2}\right)^{-1/2} \\ X = V_r S_{fc}, \quad S = S_{ms}/\left(1-\sqrt{V_r}\left(1-G_m/G_{f12}\right)\right), \quad Y = S_{mc}/\left(1-\sqrt{V_r}\left(1-E_m/E_{f2}\right)\right)\end{gathered} \tag{10.62}$$

where S_{fc} is the axial compressive strength of the fiber, S_{ms} is the shear strength of the matrix, S_{mc} is the compressive strength of the matrix, and Φ denotes the angle between the orientation of the fibers in the laminate subsystem and the simple compression loading direction in the sample reference frame. For 2D composites, Φ should be replaced by θ in Eq. (10.62). The effective strength of the composite is assumed to be given by volume averaging the failure strengths of the constituent laminate subsystems.

As described earlier, for the elementary first-order homogenization theories selected in this book, the relevant statistical description of the microstructure comprises the set $\{V_r, f(\theta)\}$. The set of possible values for V_r can be described simply as

$$M_r = \{V_r | 0 \le V_r \le 1\} \tag{10.63}$$

The complete set of all possible distributions of $f(\theta)$ and $f(\Phi,\beta)$ is significantly more difficult to specify, and here we resort to the spectral representations of these functions described earlier. The microstructure hull represents the complete set of distribution functions that correspond to all physically realizable microstructures. To delineate the microstructure hull, it is convenient to start with certain special microstructures that are called "eigen-microstructures" or "single-state" microstructures. The local state distributions for these eigen-microstructures are conveniently represented by Dirac functions:

$$f(h) = \delta(h - h_i), \quad \int_{H_i} \delta(h - h_i)dh = \begin{cases} 1 & \text{if } h_i \in H_i \\ 0 & \text{if } h_i \notin H_i \end{cases} \quad \forall\ H_i \subset H \tag{10.64}$$

The distribution functions described in Eq. (10.64) imply that the eigen-microstructures are allowed to possess only one local state. In practical terms, this means that a 2D orthorhombic composite eigen-microstructure is realized by assembling laminates with orientation θ and $-\theta$ in equal parts. Similarly, an orthorhombic 3D composite eigen-microstructure can be realized by assembling together a set of four unidirectional laminates that are selected specifically to satisfy the required elements of the orthorhombic symmetry group for the 3D composite.

The complete set of all physically realizable distributions, M, can now be identified as

$$M = \left\{ f(h) | f(h) = \sum_j \alpha_j \delta(h - h_j);\ \sum_j \alpha_j = 1;\ \alpha_j > 0;\ h_j \in H \right\} \tag{10.65}$$

Equation (10.65) reflects the fact that any physically reliable microstructure has to comprise the elements of the set of eigen-microstructures described earlier, and that this assemblage has to be accomplished in such a way that the volume fractions of the various eigen-microstructures are positive and add up to 1.

Substituting the Dirac distributions of Eq. (10.64) into Eq. (10.56) yields the Fourier coefficients for the eigen-microstructures. Let $\hat{M}$ represent the complete set of Fourier coefficients for all physically realizable eigen-microstructures. This set for the idealized orthorhombic 2D and 3D composites considered here can be expressed as

$$\begin{aligned} \text{2D:} \quad \hat{M} &= \{\hat{F}_l | \hat{F}_l = \cos(l\theta);\ \theta \in [0, \pi/2]\} \\ \text{3D:} \quad \hat{M} &= \{\hat{F}_l^{\mu} | \hat{F}_l^{\mu} = \dot{k}_l^{\mu}(\Phi, \beta);\ \Phi \in [0, \pi/2];\ \beta \in [0, \pi/2]\} \end{aligned} \tag{10.66}$$

From Eqs. (10.56) and (10.64), and making use of Eq. (10.65), one can now identify the complete set of Fourier coefficients corresponding to all physically realizable microstructures, referred to as the microstructure hull and denoted $\widetilde{M}$.

$$\begin{aligned} \text{2D:} \quad \widetilde{M} &= \left\{ F_l | F_l = \sum_j \alpha_j \hat{F}_l;\ \sum_j \alpha_j = 1;\ \alpha_j > 0;\ \hat{F}_l \in \hat{M} \right\} \\ \text{3D:} \quad \widetilde{M} &= \left\{ F_l^{\mu} | F_l^{\mu} = \sum_j \alpha_j \hat{F}_l^{\mu};\ \sum_j \alpha_j = 1;\ \alpha_j > 0;\ \hat{F}_l^{\mu} \in \hat{M} \right\} \end{aligned} \tag{10.67}$$

Equation (10.67) essentially defines $\widetilde{M}$ as a compact convex hull (region) in Fourier space, where the vertices of the convex hull are elements of $\hat{M}$. Note that this compact convex hull exists in an infinite-dimensional Fourier space. It can be shown that the projections of this convex hull in any of its finite-dimensional subspaces are also compact and convex.

Figure 10.9 depicts the projections of the convex hull defined in Eq. (10.67) for the 2D composites in the lower dimensions of Fourier space. Note that any physically realizable orthorhombic 2D composite microstructure under consideration here is guaranteed a representation either on the surface or inside of the hull depicted in Figure 10.9. Just as important, a distribution corresponding to a point outside the depicted hull in Figure 10.9 will not correspond to a physically realizable microstructure.

The locations of a few selected microstructures are also shown in Figure 10.9. The eigen-microstructures in this case happen to lie on the curved line segment ABC. Point D on the boundary of the microstructure hull is not an eigen-microstructure. However, it lies exactly midway on the line joining points A and B and can thus be realized by mixing the microstructures corresponding to these points in equal volume fractions. Therefore, point D corresponds to a 0/90 laminate composite. Extending the same logic, however, it can be seen that the interior points of the hull can be realized in a number of different ways. For example, point O (at the origin) can be realized by mixing microstructures corresponding to points D and B in equal parts or by mixing microstructures corresponding to points E and F in equal parts.

It should be recognized, then, that the interior points of the hulls shown in Figure 10.9 do not correspond to a single unique microstructure. Indeed, an entire family of microstructures can be identified corresponding to an interior point in the hulls shown in Figure 10.9, all of which have the

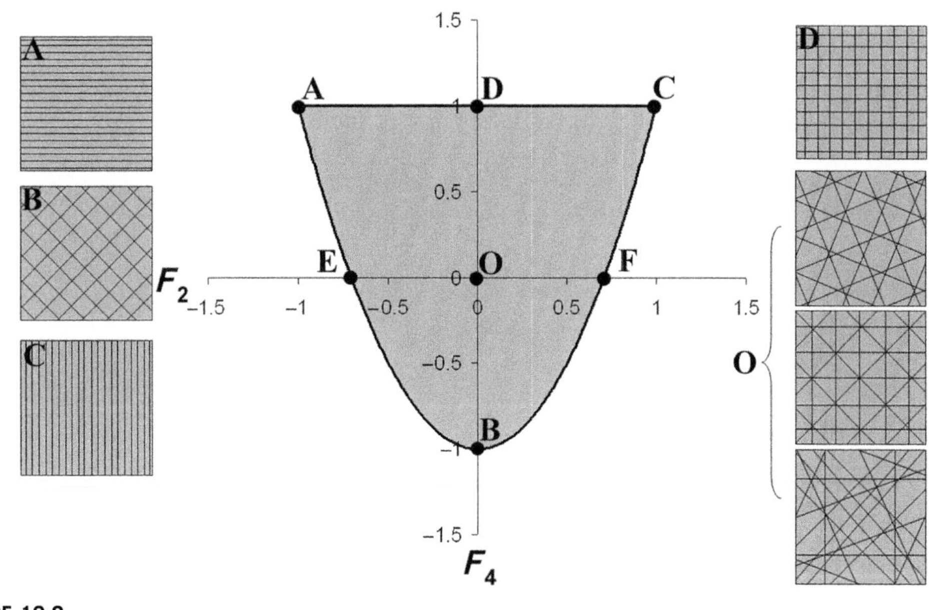

FIGURE 10.9

The microstructure hull for 2D composites in the first two dimensions of the Fourier space representing all physically realizable one-point distributions, $f(\theta)$.

exact same Fourier coefficients in the chosen subspace (the members of this family are, of course, expected to have differing Fourier coefficients in the higher dimensions of the Fourier space). It will be seen later that effective elastic properties of the composite are influenced only by a finite number of low-order Fourier coefficients. This implies that the entire class of microstructures corresponding to a single point in a properly selected lower-dimensional microstructure hull is predicted to yield the same set of effective elastic properties.

Substitution of Eq. (10.59) into Eq. (10.58) yields expressions for the components of the local stiffness and compliance tensors in the sample frame that are dependent on the local state parameters (θ in 2D composites and (Φ,β) in 3D composites). To denote this dependency, these components are denoted as $S_{abcd}(\theta)$ and $S_{abcd}(\Phi,\beta)$, respectively. Note that these functions can also be expressed in the same Fourier spaces that were used to represent the distributions characterizing the internal structure of the material.

Evaluation of the effective elastic and failure properties using the first-order theories summarized earlier is possible using the distributions described previously. In fact, the local state distributions were defined keeping in mind the requirements of the selected homogenization theories. Note that the elementary homogenization theories selected here require evaluation of ensemble averages of various quantities over the microstructure (see, for example, Eq. 10.62). The ensemble average of any quantity, A, can be evaluated using local state distributions as

$$\langle A \rangle = \int_H A(h) f(h) dh \tag{10.68}$$

It can be seen that the Fourier representation of the functions $A(h)$ and $f(h)$ will prove beneficial in evaluating the integral shown in Eq. (10.68), because of the orthonormality of the Fourier basis.

Application of the concept described before to the elastic compliance parameters yields the following results (the equations for elastic stiffness are very similar):

$$\begin{aligned} S_{abcd}(\theta) &= \Xi_0 + 2\sum_{l=2,2}^{\infty} {}_{abcd}\Xi_l \cos(l\theta), \quad {}_{abcd}\Xi_l = \frac{2}{\pi}\int_0^{\pi/2} S_{abcd}(\theta)\cos(l\theta)d\theta S_{abcd}(\Phi,\beta) \\ &= \sum_{l=0,2}^{\infty}\sum_{\mu=1}^{M(l)} {}_{abcd}\Xi_l^{\mu} \dot{k}_l^{\mu}(\Phi,\beta), \quad {}_{abcd}\Xi_l^{\mu} \\ &= \frac{2}{\pi}\int_0^{\pi/2}\int_0^{\pi/2} S_{abcd}(\Phi,\beta)\dot{k}_l^{\mu}(\Phi,\beta)\sin\Phi \, d\Phi d\beta \end{aligned} \tag{10.69}$$

In Eq. (10.69), $S_{abcd}(\theta)$ and $S_{abcd}(\Phi,\beta)$ represent the functions obtained by substituting the 2D and 3D descriptions of the transformation matrices described in Eq. (10.59) into the coordinate transformation law described in Eq. (10.58). Of particular significance is the fact that the computation of the Ξ coefficients in Eq. (10.69) reveals that these coefficients are nonzero only for $l = 0,2,4$. This is because of the specific coordinate transformation law (Eq. 10.58) that is applicable to the tensor components of the elastic stiffness and compliance tensors. Table 10.5 lists the expressions obtained from Eq. (10.69)

Table 10.5 Spectral Coefficients Calculated from Eq. (10.69)

	$l = 0$	$l = 2$	$l = 4$
$_{1111}\Xi$	$\frac{1}{8}S_{11} + \frac{3}{8}S_{22} + \frac{1}{4}S_{12} + \frac{1}{2}S_{66}$	$\frac{1}{4}S_{11} - \frac{1}{4}S_{22}$	$\frac{1}{16}S_{11} + \frac{1}{16}S_{22} - \frac{1}{8}S_{12} - \frac{1}{4}S_{66}$
$_{2222}\Xi$	$\frac{3}{8}S_{11} + \frac{3}{8}S_{22} + \frac{1}{4}S_{12} + \frac{1}{2}S_{66}$	$-\frac{1}{4}S_{11} + \frac{1}{4}S_{22}$	$\frac{1}{16}S_{11} + \frac{1}{16}S_{22} - \frac{1}{8}S_{12} - \frac{1}{4}S_{66}$
$_{3333}\Xi$	S_{22}	0	0
$_{2233}\Xi$	$\frac{1}{2}S_{23} + \frac{1}{2}S_{12}$	$\frac{1}{4}S_{23} - \frac{1}{4}S_{12}$	0
$_{1133}\Xi$	$\frac{1}{2}S_{23} + \frac{1}{2}S_{12}$	$-\frac{1}{4}S_{23} + \frac{1}{4}S_{12}$	0
$_{1122}\Xi$	$\frac{1}{8}S_{11} + \frac{1}{8}S_{22} + \frac{3}{4}S_{12} - \frac{1}{2}S_{66}$	0	$-\frac{1}{16}S_{11} - \frac{1}{16}S_{22} + \frac{1}{8}S_{12} + \frac{1}{4}S_{66}$
$_{2323}\Xi$	$S_{22} - S_{23} + \frac{1}{2}S_{66}$	$\frac{1}{2}S_{22} - \frac{1}{2}S_{23} - \frac{1}{4}S_{66}$	0
$_{1313}\Xi$	$S_{22} - S_{23} + \frac{1}{2}S_{66}$	$-\frac{1}{2}S_{22} + \frac{1}{2}S_{23} + \frac{1}{4}S_{66}$	0
$_{1212}\Xi$	$\frac{1}{8}S_{11} + \frac{1}{8}S_{22} - \frac{1}{4}S_{12} + \frac{1}{2}S_{66}$	0	$-\frac{1}{16}S_{11} - \frac{1}{16}S_{22} + \frac{1}{8}S_{12} + \frac{1}{4}S_{66}$

for the 2D composites. Exploiting the orthonormality of the Fourier basis, the ensemble averages in the expressions for bounds (Eq. 10.60) can be evaluated as

$$
\begin{aligned}
2D: \quad \langle S_{abcd} \rangle &= \Xi_0 F_0 + 2 \sum_{l=2,2}^{4} {}_{abcd}\Xi_l F_l \\
3D: \quad \langle S_{abcd} \rangle &= \sum_{l=0,2}^{4} \sum_{\mu=1}^{M(l)} {}_{abcd}\Xi_l^{\mu} F_l^{\mu}
\end{aligned}
\tag{10.70}
$$

The main advantage of using the MSD framework is realized in Eq. (10.70). First, note that the ensemble averages needed to obtain the bounds (or estimates) require the use of only a finite number of terms in the Fourier expansion of the distributions. In other words, the higher-order microstructure coefficients (F coefficients corresponding to $l > 4$) do not have any influence on the ensemble averages for the properties of interest. We expect this important feature to carry over to other physical properties as well. This is because the Fourier basis used is already optimized to represent tensorial variables in the most economical way (i.e., the coordinate transformation properties expected from tensors are captured with the least number of terms in the Fourier expansion).

Second, the microstructure-effective property linkage expressed in Eq. (10.70) is invertible. The ensemble averages described by Eq. (10.70) denote hypersurfaces in the Fourier space. The intersection of these hypersurfaces with the microstructure hulls (see Figure 10.9) identify the set of microstructure families that are predicted to yield the specified ensemble averages (these can in turn be related to bounds on effective properties). More important, these hypersurfaces dissect the microstructure hull into two regions—one region depicting the set of one-point distributions with an ensemble-averaged property inferior to a specified value; the other region, an ensemble-averaged property superior to the specified value.

As an illustration of the concepts described here, we present a few samples of our computations on the complete set of orthorhombic 2D composites in a carbon-epoxy composite material system. The relevant properties of the individual phases are taken to be as follows:

$$E_m = 2.94\,\text{GPa},\ G_m = 1.7\,\text{GPa},\ E_{f1} = 234.6\,\text{GPa},\ E_{f2} = G_{f12} = 13.8\,\text{GPa}$$

$$G_{f23} = 5.5\,\text{GPa},\ \nu_{f12} = 0.2,\ S_{mc} = 50\,\text{MPa},\ \ S_{ms} = 29\,\text{MPa},\ S_{fc} = 870\,\text{MPa}$$

Since the ensemble averages of the elastic properties need only the Fourier coefficients of the microstructure up to $l = 4$, the relevant microstructure hull is two-dimensional and has been depicted in Figure 10.9. On this microstructure hull, we have identified the set of points satisfying Eq. (10.68) for $(\langle S\rangle^{-1})_{1111} = 10$ GPa and $(\langle S\rangle^{-1})_{1212} = 10$ GPa, shown in Figure 10.10 as a curve and a line, respectively. Points on these curves that lie inside the microstructure hull identify the complete set of distributions that would be predicted to yield the prescribed ensemble averages. Just as important, we can also identify in Figure 10.10 the sets of microstructures corresponding to $(\langle S\rangle^{-1})_{1111} < 10$ GPa and $(\langle S\rangle^{-1})_{1111} > 10$ GPa, respectively. Region A, corresponding to $(\langle S\rangle^{-1})_{1111} > 10$ GPa in dark gray, is particularly significant because the bounding theories indicate that all the microstructures in this region would exhibit a minimum effective value of C^*_{1111} of 10 GPa, independent of the spatial

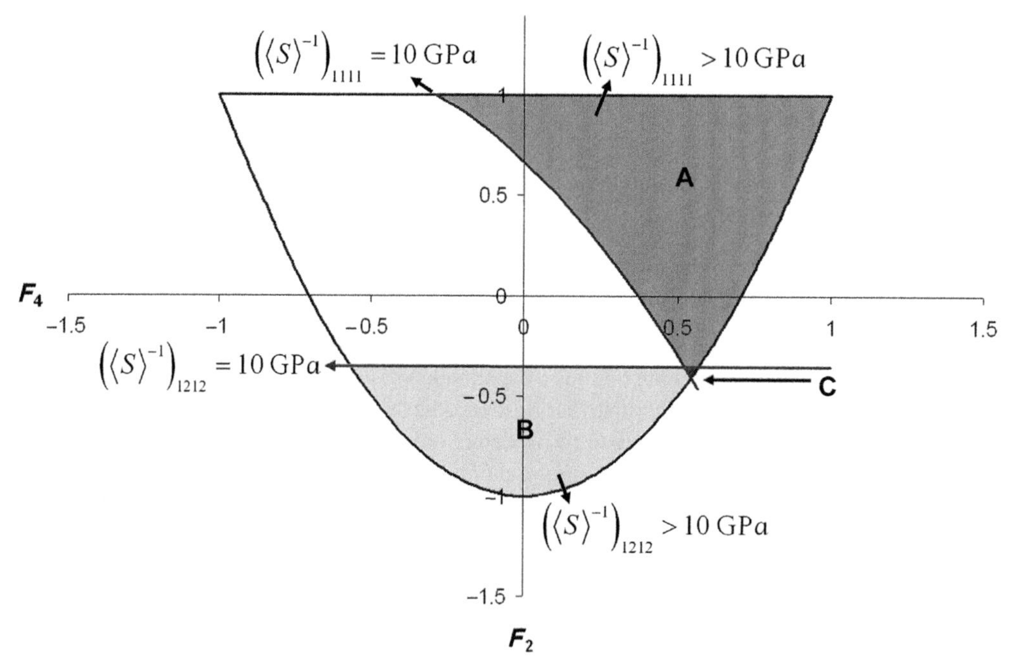

FIGURE 10.10

Iso-property hypersurfaces delineating the set of microstructures in Fourier space that meet prescribed conditions.

arrangements of the local states in the microstructure. Likewise, the region corresponding to $(\langle S\rangle^{-1})_{1212} > 10$ GPa, B, is shown in lighter gray and depicts the microstructures that are guaranteed to possess a minimum effective value of C^*_{1212} of 10 GPa.

Furthermore, the intersection of the two shaded regions within the microstructure hull (the small common area, C, identified in Figure 10.10) delineates the complete set of distribution functions that would satisfy both criteria, that is, $C^*_{1111} > 10$ GPa and $C^*_{1212} > 10$ GPa. Therefore, it can be seen that the mathematical treatment presented here can be extended easily to the consideration of multiple property/performance criteria. Further inspection of Eqs. (10.60), (10.69), and (10.70) reveals that iso-property hypersurfaces can be depicted directly for bounds or estimates of all effective properties by invoking the WAM model, instead of the ensemble averages treated in Figure 10.10.

Next, we turn our attention to the construction of property closures that delineate the complete set of theoretically feasible combinations of specified effective properties for a given composite material system and selected homogenization model (e.g., Eq. 10.61).

Construction of the property closures entails evaluation of property combinations from the microstructure-effective property linkages described before (Eqs. 10.68–10.70) while constraining the choice of microstructure coefficients to lie inside the microstructure hulls (defined by Eq. (10.67) and depicted in Figure 10.9). The numerical procedures used to delineate the property are accomplished primarily by evaluating if the intersections of the iso-property hypersurfaces corresponding to selected combinations of effective properties intersect on or inside the microstructure hull. The selected effective property combination is only feasible (as predicted by the chosen homogenization theory) if the corresponding iso-property hypersurfaces intersect on or inside the hull. Figure 10.11 presents examples of property closures computed using the methods described earlier together with the WAM model described in Eq. (10.61). The three closures correspond to three different values of the interpolation parameter α. The union of such closures for all possible values of α between 0 and 1 yields the overall closure for the selected material system.

The approach described here can now be extended to the simple failure model described in Eq. (10.62). For the 2D composites, the failure model can be expressed in the Fourier space as

$$\sigma_f(\theta) = \Psi_0 + 2\sum_{l=0,2}^{\infty} \Psi_l \cos(l\theta), \quad \psi_l = \frac{2}{\pi}\int_0^{\pi/2} \sigma_f(\theta)\cos(l\theta)d\theta$$
$$\langle\sigma_f\rangle = \Psi_0 + 2\sum_{l=2,2}^{\infty} \psi_l F_l \tag{10.71}$$

Table 10.6 provides the computed values of ψ_l for the selected material system. Note that, as expected, the values of ψ_l are converging quickly to 0. The average truncation error in the series representation of $\langle\sigma_f\rangle$ in Eq. (10.71) was estimated to be 23.9% at $l = 2$, 23% at $l = 4$, 6.6% at $l = 6$, and 0.03% at $l = 8$. Figure 10.12 shows an example of an elastic property-failure property closure computed using the methodology described before.

In design of engineering components, holes and notches are often encountered. These geometrical nonlinearities introduce stress concentrations. As an example, we consider here a thin orthotropic plate containing a circular hole and loaded in in-plane compression. Naturally, there is a stress concentration in the neighborhood of the hole. For a material exhibiting an isotropic elastic response, the stress

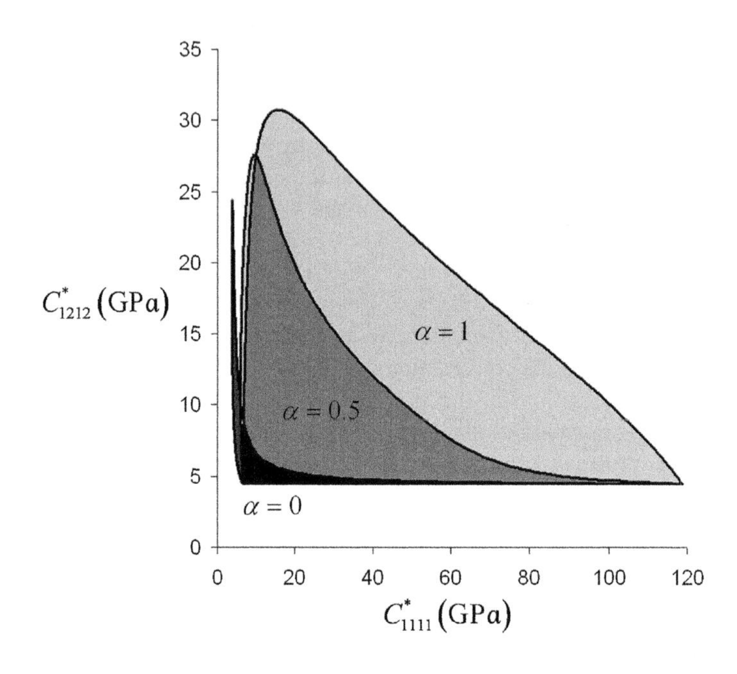

FIGURE 10.11

The property closure for the effective elastic properties C^*_{1111}(GPa) and C^*_{1212}(GPa) of the composite material system studied here. The material system studied was an AS-4 carbon fiber and epoxy system with a fiber volume fraction (V_f) of 0.5.

Table 10.6 Fourier Coefficients Summary (Eq. 10.71) in the Linkage's Spectral Representation between Microstructure and Effective Failure Strength in Uniaxial Compression for 2D Composites

	Failure Coefficients, ψ_l (MPa)						
l	0	2	4	6	8	10	12
ψ_l	250	86	17	−3.2	−5	−2.7	0.8

concentration factor is known to be 3. Therefore, design in the isotropic case essentially entails selection of the strongest possible material to sustain the increased stress at the "hot spot." One of the many appeals of using composites is that the inherent anisotropy of the properties can be exploited to improve the performance in the selected application.

In the study, we limit our consideration to the 2D AS-4 carbon fiber–epoxy material system with a fiber volume fraction of 0.5. The goal of the design exercise is to maximize the load-carrying capacity of the plate by an optimal distribution of the fiber orientations, while avoiding failure. The

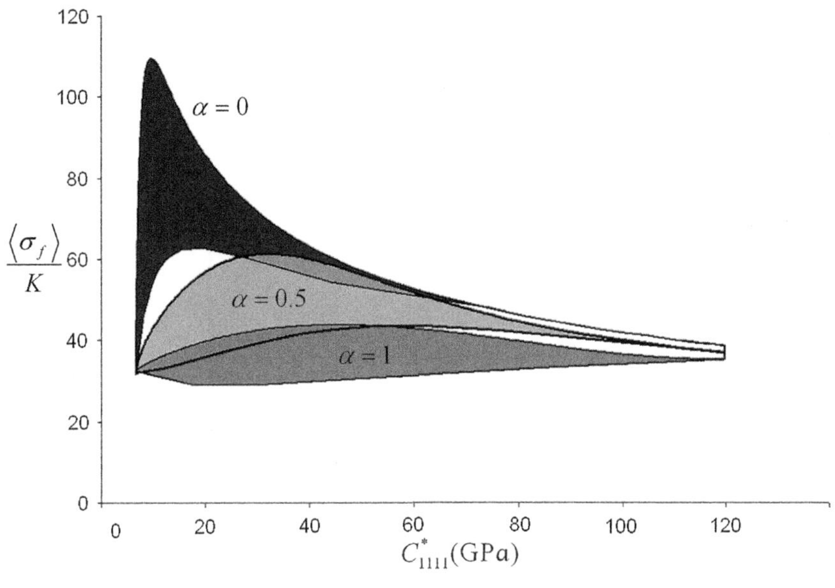

FIGURE 10.12

An example of a performance closure for a thin orthotropic plate containing a central circular hole and subjected to in-plane compression. Closures are shown for three different values of the interpolation parameter α in the WAM model.

specific property combination that controls the load-carrying capacity in this specific design is shown to be $\langle \sigma_y \rangle / K$. The stress concentration factor for this case study is obtained from the orthotropic-elasticity solution provided by Lekhnitskii (1968) as

$$K = \left(1 + \sqrt{\frac{4S^*_{66}}{S^*_{11}} + 2\sqrt{\frac{S^*_{22}}{S^*_{11}}}}\right) \tag{10.72}$$

Here, we assume that the effective elastic compliance components in Eq. (10.72) can be obtained from the WAM model (Eq. 10.61). Figure 10.12 shows a property closure for $\langle \sigma_y \rangle / K$ and C^*_{1111}, obtained using the spectral representations described in previous sections for three different values of the interpolation parameter used in the WAM models.

The points in the closures shown in Figure 10.12 that correspond to the maximum possible values of$\langle \sigma_y \rangle / K$ delineate the specific microstructures that are predicted to yield the highest load-carrying capacity based on the selected homogenization theories. For the present example, it is identified that a fiber orientation with an angle $(\theta) = 36.5°$ produces the highest load-carrying capacity for all three choices of the interpolation parameter α. The corresponding composite microstructure is depicted schematically in Figure 10.13. Note also that, in some design cases, there may be a need to trade off performance in the load-carrying capacity to ensure a certain minimum elastic stiffness in the loading

FIGURE 10.13

Schematic depiction of one of the microstructures identified by the microstructure-sensitive design methodology for the optimal performance of a 2D orthotropic thin composite plate with a central circular hole and subjected to in-plane compression.

direction. Closures presented in Figure 10.12 will allow the designer to make appropriate decisions in such cases.

10.5 HETEROGENEOUS DESIGN

In this section we briefly introduce heterogeneous design using the tools developed so far. The objective up to this point has been to characterize the material property space in such a manner that a designer might be able to search the space and thus optimize the properties of a material to meet design requirements. In traditional design the appropriate material will be chosen for a given component and the geometry then varied to meet structural or other requirements. Clearly, this loop may be traveled several times with different material specifications in an iterative attempt to optimize both material and geometry properties.

Ideally a designer would be able to optimize material and geometrical design simultaneously. To demonstrate this ideal in the extreme, we assume for the present that once the property closure for a series of properties has been found, the designer can actually manufacture material that lies at any point of this region. Furthermore, we assume that if two points are chosen in the property closure, and a corresponding two points are chosen on the geometry of the component, then the designer can process the material in such a manner that the first property is achieved at the first geometrical point, and the properties are varied in a continuous manner in both property and geometrical space until the second property is achieved at the second geometrical point. This is clearly a gross idealization, but it makes for a simpler theoretical problem that can be used to demonstrate the concepts.

The idea is that we select a set of property parameters that are critical to the design of a component; for example, the set might include particular components of the stiffness tensor, yield strength, thermal conductivity, and so forth. We then formulate the property closure for these parameters using the methods described earlier. The boundaries of the closure then become the constraints on these parameters in an optimization process that includes geometry and material properties.

We illustrate the method with a simple example. Clearly, variation of the geometry must be implemented in a manner that utilizes a finite-dimensional space; that is, we cannot simply allow any point in space to arbitrarily take a value inside or outside of the component. This issue is commonly handled by choosing key points on the geometry that can be varied. Bezier curves are subsequently defined between these points, thus resulting in a limited set of parameters that define the positions of the key points and the shape of the curves. A similar process may be defined in material space by defining Bezier curves in the property closure. One of the special properties of these curves is that once the points defining the curve are given, the curve must lie in the convex region between these points. There is no guarantee that the property closure is convex, however, so care must be taken that curves do not leave the closure.

On the other hand, the microstructure hull is guaranteed to be convex, and it is therefore straightforward to define Bezier curves in this space that are guaranteed to remain in the space. The points may then be mapped into the property closure using homogenization theories already described. The disadvantage with this approach is the potential size of the space. The introduction of spherical harmonic representations in Chapter 9 helps by reducing the hull's size. For the present illustration, we limit the (approximate) microstructure hull's size by discretization.

A second issue that arises involves the mapping between the property closure and the microstructure hull. Using the bounding techniques described earlier, each point in the microstructure hull maps to a rectangle in the closure. Thus, there may be various ways to map a point in the closure back to the hull. Take, for example, the closure from Example 9.2, which was created using two orientations in the fundamental zone. We illustrate the regions corresponding to single points in the hull both for this example and for a closure created using four orientations.

If, for example, we decide that any point on a curve in the property closure associates with the rectangle (and, hence, point in the microstructure hull) with the closest center point, then curves in Figure 10.14(a) will be mapped to a straight line in the 1D hull, while curves in Figure 10.14(b) may map to straight lines or 3D-curves.

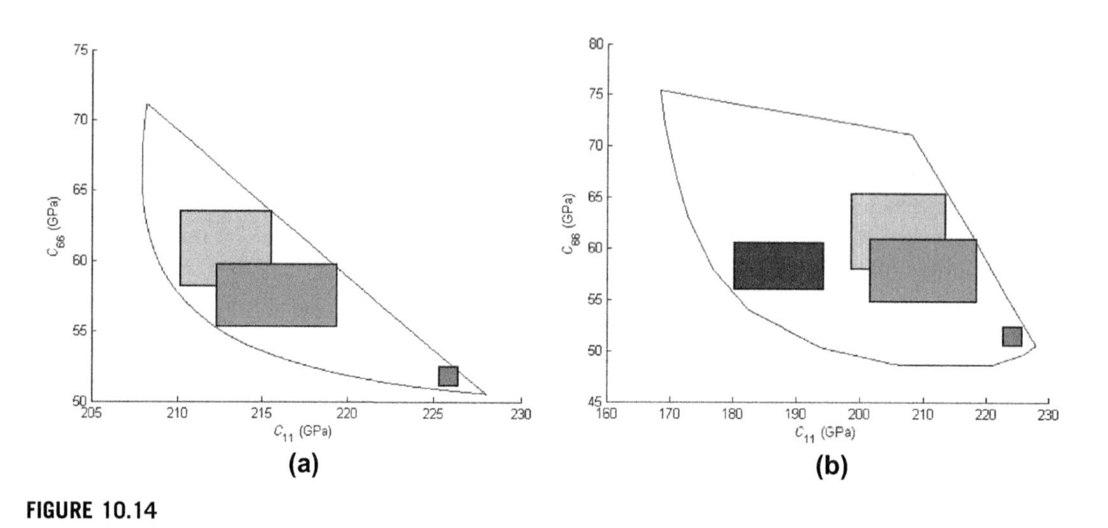

FIGURE 10.14

(a) Schematic of regions in the closure from Example 9.2 relating to single points in the microstructure hull (these are illustrations rather than actual calculated regions). (b) A closure for the same material, with four orientations used to create the closure. The boxes of the same shade (a and b) represent the same points in the microstructure hull.

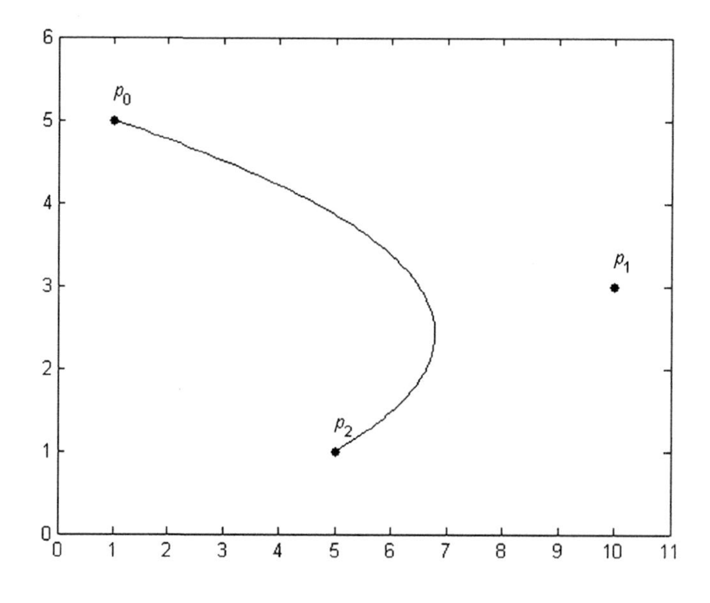

FIGURE 10.15

Second-order Bezier curve defined by p_0, p_1, and p_2. Note that the curve lies in the convex region between the points.

We next define a Bezier curve. Let t be the variable that parameterizes the curve $t \in [0, 1]$, and let n be the order of the curve. Then $n + 1$ points are required to define the curve, and the curve itself is defined in terms of combinations of these points:

$$f(t) = \sum_{i=0}^{n} \binom{n}{i} p_i (1-t)^{n-i} t^i \tag{10.73}$$

For example, a second-order curve is given by

$$f(t) = p_0(1-t)^2 + 2p_1(1-t)t + p_2 t^2 \tag{10.74}$$

and is illustrated in Figure 10.15.

Example 10.1

Here we show how to set up an example for heterogeneous design of a shaft with restrictions on deflection and twist. Assume that the bar is of length 0.7 m with a torque of 4000 Nm and an axial force of 5000 N at one end; the other end is assumed fixed. Assume that there is a target maximum twist in each section of the shaft and a target deflection for the entire shaft. The bar is assumed to be cylindrical at any cross-section, and the material is assumed to be constant in any such section.

Solution

Approximate the shape of the bar by dividing it into N equal lengths, δ, and label the sections by $i = 1:N$. Let the axial force be F and the torque be T. Let l be the length of the bar, r_i is the radius of the ith section, and I_i is the moment of inertia of that section: $I_i = \pi r_i^4/4$.

The angle of twist over each section is given by $\Delta\theta_i = Tl_i/J_iG_i$, where G_i is the shear modulus in that section and $J_i = I_i/2$.

The deflection over each section is calculated using the differential equation

$$\left. d^2y/dx^2 \right|_i = \frac{M_i}{E_i I_i}$$

where E is the usual Young's modulus, from which one may readily deduce that

$$\left. dy/dx \right|_i = \frac{M_i x}{E_i I_i} + \left. \frac{dy}{dx} \right|_{(i-1)}$$

Hence,

$$\Delta y_i = \delta\left(\frac{M_i \delta}{E_i I_i} + \Delta y_{(i-1)}/\delta\right)$$

Now $E = \dfrac{\mu(3\lambda + 2\mu)}{\lambda + \mu}$, $G = \mu$, and $\lambda = C_{1122} = C_{12}$, $\mu = C_{1212} = C_{66}$. Thus, we require the closure for these two properties. The bounds are given as (Eq. 8.2.19)

$$\overline{S}_{1212}^{-1} \leq C_{1212}^{*} \leq \overline{C}_{1212}$$

For C_{1122} we need to calculate a "degraded bound" as follows:

$$\Delta_{11} = \overline{C}_{1111} - \overline{S}_{1111}^{-1}, \quad \Delta_{22} = \overline{C}_{2222} - \overline{S}_{2222}^{-1}$$

$$\overline{C}_{1122} - \sqrt{\Delta_{11}\Delta_{22}} \leq C_{1122}^{*} \leq \overline{C}_{1122} + \sqrt{\Delta_{11}\Delta_{22}}$$

Using four orientations (see, for example, Example 13.3) the resulting closure will look something like the closure in Figure 10.14(b). The top and right boundaries may be fitted by straight lines; the left boundary may be fitted with a quadratic or cubic polynomial. These then represent the boundaries of the search in the closure.

If a target twist is assigned to each section, the aim will be to minimize the absolute value of the sum of all the differences between actual and target twists. An optimization routine may be set up in, say, Excel to achieve this. Assuming second-order Bezier curves, the variables will be three radii (for the first, last, and middle sections) and three points in the closure. The bounds on the radii will be some geometric constraints; for example, we may bound them between 0.01 and 0.1 m. The Bezier curve parameter, t, in each case will also correspond to the length along the shaft.

It is left to readers to play with this problem to get a feel for the intricacies. It will soon be noticeable that each different starting point for the optimization is likely to produce a different result; that is, there are many local minimum points. This is because for each fixed geometry there will be a material design that will minimize the problem, and vice versa. Hence, an optimization routine such as simulated annealing or genetic algorithms is likely to be the best approach for this type of problem.

Readers should also determine a way to map the final Bezier curve in the property closure back to the microstructure hull.

SUMMARY

In this chapter the full (first-order) microstructure-sensitive design framework was demonstrated via four case studies. The potential for heterogeneous design (meaning that the local structure can vary across a component) was also illustrated by a simple extension of the methodology. In Chapter 11 we look at the potential for guiding the processing of materials to arrive at superior properties. Subsequent chapters then consider a higher-order microstructure design framework (i.e., one that includes geometry in the structure description).

Exercises

Section 10.4

1. Using Eq. (10.69) and Table 10.5, verify that the Fourier representation is accurate by comparing results from the Fourier representation and direct computation from coordinate transformation laws for $S_{1111}(\theta)$ of a laminate with fibers oriented at 20 degrees to the $\mathbf{e}_1$-axis.

2. Obtain a property closure between C_{1111}^{*} and C_{2222}^{*} for the composite plate case study using the information in Table 10.5.

Section 10.5

3. Repeat Example 9.2, but plot the value of $K = \left(1 + \sqrt{\frac{4S_{66}^*}{S_{11}^*} + 2\sqrt{\frac{S_{22}^*}{S_{11}^*}}}\right)$ around the boundary (i.e., use 3D plotting). For the values of the stiffness matrix that are not being varied, use the average of the values from the two orientations. Which point will give the lowest value of K, and what is the associated texture?

4. Repeat the previous example with the four orientations of Example 13.3, finding all available bounding points.

5. Implement Example 10.1.

CHAPTER

Microstructure Evolution by Processing

11

CHAPTER OUTLINE

The focus of the mathematical framework developed thus far has been the identification of the class of microstructures that are theoretically predicted to satisfy a set of designer-specified criteria for effective properties. Success in that endeavor immediately summons us to the next essential task in the inverse methodologies of materials design—that of process design. The stated goal of process design is to find a processing route made up of arbitrary combinations of available manufacturing options that will transform a given initial microstructure into one of the elements of the set of desired microstructures (presumably identified to satisfy a certain set of property requirements or performance characteristics). The solution to the process design problem is significantly more difficult than the microstructure design problem discussed thus far. Mathematically rigorous methodologies for inverse solutions to the process design problem continue to elude us.

The overall strategy for addressing the process design problem in MSDPO follows the same logic as our earlier strategy for microstructure design. Our goal here will be to explore methodologies that permit an efficient description of the microstructure evolution in a given material system in response to a prescribed processing operation, and to formulate this linkage in such a manner that process design solutions can be sought. As noted in preceding chapters, a central feature of the MSDPO framework is to characterize the microstructure in a given sample by certain statistical distribution functions that are needed to evaluate the effective properties exhibited by the sample. Since these functions are conveniently represented in appropriate Fourier spaces, our goal here will be to represent the evolution of the microstructure in response to a prescribed processing operation in terms of the evolution of the corresponding Fourier coefficients.

As an example, consider a polycrystalline sample subjected to a deformation processing operation such as rolling, drawing, or extrusion. One important consequence of such operations is that the lattice orientations of the constituent crystals change substantially, thus changing the overall texture in the sample. As described in the preceding chapters, the first-order description of texture in the sample results in the definition of the orientation distribution function (ODF) that can be represented as a point in a Fourier space (using either the indicator functions or the generalized spherical harmonic basis). Recall also that all physically realizable textures have to have a representation inside texture hulls.

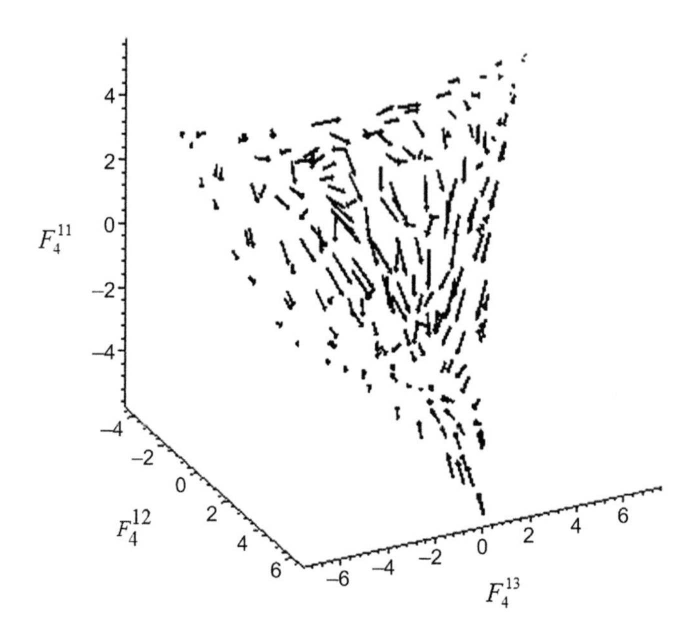

FIGURE 11.1

The vector flow field representing the crystal lattice rotations in the Fourier space for fcc crystals in a small increment of rolling reduction as predicted by the Taylor-type model.

Therefore, one should be able to represent any change in the texture of a given sample induced by a processing operation as a suitable transformation in the Fourier space—that is, as a transformation from one location to another in the texture hulls described in earlier chapters. As an example, Figure 11.1 shows the vector field that captures the transformation of points in the texture hull corresponding to a small rolling strain (5% reduction). This figure, of course, shows only a projection of the vector field in the first three dimensions of the Fourier space; a complete graphical description of this vector field needs new visualization tools that can handle a much larger-dimensional Fourier space.

Thermomechanical deformation processes are widely employed in the manufacture of various metallic engineering components. Of these, only the deformation processes in single-phase cubic metals at low homologous temperatures have thus far been successfully simulated, with satisfactory fidelity, by physics-based models that employ the one-point statistics of the microstructure (ODF); this set of physical theories was introduced in Chapter 7 as the crystal plasticity modeling framework. In the relatively short history of MSDPO, we have thus far only explored process design solutions within the confines of this well-established and extensively validated crystal plasticity theory. Extension of this class of models to include a broad range of thermomechanical processing operations and a broader class of metals continues to be the central focus of ongoing research activities worldwide.

Although we present only a limited number of case studies exploring a severely limited set of available processing operations, we feel that the central ideas presented here are extendable to a much broader set of processing operations. However, this requires the development of physics-based micromechanical models and their extensive validation. Alternatively, one can aim to get the required information for the spectral linkages described in this chapter directly from experiments. Perhaps it is desirable to devise new innovative procedures to obtain the desired information from a judicious combination of physics-based models and some direct experimental observations. All of these ideas are being pursued in our ongoing work. This chapter should be treated by readers as a "work in progress." Our main goal in this chapter is to open readers' minds to various possibilities for addressing the process design problem in the MSDPO framework.

11.1 FIRST-ORDER CRYSTAL PLASTICITY MODELS IN A SPECTRAL FRAMEWORK

Crystal plasticity models are used extensively in understanding and predicting the evolution of the underlying microstructure evolution and the concomitant anisotropic stress–strain response in polycrystalline metals subjected to large plastic strains. In the first-order approaches (e.g., Taylor model, Sachs model), each individual grain in the representative volume element (RVE) of the polycrystal is treated completely independent of all the other grains in the polycrystal. It has also been demonstrated in literature that the predictions from these first-order theories correspond to first-order bounds on the anisotropic stress–strain response of the polycrystals (see also Chapter 8). Most important, these first-order theories utilize only the one-point statistics of the microstructure (i.e., ODF). Of the various first-order crystal plasticity theories, the Taylor-type models have been demonstrated to provide good predictions of the averaged texture evolution and the overall anisotropic stress–strain behavior for single-phase, medium- to high-stacking fault energy cubic metals.

It should also be pointed out that there are indeed other approaches, distinct from those described here, to capture the essential features of crystal plasticity models in an efficient spectral framework. For example, a concept originally proposed by Bunge and Esling (1984) was revisited and expounded in substantial detail in several recent papers. Although this concept showed great promise, it was identified that the evaluation of the higher-order terms of the spectral representation in this specific methodology demanded a prohibitively high computational cost.

In the spectral linkages described in this chapter, our main goal is to establish suitable databases that capture efficiently the most important aspects of microstructure evolution in a selected class of processing operations. A review of the crystal plasticity theory presented in Chapter 7 suggests that our main focus should be to capture the dependence of stresses, the lattice spins, and the strain hardening rates in individual crystals as a function of their lattice orientation for a specified monotonic deformation path. Extension of this approach to include all deformation paths will be discussed later. We will aim to develop these linkages in the same spectral framework that was presented earlier for establishing microstructure–property linkages, since it will allow us to integrate seamlessly the microstructure design solutions with the process design solutions. In the following discussion, we will also employ the generalized spherical harmonics as the basis functions because they are far more economical in capturing the linkages we seek when compared to the indicator functions.

Specifically, we seek to establish the following representations:

$$W^*_{ij}(g) = \sum_{l=0}^{\infty} \sum_{\mu=1}^{M(l)} \sum_{\nu=1}^{2l+1} {}_{ij}A_l^{\mu\nu} \dot{T}_l^{\mu\nu}(g) \tag{11.1}$$

$$\sigma_{ij}(g) = \sum_{l=0}^{\infty}\sum_{\mu=1}^{M(l)}\sum_{\nu=1}^{2l+1} {}_{ij}S_l^{\mu\nu}\dot{T}_l^{\mu\nu}(g) \tag{11.2}$$

$$\sum_{\alpha}|\dot{\gamma}^{\alpha}|(g) = \sum_{l=0}^{\infty}\sum_{\mu=1}^{M(l)}\sum_{\nu=1}^{2l+1} G_l^{\mu\nu}\dot{T}_l^{\mu\nu}(g) \tag{11.3}$$

Note that Eqs. (11.1), (11.2), and (11.3) imply that we are seeking Fourier representations for three independent components of the skew-symmetric $W_{ij}^*(g)$, for five independent components of the symmetric and deviatoric component of $\sigma_{ij}(g)$ (the isochoric deformation modes associated with metal plasticity produce deviatoric stress components; hydrostatic stress can be superimposed as needed to recover the complete stress state), and for the sum of the absolute values of the slip rates on the different slip systems in the crystal. These variables have been selected because they constitute the essential information needed to predict deformation texture evolution and the overall anisotropic stress–strain behavior of polycrystals. In the examples presented here, all slip systems are assumed to harden equally, and in this situation, the accumulated slip is the only information needed for capturing the slip hardening description. If latent hardening is introduced, then it will be necessary to track the slip rates on the different slip systems individually.

The Fourier coefficients in Eqs. (11.1), (11.2), and (11.3) can be expressed as

$${}_{ij}A_l^{\mu\nu} = \frac{1}{2l+1}\oint W_{ij}^*(g)\overline{\dot{T}_l^{\mu\nu}}(g)dg \tag{11.4}$$

$${}_{ij}S_l^{\mu\nu} = \frac{1}{2l+1}\oint \sigma_{ij}(g)\overline{\dot{T}_l^{\mu\nu}}(g)dg \tag{11.5}$$

$$G_l^{\mu\nu} = \frac{1}{2l+1}\oint \sum_{\alpha}|\dot{\gamma}^{\alpha}|(g)\overline{\dot{T}_l^{\mu\nu}}(g)dg \tag{11.6}$$

where the overbar indicates a complex conjugate. The coefficients in Eqs. (11.4) through (11.6) can be evaluated numerically using an appropriate numerical integration scheme (e.g., the Simpson method) once a suitable model has been selected for predicting $W_{ij}^*(g)$, $\sigma_{ij}(g)$, and $\sum_{\alpha}|\dot{\gamma}^{\alpha}|(g)$.

As noted earlier, the most successful first-order approach to obtaining the response of a polycrystal from the response of the individual grains is to use the extended Taylor's assumption of iso-deformation gradient in all of the crystals comprising the polycrystal. In this model, the Cauchy stress in the polycrystal is taken as the volume average of the Cauchy stresses in the constituent grains. For the following examples, the Taylor-type model described in Chapter 7 has been employed to provide values of W_{ij}^*, σ_{ij}, and $\sum_{\alpha}|\dot{\gamma}^{\alpha}|$ for selected grain orientations that were distributed in the cubic-triclinic fundamental zone to enable the numerical integration in Eqs. (11.4) through (11.6). To compute these values a small strain step ($\varepsilon = 2\%$) was imposed on the selected crystals and the desired variables were extracted from the Taylor-type computations.

We present here some sample computations for plane strain compression, where the Fourier coefficients in Eqs. (11.4) through (11.6) were computed up to $l = 23$. It was later discovered that the contributions from the terms corresponding to $l > 10$ were negligibly small and could be ignored in the examples shown here. Using the Fourier representations obtained recursively in small strain steps and a continuous update of the crystal orientations and their slip resistances in each step, plane strain compression to $\varepsilon = -1.0$ was simulated for two different initial textures: (1) a random initial texture and (2) a <110> fiber texture. The deformed textures and the anisotropic stress–strain curves, predicted using both the Taylor-type model and the spectral representations, are shown in Figures 11.2 through 11.4. It is seen that the spectral methods accurately reproduce all of the important aspects of the Taylor-type model predictions.

The most generic isochoric deformation path to be imposed on a polycrystalline metal can be expressed in the principal frame of the stretching tensor as

$$\mathbf{L} = \frac{\dot{\varepsilon}}{\dot{\varepsilon}_o}\mathbf{D}_o + \mathbf{W}, \quad \mathbf{D}_o = \begin{bmatrix} \frac{\cos\theta}{\sqrt{2+\sin 2\theta}}\dot{\varepsilon}_o & 0 & 0 \\ 0 & \frac{\sin\theta}{\sqrt{2+\sin 2\theta}}\dot{\varepsilon}_o & 0 \\ 0 & 0 & \frac{-(\cos\theta+\sin\theta)}{\sqrt{2+\sin 2\theta}}\dot{\varepsilon}_o \end{bmatrix} \tag{11.7}$$

where $\dot{\varepsilon}_o$ is a reference value of strain rate and $\theta \in [0, 2\pi)$. This implies that to extend the spectral method previously described to the most general deformation path, one needs to create a database of Fourier coefficients in Eqs. (11.4) through (11.6) for all descriptions of $\mathbf{D}_o$. This can be accomplished by expanding the Fourier representations in Eqs. (11.4) through (11.6) to include an additional scalar variable, θ.

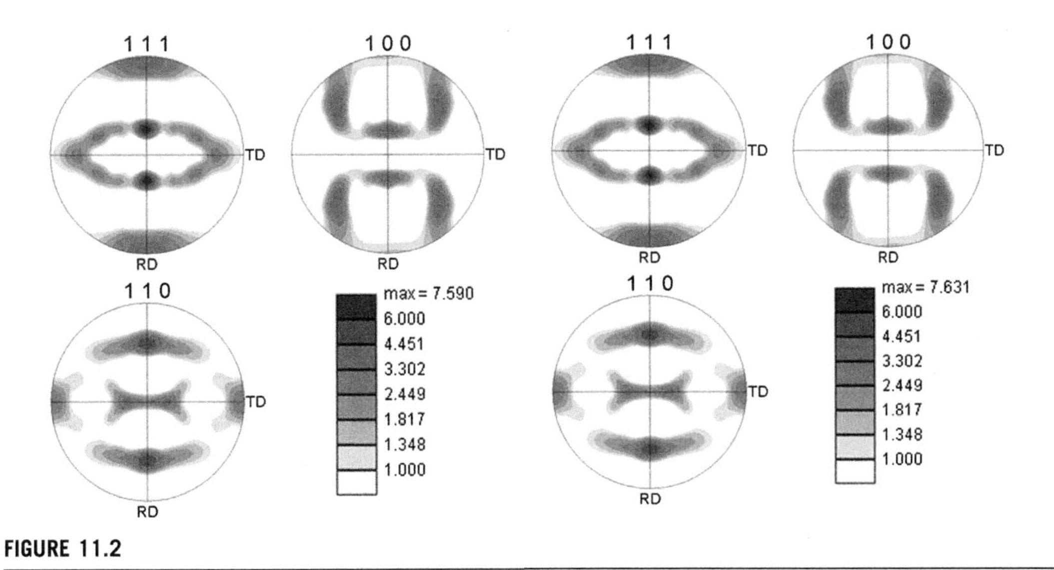

FIGURE 11.2

Comparison of the (111), (100), and (110) predicted pole figures in plane strain compression of fcc metals after a compressive true strain of −1.0. The starting texture for these simulations is a random texture.

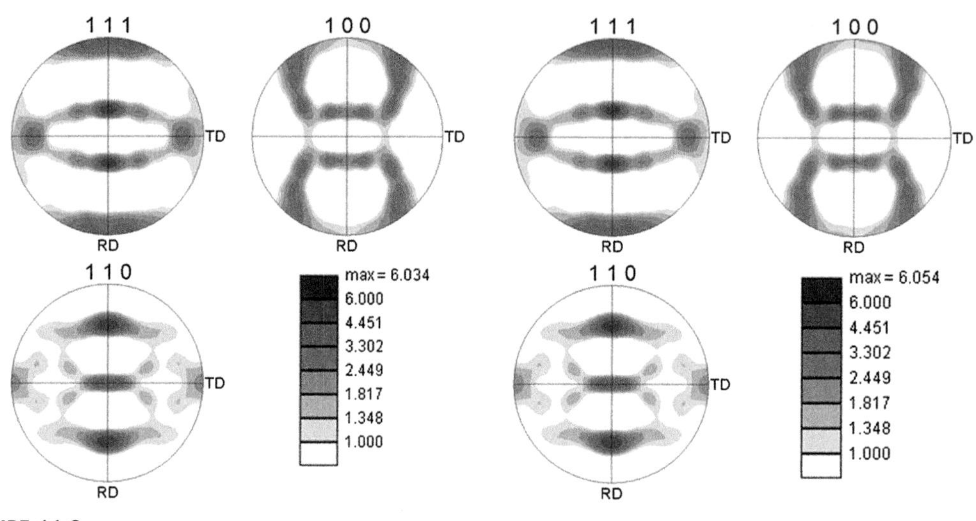

FIGURE 11.3

Comparison of the (111), (100), and (110) predicted pole figures in plane strain compression of fcc metals after a compressive true strain of −1.0. The starting texture for these simulations is a <110> fiber texture obtained by simulating simple compression to a strain of −1.0 on an initially random texture using the Taylor-type model.

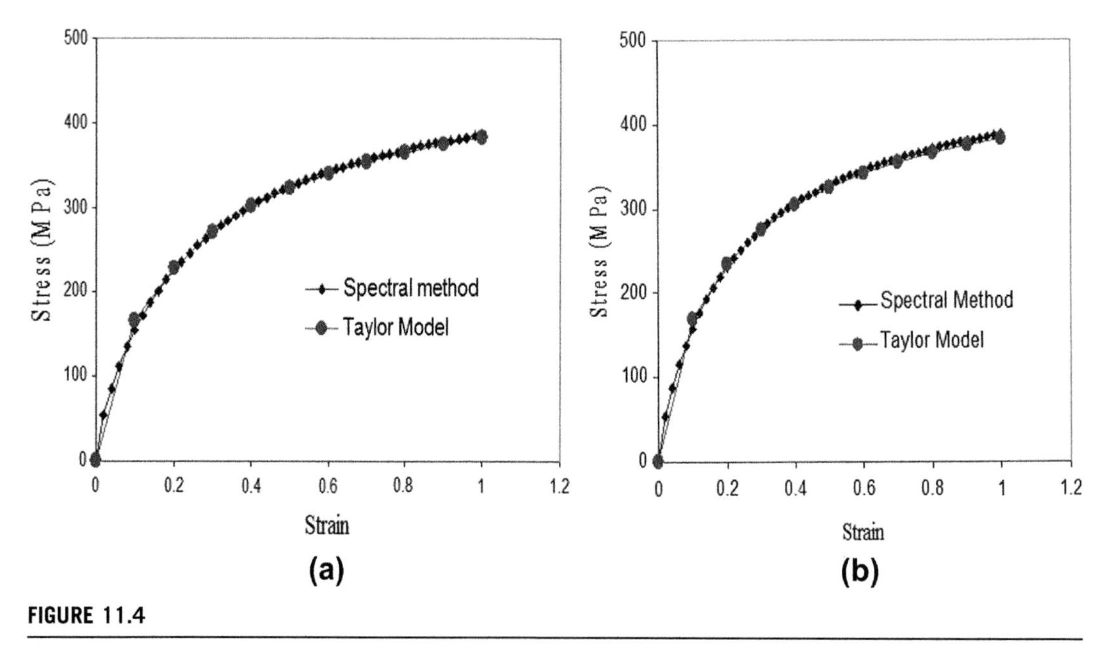

FIGURE 11.4

Predicted stress–strain response in plane strain compression: (a) random initial texture; (b) initial texture is a <110> fiber texture.

Note that the representation of the deformation path in Eq. (11.7) is in the principal frame of the imposed stretching tensor. Therefore, the lattice spins and the stresses computed by Eqs. (11.1) through (11.3) would also be in the same reference frame. These can be transformed, as needed, to the sample coordinate frame. The introduction of the reference strain rate and the superimposed spin in the deformation path in Eq. (11.7) requires us to modify Eqs. (11.1) through (11.3) as

$$W^*_{ij}(g,\theta) = \frac{\dot{\varepsilon}}{\dot{\varepsilon}_o}\sum_{l=0}^{\infty}\sum_{\mu=1}^{M(l)}\sum_{\nu=1}^{2l+1} {}_{ij}A_l^{\mu\nu}(\theta)\dot{T}_l^{\mu\nu}(g) + W_{ij} \tag{11.8}$$

$$\sigma_{ij}(g,\theta) = \left|\frac{\dot{\varepsilon}}{\dot{\varepsilon}_o}\right|^m \operatorname{sgn}\left(\frac{\dot{\varepsilon}}{\dot{\varepsilon}_o}\right)\sum_{l=0}^{\infty}\sum_{\mu=1}^{M(l)}\sum_{\nu=1}^{2l+1} {}_{ij}S_l^{\mu\nu}(\theta)\dot{T}_l^{\mu\nu}(g) \tag{11.9}$$

$$\sum_{\alpha}|\dot{\gamma}^{\alpha}|(g,\theta) = \left|\frac{\dot{\varepsilon}}{\dot{\varepsilon}_o}\right|\sum_{l=0}^{\infty}\sum_{\mu=1}^{M(l)}\sum_{\nu=1}^{2l+1} G_l^{\mu\nu}(\theta)\dot{T}_l^{\mu\nu}(g) \tag{11.10}$$

where it is implicit that the Fourier coefficients were obtained with $\dot{\varepsilon} = \dot{\varepsilon}_o$ and $\mathbf{W} = 0$.

Alternatively, one can also use a new Fourier basis defined as the product space of the spherical harmonic functions for the orientation variable g and the classical harmonic functions for the variable θ. In such a Fourier basis, the Fourier coefficients in the description of $W^*_{ij}(g,\theta)$, $\sigma_{ij}(g,\theta)$, and $\sum_{\alpha}|\dot{\gamma}^{\alpha}|(g,\theta)$ would be just constants.

11.2 PROCESS DESIGN USING DEFORMATION PROCESSING OPERATIONS

The goal of process design is to develop rigorous mathematical procedures that can identify one or more processing paths that are predicted to produce either an element of the class of optimal microstructures or a microstructure (presumably suboptimal) that is close to the optimal set, and one that could be realized within reasonable cost by a small set of available manufacturing routes.

Rigorous solution methodologies for the process design problem stated before continue to elude us presently. We describe in Figure 11.5 one strategy that has enjoyed limited success; much further work is needed in this area. The texture hull, delineating the space corresponding to all physically realizable textures, is shown in Fourier space in the first three dimensions. Every point in this hull represents a class of textures. Let the A circle in the far right represent the microstructure of the as-received material, and the B circle on the far left represent the location in Fourier space of a class of microstructures with a desired combination of properties (presumably identified using the MSDPO methodologies described in earlier chapters). The problem at hand is to find an overall processing route that will transform the given initial microstructure to an element of the family of desired microstructure(s). The processing route has to be made up of segments chosen from a set of available

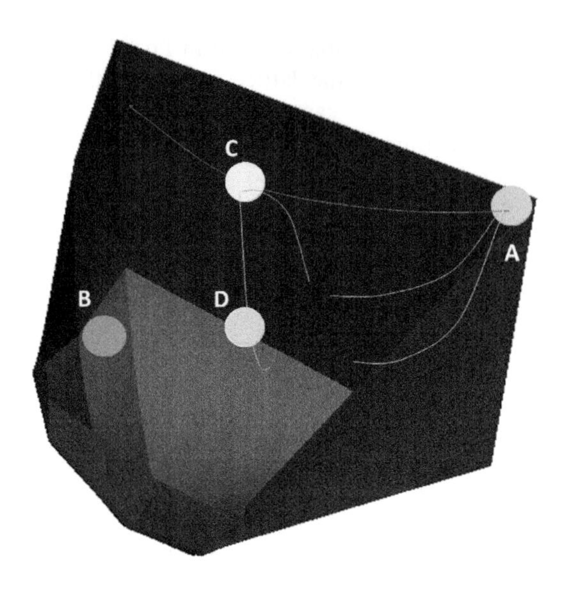

FIGURE 11.5

A schematic description of a strategy to find process design solutions. The circles indicate specific textures and the lines describe streamlines (or paths) of texture evolution for a selected processing method. The B circle represents a desired texture, while the A circle represents a starting initial texture.

manufacturing routes. If it is not possible to reach the desired microstructure(s) using the available manufacturing routes, process design aims to get as close as possible to the desired microstructure(s).

For each available manufacturing process, one can envision developing the spectral representations of the type described in Section 11.1. Figure 11.5 shows schematically three different microstructure evolution paths starting from the A circle, each corresponding to an available manufacturing route. These paths can be computed very quickly for any given location in the microstructure hull using the spectral databases described in Eqs. (11.4) through (11.6). Although none of these available manufacturing routes transform the microstructure directly into the desired microstructure, it is seen that one of the process routes brings us closest to the desired microstructure. The point on the process closest to the desired microstructure is represented by the C circle in Figure 11.5. Now one can explore again all possible manufacturing routes starting with this new location, and the three different lines shown correspond to evolutions for the same three available manufacturing processes. Applying the same philosophy recursively, one can get as close to the desired microstructure as possible. Of course, our ability to get close to the desired microstructure depends on the available manufacturing routes.

As a case study, let us revisit the design of an orthotropic thin plate with a central circular hole and subjected to in-plane tensile loads. The design objective is to maximize the load-carrying capacity of the plate without allowing plastic deformation. An optimum texture was identified in Kalidindi et al. (2004). The goal now is to find a processing solution to obtain the desired texture assuming that we are given a sample with a random initial texture. Further assume that the processing routes to be considered are limited only to those room-temperature deformation processes that preserve the orthorhombic sample symmetry. This limited set of processing paths can be defined through the following velocity gradient tensor:

$$\mathbf{L} = \begin{bmatrix} \alpha\dot{\varepsilon} & 0 & 0 \\ 0 & \beta\dot{\varepsilon} & 0 \\ 0 & 0 & -(\alpha+\beta)\dot{\varepsilon} \end{bmatrix} \quad \alpha, \beta \in [-1, 1] \tag{11.11}$$

Figure 11.6 delineates the region corresponding to all physically realizable textures, in the first three component subspaces of the Fourier space, as a gray wire-framed convex hull. However, not all textures in this convex hull can be realized by the selected set of processing paths starting with the random initial texture. The specific set of textures that can be produced can be identified using the strategy described earlier, and is depicted as a black wire-framed region in Figure 11.6. The strategy described earlier allows us to identify the complete set of textures that can be produced, starting from a given initial texture, by an arbitrary combination of selected processing techniques. It is further possible to design a processing route for any selected target texture that lies in this subset of realizable textures (the black wire-framed subspace in this case study).

This example reveals that none of the elements of the previously identified class of textures that were deemed best for the thin orthotropic plate with a circular hole can be produced starting with an initial random texture. Therefore, the MSDPO computations were repeated a second time, and this time the choice of Fourier coefficients in the MSDPO computations was restricted to the inner region

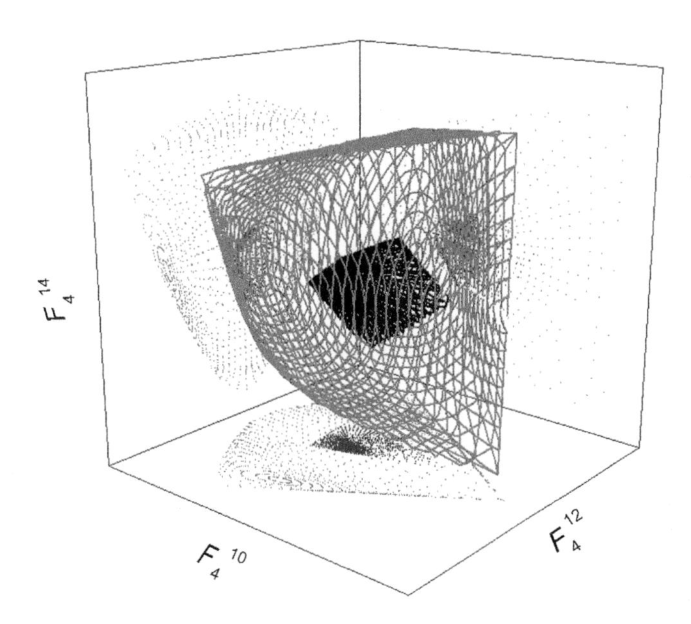

FIGURE 11.6

The gray wire-framed convex hull denotes the set of all possible textures, while the black subspace denotes the set of textures that can be produced by a combination of selected processing routes starting with a random initial texture.

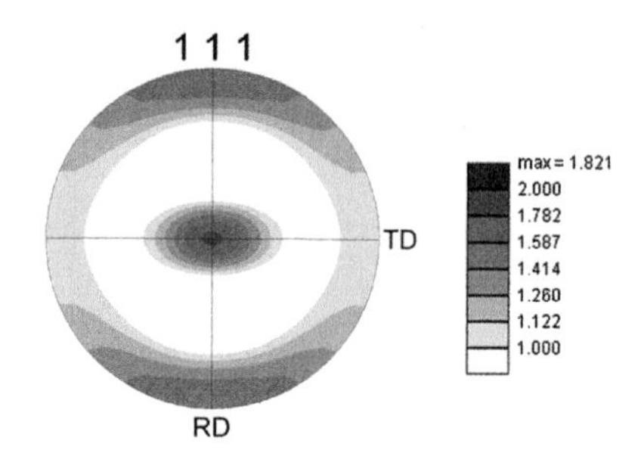

FIGURE 11.7

A (111) pole figure corresponding to the Fourier coefficients of the ODF identified as the optimized texture in the MSDPO framework, while being producible using deformation processing steps on a material with an initial random texture.

in Figure 11.6, because we know we can realize these textures by deformation processing (starting with a random initial texture). In this manner, a target texture was identified as the texture that would yield the best possible performance for the selected design case study that would also be easily producible using deformation processing steps. The texture identified is shown in Figure 11.7 as a (111) pole figure, and Taylor-type models predict that this texture can be produced in a two-step rolling process: a 50% rolling of the plate followed by a 20% rolling in a direction perpendicular to the original rolling direction.

11.3 A BRIEF OUTLINE FOR HETEROGENEOUS DESIGN

We conclude this chapter with some brief comments on how the theory introduced so far in this book might be incorporated into a heterogeneous design framework (see Section 10.5) that includes material processing.

In a traditional design process, materials are selected from a database of available stock products and introduced into a model with variable geometry. The model is optimized to meet customer requirements, with a loop that allows the material definition to be modified in an iterative manner. For polycrystalline-based materials (e.g., metals) the assumptions of homogeneity and isotropy are often integral to the design process.

In recent years heterogeneous design has emerged to allow the material definition to vary along with the geometry. This generally occurs in an iterative manner rather than including all variables in a single optimization process. Furthermore, materials are still selected from databases of producible substances.

In microstructure-sensitive design the space of all physically realizable microstructures is explored. Thus, the material design can be driven by customer requirements and geometry in a truly inverse approach. However, this does not lead to a practical design framework if the resultant material design cannot be arrived at by a known processing route. A more constrained theory must be adopted by combining the full material space with both known materials and available processing paths in the space.

Furthermore, it must be recognized that the processing route for a particular component may significantly modify the initial material structure, and should be factored into the full design process. Thus, we propose an anisotropic heterogeneous design process that incorporates a much more general framework than previous approaches. The main spaces and constructs include

1. The microstructure hull of all physically possible (anisotropic) materials.
2. An atlas of known materials and material processing paths that overlays the microstructure hull.
3. Homogenization relations that map the microstructure hull to the property closure of physical properties pertinent to the particular design.
4. The geometry space with known constraints that must be applied to the design.
5. The combined design space of material properties and geometry where the optimization process is applied. The inputs to this space are the raw material properties and initial geometry of the component. Any manufacturing operations that have a significant effect on the final material properties (rolling, stamping, etc.) must then be included in the calculations that arrive at the final component design before assessment of the design is undertaken. A heterogeneous design process is employed, where the geometry is varied over the available envelope, and the full material atlas is explored.

For example, assume that we wish to design the bar mentioned in Section 10.5. Suppose that we have two processing routes available for forming the bar (i.e., two processing routes that affect the microstructure). These might be shown schematically as in Figure 11.8.

The potential for different forming routes for different sections of the bar must be included in the optimization routine, along with the corresponding effect on the microstructure and, hence, properties. The property closure is thus constrained to those properties achievable for realizable materials and processing routes. The resultant properties must be mapped in a continuous manner onto the component for analysis, for example, by FE codes.

Thus, all aspects of microstructure-sensitive design as presented in this book are employed to arrive at a design space that is much larger than traditionally used, and that has the potential to achieve dramatic improvements in performance in many areas of engineering.

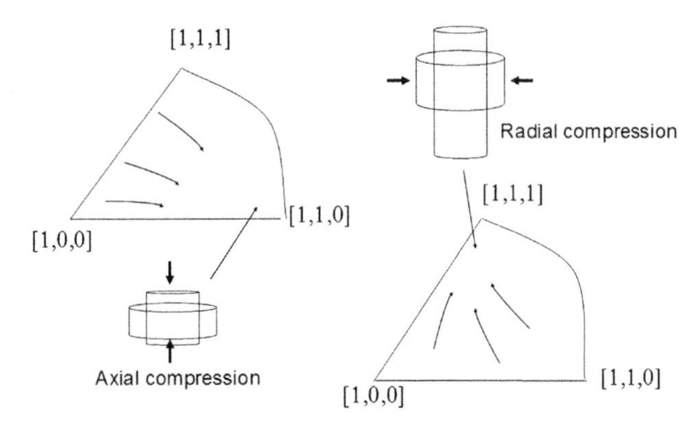

FIGURE 11.8

Schematic of two forming routes that have a known effect on the microstructure of an axially symmetric bar. The evolution of the texture is sketched in the rough inverse pole figures.

SUMMARY

In this chapter, readers were introduced to the very difficult inverse problem of process design. The design space here is the very large space of all possible hybrid process paths that combine available manufacturing options in arbitrary sequences. In the chapter, readers were taken through the various steps involved in accurately capturing the details of microstructure evolution in computationally efficient spectral databases. It is hoped that readers can appreciate the tremendous value of these spectral representations in obtaining inverse solutions for process design. Although the steps involved were demonstrated only for first-order models, they are extendable to high-order homogenization theories presented later in this book.

CHAPTER

Higher-Order Microstructure Representation

12

CHAPTER OUTLINE

12.1 CORRELATION FUNCTIONS AND MICROSTRUCTURE REPRESENTATION

Up to this point, the description of the microstructure has been purely in terms of the (first-order) distribution functions. These give information about the proportion of each local state in the overall microstructure. However, as early as Section 1.3 it became apparent that for a given distribution, the local states could be arranged physically to produce widely varying properties (e.g., see Figure 1.9).

This was also demonstrated in the first-order homogenization relations of Chapter 8. It is evident that better estimates of material property values must rely on some information regarding the spatial position of the local states, rather than simply the proportion of each state.

In materials science, it is a standard practice to sample the material internal structure by obtaining representative (2D or 3D) microstructure data sets. Since the sampled microstructure data sets have no natural origin, the only meaningful information one can extract from each data set has to be concerning the relative spatial arrangement or spatial correlations of local material states (e.g., phases, orientations) in the structure. In other words, we should focus our effort on extracting statistically meaningful data regarding relative positions of components rather than their absolute positions. Such information is embodied in *correlation functions*.

12.1.1 Two-point local state correlation functions

The simplest example of a correlation function is the *two-point local state correlation function*. For the individual microstructure function (MF) this correlation is obtained by integrating the product of the MF at points separated by the vector **r**:

$$^{(k)}f_2(h, h'|\mathbf{r}) = \frac{1}{vol(\Omega|\mathbf{r})} \iiint_{x \in \Omega|\mathbf{r}} M(x,h)M(x+\mathbf{r},h')dx \tag{12.1}$$

Note that the integration is over region $\Omega|\mathbf{r} \subset \Omega$ (Ω is the kth region of an ensemble), which is defined to be

$$\Omega|\mathbf{r} = \{x | x \in \Omega \text{ and } x + \mathbf{r} \in \Omega\} \tag{12.2}$$

Physically, $^{(k)}f_2(h, h'|\mathbf{r})$ is the joint volume density of occurrence of local state h at the tail of a randomly placed vector **r**, and of local state h' at the head of **r** in the kth element of the ensemble (Figure 12.1). Note that vector **r** is restricted in length and direction to having its head and tail both lay

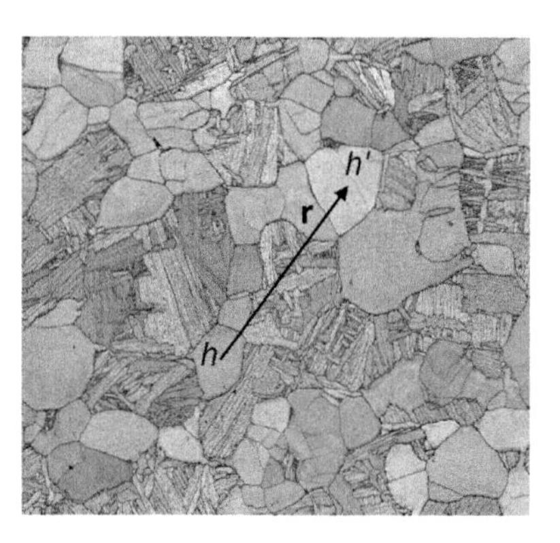

FIGURE 12.1

Schematic of an arbitrary vector, **r**, placed in region Ω to determine $^{(k)}f_2(h, h'|\mathbf{r})$. OIM image of titanium courtesy of EDAX-TSL.

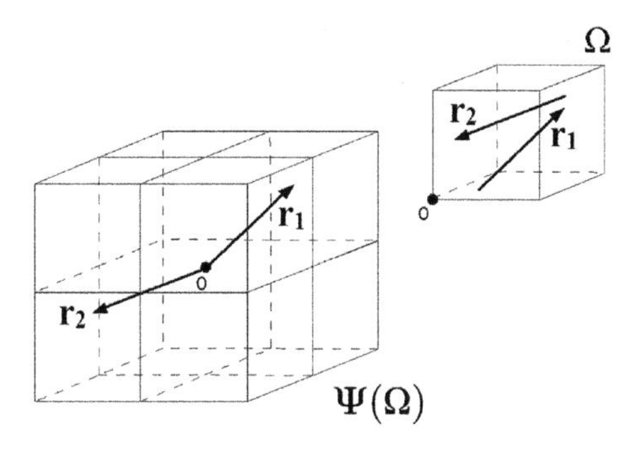

FIGURE 12.2

Representation of $\Psi(\Omega)$ with two arbitrary vectors shown.

in region Ω. The set of all possible $\mathbf{r}$ vectors, say $\Psi(\Omega)$, occupies a volume of space eight times the volume of region Ω (Figure 12.2). Formally,

$$\Psi(\Omega) = \{\mathbf{r}|x,\ x+\mathbf{r} \in \Omega\ \forall\ x \in \Omega\} \tag{12.3}$$

The two-point function ${}^{(k)}f_2(h,h'|\mathbf{r})$ satisfies the following symmetry relationship:

$$ {}^{(k)}f_2(h,h'|\mathbf{r}) = {}^{(k)}f_2(h',h|-\mathbf{r}) \tag{12.4}$$

Expressions for the two-point local state correlation functions for the ensemble follow directly from the definitions

$$f_2(h,h'|\mathbf{r}) = \langle {}^{(k)}f_2(h,h'|\mathbf{r})\rangle \tag{12.5}$$

In statistically homogeneous ensembles of microstructure, from Eq. (12.1) we see that

$$f_2(h,h'|\mathbf{r}) = \langle M(x,h)M(x+\mathbf{r},h')\rangle \quad (\text{for all } x \in \Omega|\mathbf{r}) \tag{12.6}$$

12.1.2 Generalized *n*-point local state correlation functions

Extension of the developments of the previous section to n-point local state correlation functions is straightforward:

$$ {}^{(k)}f_n(h_0,h_1,\ldots,h_{n-1}|\mathbf{r}_1,\mathbf{r}_2,\ldots,\mathbf{r}_{n-1}) = \frac{1}{vol(\Omega|\mathbf{r}_1,\mathbf{r}_2,\ldots,\mathbf{r}_{n-1})}\iiint\limits_{x\in\Omega|\mathbf{r}_1,\mathbf{r}_2,\ldots,\mathbf{r}_{n-1}} M(x,h_0)M(x+\mathbf{r}_1,h_1)M(x+\mathbf{r}_2,h_2)\ldots M(x+\mathbf{r}_{n-1},h_{n-1})dx \tag{12.7}$$

where

$$\Omega|\mathbf{r}_1,\mathbf{r}_2,\ldots,\mathbf{r}_{n-1} = \bigcap_{i=1}^{n-1}\Omega|\mathbf{r}_i \tag{12.8}$$

which comprises the intersection of the restricted sets $\Omega|\mathbf{r}_i$. The ensemble-averaged n-point functions follow the established pattern:

$$f_n(h_0, h_1, \ldots, h_{n-1}|\mathbf{r}_1, \mathbf{r}_2, \ldots, \mathbf{r}_{n-1}) = \left\langle {}^{(k)}f_n(h_0, h_1, \ldots, h_{n-1}|\mathbf{r}_1, \mathbf{r}_2, \ldots, \mathbf{r}_{n-1}) \right\rangle \tag{12.9}$$

When the ensemble of microstructure functions is statistically homogeneous we must have

$$\begin{aligned} & f_n(h_0, h_1, \ldots, h_{n-1}|\mathbf{r}_1, \mathbf{r}_2, \ldots, \mathbf{r}_{n-1}) \\ & = \langle M(x, h_0)M(x+\mathbf{r}_1, h_1)M(x+\mathbf{r}_2, h_2)\ldots M(x+\mathbf{r}_{n-1}, h_{n-1})\rangle \\ & \text{for all } x \in \Omega|\mathbf{r}_1, \mathbf{r}_2, \ldots, \mathbf{r}_{n-1}) \end{aligned} \tag{12.10}$$

12.2 REPRESENTATION OF CORRELATION FUNCTIONS IN THE PRIMITIVE BASIS

In Section 5.4.1 we represented the local state distribution function in terms of a simple Fourier series. Such a representation is now required for the correlation functions to allow manipulation of problems that would otherwise be overwhelming. The benefits of this decomposition process will become apparent in Chapters 13 and 14.

12.2.1 Fundamental relation in Fourier space

Consider the representation of the fundamental relation between the microstructure function and the two-point correlations of local state. Recall that the range of vector $\mathbf{r}$ is over all possible pairs of points lying in Ω. This region is rectangular and exactly twice the size of Ω in each of its dimensions, or eight times the volume of Ω in the 3D real space. Label this set of possible $\mathbf{r}$ vectors $\Psi(\Omega)$. It is defined as

$$\Psi(\Omega) = \{(r_1, r_2, r_3) \mid -D_1 \leq r_1 < D_1, \ \ -D_2 \leq r_2 < D_2, \ \ -D_3 \leq r_3 < D_3\} \tag{12.11}$$

This region can also be subdivided into subcells, in a manner similar to the partitioning of Ω. Let these subcells be $\psi_{t_1t_2t_3} \leftrightarrow \psi_t$, where the indices t_1, t_2, $t_3 \leftrightarrow t$ are integers that enumerate in the three dimensions, as before. The real interval $[-D_k, D_k)$ contains all possible values of the variable r_k. Acting as before, with an eye to compatibility with Haar wavelet representations, partition this interval into $2^{P'_k}$ subintervals of size $\delta' = D_k/2^{(P'_k-1)}$. Note that the subcell size δ' is not necessarily dictated by δ; however, given the fact that spatial resolution of the approximate representation of the sample microstructure is limited to δ, for the following we stipulate $\delta' = \delta$. It follows that $P'_k = P_k + 1$. Consistent with this scheme, the subcells themselves are defined by

$$\psi_t \leftrightarrow \psi_{t_1t_2t_3} = \left\{ (r_1, r_2, r_3) \left| \begin{array}{l} \left(t_1 - 2^{P'_1-1} - 1\right)\delta' \leq r_1 < \left(t_1 - 2^{P'_1-1}\right)\delta' \\ \left(t_2 - 2^{P'_2-1} - 1\right)\delta' \leq r_2 < \left(t_2 - 2^{P'_2-1}\right)\delta' \\ \left(t_3 - 2^{P'_3-1} - 1\right)\delta' \leq r_3 < \left(t_3 - 2^{P'_3-1}\right)\delta' \\ 1 \leq t_k \leq 2^{P'_k} \text{ for } k = 1, 2, 3 \end{array} \right. \right\} \tag{12.12}$$

Indicator functions on these subcells ψ_t are defined as before:

$$\chi_t(\mathbf{r}) = \begin{cases} 1 & \text{if } \mathbf{r} \in \psi_t \\ 0 & \text{otherwise} \end{cases} \tag{12.13}$$

The two-point local state correlation function, or *pair correlation function* (PCF), then accepts the following representation using the primitive basis for the $\vec{r}$-dependence:

$$f_2(h,h'|\mathbf{r}) \approx F_t^{nn'} \chi^n(h)\chi^{n'}(h')\chi_t(\mathbf{r}) \tag{12.14}$$

Introducing the Fourier representation of the microstructure function (5.45) into the fundamental equation (12.1), we obtain

$$f_2(h,h'|\mathbf{r}) \approx D_s^n D_{s'}^{n'} \chi^n(h)\chi^{n'}(h') \left[\frac{1}{vol(\Omega|\mathbf{r})} \iiint_{\Omega} \chi_s(x)\chi_{s'}(x+\mathbf{r})dx \right] \tag{12.15}$$

The last term in (12.15), involving the integration over Ω, carries the $\mathbf{r}$ dependence of the PCF. It is convenient to approximate this function with its own Fourier series, with a spatial indicator function basis over the subcells ψ_t. Proceeding in this direction,

$$h_{ss'}(\mathbf{r}) = \frac{1}{vol(\Omega|\mathbf{r})} \iiint_{\Omega} \chi_s(x)\chi_{s'}(x+\mathbf{r})dx \approx H_{ss't'}\chi_{t'}(\mathbf{r}) \tag{12.16}$$

Multiplying both sides of this equation by $\chi_t(r)$, and integrating over all $\mathbf{r} \in \Psi(\Omega)$, we obtain

$$\begin{aligned} H_{ss't} &= \iiint_{\Omega} \chi_s(x) \iiint_{\Psi(\Omega)} \frac{1}{\delta'^3 vol(\Omega|\mathbf{r})} \chi_t(\mathbf{r})\chi_{s'}(x+\mathbf{r})d\mathbf{r}dx \\ &= \iiint_{\omega_s} \chi_s(x) \iiint_{\psi_t} \frac{1}{\delta'^3 vol(\Omega|\mathbf{r})} \chi_t(\mathbf{r})\chi_{s'}(x+\mathbf{r})d\mathbf{r}dx \end{aligned} \tag{12.17}$$

The subvolume sampled by vector $\mathbf{r}$ is readily expressed for rectangular models as

$$vol(\Omega|\mathbf{r}) = (D_1 - |r_1|)(D_2 - |r_2|)(D_3 - |r_3|) \tag{12.18}$$

Notice that for fixed s_k and t_k, $H_{ss't}$ has value only when $s'_k = s_k + t_k - 2^{P_k} - 1$ or $s'_k = s_k + t_k - 2^{P_k}$.

The final expression for the local PCF is obtained by incorporating (12.17) into (12.15):

$$f_2(h,h'|\mathbf{r}) \approx H_{ss't} D_s^n D_{s'}^{n'} \chi^n(h)\chi^{n'}(h')\chi_t(\mathbf{r}) \tag{12.19}$$

Recalling relation (12.14), with Fourier coefficients of the PCF, $F_t^{nn'}$, and equating to relation (12.19), we obtain the fundamental relation in terms of the primitive Fourier coefficients:

$$F_t^{nn'} = H_{ss't}\, D_s^n D_{s'}^{n'} \tag{12.20}$$

Thus, the coefficients of the local PCF, $F_t^{nn'}$, are related through (12.20) to the coefficients of the specified sample MF, D_s^n. Transition from local- to ensemble-level primitive Fourier coefficients is straightforward, with the ensemble-level PCF taking coefficients $\langle F_t^{nn'} \rangle$ related to the ensemble average of the quadratic product of coefficients of the microstructure functions:

$$\left\langle F_t^{nn'} \right\rangle = H_{ss't} \left\langle D_s^n D_{s'}^{n'} \right\rangle \tag{12.21}$$

12.2.2 Evaluation of the weighting coefficients, $H_{ss't}$

The values of the components of $H_{ss't}$ are determined by substituting relation (12.18) into relation (12.17) and evaluating the integral in the expression. As noted earlier, this integral produces nonzero values only when $s'_k = s_k + t_k - 2^{P_k} - 1$ or $s'_k = s_k + t_k - 2^{P_k}$. Furthermore, in evaluating this integral, one needs to pay attention to the sign of each of the components of the variable $\vec{r}$. It can be shown that

$$H_{ss't} \leftrightarrow {}^{s'_1 s'_2 s'_3}_{s_1 s_2 s_3} H_{t_1 t_2 t_3} \approx h(s_1, s'_1, t_1 | P_1) h(s_2, s'_2, t_2 | P_2) h(s_3, s'_3, t_3 | P_3) \tag{12.22}$$

where

$$h(s, s', t\,|P) = \begin{cases} -1 - \left(2^{P+1} - t + 1\right) \ln\left(\dfrac{2^{P+1} - t}{2^{P+1} - t + 1}\right) & \text{if } t = s' - s + 2^P \text{ and } s' > s \\ 1 + \left(2^{P+1} - t\right) \ln\left(\dfrac{2^{P+1} - t}{2^{P+1} - t + 1}\right) & \text{if } t = s' - s + 2^P + 1 \text{ and } s' \geq s \\ 1 - (t - 1) \ln\left(\dfrac{t}{t - 1}\right) & \text{if } t = s' - s + 2^P \text{ and } s' \leq s \\ -1 + t \ln\left(\dfrac{t}{t - 1}\right) & \text{if } t = s' - s + 2^P + 1 \text{ and } s' < s \\ 0 & \text{if } t \neq s' - s + 2^P \text{ and } t \neq s' - s + 2^P + 1 \end{cases} \tag{12.23}$$

Note that relations (12.23) possess a singularity at the edges of the $\Psi(\Omega)$, when $t = 1$ or $t = 2^{P+1}$. Furthermore, in most cases, one might find the following approximation of relation (12.23) useful:

$$h(s, s', t\,|P) = \begin{cases} \dfrac{1}{2\left(2^P - \left|t - 2^P - \dfrac{1}{2}\right|\right)} & \text{if } t = s' - s + 2^P \quad \text{or} \quad t = s' - s + 2^P + 1 \\ 0 & \text{if } t \neq s' - s + 2^P \quad \text{and} \quad t \neq s' - s + 2^P + 1 \end{cases} \tag{12.24}$$

12.2.3 The r interdependence of the fundamental relation

At this juncture an important observation can be made with reference to the fundamental relation (12.20). Approximations to the microstructure function of resolution $1/S$ in the spatial variables and resolution $1/N$ in the orientation variables give rise to NS Fourier coefficients, D_s^n. On the other hand, the same resolution in both types of variables requires $8N^2S$ coefficients $F_t^{nn'}$ of the two-point correlation function. Thus, relations (12.20) and (12.21) reflect the fact that only a small fraction of the $F_t^{nn'}$ coefficients can be independent (the same observations hold whether our consideration is of individual sample microstructures or those pertaining to the ensemble). *Indeed, the* **r** *interdependence of the pair correlation function is massive, and must be dealt with in second-order treatments of microstructure and microstructure–properties relations.* This issue is tackled in some detail for the DFT representation of the two-point functions in Section 12.3.

12.2.4 Representation of the *n*-point correlation functions

The methods illustrated for expressing the fundamental equation in the primitive basis (12.20 and 12.21), linking microstructure functions to the pair correlation functions, can be easily extended to consider the *n*-point correlation functions. For example, relation (12.14) can be extended to consider three-point correlations by appending additional basis functions:

$$f_3(h_0, h_1, h_2|\mathbf{r}_1, \mathbf{r}_2) \approx F_{t_1 t_2}^{n_0 n_1 n_2} \chi^{n_0}(h_0)\chi^{n_1}(h_1)\chi^{n_2}(h_2)\chi_{t_1}(\mathbf{r}_1)\chi_{t_2}(\mathbf{r}_2) \tag{12.25}$$

Considerations of the relationship of the three-point correlation functions to the microstructure functions, as described in (12.7), naturally lead to an extended fundamental equation of the form

$$F_{t_1 t_2}^{n_0 n_1 n_2} = H_{s_0 s_1 s_2 t_1 t_2} D_{s_0}^{n_0} D_{s_1}^{n_1} D_{s_2}^{n_2} \tag{12.26}$$

or by

$$\left\langle F_{t_1 t_2}^{n_0 n_1 n_2} \right\rangle = H_{s_0 s_1 s_2 t_1 t_2} \left\langle D_{s_0}^{n_0} D_{s_1}^{n_1} D_{s_2}^{n_2} \right\rangle \tag{12.27}$$

in the ensemble. Evaluation of the $H_{s_0 s_1 s_2 t_1 t_2}$ coefficients follows from (12.7). Extensions of these to the higher-order local state correlation functions can be realized by continuing the same process.

12.2.5 Variance in the local state distribution function

We presently limit our discussion to materials possessing statistical homogeneity, and our focus will be on the ${}^{(k)}M(x,h)$ that was defined previously by the relation

$${}^{(k)}M(x,h)dh = \frac{dV}{V} \tag{12.28}$$

The interpretation of ${}^{(k)}M(x,h)$ is the volume density of material lying in the neighborhood of position *x*, within the kth sample of the ensemble, having local state lying within (invariant) measure *dh* of local state *h*. For our purposes here we integrate ${}^{(k)}M(x,h)$ over some specified region Ω, and normalize to its volume $V(\Omega)$, to obtain

$$\frac{1}{V(\Omega)} \iiint_{x \in \Omega} {}^{(k)}M(x,h)dhdx = {}^{(k)}f(h|\Omega)dh \tag{12.29}$$

where ${}^{(k)}f(h|\Omega)$ is the local state distribution in the kth element of the ensemble, occupying specified region Ω. Note that if relation (12.29) is ensemble averaged, then the conventional local state distribution function for the ensemble is retrieved:

$$\frac{1}{V(\Omega)}\iiint_{\vec{x}\in\Omega}\langle{}^{(k)}M(x,h)dh\rangle dx = f(h|\Omega)dh \tag{12.30}$$

and, if the microstructure is statistically homogeneous, it will be true that

$$f(h|\Omega) = f(h) \tag{12.31}$$

Or, in other words, the ordinary local state distribution function is equivalent to the local state distribution function confined to region Ω.

Next, consider the variance of ${}^{(k)}f(h|\Omega)$ in the ensemble. This is readily expressed as

$$\begin{aligned}\sigma^2_{f(h|\Omega)} &= \frac{1}{K}\sum_{k=1}^{K}\left[{}^{(k)}f(h|\Omega) - \langle{}^{(k)}f(h|\Omega)\rangle\right]^2 \\ &= \left\langle[{}^{(k)}f(h|\Omega) - \langle{}^{(k)}f(h|\Omega)\rangle]^2\right\rangle = \left\langle{}^{(k)}f(h|\Omega)^2\right\rangle - \left\langle{}^{(k)}f(h|\Omega)\right\rangle^2\end{aligned} \tag{12.32}$$

where K denotes the number of samples in the ensemble. In statistically homogeneous microstructures, relation (12.31) is valid and (12.32) can be rewritten as

$$\sigma^2_{f(h|\Omega)} = \left\langle{}^{(k)}f(h|\Omega)^2\right\rangle - f(h)^2 \tag{12.33}$$

Introduce (12.29) into the first term on the right side of (12.33) to obtain

$$\left\langle{}^{(k)}f(h|\Omega)^2\right\rangle = \left\langle\frac{1}{V(\Omega)}\iiint_{x\in\Omega}{}^{(k)}M(x,h)dx\,\frac{1}{V(\Omega)}\iiint_{x'\in\Omega}{}^{(k)}M(x',h)dx'\right\rangle \tag{12.34}$$

For purposes that will become clear later, it is convenient to define an indicator function $\theta(x)$ over region Ω such that

$$\theta(x) = \begin{cases}1 & \text{if } x\in\Omega \\ 0 & \text{otherwise}\end{cases} \tag{12.35}$$

With this definition (12.34) can be rewritten as

$$\left\langle{}^{(k)}f(h|\Omega)^2\right\rangle = \frac{1}{V(\Omega)^2}\left\langle\iiint_{\infty}{}^{(k)}M(x,h)\theta(x)dx\iiint_{\infty}{}^{(k)}Mx',h)\theta(x')dx'\right\rangle \tag{12.36}$$

Next, introduce a change of variables, $x' = x+\mathbf{r}$, where (12.36) can be written as

$$\left\langle{}^{(k)}f(h|\Omega)^2\right\rangle = \frac{1}{V(\Omega)^2}\iiint_{\infty}\iiint_{\infty}\langle{}^{(k)}M(x,h){}^{(k)}M(x+\mathbf{r},h)\rangle\theta(x)\theta(x+\mathbf{r})dxdr \tag{12.37}$$

However, in statistically homogeneous microstructures $\langle {}^{(k)}M(x,h)^{(k)}M(x+\mathbf{r},h)\rangle$ is independent of position x, and hence (12.37) can be expressed as

$$\left\langle {}^{(k)}f(h|\Omega)^2 \right\rangle = \iiint_{\infty} \langle {}^{(k)}M(x,h)^{(k)}M(x+\mathbf{r},h)\rangle \Theta(\mathbf{r}|\Omega)d\mathbf{r} \tag{12.38}$$

where

$$\Theta(\mathbf{r}|\Omega) = \frac{1}{V(\Omega)^2} \iiint_{\infty} \theta(x)\theta(x+\mathbf{r})dx \tag{12.39}$$

Geometrically, $\Theta(\mathbf{r}|\Omega)$ is understood to be a normalized volume of overlap between two regions of the specified size and shape, Ω, separated from one another by a translation $\mathbf{r}$. When $\mathbf{r} = 0$, $\Theta(\mathbf{r}|\Omega) = 1/V(\Omega)$, which is its maximum. When $\mathbf{r}$ is larger than the maximum caliper length of Ω, say $H(\Omega)_{\text{max}}$, since there will be no overlap, $\Theta(\mathbf{r}|\Omega) = 0$; this is its minimum. Hence, $0 \leq \Theta(\mathbf{r}|\Omega) \leq 1/V(\Omega)$.

Recall that $\langle {}^{(k)}M(x,h)^{(k)}M(x+\mathbf{r},h)\rangle$ is just the two-point local state auto-correlation function $f_2(h,h|\mathbf{r})$ (see relation 12.6). Thus, the relationship for the variance of the local state distribution function becomes

$$\sigma^2_{f(h|\Omega)} = \iiint_{\infty} f_2(h,h|\mathbf{r})\Theta(\mathbf{r}|\Omega)d\mathbf{r} - f(h)^2 \tag{12.40}$$

This is the key result. Note that the variance of the local state distribution function is closely related to the two-point auto-correlation functions, evaluated over the domain of $\Theta(\mathbf{r}|\Omega)$, which is only nonzero for $\mathbf{r}$ vectors that give rise to nonzero overlap of a pair of regions Ω that are separated in their centroids by vector $\mathbf{r}$. Roughly speaking, this excludes $\mathbf{r}$ vectors larger than the largest caliper length of Ω, $H(\Omega)_{\text{max}}$. And therefore only the two 2-point auto-correlation functions of size admitted by Ω enter into the variance $\sigma^2_{f(h|\Omega)}$.

The dependence of the variance $\sigma^2_{f(h|\Omega)}$ on the size and shape of Ω is obviously nontrivial, as it depends on the two-point correlations of local state. It is worthwhile to note the two limits, however. If $H(\Omega)_{\text{max}}$ shrinks to 0, then $\mathbf{r}\rightarrow 0$ and the two-point correlation function $f_2(h,h'|\mathbf{r})\rightarrow f(h)\delta(h'-h)$. Considering that in this limit $\Theta(\mathbf{r}|\Omega)\rightarrow 1/V(\Omega)$, it follows that

$$\lim_{\mathbf{r}\rightarrow 0} \sigma^2_{f(h|\Omega)} = f(h)(1-f(h)) \tag{12.41}$$

In the limit that $|\mathbf{r}| \rightarrow \infty$, the two-point correlation functions, over the vast majority of the relevant space, go to $f_2(h,h'|\mathbf{r})\rightarrow f(h)^2$, from which it follows that

$$\lim_{|\mathbf{r}|\rightarrow \infty} \sigma^2_{f(h|\Omega)} = 0 \tag{12.42}$$

Similar expressions for two-phase composites were obtained by Torquato (2002).

12.3 DISCRETE FOURIER TRANSFORM REPRESENTATION OF CORRELATION FUNCTIONS

One big advantage of using DFT representations of correlation functions is that FFTs allow for the efficient calculation of the correlations. For example, to calculate the full set of two-point autocorrelations for a given phase, on an RVE of S spatial positions, requires an order of S^2 operations. Using FFTs for the same calculation requires an order of S*log(S) operations due to the convolutions theorem (see Section 3.3.4). We demonstrate these efficiencies before discussing the DFT representation in more detail.

12.3.1 Calculation of two-point correlation functions via FFTs

The definition of the two-point correlations presented in Eq. (12.1) allows their efficient computation using established FFT algorithms on a discrete space. Let S be the number of points in the finite lattice that covers the RVE. We will momentarily assume that the RVE is a 1D space for ease of index notation, but it will be obvious how to generalize to higher dimensions. Using the primitive basis form (see Section 5.4), Eq. (12.20) may be rewritten as

$$f_2^{np}(\mathbf{r}_t) = f_t^{np} = \frac{1}{S}\sum_{s=0}^{S-1} m_s^n m_{s+t}^p \tag{12.43}$$

where $m_s^n = D_s^n/N$ and $\mathbf{r}_t$ take values on a lattice of the same spacing as $\mathbf{x}_s$. The primitive representation of the microstructure function requires that

$$\sum_{n=1}^{N} m_s^n = 1, \quad 0 \leq m_s^n, \quad \sum_{s=0}^{S-1} m_s^n = V^n S \tag{12.44}$$

where V^n is the volume fraction of phase n in the sample. The DFT (in the spatial dimensions) of the microstructure function is expressed as

$$\widetilde{M}_k^n = \Im(m_s^n) = \sum_{s=0}^{S-1} m_s^n e^{2\pi isk/S} = \left|\widetilde{M}_k^n\right| e^{i\theta_k^n} \tag{12.45}$$

The term $|\widetilde{M}_k^n|$ is referred to as the amplitude of the Fourier transform and θ_k^n as the phase. Taking the FFT of Eq. (12.43) gives

$$\begin{aligned} \Im[f_t^{np}]_k &= \frac{1}{S}\sum_{t=0}^{S-1}\sum_{s=0}^{S-1} m_s^n m_{s+t}^p e^{-2\pi itk/S} \\ &= \frac{1}{S}\sum_{s=0}^{S-1} m_s^n \sum_{t=0}^{S-1} m_{s+t}^p e^{-2\pi itk/S} \end{aligned} \tag{12.46}$$

Now let $s + t = z$. Then

$$\begin{aligned}\Im[f_t^{np}]_k &= \frac{1}{S}\sum_{s=0}^{S-1} m_s^n \sum_{z=s}^{S-1+s} m_z^p e^{-2\pi i(z-s)k/S} \\ &= \frac{1}{S}\sum_{s=0}^{S-1} m_s^n e^{-2\pi i(-s)k/N} \sum_{z=s}^{S-1+s} m_z^p e^{-2\pi izk/S}\end{aligned} \tag{12.47}$$

Now let us assume periodicity (cyclicity), that is, in a single dimension we have $m_z^p = m_{z+S}^p$. Since $e^{-2\pi iz/S}$ is also cyclic with the same period, we may renumber the second summation to obtain

$$\begin{aligned}\Im[f_t^{np}]_k &= \frac{1}{S}\sum_{s=0}^{S-1} m_s^n e^{2\pi isk/S} \sum_{z=0}^{S-1} m_z^p e^{-2\pi izk/S} \\ &= \frac{1}{S}\widetilde{M}_k^{*n}\widetilde{M}_k^p\end{aligned} \tag{12.48}$$

where $\widetilde{M}_k^{*n}$ is the complex conjugate of $\widetilde{M}_k^n$ (since m is real valued, thus $m_s^{*n} = m_s^n$). Hence, finally we have

$$f_t^{np} = \frac{1}{S}\Im^{-1}\left[\widetilde{M}_k^{*n}\widetilde{M}_k^p\right] \tag{12.49}$$

The result is that in practice we may efficiently obtain the two-point correlation function using pointwise multiplication of the microstructure function's FFTs, and an inverse FFT. This is dramatically more efficient than Eq. (12.43).

12.3.2 Calculation of *n*-point correlation functions via FFTs

We now repeat the ideas in the previous section to obtain efficient methods of finding higher-order n-point correlation functions. We give the details only for the three-point function; extension to higher orders will be clear. Again, we assume that $\mathbf{x}_s$ is in a 1D space for ease of index notation:

$$f_3^{npq}(\mathbf{r}_t, \mathbf{r}_u) = f_{tu}^{npq} = \frac{1}{S}\sum_{s=0}^{S-1} M_s^n M_{s+t}^p M_{s+t+u}^q \tag{12.50}$$

where $\mathbf{r}_t$, $\mathbf{r}_u$ take values on a lattice of the same basis parameters as $\mathbf{x}_s$. Take the DFT of both sides:

$$\begin{aligned}\Im[f_{tu}^{npq}]_{k_1k_2} &= \frac{1}{S}\sum_{u=0}^{S-1}\sum_{t=0}^{S-1}\sum_{s=0}^{S-1} m_s^n m_{s+t}^p m_{s+t+u}^q e^{-2\pi itk_1/S} e^{-2\pi iuk_2/S} \\ &= \frac{1}{S}\sum_{s=0}^{S-1} m_s^n \sum_{t=0}^{S-1} m_{s+t}^p e^{-2\pi itk_1/S} \sum_{u=0}^{S-1} m_{s+t+u}^q e^{-2\pi iuk_2/S}\end{aligned} \tag{12.51}$$

Now let $s+t+u = y$. Then,

$$\begin{aligned}\Im[f_{tu}^{npq}]_{k_1k_2} &= \frac{1}{S}\sum_{s=0}^{S-1} m_s^n \sum_{t=0}^{S-1} m_{s+t}^p e^{-2\pi i t k_1/S} \sum_{y=s+t}^{S-1+s+t} m_y^q e^{-2\pi i (y-s-t)k_2/S} \\ &= \frac{1}{S}\sum_{s=0}^{S-1} m_s^n \sum_{t=0}^{S-1} m_{s+t}^p e^{-2\pi i (tk_1-sk_2-tk_2)/S} \sum_{y=s+t}^{S-1} m_y^q e^{-2\pi i y k_2/S}\end{aligned} \tag{12.52}$$

Again, assume periodicity and renumber the third summation to obtain

$$\Im[f_{tu}^{npq}]_{k_1k_2} = \widetilde{M}_{k_2}^q \frac{1}{S}\sum_{s=0}^{S-1} m_s^n \sum_{t=0}^{S-1} m_{s+t}^p e^{-2\pi i (tk_1-sk_2-tk_2)/S} \tag{12.53}$$

Let $s+t = z$:

$$\Im[f_{tu}^{npq}]_{k_1k_2} = \widetilde{M}_{k_2}^q \frac{1}{S}\sum_{s=0}^{S-1} m_s^n e^{2\pi i s k_1/S} \sum_{z=s}^{S-1+s} m_z^p e^{-2\pi i z(k_1-k_2)/S} \tag{12.54}$$

Again, renumbering the final summation, we have

$$\begin{aligned}\Im[f_{tu}^{npq}]_{k_1k_2} &= \widetilde{M}_{k_1-k_2}^p \widetilde{M}_{k_2}^q \frac{1}{S}\sum_{s=0}^{S-1} m_s^n e^{2\pi i s k_1/S} \\ &= \frac{1}{S}\widetilde{M}_{k_1}^{*n} \widetilde{M}_{k_1-k_2}^p \widetilde{M}_{k_2}^q\end{aligned} \tag{12.55}$$

Hence, finally we have

$$f_{tu}^{npq} = \frac{1}{S}\Im^{-1}\left[\widetilde{M}_{k_1}^{*n} \widetilde{M}_{k_1-k_2}^p \widetilde{M}_{k_2}^q\right] \tag{12.56}$$

By a similar procedure we obtain

$$f_{tuv}^{npqr} = \frac{1}{S}\Im^{-1}\left[\widetilde{M}_{k_1}^{*n} \widetilde{M}_{k_1-k_2}^p \widetilde{M}_{k_2-k_3}^q \widetilde{M}_{k_3}^r\right] \tag{12.57}$$

And so on.

12.3.3 Boundary conditions in the DFT representation

As indicated in Eq. (12.48), using FFTs to calculate correlation functions implicitly assumes periodic boundary conditions. A different boundary condition may be applied by using padding (see, for example, Walker (1996), for a discussion of padding). Continuing with a 1D example, assume a discrete domain of S points where $\mathbf{x}$ takes the values from 0 to $S-1$. One commonly used definition of the two-point correlation function that does not involve periodic boundary conditions is given by Adams et al. (2005):

$$f_2(h, h'|\mathbf{r}) = Q_{\Omega|\mathbf{r}} \int_{\Omega|\mathbf{r}} M(\mathbf{x}, h) M(\mathbf{x} + \mathbf{r}, h') d\mathbf{x} \tag{12.58}$$

where $Q_{\Omega|\mathbf{r}}$ is a normalization factor, and $\Omega|\mathbf{r} = \{\mathbf{x} \in \Omega | \mathbf{x} + \mathbf{r} \in \Omega\}$. In the continuous case the normalization factor is given by $Q_{\Omega|\mathbf{r}} = \frac{1}{V_{\Omega|\mathbf{r}}}$. In the discrete 1D case of S points, the normalization factor is given by 1/(number of points sampled); that is, $Q_{\Omega|\mathbf{r}} = 1/(S - r)$ where r takes the integer values from 0 to $S - 1$. If we pad our discrete 1D example with $S - 1$ zeros, then the resultant calculation for f_2 effectively calculates the integral in Eq. (12.58) without the normalization factor; hence, the FFT results must be normalized by $1/(S - r)$.

Note that one issue with these boundary conditions is that the values of $f_2(h, h'|\mathbf{r})$ for a larger value of $|\mathbf{r}|$ are sampled with significantly less frequency than those for smaller values of $|\mathbf{r}|$; nevertheless, they are given equal importance. To deal with this issue, it may be assumed, for example, that values of $|\mathbf{r}|$ above one-half the size of the domain should be ignored as being particularly biased. If this same approach is taken in the case of the unpadded FFT, it will reduce the effect from the assumed periodicity, since the smaller **r** vectors will cross the boundary of the domain for a smaller proportion of the samples.

Another approach, if our sample size is not large enough to comfortably ignore the larger **r** vectors, is to artificially increase the sample size using a collage of the original RVE created from random translations. This effectively increases the size of the sample with a layer of material having the same statistics as the original sample, but not in a periodic manner. The approach will, of course, not be valid if the assumption of homogeneity is not correct.

12.3.4 Interrelations between two-point correlation functions

The DFT of f_t^{np} is computed as (no summation on the repeated k indices)

$$F_k^{np} = \Im(f_t^{np}) = \frac{1}{S} \widetilde{M}_k^{*n} \widetilde{M}_k^{p} = \frac{1}{S} \left|\widetilde{M}_k^{n}\right| \left|\widetilde{M}_k^{p}\right| e^{-i\,\theta_k^n} e^{i\,\theta_k^p} \tag{12.59}$$

where $\widetilde{M}_k^{*n}$ denotes the complex conjugate of $\widetilde{M}_k^{n}$. If we take $n = p$ in Eq. (12.59), we arrive at a special set of correlations termed *autocorrelations*. Note that the autocorrelation is a real-valued even function and therefore its DFTs are also real. On the other hand, $n \neq p$ leads to cross-correlations, and their DFTs are, in general, complex.

The complete set of two-point correlations for any given microstructure constitutes a very large data set. If a microstructure of interest comprises N distinct local states, the complete set of its two-point statistics contains N^2 correlations for each value of the 3D vector, **r**. The full set of correlations is conveniently visualized as an $N \times N$ array:

$$f_t^{np} = \begin{bmatrix} f_t^{11} & f_t^{12} & \cdots & \cdots & f_t^{1N} \\ f_t^{21} & \ddots & & & \vdots \\ \vdots & & \ddots & & \vdots \\ \vdots & & & \ddots & \vdots \\ f_t^{N1} & \cdots & \cdots & \cdots & f_t^{NN} \end{bmatrix} \tag{12.60}$$

It has long been known that this set exhibits many interdependencies. Frisch and Stillinger (1963) showed that only one of the four correlations is independent for a two-phase material (i.e., if f_t^{11} is known, then f_t^{12}, f_t^{21}, and f_t^{22} can be calculated). More recently, Gokhale et al. (2005) showed that at most $\frac{1}{2}(N(N-1))$ correlations are independent for a material system comprising N distinct local states. In this book, exploiting the known properties of DFTs, we demonstrate that the number of independent correlations is only $N-1$. This constitutes a dramatic reduction in the size of the data set needed to specify uniquely the complete set of two-point correlations for any given microstructure.

The definition of the discrete two-point correlations in Eq. (12.43) requires that $f_t^{np} = f_{-t}^{pn}$. In the Fourier space, this requirement translates to

$$F_k^{np} = \frac{1}{S}\left|\widetilde{M}_k^n\right|\left|\widetilde{M}_k^p\right| e^{-i\theta_k^n} e^{i\theta_k^p} = F_k^{*pn} \tag{12.61}$$

Furthermore, the product of two of the DFTs of the two-point correlations exhibits the following property:

$$F_k^{np} F_k^{pq} = \frac{1}{S^2}\left|\widetilde{M}_k^p\right|\left|\widetilde{M}_k^p\right|\left|\widetilde{M}_k^n\right| e^{-i\theta_k^n}\left|\widetilde{M}_k^q\right| e^{i\theta_k^q} = F_k^{pp} F_k^{nq} \tag{12.62}$$

Using Eq. (12.61), Eq. (12.62) can be recast into a more convenient form as

$$F_k^{nq} = \frac{F_k^{*pn} F_k^{pq}}{F_k^{pp}} \tag{12.63}$$

Equation (12.63) is a key result. This equation implies that if F_k^{pn} (and by extension f_t^{pn}) are known for any one choice of p and all n or vice versa (i.e., either a column or a row in the array in Eq. (12.60) is known), then all of the other correlations can be calculated. In addition to the redundancies described before between the various two-point correlations, there exist a number of other constraints or bounds on the values of F_k^{np}. The following relationships are easily seen from the definitions in Eqs. (12.43) through (12.59):

$$F_0^{np} = S\, V^n\, V^p,\ 0 \leq F_0^{np} \leq S,\ \left|F_k^{np}\right| \leq S^2 \tag{12.64}$$

These bounds allow us to define an additional redundancy; therefore, Eq. (12.64) (together with Eqs. (12.43) and (12.59)) requires that

$$\sum_{n=1}^{N} F_k^{np} = \begin{cases} S\,V^p, & \text{if } k = 0 \\ 0, & \text{if } k \neq 0 \end{cases} \tag{12.65}$$

Since V^p is already specified by F_0^{pp} (Eq. 12.64), it follows that there are only $(N-1)$ independent correlations in the complete set shown in Eq. (12.60). Although the interdependencies shown in Eqs. (12.63) and (12.65) are formulated between the DFTs of the two-point correlations, the same relationships can be expressed in terms of their continuous Fourier transforms. However, since we plan to exploit the significant advantages associated with FFT calculation of DFTs, all further discussion in this book is restricted to DFTs.

Substituting Eq. (12.64) into Eq. (12.65) allows us to recast Eq. (12.65) completely in terms of the DFTs as

$$\sum_{n=1}^{N} F_k^{np} = \begin{cases} \sqrt{S F_0^{pp}}, & \text{if } k = 0 \\ 0, & \text{if } k \neq 0 \end{cases} \tag{12.66}$$

For eigen-microstructures ($m_s^n \in \{0, 1\}$), one additional redundancy can be established by summing over all frequencies in Fourier space as

$$\sum_k F_k^{np} = \delta_{np} \sqrt{S} \sqrt{F_0^{nn}} \tag{12.67}$$

where δ_{np} denotes the Kronecker-delta.

As a final note we remind readers, since all f_t^{np} are real valued, there is an implicit symmetry in their DFTs that may be expressed as

$$F_k^{np} = F_{S-k}^{*np} \tag{12.68}$$

for all $k \neq 0$. In other words, about one-half of the transforms are simply complex conjugates of the other half.

As a final note we mention that discrete higher-order n-point correlation functions can also be obtained efficiently using DFTs in an analogous manner to Eq. (12.59) (Fullwood et al., 2008):

$$F_{k_1 k_2}^{npq} = \Im\left(f_{t_1 t_2}^{npq}\right) = \frac{1}{S} \widetilde{M}_{k_1}^{*n} \widetilde{M}_{k_1 - k_2}^{p} \widetilde{M}_{k_2}^{q} \tag{12.69}$$

$$F_{k_1 k_2 k_3}^{npqr} = \Im\left(f_{t_1 t_2 t_3}^{npqr}\right) = \frac{1}{S} \widetilde{M}_{k_1}^{*n} \widetilde{M}_{k_1 - k_2}^{p} \widetilde{M}_{k_2 - k_3}^{q} \widetilde{M}_{k_3}^{r} \tag{12.70}$$

where $f_{t_1 t_2}^{npq}$ and $f_{t_1 t_2 t_3}^{npqr}$ are the three-point and four-point correlation functions, respectively.

12.4 QUANTITATIVE REPRESENTATIONS OF INTERFACE MICROSTRUCTURE

Thus far, our focus has been on the volume density of local state and its spatial correlations in the microstructure. Representations in terms of the local state distribution functions and their spatial correlations contain sufficient information to predict effective properties where the microstructure–properties model belongs to the class of mean-field constitutive relations. Examples of this class include estimates for linear properties, such as elastic and conductive properties, thermal expansion, and others. Another important example is the theory for initial yielding. Other types of properties, however, are sensitive to a microstructure's internal interfacial structure, and these require a different approach. Theory relating the character of interfaces to the properties of material points lying on the interface is rather limited at this time. Also, theories of homogenization, linking the distribution of interfacial properties to effective properties, are also very limited. Some progress has been made in terms of a mesoscale description of the local state of interfaces and the distribution of the interfacial component within the microstructure. Here we present the main elements of low-order representation of interfaces, consistent with the previous representations of distribution of local state.

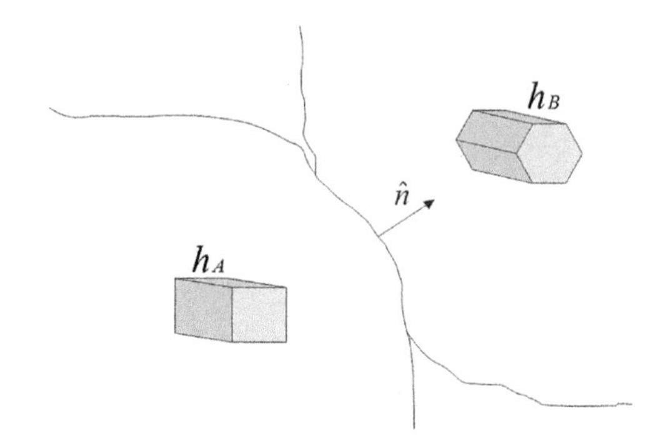

FIGURE 12.3

Schematic of local state $\partial h = (h_A, \hat{n}, h_B)$ at interface between two grains.

12.4.1 Mesoscale interface character and the local state space, ∂H

Consider material points that lie at the interface between two regions of differing local state, say h_A and h_B. The structure of the interface is known to depend not only on h_A and h_B but also on the *inclination* of the interface, which we here denote as $\hat{n}$ (Figure 12.3). We adopt the convention that $\hat{n}$ begins in the region identified with h_A and ends in the region associated with h_B. Thus, the local state of material points lying on the semi-continuous interface network consists in the ordered triple $\partial h = (h_A, \hat{n}, h_B)$. Since which local region is labeled A and which is labeled B is arbitrary, it follows that

$$\partial h = (h_A, \hat{n}, h_B) = (h_B, -\hat{n}, h_A) \tag{12.71}$$

Since $\hat{n} \in S^2$, the unit sphere of all possible inclinations of $\hat{n}$, taking the symmetry relation (12.71) into account, the complete mesoscale local state space for material points on the interface, ∂H, must be

$$\partial h \in \partial H = H \times RP^2 \times H \tag{12.72}$$

for polyphase polycrystalline microstructures where H is defined according to relation (4.10). The set RP^2, the *real projective plane*, is just the set of all distinct lines passing through a common origin, through the surface of a unit hemisphere.

12.4.2 Interface character distribution

The distribution of interface types, within the mesoscale description of interface character, is described as the *interface character distribution* (ICD). $S_V(\partial h)$ describes the density of surface area per unit volume of interface type ∂h. More precisely, the increment of interface surface area per unit volume of microstructure that associates with an inclination lying within an infinitesimal neighborhood of (invariant) measure $d\hat{n}$ of inclination $\hat{n}$, the tail of which lies in local state within a neighborhood of (invariant) measure dh_A of local state h_A, and the head of which lies in local state within a neighborhood of (invariant) measure dh_B of local state h_B, is

$$dS_V = S_V(\partial h)d\partial h = S_V(h_A, \hat{n}, h_B)dh_A d\hat{n} dh_B \tag{12.73}$$

The ICD is normalized such that the total surface area per unit volume of interfaces, $\overline{S}_V$, is

$$\overline{S}_V = \oint_{\partial H} S_V(\partial h)d\partial h = \int_H \int_{RP^2} \int_H S_V(h_A, \hat{n}, h_B)dh_A d\hat{n} dh_B \tag{12.74}$$

The main application of the ICD has been in terms of single-phase polycrystals, where the local state is the ordered triplet $\partial h = (R_A, \hat{n}, R_B)$, $R_A, R_B \in SO(3)/G$. In this case, the function of interest, $S_V(R_A, \hat{n}, R_B)$, is called the *grain boundary character distribution* (GBCD). In some cases, it is reduced even further, to the *misorientation distribution function* (MDF) by focusing on the *lattice misorientation* $\Delta R = R_A^T R_B$ by further integration of (12.73):

$$S_V(\Delta R) = \iiint_{SO(3)/G} \iint_{RP^2} S_V(R_A, \hat{n}, R_A \Delta R)d\hat{n} dR_A \tag{12.75}$$

Many other examples of the class of interface character distribution functions can be considered following the same general principles. Recovery of these functions from experimental data is a complex matter involving well-established principles of stereology. We will return to this question in Chapter 16.

12.4.3 Representation of the ICD in the primitive basis

Since the local state space associated with interfaces is $\partial H = H \times S^2 \times H$, representation of the ICD in the primitive basis requires a partitioning of S^2 (the set of all physically distinct unit vectors) into subregions. There are many ways to accomplish this. Here we assume that S^2 has been partitioned into P subregions $\eta_{pq} \leftrightarrow \eta_p$ satisfying the following relationships:

$$\bigcup_{p=1}^{P} \eta_p = RP^2, \quad \eta_p \cap \eta_{p'} = \varnothing \ (p \neq p'), \quad \iint_{\eta_p} d\hat{n} = \frac{1}{P} \ (\text{for all } p) \tag{12.76}$$

Furthermore, define indicator functions on these subregions according to

$$\chi^P(\hat{n}) = \begin{cases} 1 \ \text{if} \ \hat{n} \in \eta_p \\ 0 \ \text{otherwise} \end{cases} \tag{12.77}$$

It follows that the ICD can be expressed in the primitive basis as

$$S_V(h_A, \hat{n}, h_B) \approx S^{npn'} \chi^n(h_A) \chi^P(\hat{n}) \chi^{n'}(h_B) \tag{12.78}$$

Expressions for the GBCD or the MDF can be readily formulated by restricting the ICD's domain to the appropriate subspaces, partitioning these as required, forming the associated primitive indicator basis in terms of products, and then describing the primitive Fourier coefficients in the same manner described in obtaining (12.78). Since it is much more common to express the ICD and its subspaces in terms of classic basis functions, we will not deal further with the primitive basis in describing interface distributions.

12.4.4 Variance in the interface character distribution

We now consider variance of the ICD. The basic definition of the function $S_V(\partial h)$ was described earlier. For our purposes we require a new interface function, $\partial M(x, \partial h)$, that is defined as follows:

$$\partial M(x, \partial h) d\partial h = \begin{cases} 1/t \text{ if } x \text{ lies within } t/2 \text{ of interface lying within measure } d\partial h \text{ of } \partial h \\ 0 \text{ otherwise} \end{cases} \tag{12.79}$$

Recall that ∂h is the ordered set of parameters $(h, \hat{n}, h')$ defining the local state of interface. As in the previous section we enumerate samples of the ensemble with the index k, and the corresponding interface function as ${}^{(k)}\partial M(x, \partial h)$. Note that if this function is integrated over specified region Ω, and normalized to the volume of that region, we must have, in the limit that $t \to 0$,

$$\lim_{t \to 0} \frac{1}{V(\Omega)} \iiint_{\infty} {}^{(k)}\partial M(x, \partial h) \theta(x) dx = {}^{(k)}S_V(\partial h) \tag{12.80}$$

Taking the ensemble average of this relation we obtain the following:

$$\lim_{t \to 0} \langle {}^{(k)}\partial M(\vec{x}, \partial h) \rangle = \langle {}^{(k)}S_V(\partial h) \rangle = S_V(\partial h) \tag{12.81}$$

Our interest now is the variance of the ICD, which is defined as

$$\sigma^2_{S_V(\partial h)} = \left\langle {}^{(k)}S_V(\partial h)^2 \right\rangle - S_V(\partial h)^2 \tag{12.82}$$

Introducing (12.80) into (12.82), we obtain

$$\begin{aligned} \sigma^2_{S_V(\partial h)} &= \lim_{t \to 0} \iiint_{\infty} \left\langle {}^{(k)}\partial M(x, \partial h) {}^{(k)}\partial M(x + \mathbf{r}, \partial h) \right\rangle \Theta(\mathbf{r}|\Omega) d\mathbf{r} - S_V(\partial h)^2 \\ \sigma^2_{S_V(\partial h)} &= \lim_{t \to 0} \iiint_{\infty} \partial f_2(\partial h, \partial h|\mathbf{r}) \Theta(\mathbf{r}|\Omega) d\mathbf{r} - S_V(\partial h)^2 \end{aligned} \tag{12.83}$$

Thus, as with the previous example, it is illustrated that the variance of the ICD is dependent on the two-point autocorrelation functions of interface character, denoted here by $\partial f_2(\partial h, \partial h|\mathbf{r})$. The limiting relations corresponding to (12.41) and (12.42) are obtained by substituting $S_V(\partial h)$ for $f(h)$.

12.5 RELATIONSHIP BETWEEN TWO-POINT CORRELATION FUNCTIONS AND THE ICD

A formal connection between short-range two-point local state correlation functions, $f_2(h, h'|\mathbf{r})$, and the ICD, $S_V(\partial h) = S_V(h, \hat{n}, h')$, is now considered. (As in the previous section, for pedagogical clarity, we illustrate with the GBCD pertinent to single-phase polycrystals.) Imagine that we partition the grain boundary interface network into small patches, each of which is characterized by a vector $\mathbf{A}_j = dA_j\hat{n}_j$, where dA_j is the measure of area of the jth patch and $\hat{n}_j$ is its unit normal. Vectors $\mathbf{r}$, placed randomly in Ω, can only cross a particular patch $\mathbf{A}_j$ when their tails fall in a subvolume of measure $|\mathbf{A}_j \cdot \mathbf{r}|$. Thus, an estimate of the fraction of randomly placed vectors $\mathbf{r}$ that cut across interfaces can be obtained from the volume fraction $\sum_j |\mathbf{A}_j \cdot \mathbf{r}|/V(\Omega)$ where the summation is over all patches. Two sources of error can occur in this estimate. For patches lying near the surface of Ω, $\partial\Omega$, the volume considered in $|\mathbf{A}_j \cdot \mathbf{r}|$ may extend outside of Ω, leading to an overestimate of volume fraction. Also, for those patches participating in triple junctions, the volume $|\mathbf{A}_j \cdot \mathbf{r}|$ will overlap for adjacent patches, again giving rise to an overestimate of volume fraction.

We can bound the total amount of error occurring from these two sources by the sum $\sum_j \pi r^2 P_j/V(\Omega)$, where P_j denotes the perimeter of $\mathbf{A}_j$ including its intersection with $\partial\Omega$. This is just the volume of a cylinder of radius $r = |\mathbf{r}|$ and of length equal to the perimeter of the grain boundary network (including the surface terms), normalized to the total volume of Ω. It should be evident that the magnitude of the error will always be smaller than this estimate.

Next, we apply these ideas to the density functions. Partition the domain S^2 into two real-projective planes $S^2 = RP^{2+} \cup RP^{2-}$, according to whether $\mathbf{r} \cdot \hat{n}$ is positive or negative. Assume that r is small in comparison with the grain size d. Let $P(g, dg; g', dg')$ denote the combined perimeter of all (planar) interface patches exhibiting lattice orientation in a neighborhood of measure dg of orientation g at the tail, and lattice orientation in a neighborhood of measure dg' of orientation g' at the head. Following the previous discussion, it is evident that the following inequality can be established:

$$\left| \iint_{RP^{2+}} S_V(g, \hat{n}, g')\mathbf{r} \cdot \hat{n} dg d\hat{n} dg' - \iint_{RP^{2-}} S_V(g, -\hat{n}, g')\mathbf{r} \cdot \hat{n} dg d\hat{n} dg' - f_2(g, g'|\mathbf{r}) dg dg' \right| \\ \leq \alpha\pi r^2 P(g, dg; g', dg')/V(\Omega) \tag{12.84}$$

Here α is a scalar parameter approximately equal to ½. A similar expression is obtained for the volume fraction of Ω in which the tail of random vectors $\mathbf{r}$ can reside such that they have lattice orientation in a region of measure dg about g at the tail and lattice orientation within dg' of orientation g' at the head:

$$\left| \iint_{RP^{2+}} S_V(g, -\hat{n}, g')\mathbf{r} \cdot \hat{n} dg d\hat{n} dg' - \iint_{RP^{2-}} S_V(g, \hat{n}, g')\mathbf{r} \cdot \hat{n} dg d\hat{n} dg' \right. \\ \left. - f_2(g, g'|\mathbf{r}) dg dg' \right| \leq \beta\pi r^2 P(g, dg; g', dg')/V(\Omega) \tag{12.85}$$

where $\alpha + \beta = 1$. These two equations can be combined into a single one as follows:

$$\left| \left[\iint_{S^2} [S_V(g, \hat{n}, g') + S_V(g, -\hat{n}, g')] |\mathbf{r} \cdot \hat{n}| d\hat{n} - [f_2(g, g'|\mathbf{r}) + f_2(g, g'|-\mathbf{r})] \right] dg dg' \right| \\ \leq \pi r^2 P(g, dg; g', dg')/V(\Omega) \tag{12.86}$$

Notice that in the limit that r approaches 0, clearly the right side of (12.86) goes to 0 faster than the left side (i.e., r^2 versus r); therefore,

$$\lim_{r\to 0}\left[\iint_{S^2} [S_V(g,\hat{n},g') + S_V(g,-\hat{n},g')]|\mathbf{r}\cdot\hat{n}|d\hat{n} = [f_2(g,g'|\mathbf{r}) + f_2(g,g'|-\mathbf{r})]\right] \tag{12.87}$$

We find it most convenient to define a nonsided GBCD,

$$\widetilde{S}_V(g,\hat{n},g') = [S_V(g,\hat{n},g') + S_V(g,-\hat{n},g')]/2 \tag{12.88}$$

in conjunction with an averaged two-point orientation correlation function,

$$\widetilde{f}_2(g,g'|\mathbf{r}) = [f_2(g,g'|\mathbf{r}) + f_2(g,g'|-\mathbf{r})]/2 \tag{12.89}$$

such that relation (12.87) becomes

$$\lim_{r\to 0}\left[\iint_{S^2} \widetilde{S}_V(g,\hat{n},g')|\mathbf{r}\cdot\hat{n}|d\hat{n} = \widetilde{f}_2(g,g'|\mathbf{r})\right] \tag{12.90}$$

Next, consider the meaning of the relation $\widetilde{f}_2(g,g'|\mathbf{r})dgdg'/r$. From (12.89) it is evident that it denotes the probability that randomly placed vectors $\mathbf{r}$ cross grain boundaries, with either their head lying within measure dg of orientation g and their tail lying within measure dg' of orientation g', or vice versa, normalized to the length r. Having normalized this expression, it must be equivalent to the familiar $P_L(g,g'|\hat{r})$ of the P_L/S_V stereology, where $\hat{r}$ here suggests test lines in the direction signified by $\hat{r}$. Thus, when r is sufficiently small we expect

$$\widetilde{f}_2(g,g'|\mathbf{r})dgdg'/r = P_L(g,g'|\hat{r})dgdg' \tag{12.91}$$

This expression reflects the fact that sampling with small, randomly placed vectors is completely equivalent to sampling with randomly placed lines of substantial length, when both are normalized to their respective lengths. Thus, in the conventional P_L/S_V stereology, one can replace the traditional measurement of $P_L(g,g'|\hat{r})$ with the nonsided two-point correlation functions $\widetilde{f}_2(g,g'|\mathbf{r})/r$ normalized to r.

SUMMARY

In this chapter readers were introduced to the fundamental concepts needed for higher-order description of microstructure using spatial correlation functions. Although much of the focus was on two-point correlation functions, we hope readers can see the extension of these ideas to three-point and other higher-order correlations. It is also hoped that readers appreciate the value of Fourier representations in seeking computationally efficient descriptions of these higher-order correlation functions. These higher-order representations of microstructure form the basis for MSDPO using higher-order composite theories presented in Chapter 13.

CHAPTER

Higher-Order Homogenization

13

CHAPTER OUTLINE

The n-point (and specifically, the two-point) correlation functions introduced in the previous chapter form the basis for higher-order descriptions of microstructure. However, to incorporate them fully into the MSDPO framework, they must also be embedded in structure–property relations; that is, higher-order homogenization methods are needed that take the n-point functions as inputs and return effective property estimates. This chapter develops the required homogenization methods in the settings of both the primitive basis and discrete Fourier transforms. Perturbation expansions are used to provide the homogenization formulation, and various issues with implementing the method numerically are discussed. A brief discussion then demonstrates how the same underlying methodology can be applied to localization of properties, such as must be considered when dealing with fracture and fatigue. However, this topic is not discussed in any significant detail in this book. Finally, an alternative homogenization formulation (the strong-contrast expansion) is developed that provides more accurate estimates of effective properties when there is high contrast between the properties of the constituent phases.

13.1 HIGHER-ORDER PERTURBATION ESTIMATES FOR ELASTIC PROPERTIES

All of the bounds discussed in Chapter 8 can be evaluated using only the information on the volume fractions of the distinct local states in the internal structure of the material system. Often, one might find that the bounds are too far apart, requiring the need for the use of higher-order theory in obtaining either narrower bounds or more precise estimates. The higher-order theories should necessarily use more information regarding the spatial distribution of the local states in the material internal structure. In this chapter we focus on higher-order bounds on the effective elastic properties, obtained from the use of perturbation theory.

We start with the representative volume shown in Figure 8.2 and employ an additive decomposition of the strain field in the element into an average quantity and a perturbation from the average as

$$\boldsymbol{\varepsilon}(\mathbf{x}) = \overline{\boldsymbol{\varepsilon}} + \boldsymbol{\varepsilon}'(\mathbf{x}), \quad \overline{\boldsymbol{\varepsilon}'} = \mathbf{0} \tag{13.1}$$

The local perturbation in the strain field can be expressed in terms of a fourth-rank polarization tensor, **a**, as

$$\boldsymbol{\varepsilon}'(\mathbf{x}) = \mathbf{a}(\mathbf{x})\overline{\boldsymbol{\varepsilon}}, \quad \overline{\mathbf{a}} = \mathbf{0} \tag{13.2}$$

The fact that $\mathbf{S}^{*-1} = \mathbf{C}^*$ (see Eq. 7.53) permits equating two definitions of the macroscale stress and the derivation of a relationship for the effective stiffness tensor. In deriving this expression next, we have not shown explicitly the functional dependence of the various field quantities with the position in the representative volume element:

$$\mathbf{C}^*\overline{\boldsymbol{\varepsilon}} = \overline{\mathbf{C}\boldsymbol{\varepsilon}} = \overline{\mathbf{C}(\boldsymbol{\varepsilon}+\overline{\boldsymbol{\varepsilon}})} = \overline{\mathbf{C}\boldsymbol{\varepsilon}'} + \overline{\mathbf{C}}\overline{\boldsymbol{\varepsilon}} = \overline{\mathbf{C}\mathbf{a}}\overline{\boldsymbol{\varepsilon}} + \overline{\mathbf{C}}\overline{\boldsymbol{\varepsilon}} \Rightarrow \mathbf{C}^* = \overline{\mathbf{C}\mathbf{a}} + \overline{\mathbf{C}} \tag{13.3}$$

The local stiffness can also be additively decomposed into a reference value and a perturbation from this reference stiffness tensor as

$$\mathbf{C}(\mathbf{x}) = \mathbf{C}^r + \mathbf{C}'(\mathbf{x}), \quad \overline{\mathbf{C}'} = \overline{\mathbf{C}} - \mathbf{C}^r \tag{13.4}$$

Substitution of Eq. (13.4) into Eq. (13.3) yields

$$\mathbf{C}^* = \overline{\mathbf{C}} + \overline{\mathbf{C}\mathbf{a}} = \mathbf{C}^r + \overline{\mathbf{C}'} + \overline{\mathbf{C}\mathbf{a}} \tag{13.5}$$

It is clear from Eq. (13.5) that the computation of $\mathbf{C}^*$ requires the derivation of a suitable expression for the polarization tensor **a** relating the averaged strain tensor to the perturbation of the local strain tensor from the averaged quantity (Eq. 13.2). The perturbation in the strain field can be related to a perturbation in the displacement field:

$$\varepsilon'_{kl} = \frac{1}{2}\left(u'_{k,l} + u'_{l,k}\right) \tag{13.6}$$

The field equation for the perturbed displacement field can be derived from the equilibrium conditions as

$$\begin{gathered}\sigma_{ij,j} = \left[\left(C^r_{ijkl} + C'_{ijkl}\right)\left(\overline{\varepsilon}_{kl} + \varepsilon'_{kl}\right)\right]_{,j} = C^r_{ijkl}\varepsilon'_{kl,j} + C'_{ijkl,j}\,\overline{\varepsilon}_{kl} + \left(C'_{ijkl}\,\varepsilon'_{kl}\right)_{,j} = 0 \\ \Rightarrow C^r_{ijkl}u'_{k,lj} + F_i = 0, \qquad F_i = C'_{ijkl,j}\,\overline{\varepsilon}_{kl} + \left(C'_{ijkl}\,\varepsilon'_{kl}\right)_{,j}\end{gathered} \tag{13.7}$$

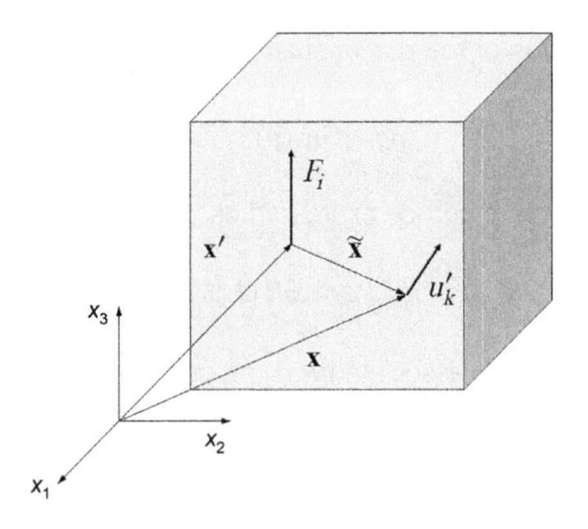

FIGURE 13.1

Physical meaning of the Green's function, G_{ki} (see Eq. 13.9). Green's function represents the displacement u'_k at location $\mathbf{x}$ due to the application of a force F_i at location $\mathbf{x}'$.

It is convenient to interpret the governing field equation in (13.7) as a stress analysis problem with a body force per unit volume of $F_i = C'_{ijkl,j}\bar{\varepsilon}_{kl} + (C'_{ijkl}\varepsilon'_{kl})_{,j}$. A solution for u'_i can be obtained from the field equation (13.7) using Green's function method. The main strategy in this method will be to first solve the governing partial differential equation in (13.7) for a unit body force and then obtain the solution for u'_i as a convolution integral over the body force distribution defined in the governing field equation (13.7). In other words, the strategy here can be summarized through the following equations (see Figure 13.1):

$$C^r_{ijkl}G_{km,lj}(\mathbf{x}-\mathbf{x}') + \delta\,(\mathbf{x}-\mathbf{x}')\delta_{im} = 0 \tag{13.8}$$

$$u'_k(\mathbf{x}) = \int_V G_{ki}(\mathbf{x}-\mathbf{x}')F_i(\mathbf{x}')d\mathbf{x}' \tag{13.9}$$

Eq. (13.8) defines the fundamental Green's function solution, G_{km}, as the kth component of the displacement field at location $\mathbf{x}$ in a material of which the elastic stiffness is defined by C^r_{ijkl}, when a unit body force defined with a single mth component is applied at position $\mathbf{x}'$.

Example 13.1

Derive the solution for Green's function, G_{km}, if C^r_{ijkl} is known to be isotropic.

Solution

Let $\tilde{\mathbf{x}} = \mathbf{x} - \mathbf{x}'$. To solve Eq. (13.1.8), the 3D Fourier transform of $G_{km}(\tilde{\mathbf{x}})$ is expressed as

$$G_{km}(\tilde{\mathbf{x}}) = \frac{1}{8\pi^3}\int_{\mathbf{p}\in P^3} \tilde{G}_{km}(\mathbf{p})e^{\,i\mathbf{p}.\tilde{\mathbf{x}}}d\mathbf{p}$$

The corresponding Fourier transform of the partial differential equation (13.8) yields

$$-C^r_{ijkl}p_jp_l\widetilde{G}_{km}(\mathbf{p})+\delta_{im}=0$$

The isotropic elastic stiffness tensor can be expressed as

$$C^r_{ijkl}=\lambda^r\delta_{ij}\delta_{kl}+\mu^r\left(\delta_{ik}\delta_{jl}+\delta_{il}\delta_{jk}\right)$$

where λ^r and μ^r are the elastic constants associated with the isotropic stiffness tensor C^r_{ijkl}. Substitution of this expression in the Fourier transform of the governing partial differential equation yields

$$-(\lambda^r+\mu^r)p_ip_k\widetilde{G}_{km}(\mathbf{p})-\mu^r|\mathbf{p}|^2\widetilde{G}_{im}(\mathbf{p})+\delta_{im}=0$$

Multiplication of both sides of the previous equation by p_i produces

$$p_k\widetilde{G}_{km}(\mathbf{p})=\frac{p_{\mathrm{m}}}{(\lambda^r+2\mu^r)|\mathbf{p}|^2}$$

Solving for $G_{km}(\widetilde{\mathbf{x}})$ from the previous two equations yields

$$\widetilde{G}_{km}(\mathbf{p})=\frac{-(\lambda^r+\mu^r)p_kp_{\mathrm{m}}+(\lambda^r+2\mu^r)|\mathbf{p}|^2\delta_{km}}{\mu^r(\lambda^r+2\mu^r)|\mathbf{p}|^4}$$

The corresponding inverse Fourier transform yields the desired solution (this is not a trivial step, as comprehensive 3D Fourier transform tables are not readily available; see, for example, Champeney (1973) and Bracewell (1978) and Example 15.1):

$$G_{km}(\widetilde{\mathbf{x}})=\frac{1}{8\pi\,\mu^r(\lambda^r+2\mu^r)}\left\{\frac{(\lambda^r+3\mu^r)}{|\tilde{\mathbf{x}}|}\delta_{km}+\frac{(\lambda^r+\mu^r)}{|\tilde{\mathbf{x}}|^3}\widetilde{x}_k\widetilde{x}_{\mathrm{m}}\right\} \tag{13.10}$$

The perturbation in the strain field can be computed from (13.9) as

$$\begin{aligned}\varepsilon'_{kl}(\mathbf{x})&=\int_V\frac{1}{2}(G_{ki,l}(\mathbf{x}-\mathbf{x}')+G_{li,k}(\mathbf{x}-\mathbf{x}'))F_i(\mathbf{x}')d\mathbf{x}'\\&=\int_V\frac{1}{2}(G_{ki,l}(\mathbf{x}-\mathbf{x}')+G_{li,k}(\mathbf{x}-\mathbf{x}'))\left[C'_{ijmn}(\mathbf{x}')_{,j}\bar{\varepsilon}_{mn}+\left(C'_{ijmn}(\mathbf{x}')\varepsilon'_{mn}(\mathbf{x}')\right)_{,j}\right]d\mathbf{x}'\end{aligned} \tag{13.11}$$

Substituting Eq. (13.2) into Eq. (13.11) and $\widetilde{\mathbf{x}}=\mathbf{x}-\mathbf{x}'$ permits the extraction of the following expression for a:

$$a_{klmn}(\mathbf{x})=\int_V\frac{1}{2}\left(G_{ki,l}(\widetilde{\mathbf{x}})+G_{li,k}(\widetilde{\mathbf{x}})\right)\left[C'_{ijmn}(\mathbf{x}')+C'_{ijpq}(\mathbf{x}')a_{pqmn}(\mathbf{x}')\right]_{,j}d\mathbf{x}' \tag{13.12}$$

Note that Green's function has a singularity ($G_{km}(\widetilde{\mathbf{x}})\to\infty$ as $\widetilde{\mathbf{x}}\to 0$), and therefore the integral in Eq. (13.12) has a principal value in the neighborhood of $\widetilde{\mathbf{x}}=0$. The approach adopted here would be to evaluate the integral in Eq. (13.12) over a volume $\widetilde{V}$ that lies between two spherical surfaces centered at $\widetilde{\mathbf{x}} = 0$ with radii approaching 0 and ∞, respectively. Expansion of Eq. (13.12) using integration by parts yields

$$a_{klmn}(\mathbf{x}) = \left[\int_S \frac{1}{2}\left(G_{ki,l}(\widetilde{\mathbf{x}}) + G_{li,k}(\widetilde{\mathbf{x}})\right)\left[C'_{ijmn}(\mathbf{x}') + C'_{ijpq}(\mathbf{x}')\, a_{pqmn}(\mathbf{x}')\right] n_j dS \right]_{\substack{\text{sphere with} \\ \widetilde{\mathbf{x}}\to 0}}^{\substack{\text{sphere with} \\ \widetilde{\mathbf{x}}\to\infty}} - \int_{\widetilde{V}} \frac{1}{2}\left(G_{ki,lj}(\widetilde{\mathbf{x}}) + G_{li,kj}(\widetilde{\mathbf{x}})\right)\left[C'_{ijmn}(\mathbf{x}^*) + C'_{ijpq}(\mathbf{x}')a_{pqmn}(\mathbf{x}')\right] d\mathbf{x}' \tag{13.13}$$

For clarity, let us consider only the terms on the right side, without tensor a, and group them into a single term, with the integral operator, $\mathbf{\Gamma}$, acting on the perturbation of the stiffness tensor:

$$-\mathbf{\Gamma}\mathbf{C}' = \left[\int_S \frac{1}{2}\left(G_{ki,l}(\widetilde{\mathbf{x}}) + G_{li,k}(\widetilde{\mathbf{x}})\right)\left[C'_{ijmn}(\mathbf{x}')\right] n_j dS \right]_{\substack{\text{sphere with} \\ |\widetilde{\mathbf{x}}|\to 0}}^{\substack{\text{sphere with} \\ |\widetilde{\mathbf{x}}|\to\infty}} - \int_{\widetilde{V}} \frac{1}{2}\left(G_{ki,lj}(\widetilde{\mathbf{x}}) + G_{li,kj}(\widetilde{\mathbf{x}})\right)\left[C'_{ijmn}(\mathbf{x}')\right] d\mathbf{x}' \tag{13.14}$$

Of the two spherical surface integrals in the first term of Eq. (13.14) the second integral corresponding to $\widetilde{\mathbf{x}}\to 0$ is the principal value term. Clearly, when $\mathbf{x}\approx\mathbf{x}'$ on the surface of this infinitesimal sphere, $C'_{ijmn}(\mathbf{x}') = C'_{ijmn}(\mathbf{x})$, which is a constant, and can be taken outside of the integral. Regrouping the terms and writing the kernel of the integral operator as $\mathbf{\Gamma}_{klij}(\widetilde{\mathbf{x}})$, Eq. (13.14) leads to

$$\begin{aligned} \mathbf{\Gamma}_{klij}(\widetilde{\mathbf{x}}) &= E_{klij}\delta(\widetilde{\mathbf{x}}) + I^{\infty}_{klij}\delta(\widetilde{\mathbf{x}}) + \Phi_{klij}(\widetilde{\mathbf{x}}) \\ E_{klij} &= \lim_{|\mathbf{x}-\mathbf{x}'|\to 0}\left(\int_S \frac{1}{2}(G_{ki,l}(\widetilde{\mathbf{x}}) + G_{li,k}(\widetilde{\mathbf{x}}))n_j dS \right) \\ I^{\infty}_{klij} &= \lim_{|\mathbf{x}-\mathbf{x}'|\to\infty}\left(\int_S \frac{1}{2}(G_{ki,l}(\widetilde{\mathbf{x}}) + G_{li,k}(\widetilde{\mathbf{x}}))\left[C'_{ijmn}(\mathbf{x}')\right] n_j dS \right) \\ \Phi_{klij}\left(\widetilde{\mathbf{x}}\right) &= \frac{1}{2}(G_{ki,lj}(\widetilde{\mathbf{x}}) + G_{li,kj}(\widetilde{\mathbf{x}})) \end{aligned} \tag{13.15}$$

where $\delta(\widetilde{\mathbf{x}})$ is 1 if $\mathbf{x} = \mathbf{x}'$ and 0 otherwise.

The term I^{∞}_{klij} is typically assumed to equal 0; however, this assumption is generally not valid. This is easy to see if we take an isotropic reference tensor and the resultant analytical values of $G_{ki,l}$ (Eq. 13.10). Since it is proportional to $1/|r|^2$, the integral over the surface of a sphere remains approximately constant. In fact, if the stiffness tensor is constant throughout the material (and hence can be taken outside the integral), then clearly the integral over the outer sphere is simply the negative of the integral over the inner sphere. If we make the common assumption that the material is random at infinity (completely uncorrelated), then the stiffness tensor on a patch of the outer sphere is approximately $\overline{\mathbf{C}}'$ —the average value of the perturbation stiffness. Then the integral on the outer sphere may be approximated by

$$\mathbf{I}^{\infty} \approx \mathbf{E}\overline{\mathbf{C}}' \tag{13.16}$$

Hence, we may rewrite Eq. (13.13) truncated to the first term (not including the terms that have *a* in them) as

$$\mathbf{a}(\mathbf{x}) = \mathbf{E}\overline{\mathbf{C}}' - \mathbf{E}\mathbf{C}'(\mathbf{x}) - \int_{\tilde{V}} \mathbf{\Phi}(\tilde{\mathbf{x}})\mathbf{C}'(\mathbf{x}')d\mathbf{x}' \tag{13.17}$$

Clearly, if $\mathbf{C}^r = \overline{\mathbf{C}}$, then $\overline{\mathbf{C}}' = 0$, and $\mathbf{I}^{\infty}$ as calculated from Eq. (13.16) tends to 0, in which case we arrive at the more common first-order form of the localization equation:

$$a_{klmn}(\mathbf{x}) = -\mathrm{E}_{klij}C'_{ijmn}(\mathbf{x}) - \int_{\tilde{V}} \mathbf{\Phi}_{klij}(\tilde{\mathbf{x}})C'_{ijmn}(\mathbf{x}')d\mathbf{x}' \tag{13.18}$$

In general, we will not restrict the reference tensor to equal the average stiffness tensor; for example, it may be most desirable to choose an isotropic stiffness tensor that is closest to the average of several anisotropic tensors corresponding to the phases/crystal orientations that are present (Norris, 2006). Hence, we will use Eq. (13.17) rather than Eq. (13.18).

Equation (13.13) can now be recast as

$$a_{klmn}(\mathbf{x}) = -\int_{V} \Gamma_{klij}\left(\tilde{\mathbf{x}}\right)\left[C'_{ijmn}(\mathbf{x}') + C'_{ijpq}(\mathbf{x}')a_{pqmn}(\mathbf{x}')\right]d\mathbf{x}' \tag{13.19}$$

Substituting $a_{klmn}(\mathbf{x})$ into itself in Eq. (13.19), we obtain a series solution for this tensor as

$$a_{klmn}(\mathbf{x}) = -\int_{V} \Gamma_{klij}\left(\tilde{\mathbf{x}}\right)\left[\begin{array}{l} C'_{ijmn}(\mathbf{x}')+ \\ C'_{ijpq}(\mathbf{x}')\left(-\int_{V} \Gamma_{pquv}\left[\begin{array}{l} C'_{uvmn}(\mathbf{x}'')+ \\ C'_{uvrs}(\mathbf{x}'')\int_{V}\ldots \end{array}\right]d\mathbf{x}''\right) \end{array}\right]d\mathbf{x}' \tag{13.20}$$

For the special case that $\boldsymbol{C}^r = \overline{\boldsymbol{C}}$, then $\overline{\boldsymbol{C}}' = 0$. Hence, substitution of Eq. (13.20) in Eq. (13.5) allows one to express the effective stiffness as a series:

$$\mathbf{C}^{*} = \overline{\mathbf{C}} - \overline{\mathbf{C}'\mathbf{\Gamma}\mathbf{C}'} + \overline{\mathbf{C}'\mathbf{\Gamma}\mathbf{C}'\mathbf{\Gamma}\mathbf{C}'} - \ldots \tag{13.21}$$

It is also worth noting that Eq. (13.21) can be recast as (Torquato, 2002)

$$\mathbf{C}^{*} = \mathbf{C}^{r} + \overline{\mathbf{C}'\left(\mathbf{I}+\mathbf{\Gamma}\mathbf{C}'\right)^{-1}} \tag{13.22}$$

In closing this section, we present some details of the computation of the terms in Eq. (13.21), which will become clearer in later chapters. For statistically homogeneous microstructures, it will be shown later that the second term in Eq. (13.21) can be evaluated as

$$\overline{\mathbf{C}'\mathbf{\Gamma}\mathbf{C}'} = \left\langle \mathbf{C}'\mathbf{\Gamma}\mathbf{C}' \right\rangle = \int_H \int_H \int_{\Psi(\mathbf{r})} f_2\left(h, h'|\mathbf{r}\right)\mathbf{C}(h)\mathbf{\Gamma}(-\mathbf{r})\mathbf{C}\left(h'\right) dh dh' d\mathbf{r} \tag{13.23}$$

where $f_2(h, h'|\mathbf{r})$ denotes the two-point correlation function that describes the conditional probability density of realizing local states h and h' in the immediate neighborhood of an ordered pair of points $(\mathbf{x}, \mathbf{x}' = \mathbf{x} + \mathbf{r})$ placed randomly in the microstructure. H denotes the local state space for the composite material system of interest, and $\Psi(\mathbf{r})$ denotes the space of all vectors $\mathbf{r}$ that can be accommodated in the representative volume element of the microstructure. Note also that $\mathbf{\Gamma}(-\mathbf{r}) = \mathbf{\Gamma}(\mathbf{x} - \mathbf{x}') = \mathbf{\Gamma}(\widetilde{\mathbf{x}})$, consistent with the notation in the earlier equations (e.g., Eq. 13.15)

Example 13.2

Derive an expression for the fourth-rank tensor $\mathbf{\Gamma}$ building on Green's function solution obtained in Example 13.1 for an isotropic reference medium.

Solution

We need to evaluate the tensors E_{klij} and $\Phi_{klij}(\widetilde{\mathbf{x}})$ from Eq. (13.19), and for this we need to evaluate the derivatives of Green's the function. The derivatives of the Green's function derived in Example 13.1 (for an arbitrary vector, $\mathbf{x}$) are as follows:

$$G_{ki,l}(\mathbf{x}) = \frac{1}{8\pi\mu^r(\lambda^r + 2\mu)}\left\{\frac{(\lambda^r + \mu^r)(\delta_{kl}x_i + \delta_{il}x_k) - (\lambda^r + 3\mu^r)\delta_{ki}x_l}{|\mathbf{x}|^3} - \frac{3(\lambda^r + \mu^r)x_i x_k x_l}{|\mathbf{x}|^5}\right\}$$

The tensor E_{klij} is then

$$\frac{1}{2}\left(G_{ki,l}(\mathbf{x}) + G_{li,k}(\mathbf{x})\right) =$$

$$\frac{1}{16\pi\mu^r(\lambda^r + 2\mu^r)}\left\{\begin{array}{l}\dfrac{(\lambda^r + \mu^r)(2\delta_{kl}x_i + \delta_{il}x_k + \delta_{ik}x_l) - (\lambda^r + 3\mu^r)(\delta_{ki}x_l + \delta_{li}x_k)}{|\mathbf{x}|^3} \\ \\ -\dfrac{6(\lambda^r + \mu^r)x_i x_k x_l}{|\mathbf{x}|^5}\end{array}\right\}$$

The surface integral over a small sphere in proximity of the origin becomes

$$E_{klij} = \left[\int_S \frac{1}{2}\left(G_{ki,l}(\mathbf{x}) + G_{li,k}(\mathbf{x})\right)\frac{x_j}{|\mathbf{x}|}dS\right]_{|\mathbf{x}|\to 0} = \frac{1}{15\mu^r}\left\{\frac{\lambda^r + \mu^r}{\lambda^r + 2\mu^r}\delta_{il}\delta_{ik} - \frac{3\lambda^r + 8\mu^r}{\lambda^r + 2\mu^r}\mathbf{I}_{ijkl}\right\}$$

where

$$\mathbf{I}_{ijkl} = \frac{1}{2}\left(\delta_{ik}\delta_{jl} + \delta_{il}\delta_{jk}\right)$$

The second derivatives of the Green's function result are as follows:

$$G_{ki,lj}(\mathbf{x}) = \frac{1}{8\pi\mu^r(\lambda^r + 2\mu^r)} \left\{ \begin{array}{l} \dfrac{(\lambda^r + \mu^r)\left(\delta_{kl}\delta_{ij} + \delta_{il}\delta_{kj}\right) - (\lambda^r + 3\mu^r)\delta_{ki}\delta_{lj}}{|\mathbf{x}|^3} + \dfrac{3(\lambda^r + 3\mu^r)\delta_{ki}x_l x_j}{|\mathbf{x}|^5} \\ -\dfrac{3(\lambda^r + \mu^r)\left(\delta_{kl}x_i x_j + \delta_{il}x_k x_j + \delta_{kj}x_i x_l + \delta_{ij}x_k x_l + \delta_{lj}x_k x_i\right)}{|\mathbf{x}|^5} \\ +\dfrac{15(\lambda^r + \mu^r)x_i x_j x_k x_l}{|\mathbf{x}|^7} \end{array} \right\}$$

$$\Phi_{klij}(\mathbf{x}) = \frac{1}{2}\left(G_{ki,lj}(\mathbf{x}) + G_{li,kj}(\mathbf{x})\right)$$

13.2 CALCULATION OF SECOND-ORDER PROPERTIES IN THE PRIMITIVE BASIS

The calculations in the previous section may be combined with work presented in prior chapters to arrive at a practical method for calculating effective material properties. We begin with Eq. (13.21), which is obtained from Eqs. (13.20) and (13.5). Since we are interested in second-order terms only, we omit those of higher order. We assume statistical homogeneity; hence, we may swap volume averages for ensemble averages:

$$\mathbf{C}^* = \langle \mathbf{C} \rangle - \langle \mathbf{C}' \boldsymbol{\Gamma} \mathbf{C}' \rangle \tag{13.24}$$

In the Fourier primitive basis, D_s^n gives the proportion of state parameter $h(n)$ (associated with cell n in state space and normalized by factor N) in cell s of real space; hence, the first term on the right side may be written as

$$\langle \mathbf{C} \rangle = \frac{1}{NS} \sum_{s=1}^{S} \mathbf{C}^n D_s^n \tag{13.25}$$

where summation is assumed over the values of n, and $\mathbf{C}^n$ implies the value of $\mathbf{C}$ for the state space parameter represented by n (not a power of $\mathbf{C}$).

The second term is significantly more complicated, involving the ensemble average of the convolution of Green's function with the polarization of tensor $\mathbf{C}$. Determining a single value of the property $\mathbf{C}^*$ for a given microstructure appears a significant task. Searches or optimizations over the property space would appear to be enormous undertakings. However, by a clever rearrangement of the problem in the primitive basis we arrive at an ideal formulation for future optimization tasks.

We begin by restating this part of the problem using Eqs. (13.5) and (13.20):

$$\langle C'\Gamma C'\rangle_{ijkl} = \langle C'_{ijmn}(\mathbf{x})\int_{\Omega}\Gamma_{mnop}(\mathbf{x}-\mathbf{x}')C'_{opkl}(\mathbf{x}')d\mathbf{x}'\rangle \tag{13.26}$$

where the average is over $\mathbf{x}$; that is,

$$\langle C'\Gamma C'\rangle_{ijkl} = \frac{1}{V_\Omega}\int_{\Omega}C'_{ijmn}(\mathbf{x})\int_{\Omega}\Gamma_{mnop}(\mathbf{x}-\mathbf{x}')C'_{opkl}(\mathbf{x}')d\mathbf{x}'d\mathbf{x} \tag{13.27}$$

Now $C'_{ijkl}(x) = \int_{FZ} C'_{ijkl}(h)M(\mathbf{x},h)dh$ from the definition of the microstructure function, M; thus,

$$\langle C'\Gamma C'\rangle_{ijkl} = \frac{1}{V_\Omega}\int_{\Omega}\int_{\Omega}\Gamma_{mnop}(\mathbf{x}-\mathbf{x}')\int_{FZ}\int_{FZ}C'_{ijmn}(h)C'_{opkl}(h')M(\mathbf{x},h)M(\mathbf{x}',h')dhdh'd\mathbf{x}'d\mathbf{x} \tag{13.28}$$

Let $\mathbf{x}\prime - \mathbf{x} = \mathbf{r}$, and $\theta(\mathbf{r},\mathbf{x}) = 1$ if $\mathbf{x}+\mathbf{r}\in\Omega$ and 0 otherwise, and note that by statistical homogeneity the integral

$$\int_{\Omega}\Gamma_{mnop}(\mathbf{x}-\mathbf{x}')C'_{ijmn}(h')M(\mathbf{x}',h')'d\mathbf{x}' \tag{13.29}$$

is independent of the value of $\mathbf{x}$. Then

$$\langle C'\Gamma C'\rangle_{ijkl} = \int_{R(\Omega)}\int_{FZ}\int_{FZ}C'_{ijmn}(h)C'_{opkl}(h')\Gamma_{mnop}(-\mathbf{r})\frac{1}{V_\Omega}\int_{\mathbf{x}\in\Omega}M(\mathbf{x},h)M(\mathbf{x}+\mathbf{r},h')\theta(\mathbf{r},\mathbf{x})d\mathbf{x}dhdh'd\mathbf{r}$$

$$= \int_{R(\Omega)}\int_{FZ}\int_{FZ}C'_{ijmn}(h)C'_{opkl}(h')\Gamma_{mnop}(-\mathbf{r})\frac{1}{V_\Omega}\int_{\mathbf{x}\in\Omega|\mathbf{r}}M(\mathbf{x},h)M(x+\mathbf{r},h')d\mathbf{x}dhdh'd\mathbf{r}$$

$$= \int_{R(\Omega)}\int_{FZ}\int_{FZ}C'_{ijmn}(h)C'_{opkl}(h')\Gamma_{mnop}(-\mathbf{r})\frac{V_{\Omega|\mathbf{r}}}{V_\Omega}f_2(h,h'|\mathbf{r})dhdh'd\mathbf{r} \tag{13.30}$$

where $\Omega|\mathbf{r}$ is the subset of points in Ω such that a vector, $\mathbf{r}$, initiating from a given point remains completely within Ω. From Chapter 12, Eq. (12.19), we already have a Fourier representation for the final term in the integrand. If we include the volume term we may define

$$\frac{V_{\Omega|\mathbf{r}}}{V_\Omega}f_2(h,h'|\mathbf{r}) = \widetilde{f}_2(h,h'|\mathbf{r}) \approx \widetilde{H}_{ss't}D^n_s D^{n'}_{s'}\chi^n(h)\chi^{n'}(h')\chi_t(\mathbf{r}) \tag{13.31}$$

where the coefficients of $\widetilde{H}_{ss't}$ are much simpler than those previously defined in Section 12.2.2. Starting from the definition of $\widetilde{H}_{ss't}$ as given in Eq. (13.31), suppose that the cells in Ω are renumbered such that $s \leftrightarrow \{s_1, s_2, s_3\}$ (see Figure 5.4), and similarly for $\Psi(\Omega) : t \leftrightarrow \{t_1, t_2, t_3\}$ (t enumerates the cells in the space $\Psi(\Omega)$ of all vectors $\mathbf{r}$ that fit in Ω). We may write

$$\widetilde{H}_{ss't} = h(s_1, s'_1, t_1)h(s_2, s'_2, t_2)h(s_3, s'_3, t_3) \tag{13.32}$$

for some function h given by

$$h(s_i, s'_i, t_i) = \begin{cases} \delta/2 \text{ if } t_i = s'_i - s_i + 2^P \text{ or } t_i = s'_i - s_i + 2^P + 1 \\ 0 \quad \text{otherwise} \end{cases} \tag{13.33}$$

where $\delta = D/2^P$ is the cell dimension for a cubic sample region of length D. The cells in $\Psi(\Omega)$ are chosen to have the same dimensions as the cells in Ω, and $\Psi(\Omega)$ is eight times as large as Ω from its definition.

Now let us define a Fourier representation for the first part of the integrand in Eq. (13.30):

$$C'_{ijmn}(h)C'_{opkl}(h')\Gamma_{mnop}(-\mathbf{r}) \approx \Xi_t^{nn'}\chi^n(h)\chi^{n'}(h')\chi_t(\mathbf{r}) \tag{13.34}$$

(See Eq. (5.41) for a definition of χ; note that the Greek letter Ξ is "xi," pronounced "sigh"). Thus,

$$\Xi_t^{nn'} = \frac{1}{V_{\omega_n}V_{\omega_{n'}}V_{\psi_t}} \int_{\omega_n}\int_{\omega_{n'}}\int_{\psi_t} C'_{ijmn}(h)C'_{opkl}(h')\Gamma_{mnop}(-\mathbf{r})d\mathbf{r}dhdh' \tag{13.35}$$

where ω_n, $\omega_{n'}$, ψ_t are the relevant cells defined for the primitive basis in Section 5.4; that is, it is the average value of the integrand in the associated cells. This integral may be calculated numerically.

Then, combining Eqs. (5.41), (5.44), (13.30), (13.31), and (13.34), we arrive at

$$\langle C'\Gamma C'\rangle_{ijkl} = \frac{\delta^3}{N^2}\Xi_t^{nn'}\widetilde{H}_{sst}D_s^nD_{s'}^{n'} \tag{13.36}$$

where the summation convention is assumed (thus replacing the integral), and D_s^n are assumed to be the statistical averaged values, by a slight abuse of notation. Note that $\Xi_t^{nn'}$ on the right side are dependent on the choice of $ijkl$ (see Eq. 13.35), even though this is not explicitly shown.

In the foregoing derivation, we omitted two significant details to maintain the flow. First, for elasticity there is a singularity at the origin for Green's function as detailed in the previous section; second, the Green's function integral taken in the previous section is over an infinite sphere, whereas $R(\Omega)$ is rectangular. We address these issues in the following sections.

13.2.1 Contribution from Green's function singularity

Let us start from the integral already obtained in Eq. (13.30) and split the Green's function integral as described in Section 13.1:

$$\begin{aligned}\langle C'\Gamma C'\rangle_{ijkl} &= \int_{R(\Omega)}\int_{FZ}\int_{FZ} C'_{ijmn}(h)C'_{opkl}(h')\Gamma_{mnop}(-\mathbf{r})\widetilde{f}_2(h,h'|\mathbf{r})dhdh'd\mathbf{r} \\ &= \int_{R(\Omega)}\int_{FZ}\int_{FZ} C'_{ijmn}(h)C'_{opkl}(h')\Phi_{mnop}(-\mathbf{r})\widetilde{f}_2(h,h'|\mathbf{r})dhdh'd\mathbf{r} \\ &\quad + \int_{FZ}\int_{FZ} C'_{ijmn}(h)C'_{opkl}E_{mnop}f_2(h,h'|0)dhdh'\end{aligned} \tag{13.37}$$

In the final integral $\widetilde{f}_2 = f_2$ since $\mathbf{r} = 0$, and hence $\Omega|_{\mathbf{r}} = \Omega$. Note that $f_2(h, h'|0) = 0$ unless $h = h'$, in which case it equals the distribution function for state h, that is, *f(h)*. In the discretized Fourier representation, $f(h) \approx \frac{1}{S}\sum_{s=1}^{S} D_s^n \chi^n(h)$ (summation over n; see Eq. 5.49). We also write a Fourier representation of the remainder of the integrand in this final integral:

$$C'_{ijmn}(h) C'_{opkl} E_{mnop} \approx (C')^n_{ijmn} (C')^n_{opkl} \chi^n(h) \chi^n(h') E_{mnop} \tag{13.38}$$

Hence, we may rewrite this integral as approximately equal to

$$\begin{aligned}
&\int_{FZ}\int_{FZ} C'_{ijmn}(h) C'_{opkl} E_{mnop} \widetilde{f}_2\left(h, h'\middle|0\right) dh dh' \\
&= \int_{FZ}\int_{FZ} (C')^u_{ijmn} (C')^u_{opkl} \chi^u(h) \chi^u(h) E_{mnop} \frac{1}{S} \sum_{s=1}^{S} D_s^v \chi^v(h) dh dh \\
&= \frac{E_{mnop}}{S} \sum_{s=1}^{S} D_s^u (C')^u_{ijmn} (C')^u_{opkl} \int_{FZ}\int_{FZ} \chi^u(h) \chi^u(h) \chi^u(h) dh dh \\
&= \frac{E_{mnop}}{S} \sum_{s=1}^{S} D_s^u (C')^u_{ijmn} (C')^u_{opkl} \int_{FZ} \frac{1}{N} dh \\
&= \frac{E_{mnop}}{SN} \sum_{s=1}^{S} D_s^u (C')^u_{ijmn} (C')^u_{opkl}
\end{aligned} \tag{13.39}$$

where $\chi^u(h)\chi^v(h) = 0$ unless $u = v$. See Eqs. (5.41) and (5.44) for integrals of χ (no summation over u in the equation). Then, Eq. (13.36) becomes

$$\langle C' \Gamma C' \rangle = \aleph_{ss'}^{nn'} D_s^n D_{s'}^{n'} + Z_s^n D_s^n \tag{13.40}$$

with the "Aleph" symbol,

$$\aleph_{ss'}^{nn'} = \frac{\delta^3}{N^2} \Xi_t^{nn'} H_{sst} \tag{13.41}$$

and

$$Z_s^u = \frac{E_{mnop} (C')^u_{ijmn} (C')^u_{opkl}}{SN} \tag{13.42}$$

Note that Z_s^n is in fact only a function of n; however, the index s carries the summation requirement. We have also used the same symbol Ξ for the Fourier representations of both $\mathbf{\Gamma}$ and $\mathbf{\Phi}$, since no confusion is likely.

In the evaluation of $\mathbf{\Phi}$ in Eq. (13.37) special care must be taken in the vicinity of the origin. The main issue here is calculation of the integral of $\mathbf{\Phi}$ in the cell about the origin $\mathbf{r} = 0$. Various previous studies have either ignored the integral of Φ in the central cell, assuming that it was negligible, or have evaluated it using a regular grid of integration points. Accurate calculations using Monte Carlo integration have established that the value of the integral in the central cell accounts for approximately 15% of the total Φ integral, and this is almost independent of the size of the cell. To qualify this last statement, we mean that the value of the integral is independent of the size of the central cell if there is only a single material state within the cell. This is because the integral of Φ between two cubes is 0; hence, refining the grid or including points closer to the origin in a given discretization of space will not improve the accuracy. The integral has been calculated to high accuracy by the authors using Monte Carlo integration (see Caflisch, 1998; Fullwood et al., 2009). Other approaches are obviously possible.

It is worth noting that an equivalent approach would be to calculate **E** on the surface of a cube about the origin rather than on a sphere. In this case the calculation for Φ in the inner cell would be 0 (since it would be over a volume between two cubes), and the correction from previous work would be carried by the new value of **E**. This method was also tested as an alternative to that mentioned earlier. Since there is no known analytical solution for the calculation of **E** on the surface of a cube, a Monte Carlo integration was again employed, thus giving no significant advantage to the alternative.

13.2.2 Adjustment for a nonspherical domain

The physical space, Ω, occupied by our sample of material, is a simple 3D cube or rectangle. The volume is divided in S equal cells, and the local state space is divided into N cells of equal measure. Associated with domain Ω is a space $R(\Omega)$ of vectors, **r**, that fit inside Ω. The **r** space represents the domain where the $\boldsymbol{\Gamma}$ operator is defined. Note that each dimension in the rectangular $R(\Omega)$ space is twice that of the Ω space, as shown in Figure 13.2. For ease of illustration we will assume that Ω and $R(\Omega)$ are cubic in the following.

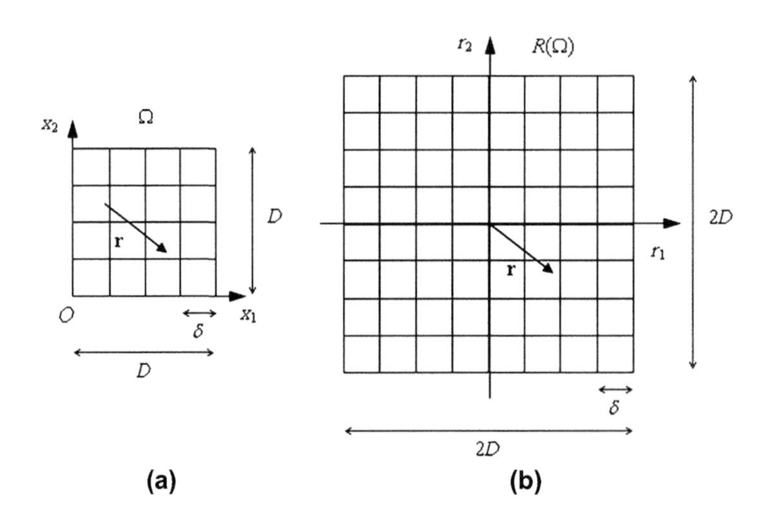

FIGURE 13.2

A 2D representation of the 3D domain Ω and associated **r** space domain $R(\Omega)$.

The analytical solutions for the $\mathbf{\Gamma}$ operator are obtained when it is integrated over a spherical domain; hence, we define a spherical region surrounding the $R(\Omega)$ domain, called $R(S)$, as shown in Figure 13.3. It is assumed that points in the cubic region $R(\Omega)$ are correlated, while outside this region they are totally uncorrelated. In other words, it is assumed that for a vector lying outside $R(\Omega)$, the probability of the local state being one value at the base of the vector is completely independent of the value of the local state at the tip of the vector. Thus, all vectors with magnitude $\mathbf{r} \leq \mathbf{r}_c$ are contained inside $R(\Omega)$, $\mathbf{r}_c$ being the "*coherence radius*."

The integral of interest is taken from Eq. (13.37):

$$\int\limits_{R(\Omega)} \int\limits_{FZ} \int\limits_{FZ} \mathbf{C}'(g)\mathbf{C}'(g')\Phi(-\mathbf{r})\widetilde{f}_2(g, g'|\mathbf{r})dgdg'd\mathbf{r} \tag{13.43}$$

Then the integral of a spherical region containing $R(\Omega)$ is obtained from summation of a first integral inside $R(\Omega)$ and a second integral over $R(S) - R(\Omega)$. The latter region of space is completely uncorrelated and the integral over this region can be determined by considering an uncorrelated

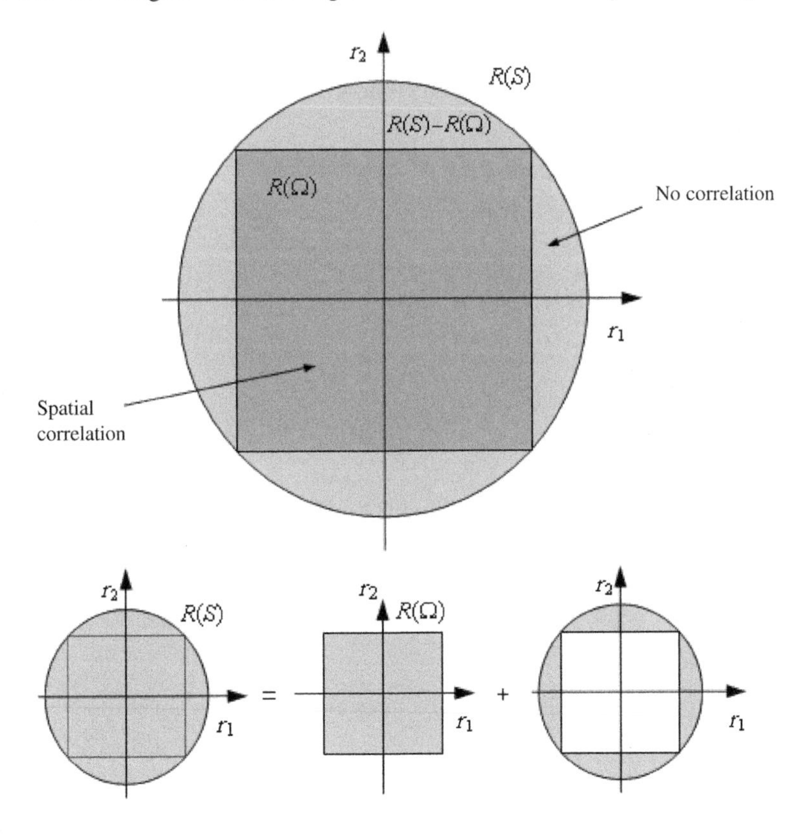

FIGURE 13.3

A 2D projection of $R(S)$ and $R(\Omega)$. The lighter-shaded region corresponds to the uncorrelated space (i.e., it is outside the coherence radius).

microstructure of the same volume Ω and $\mathbf{r}$ space $R(S)$. For an uncorrelated microstructure the integral (Eq. 13.43) on a spherical volume is 0 (f_2 is a correlation function and so is constant on an uncorrelated space). Hence, one may write

$$\iiint_{R(S)-R(\Omega)|unc} \int_{FZ}\int_{FZ} \mathbf{C}'(g)\mathbf{C}'(g')\Phi(-\mathbf{r})\widetilde{f}_2(g,g'|\mathbf{r})dgdg'd\mathbf{r} = - \iiint_{R(\Omega)|unc} \int_{FZ}\int_{FZ} \mathbf{C}'(g)\mathbf{C}'(g')\Phi(-\mathbf{r})\widetilde{f}_2(g,g'|\mathbf{r})dgdg'd\mathbf{r} \tag{13.44}$$

where *unc* is the uncorrelated microstructure.

In the uncorrelated microstructure the probability of finding a certain state n is the same in any cell s, so the microstructure is defined as

$$D_s^n = \frac{1}{S}\sum_{k=1}^{S} D_k^n \quad \text{with} \quad \sum_{n=1}^{N} D_k^n = N \text{ for all } k \tag{13.45}$$

Using the previously defined Fourier representation (Eq. (13.40), ignoring the second term that relates to the singularity), the second-order correction term is then (writing out in full summation notation to illustrate the manipulation)

$$\begin{aligned} \langle C'\Gamma C'\rangle_{R(S)-R(\Omega)} &= -\sum_{n}^{N}\sum_{n'}^{N}\sum_{s}^{S}\sum_{s'}^{S} \aleph_{ss'}^{nn'} D_s^n D_{s'}^{n'} \\ &= -\sum_{n}^{N}\sum_{n'}^{N}\sum_{k}^{S}\sum_{k'}^{S}\sum_{s}^{S}\sum_{s}^{S} \frac{\aleph_{ss'}^{nn'}}{S^2} D_k^n D_{k'}^{n'} \\ &= \sum_{n}^{N}\sum_{n'}^{N}\sum_{k}^{S}\sum_{k'}^{S} A_{kk'}^{nn'} D_k^n D_{k'}^{n'} \quad \text{with} \quad A_{kk'}^{nn'} = -\sum_{s}^{S}\sum_{s'}^{S}\frac{\aleph_{ss'}^{nn'}}{S^2} \end{aligned} \tag{13.46}$$

Finally, summation of Eqs. (13.25), (13.40), and (13.46) gives

$$\begin{aligned} C^* = \langle C\rangle - \langle C'\Gamma C'\rangle &= \sum_{s}^{S}\sum_{n}^{N} \hat{J}_s^n D_s^n - \sum_{s}^{S}\sum_{s'}^{S}\sum_{n}^{N}\sum_{n'}^{N} \hat{\aleph}_{ss'}^{nn'} D_s^n D_{s'}^{n'} \\ &= \hat{J}_s^n D_s^n - \hat{\aleph}_{ss'}^{nn'} D_s^n D_{s'}^{n'} \end{aligned} \tag{13.47}$$

where the final version uses the usual summation convention with

$$\hat{J}_s^n = J_s^n + Z_s^n = \frac{C_{ijkl}^n}{SN} - E_{mpqr}\frac{(C')_{ijmp}^n (C')_{qrkl}^n}{SN} \tag{13.48}$$

$$\hat{\aleph}_{ss'}^{nn'} = \aleph_{ss'}^{nn'} + A_{ss'}^{nn'} = \aleph_{ss'}^{nn'} - \sum_{s}^{S}\sum_{s}^{S}\frac{\aleph_{ss'}^{nn'}}{S^2} \tag{13.49}$$

Eq. (13.47) is a quadratic relationship representing the second-order effective property relation.

Example 13.3

Determine the tensors $\hat{J}_s^n$ and $\hat{\aleph}_{ss'}^{nn'}$ for the property C_{11}^* for a cubic sample made up of eight cells (i.e., each side is divided into two even lengths). The material is assumed to be composed of a single anisotropic phase of copper with the following parameters: $C_{11}=168.4$ GPa, $C_{12}=121.4$ GPa, $C_{44}=75.4$ GPa, $C_{66}=100.6$ GPa. The orientation space has four cells with orientations $(\varphi_1, \Phi, \varphi_2) = (3.51, 0.95, 0.58)$, $(1.38, 1.07, 0.38)$, (4.95, 1.28, 0.19), and (6.28, 0, 1.57) (see Example 9.2).

Solution

We first need to choose an appropriate reference tensor, C^r. In a series expansion, a logical point about which to take the expansion is halfway between the maximal points being estimated, thus reducing the error at the extreme points of the domain. In our case we are trying to approximate an anisotropic tensor with an isotropic tensor; hence, there is no obvious choice for the domain's midpoint. Since we are calculating C_{11}^*, we may base the choice of isotropic tensor on this particular component. We calculate C_{11}^n for the two orientations using the method in Example 9.2: $C_{11}^1 = 228$ GPa and $C_{11}^2 = 208.2$ GPa.

For an isotropic stiffness tensor, $C_{11} = \lambda + 2\mu$. Taking the average of the C_{11}^n values gives us one equation for the λ and μ that determines the reference tensor. To obtain a second equation, we may take, for example, the average values of the C_{66}^n—in which case, this value is equal to μ. From the values calculated in Example 9.2, we obtain $\mu = 60.9$, $\lambda = 96.3$. Then

$$C_{ijkl}^r = \mu\left(\delta_{ik}\delta_{jl} + \delta_{il}\delta_{jk}\right) + \lambda\delta_{ij}\delta_{kl}$$

We first calculate $\hat{J}_s^n$ using (Eq. 13.48):

$$\hat{J}_s^n = J_s^n + Z_s^n = \frac{C_{ijkl}^n}{SN} - E_{mpqr}\frac{(C')_{ijmp}^n (C')_{qrkl}^n}{SN}$$

As mentioned, C_{ijkl}^n can be calculated as in Example 9.2; then $(C')_{ijkl}^n = C_{ijkl}^r - C_{ijkl}^n$. For this example, $N=2$ and $S=8$. The values of E_{mpqr} are calculated using the solution given in Example 13.2. The final term in the equation is then summed over all values of *m,p,q,r*. Note that $\hat{J}_s^n$ is independent of *s*, so there are only two distinct values.

The calculation for $\hat{\aleph}_{ss'}^{nn'}$ is somewhat more complicated. We start from Eq. (13.49):

$$\hat{\aleph}_{ss'}^{nn'} = \aleph_{ss'}^{nn'} + A_{ss'}^{nn'} = \aleph_{ss'}^{nn'} - \sum_s^S \sum_s^S \frac{\aleph_{ss'}^{nn'}}{S^2}$$

and use Eq. (13.41):

$$\aleph_{ss'}^{nn'} = \frac{\delta^3}{N^2}\,\Xi_t^{nn'} H_{sst}$$

The equation for $\Xi_t^{nn'}$ is modified from its original form to take into account the singularity discussed in the previous section:

$$\Xi_t^{nn'} = \frac{1}{V_{\omega_n} V_{\omega_{n'}} V_{\psi_t}} \int_{\omega_n}\int_{\omega_{n'}}\int_{\psi_t} C'_{ijmp}(g)\, C'_{klqr}(g')\, \Phi_{mpqr}(-\mathbf{r})\, dr\, dg\, dg'$$

This integral is evaluated numerically. Since we are assuming that the values of $\mathbf{C}'$ are constant in each cell, ω_n (we only have one value of g for each cell), the integral is purely over the domain of $\mathbf{r}$. In our implementation we split each cell in the $\mathbf{r}$ domain into 8×8 ×8 subcells, that is, a total of 9×9×9 = 729 points in each subcell. Hence, in our case

$$\Xi_t^{nn'} = \frac{1}{729} C'_{ijmq}(g) C'_{prkl}(g') \sum_{v=1}^{729} \Phi_{mpqr}(-\mathbf{r}(v))$$

Note that each term must also be summed over all values of m,p,q,r. Recall that for this problem $ijkl = 1111$ (for the required property, $C_{11}^* = C_{1111}^*$). Note also that the subcells in the $\mathbf{r}$ domain must be those defined in the integration of the $H_{ss't}$ term. The values of $H_{ss't}$ are given in Eq. (13.32).

13.2.3 Reduction of the problem order

It is possible to lower the dimension of the space to $SN - S$ by considering S relations of the following form:

$$D_k^N = N - \sum_n^{N-1} D_k^n \text{ for all } k \tag{13.50}$$

By expanding the summations of Eq. (13.47) over the indexes n and n', and substituting Eq. (13.47) where possible, we get

$$\begin{aligned}
\mathbf{C}^* &= \sum_s^S \sum_n^N \hat{J}_s^n D_s^n - \sum_s^S \sum_{s'}^S \sum_n^N \left(\sum_{n'}^{N-1} \hat{\aleph}_{ss'}^{nn'} D_s^n D_{s'}^{n'} + \hat{\aleph}_{ss'}^{nN} D_s^n D_{s'}^N \right) \\
&= \sum_s^S \sum_n^N \hat{J}_s^n D_s^n - \sum_s^S \sum_{s'}^S \sum_n^N \left(\sum_{n'}^{N-1} \hat{\aleph}_{ss'}^{nn'} D_s^n D_{s'}^{n'} + \hat{\aleph}_{ss'}^{nN} D_s^n \left(N - \sum_k^{N-1} D_{s'}^k \right) \right) \\
&= \sum_s^S \sum_n^N \hat{J}_s^n D_s^n - \sum_s^S \sum_{s'}^S \sum_n^N \left(\sum_{n'}^{N-1} \hat{\aleph}_{ss'}^{nn'} D_s^n D_{s'}^{n'} + N \hat{\aleph}_{ss'}^{nN} D_s^n - \sum_k^{N-1} \hat{\aleph}_{ss'}^{nN} D_s^n D_{s'}^k \right) \\
&= \sum_s^S \sum_n^N \hat{J}_s^n D_s^n - \sum_s^S \sum_{s'}^S \left(\sum_n^N \sum_{n'}^{N-1} \left(\hat{\aleph}_{ss'}^{nn'} - \hat{\aleph}_{ss'}^{nN} \right) D_s^n D_{s'}^{n'} + N \sum_n^N \hat{\aleph}_{ss'}^{nN} D_s^n \right) \text{ for } k = n' \\
&= \sum_s^S \sum_n^{N-1} \hat{J}_s^n D_s^n + \sum_s^S \hat{J}_s^N D_s^N - \sum_s^S \sum_{s'}^S \left(\sum_n^{N-1} \sum_{n'}^{N-1} \left(\hat{\aleph}_{ss'}^{nn'} - \hat{\aleph}_{ss'}^{nN} \right) D_s^n D_{s'}^{n'} + N \sum_n^{N-1} \hat{\aleph}_{ss'}^{nN} D_s^n \right) \\
&\quad - \sum_s^S \sum_{s'}^S \left(\sum_{n'}^{N-1} \left(\hat{\aleph}_{ss'}^{Nn'} - \hat{\aleph}_{ss'}^{NN} \right) D_s^N D_{s'}^{n'} + N \hat{\aleph}_{ss'}^{NN} D_s^N \right)
\end{aligned} \tag{13.51}$$

Therefore,

$$\begin{aligned}
\mathbf{C}^* &= \sum_s^S \sum_n^{N-1} \hat{J}_s^n D_s^n + \sum_s^S \hat{J}_s^N \left(N - \sum_i^{N-1} D_s^i \right) - \sum_s^S \sum_{s'}^S \left(\sum_n^{N-1} \sum_{n'}^{N-1} \left(\hat{\aleph}_{ss'}^{nn'} - \hat{\aleph}_{ss'}^{nN} \right) D_s^n D_{s'}^{n'} \right. \\
&\quad \left. + N \sum_n^{N-1} \hat{\aleph}_{ss'}^{nN} D_s^n \right) - \sum_s^S \sum_{s'}^S \left(\sum_{n'}^{N-1} \left(\hat{\aleph}_{ss'}^{Nn'} - \hat{\aleph}_{ss'}^{NN} \right) \left(N - \sum_k^{N-1} D_s^k \right) D_{s'}^{n'} + N \hat{\aleph}_{ss'}^{NN} \left(N - \sum_j^{N-1} D_s^j \right) \right)
\end{aligned}$$

$$
\begin{aligned}
\mathbf{C}^* = {} & \sum_{s}^{S}\sum_{n}^{N-1}\hat{J}_s^n D_s^n + N\sum_{s}^{S}\hat{J}_s^N - \sum_{s}^{S}\sum_{i}^{N-1}\hat{J}_s^N D_s^i - \sum_{s}^{S}\sum_{s'}^{S}\sum_{n}^{N-1}\sum_{n'}^{N-1}\left(\hat{\aleph}_{ss'}^{nn'} - \hat{\aleph}_{ss'}^{nN}\right)D_s^n D_{s'}^{n'} \\
& - N\sum_{s}^{S}\sum_{s'}^{S}\sum_{n}^{N-1}\hat{\aleph}_{ss'}^{nN} D_s^n + \sum_{s}^{S}\sum_{s'}^{S}\sum_{k}^{N-1}\sum_{n'}^{N-1}\left(\hat{\aleph}_{ss'}^{Nn'} - \hat{\aleph}_{ss'}^{NN}\right)D_s^k D_{s'}^{n'} \\
& - N\sum_{s}^{S}\sum_{s'}^{S}\sum_{n'}^{N-1}\left(\hat{\aleph}_{ss'}^{Nn'} - \hat{\aleph}_{ss'}^{NN}\right)D_{s'}^{n'} + N\sum_{s}^{S}\sum_{s'}^{S}\sum_{j}^{N-1}\hat{\aleph}_{ss'}^{NN} D_s^j - \sum_{s}^{S}\sum_{s'}^{S}N^2\hat{\aleph}_{ss'}^{NN} \\
= {} & \sum_{s}^{S}\sum_{n}^{N-1}\hat{J}_s^n D_s^n + N\sum_{s}^{S}\hat{J}_s^N - \sum_{s}^{S}\sum_{n}^{N-1}\hat{J}_s^N D_s^n - \sum_{s}^{S}\sum_{s'}^{S}\sum_{n}^{N-1}\sum_{n'}^{N-1}\left(\hat{\aleph}_{ss'}^{nn'} - \hat{\aleph}_{ss'}^{nN} - \hat{\aleph}_{ss'}^{Nn'}\right. \\
& \left. + \hat{\aleph}_{ss'}^{NN}\right)D_s^n D_{s'}^{n'} - N\sum_{s}^{S}\sum_{s'}^{S}\sum_{n'}^{N-1}\left(\hat{\aleph}_{ss'}^{Nn'} - \hat{\aleph}_{ss'}^{NN}\right)D_{s'}^{n'} - N\sum_{s}^{S}\sum_{s'}^{S}\sum_{n}^{N-1}\left(\hat{\aleph}_{ss'}^{nN} - \hat{\aleph}_{ss'}^{NN}\right)D_s^n \\
& - N^2\sum_{s}^{S}\sum_{s'}^{S}\hat{\aleph}_{ss'}^{NN} \text{ for } i = j = k = n
\end{aligned}
$$

Using the symmetry of the matrix $\hat{\aleph}_{ss'}^{nn'} = \hat{\aleph}_{s's}^{n'n}$, the effective tensor becomes

$$
\begin{aligned}
\mathbf{C}^* = {} & \sum_{s}^{S}\sum_{n}^{N-1}\left(\hat{J}_s^n - \hat{J}_s^N\right)D_s^n + N\sum_{s}^{S}\hat{J}_s^N - \sum_{s}^{S}\sum_{s'}^{S}\sum_{n}^{N-1}\sum_{n'}^{N-1}\left(\hat{\aleph}_{ss'}^{nn'} - 2\hat{\aleph}_{ss'}^{nN} + \hat{\aleph}_{ss'}^{NN}\right)D_s^n D_{s'}^{n'} \\
& - 2N\sum_{s}^{S}\sum_{s'}^{S}\sum_{n}^{N-1}\left(\hat{\aleph}_{ss'}^{nN} - \hat{\aleph}_{ss'}^{NN}\right)D_s^n - N^2\sum_{s}^{S}\sum_{s'}^{S}\hat{\aleph}_{ss'}^{NN}
\end{aligned}
$$

Finally, the following equations are obtained:

$$
\hat{\mathbf{C}}^* = \sum_{s}^{S}\sum_{n}^{N-1}\hat{B}_s^n D_s^n - \sum_{s}^{S}\sum_{s'}^{S}\sum_{n}^{N-1}\sum_{n'}^{N-1}\hat{A}_{ss'}^{nn'} D_s^n D_{s'}^{n'} \tag{13.52}
$$

with

$$
\hat{A}_{ss'}^{nn'} = \hat{\aleph}_{ss'}^{nn'} - \hat{\aleph}_{ss'}^{nN} - \hat{\aleph}_{ss'}^{Nn'} + \hat{\aleph}_{ss'}^{NN} \quad \hat{B}_s^n = \hat{J}_s^n - \hat{J}_s^N - N\sum_{s'}^{S}\left(\hat{\aleph}_{ss'}^{nN} - 2\hat{\aleph}_{ss'}^{NN} + \hat{\aleph}_{ss'}^{Nn}\right)
$$

$$
\hat{\mathbf{C}}^* = \mathbf{C}^* - N\sum_{s}^{S}\hat{J}_s^N + N^2\sum_{s}^{S}\sum_{s'}^{S}\hat{\aleph}_{ss'}^{NN} \tag{13.53}
$$

Eq. (13.52) represents a simplified form of the problem as it operates with an inferior number of dimensions. It is always possible to reduce the local state space to $N - 1$ dimensions. Note that the benefits are minimal when the number of orientations under consideration (N) is large. However, a particular solution technique described later will make use of the order reduction.

13.3 HOMOGENIZATION IN DISCRETE FOURIER TRANSFORM SPACE

The previous section presented the homogenization approach in the primitive basis form. A discrete Fourier formulation has additional advantages; for example, convolutions required to determine the two-point statistics can be efficiently calculated using fast Fourier transforms (see Section 12.3). Furthermore, it may be possible to use fewer terms of the Fourier expansion to arrive at the same accuracy in the structure–property relations. To convert to Fourier space we first rewrite Eq. (13.21) in discrete form. If values of $\mathbf{x}$ are defined on a discrete grid and enumerated by s, then

$$\mathbf{C}^* = \frac{1}{S}\sum_s \mathbf{C}_s - \frac{1}{S}\sum_s \mathbf{C}'_s \sum_{s'} \mathbf{\Gamma}_{s-s'}\mathbf{C}'_{s'} + \ldots \tag{13.54}$$

where $\mathbf{C}_s$ indicates the appropriate stiffness tensor relating to the material at the position/cell enumerated by s in the sample. To introduce correlation functions, a variable, t, is defined, $t = s - s'$. The summation over s' in the second term is replaced by the value $s - t$; however, if periodic boundary conditions are employed, the sum becomes independent of s and one may write

$$\begin{aligned}\mathbf{C}^* &= \frac{1}{S}\sum_s \mathbf{C}_s - \frac{1}{S}\sum_s \mathbf{C}'_s \sum_t \mathbf{\Gamma}_t\mathbf{C}'_{s-t} + \ldots \\ &= \frac{1}{S}\sum_s \mathbf{C}_s - \frac{1}{S}\sum_t \mathbf{\Gamma}_t \sum_s \mathbf{C}'_s\mathbf{C}'_{s-t} + \ldots\end{aligned} \tag{13.55}$$

We wish to separate the material data from the structure data in our formulation for the localization tensor. Therefore, we utilize the microstructure function $M(\mathbf{x}, h)$ (see Sections 4.4 and 5.4). We wish to discretize the function $M(\mathbf{x}, h)$ in both the real space, $\Re^n$, and the local state space, H. As before, the real space is discretized into a regular grid, with vertices enumerated by $s = 1{:}S$, and the local state space is enumerated by $n = 1{:}N$. We will write $m_s^n = D_s^n/N = M(\mathbf{x}_s, h_n)$.

Then we may write $\mathbf{C}'_{s-t} = \sum_n \mathbf{C}'^n m_{s-t}^n$. And

$$\mathbf{C}^* = \frac{1}{S}\sum_s \sum_n \mathbf{C}^n m_s^n - \frac{1}{S}\sum_s \sum_n \mathbf{C}'^n m_s^n \sum_t \mathbf{\Gamma}_t \sum_{n'} \mathbf{C}'n' m_{s-t}^{n'} + \ldots \tag{13.56}$$

$$\mathbf{C}^* = \frac{1}{S}\sum_s \sum_n \mathbf{C}^n m_s^n - \frac{1}{S}\sum_n \mathbf{C}'^n \sum_{n'} \mathbf{C}'n' \sum_t \mathbf{\Gamma}_t \sum_s m_s^n m_{s-t}^{n'} + \ldots \tag{13.57}$$

The final summation in the second term is the definition of the two-point correlation function (see Section 12.3). In terms of notation we write the n-point correlation functions as

$$f^n = \frac{1}{S}\sum_s m_s^n, \quad f_t^{nn'} = \frac{1}{S}\sum_s m_s^n m_{s-t}^{n'}, \text{ and so on.} \tag{13.58}$$

These may be evaluated using FFTs, as described in Section 12.3 (Fullwood et al., 2008a; Fullwood et al., 2008b):

$$f_t^{nn'} = \frac{1}{S}\Im^{-1}\left(\left(\Im\left(m_s^n\right)\right)^* \Im\left(m_s^{n'}\right)\right) \tag{13.59}$$

$$f^n = f_0^{nn} \tag{13.60}$$

Hence, we may write Eq. (13.57) as

$$\mathbf{C}^* = \sum_n \mathbf{C}^n f^n - \sum_n \mathbf{C}'^n \sum_{n'} \mathbf{C}'^{n'} \sum_t \mathbf{\Gamma}_t f_t^{nn'} + \ldots \tag{13.61}$$

Taking the Fourier transforms of both sides and using Plancherel's theorem,

$$\mathbf{C}^* = \frac{1}{N}\sum_j (\widetilde{\mathbf{C}}^j)^* F^j - \frac{1}{N^2 S}\sum_j (\widetilde{\mathbf{C}}'_j)^* \sum_{j'} (\widetilde{\mathbf{C}}'_{j'})^* \sum_k (\widetilde{\mathbf{\Gamma}}_k)^* F_k^{jj'} + \ldots \tag{13.62}$$

where $\widetilde{\mathbf{C}}^j$ is the FFT of $\mathbf{C}'^n$, $\widetilde{\mathbf{\Gamma}}_k$ is the FFT of $\mathbf{\Gamma}_t$, $F_k^{jj'} = \Im_t(\Im_n(f_t^{nn'}))$, and so forth; $\Im_n$ indicates the FFT with respect to variable n, and the asterisk indicates complex conjugation for terms on the right side of the equation. We may combine terms on the right side of the equation to give (using the usual summation convention on repeated indices of a given term)

$$\begin{aligned} \mathbf{C}^* &= \frac{1}{N}(\widetilde{\mathbf{C}}^j)^* F^j - Z_k^{jj'} F_k^{jj'} + \ldots \\ Z_k^{jj'} &= \frac{1}{N^2 S}(\widetilde{\mathbf{C}}'_j)^* (\widetilde{\mathbf{C}}'_{j'})^* (\widetilde{\mathbf{\Gamma}}_k)^* \end{aligned} \tag{13.63}$$

Thus, we have effectively split the structural information of the material contained in the microstructure function from the property information contained in the local stiffness tensor. This approach dramatically reduces the summation in Eq. (13.63), leading to even higher efficiency. Since these terms are only calculated once for a given reference tensor, the significant terms may be stored in a concise database for economical material analysis and design.

One consequence of the use of FFTs to calculate the convolution in Eq. (13.56) is the introduction of periodic boundary conditions. While different boundary conditions may be utilized by padding in the various dimensions (Fullwood et al., 2008a), periodic boundaries were assumed for this chapter. Since the integral is centered at the origin (about **x**), the vectors represented by t in Eq. (13.56) have a maximum length in each dimension of half the size of the sample, to represent a cube that is centered at the origin, and the same side dimensions as the sample. Care must be taken in setting up the relevant tensors to reflect this. Note also that the calculations are expedited by writing the stiffness tensors in 9×9 matrix form.

13.4 EXTENSION OF THE HOMOGENIZATION METHOD TO LOCALIZATION PROBLEMS

The polarization tensor introduced in Eq. (13.2) is sometimes also termed the *localization tensor*, in that it provides a link between macro-strain and local strain. The variations in local stress and strain are key to material response in the realm of fracture and fatigue, for example. Hence, we provide a brief summary of how the homogenization method described before may be modified to deal with localization issues. We wish to calculate Eq. (13.2) using a spectral approach. Thus, we rewrite Eq. (13.14) (see also Eq. (13.17) for details of Green's function term) in an analogous way to Eq. (13.54):

$$\mathbf{a}_s = -\sum_t \mathbf{\Gamma}_t \mathbf{C}'_{s-t} \tag{13.64}$$

Then, as before, we may write $\mathbf{C}'_{s-t} = \sum_n \mathbf{C}'^n m^n_{s-t}$. And

$$\mathbf{a}_s = \sum_t \mathbf{\Gamma}_t \sum_n \mathbf{C}'^n m^n_{s-t} \tag{13.65}$$

Taking the Fourier transforms of both sides and using Plancherel's theorem and the convolution theorem,

$$\mathbf{A}_k = \frac{1}{N} \sum_{j=1:N} \widetilde{\mathbf{\Gamma}}_k \left(\widetilde{\mathbf{C}}^j\right)^* \widetilde{M}^j_k \tag{13.66}$$

where $\widetilde{\mathbf{C}}^j$ is the FFT of $\mathbf{C}'^n$, $\widetilde{\mathbf{\Gamma}}_k$ is the FFT of $\mathbf{\Gamma}_t$, and $\widetilde{M}^j_k = \Im_s(\Im_n(m^n_s))$; $\Im_n$ indicates the FFT with respect to variable n, and the asterisk indicates complex conjugation. We may combine the first two terms on the right side of the equation to give

$$\mathbf{A}_k = \frac{1}{N} \sum_{j=1:N} Z^j_k \widetilde{M}^j_k \tag{13.67}$$

The inverse Fourier transform of $\mathbf{A}$ can then be taken and inserted into Eq. (13.2) to determine the local strain (and thereby the stress) from the macro-strain condition.

To test the validity of the formulation previously given, various material structures were evaluated using finite-element analysis to determine the local strains based on an applied global strain (Fullwood et al., 2009). The results were compared with calculations using the method outlined before. A hypothetical material was constructed of two isotropic phases of varying contrast in their stiffness coefficients. For these tests the reference tensor was assumed to be given by the mean of that for the two constituents, independent of the overall volume fractions of these constituents.

Example structures are shown in Figure 13.4 for 75% of the stiff phase. The structures are randomly dispersed single cells, unidirectional fibers, and laminates. The Lamé constants for the two materials are $\mu = 35$, $\lambda = 75$ and $\mu = 55$, $\lambda = 100$, respectively; a global strain of 1% was applied in the

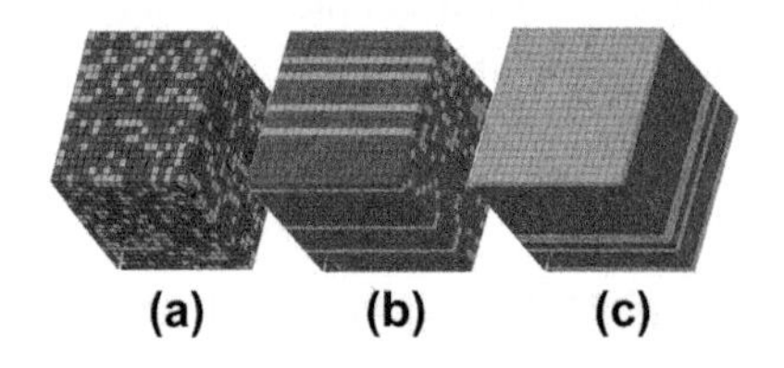

FIGURE 13.4

Schematic of two-phase cube with darker areas representing stiffer phase (75% volume fraction) for (a) random, (b) fiber, and (c) laminate structures (see Fullwood et al., 2009).

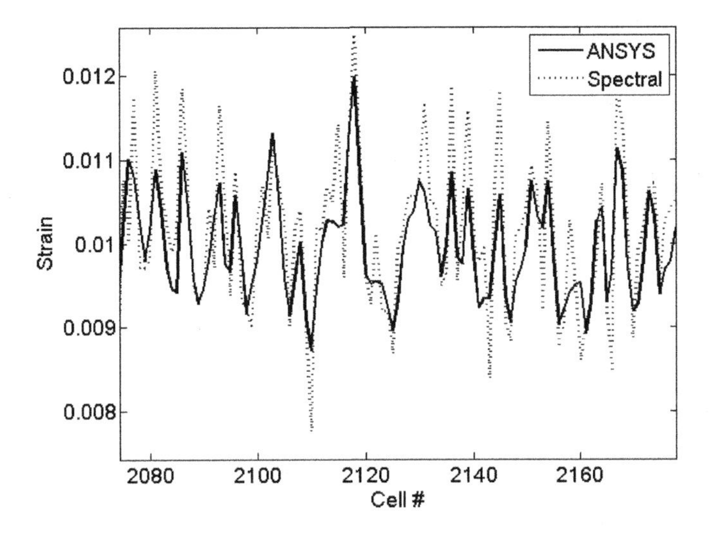

FIGURE 13.5

Typical graph of strain versus cell number as calculated by the spectral method (*dotted*) versus the finite-element method (*solid*). A macroscopic strain of 1% is applied to the cube of Figure 13.4a in the *z*-direction (see Fullwood et al., 2009).

3-direction. The average relative error between the FEM, results and the analytical framework is 4.2% (for recent work on homogenization using FEM, see Xu et al. (2008)). A plot of strain versus cell number for the analytical and FE results is shown in Figure 13.5. The analytical results correlate well with the FEM calculations.

In this example the reference tensor was chosen to be the mean of the two isotropic constituents, and not the average stiffness tensor. This choice highlights the importance of the first term on the right side of Eq. (13.17). Without this term the average strain as calculated by the spectral method was 4% different from the applied (macro) strain; with this term the difference was less than 0.2%. The differences would be more apparent with phases of higher contrast. The ability to choose a reference tensor that is not the mean of the constituent tensors is critical to efficient material design. If the calculations use the mean tensor, then Green's function terms in the integrals must be reevaluated for each choice of material structure (since the mean would change). A better choice is to choose a reference tensor that lies between the stiffness tensors of the constituent phases (Kalidindi et al., 2006). See Norris (2006) for details on how to choose the closest isotropic reference tensor to a given anisotropic tensor (calculated from that of the phases). Then only the values of M_k^j change in Eq. (13.67), leading to rapid optimization/analysis of material design.

13.5 A FORMULATION FOR STRONG-CONTRAST MATERIALS

Relationships between microstructure (local) properties and effective (global) properties of a polycrystal are often achieved through perturbation expansions. While these expansions are formally exact, the convergence depends on the "contrast" exhibited between the phases and hence the polarization of the local state from the chosen reference state. Furthermore, the integral equations present in the terms of the expansion are generally conditionally convergent and therefore depend on the shape of the sample region. Since an effective property is not dependent on the shape of the sample, absolutely convergent integrals should be used.

In 1955 Brown (Brown, 1955) suggested an expansion for effective conductivity that resulted in absolutely convergent integrals. Torquato (2002) subsequently expanded on the technique and applied it to elasticity. The resultant expansions have been termed "strong contrast" due to their ability to achieve good convergence for materials with a high degree of contrast between the properties of their phases.

These previous results were established for a material with two isotropic phases. Limiting the number of phases to two results in a more elegant form of the solution. Furthermore, if the effective property is assumed to be isotropic thc solution proccss is significantly simplified (Torquato, 2002; note that Torquato also briefly considers globally anisotropic materials). Here, we extend the previous elasticity results to multiple phases and consider both isotropic and anisotropic phases and global solutions. The form of the results is then adapted to fit with recent microstructure-sensitive design (MSD) techniques that are aimed at facilitating inversion of the problem (i.e., design of the microstructure to suit a chosen effective property).

In the next section we repeat the results of Brown and Torquato for two phases. This both introduces the required notation and allows a transparent extension of the theory to multiple phases. We subsequently validate the technique for two isotropic and anisotropic phases and then present results for higher numbers of (anisotropic) phases.

13.5.1 Strong-contrast expansion for two phases

Take a material with two isotropic phases, A and B, with stiffness tensors $\boldsymbol{C}_A$ and $\boldsymbol{C}_B$. Introduce the polarization field:

$$p_{ij}(\mathbf{x}) = \left[C(\mathbf{x}) - C^R\right]_{ijkl}\varepsilon_{kl}(\mathbf{x}) \tag{13.68}$$

So

$$\boldsymbol{\sigma}(\mathbf{x}) = \mathbf{C}(\mathbf{x})\boldsymbol{\varepsilon}(\mathbf{x}) = \mathbf{C}^R\boldsymbol{\varepsilon}(\mathbf{x}) + \mathbf{p}(\mathbf{x}) \tag{13.69}$$

Then, using $\varepsilon_{ij} = (u_{i,j} + u_{j,i})/2$, the conservation principle $\nabla\cdot\boldsymbol{\sigma} = \sigma_{ij,j} = 0$ implies

$$C^R_{ijkl}u_{k,lj} + p_{ij,j} = 0 \tag{13.70}$$

$$p_{ij,j}(\mathbf{x}) = \left\{\left[C(\mathbf{x}) - C^R\right]_{ijkl}\varepsilon_{kl}(\mathbf{x})\right\}_{,j} = C_{ijkl,j}(\mathbf{x})\varepsilon_{kl}(\mathbf{x}) + C_{ijkl}(\mathbf{x})\varepsilon_{kl,j}(\mathbf{x}) - {C^R}_{ijkl}\varepsilon_{kl,j}(\mathbf{x}) \tag{13.71}$$

We require boundary conditions for the differential equation. Thus, we consider the polarized displacement, $u'(\mathbf{x}) = u(\mathbf{x}) - u_\infty$, such that $u'(|\mathbf{x}| \to \infty) = 0$.

Now consider Green's function that is the solution of

$$C^R_{ijkl}G_{im,lj}(\mathbf{x},\mathbf{x}') + \delta_{km}\delta(\mathbf{x} - \mathbf{x}') = 0 \tag{13.72}$$

with the same boundary conditions, $\mathbf{G}(|\mathbf{x}| \to \infty, \mathbf{x}') = 0$ and using the usual delta function notation:

$$\delta_{km} = \begin{cases} 1 & \text{if } k = m \\ 0 & \text{otherwise} \end{cases} \tag{13.73}$$

$$\int_{\Omega} \delta(\mathbf{x} - \mathbf{x}')dx' = \begin{cases} 1 & \text{if } \mathbf{x} \in \Omega \\ 0 & \text{otherwise} \end{cases} \tag{13.74}$$

Multiplying both sides of (13.70) by $\mathbf{G}$ and using integration by parts (noting that the boundary conditions are 0 at infinity),

$$u_k(\mathbf{x}) = u_{k\,\infty} + \int_V G_{ik}(\mathbf{x}, \mathbf{x}')p_{ij,j}(\mathbf{x}')d\mathbf{x}' \tag{13.75}$$

One may again use integration by parts on the right side as follows:

$$u_k(\mathbf{x}) = u_{k\,\infty} + \int_V G_{ik,j}(\mathbf{x}, \mathbf{x}')p_{ij}(\mathbf{x}')d\mathbf{x}' \tag{13.76}$$

Since one may not differentiate under the integral sign to arrive at the strain (due to the singularity at the origin), a sphere around the origin is excluded and, using integration by parts and the divergence theorem (see Appendix A in Torquato, 1997),

$$\varepsilon_{kl}(\mathbf{x}) = \varepsilon_{kl\,\infty} - E_{ikjl}p_{ij}(\mathbf{x}) + \int_V G_{[ik,jl]}(\mathbf{x}, \mathbf{x}')p_{ij}(\mathbf{x}')d\mathbf{x}' \tag{13.77}$$

where E is the contribution from the singularity, and $G_{[ik,jl]}$ ($= \Phi_{ikjl}$, say) is the symmetrized double gradient of $\mathbf{G}$. Note that all of these terms depend on the choice of reference tensor.

For an isotropic reference tensor defined by parameters μ and λ, it may be shown that Φ is given by (Torquato, 2002, p. 533; Kröner, 1986, p. 260):

$$\Phi_{ijkl} = \frac{1}{8\pi\mu(\lambda + 2\mu)r^3} \begin{bmatrix} (\lambda + \mu)\left(\delta_{ij}\delta_{kl} - 3\delta_{ij}n_k n_l - 3\delta_{kl}n_i n_j + 15 n_i n_j n_k n_l\right) \\ -\mu\left(\delta_{ik}\delta_{jl} + \delta_{il}\delta_{jk}\right) - 3\lambda\left(\delta_{ik}n_j n_l + \delta_{il}n_j n_k + \delta_{jk}n_i n_l + \delta_{jl}n_i n_k\right)/2 \end{bmatrix} \tag{13.78}$$

where $\mathbf{r} = (\mathbf{x} - \mathbf{x}')$ and $\mathbf{n} = \mathbf{r}/|\mathbf{r}|$. Note that, following Torquato, the definition of Φ here is the negative of the more common definition, so the integral is added (rather than subtracted) in Eq. (13.77). Also, $\mathbf{E}$ is given by

$$E_{ijkl} = \frac{1}{15\mu}\left\{\frac{\lambda + \mu}{\lambda + 2\mu}\delta_{ij}\delta_{kl} - \frac{3\lambda + 8\mu}{\lambda + 2\mu}I_{ijkl}\right\} \tag{13.79}$$

where $\boldsymbol{I}$ is the fourth-order symmetrized tensor identity:

$$I_{ijkl} = \frac{1}{2}\left(\delta_{ik}\delta_{jl} + \delta_{il}\delta_{jk}\right) \tag{13.80}$$

Define the cavity strain field, **f**, after Torquato, which is the usual strain plus a contribution from the "cavity," or infinitesimal sphere, about point **x**, in analogy to the Lorentz strain field (as used by Brown, 1955):

$$f_{ij}(\mathbf{x}) = \left\{ I_{ijkl} + E_{ijmn}\left[C(\mathbf{x}) - C^R\right]_{mnkl} \right\} \varepsilon_{kl} \tag{13.81}$$

Using the definition of **p**, one may rearrange (13.77):

$$f_{kl}(\mathbf{x}) = \varepsilon_{kl\,\infty} + \int_V \Phi_{ikjl}(\mathbf{x}, \mathbf{x}')p_{ij}(\mathbf{x}')d\mathbf{x}' \tag{13.82}$$

Then, from (13.68.),

$$p_{ij}(\mathbf{x}) = L_{ijkl}f_{kl}(\mathbf{x}) \tag{13.83}$$

where

$$L_{ijkl}(\mathbf{x}) = \left[C(\mathbf{x}) - C^R\right]_{ijmn} \left\{ I + E\left[C(\mathbf{x}) - C^R\right] \right\}^{-1}{}_{mnkl}$$

Note that if the reference material is one of the phases, say B, and the phases are isotropic, the tensor **L** can be written as a constant tensor times the indicator function for the nonreference phase:

$$p_{ij}(\mathbf{x}) = L^B_{ijkl}\chi^A(\mathbf{x})f_{kl}(\mathbf{x}) \tag{13.84}$$

The indicator function, χ^A, is used to "pick out" the local state at each point of the material:

$$\chi^A(\mathbf{x}) = \begin{cases} 1 \text{ if the material is phase } A \text{ at point } \mathbf{x} \\ 0 \text{ otherwise} \end{cases} \tag{13.85}$$

We also define ensemble averaging, denoted by $\langle\ \ \rangle$, as the average over a large number of representative samples (of identical dimensions) of the material. We assume a homogenous material and apply the ergodic hypothesis, which requires that the ensemble average of a property be equal to the volume average over a representative volume.

Then the volume fraction of phase A may be written in the following three ways:

$$\langle \chi^A(\mathbf{x}) \rangle = P(A) = f_1(A)dA \tag{13.86}$$

where $P(A)$ is the probability of finding phase A at a randomly selected position in the material, and $f_1(A)$ is the distribution function for the phases; the volume of phase space is 1, and therefore $dA = 1/2$ for two phases.

Choose two arbitrary points, $\mathbf{x}_1$, $\mathbf{x}_2$, in the representative volume, and define the vector $\mathbf{r}_{12} = \mathbf{x}_2 - \mathbf{x}_1$. Let $P(A, A|\mathbf{r}_{12})$ be the probability that for randomly placed $\mathbf{r}_{12}$ in the representative volume element the material is phase A at both the head and the tail. The probability distribution function for this probability is the usual two-point correlation function, $f_2(A, A|\mathbf{r}_{12})$. Hence,

$$\langle \chi^A(\mathbf{x}_1)\chi^A(\mathbf{x}_2) \rangle = P(A, A|\mathbf{r}_{12}) = f_2(A, A|\mathbf{r}_{12})dA^2 \tag{13.87}$$

since the material is homogeneous. This is similar for higher distribution functions.

Define the effective tensor, $\mathbf{L}^e$, by

$$\langle p_{ij}(\mathbf{x}) \rangle = L^e_{ijkl}\langle f_{kl}(\mathbf{x}) \rangle \tag{13.88}$$

Then

$$L^{e}{}_{ijkl} = \left[C^{e} - C^{R}\right]_{ijmn}\left\{I + E\left[C^{e} - C^{R}\right]\right\}^{-1}{}_{mnkl} \tag{13.89}$$

Since we require $\mathbf{C}^{e}$, we require a solution for $\mathbf{L}^{e}$. Rewrite (13.82) as

$$\mathbf{f} = \boldsymbol{\varepsilon}_{\infty} + \boldsymbol{\Phi}\mathbf{p} \tag{13.90}$$

Substituting into (13.83):

$$\mathbf{p} = \mathbf{L}\boldsymbol{\varepsilon}_{\infty} + \mathbf{L}\boldsymbol{\Phi}\mathbf{p} \tag{13.91}$$

Now reinsert (13.91) into itself repeatedly to obtain

$$\mathbf{p} = \mathbf{L}\boldsymbol{\varepsilon}_{\infty} + \mathbf{L}\boldsymbol{\Phi}\mathbf{L}\boldsymbol{\varepsilon}_{\infty} + \mathbf{L}\boldsymbol{\Phi}\mathbf{L}\boldsymbol{\Phi}\mathbf{L}\boldsymbol{\varepsilon}_{\infty} + \ldots \tag{13.92}$$

Or, more explicitly:

$$\begin{aligned}\mathbf{p}(\mathbf{x}_1) = \mathbf{L}^{B}\chi^{A}(\mathbf{x}_1)\boldsymbol{\varepsilon}_{\infty} + \int \mathbf{L}^{B}\chi^{A}(\mathbf{x}_1)\boldsymbol{\Phi}(\mathbf{x}_1,\mathbf{x}_2)\mathbf{L}^{B}\chi^{A}(\mathbf{x}_2)\boldsymbol{\varepsilon}_{\infty}d\mathbf{x}_2 \ldots \\ + \int\int \mathbf{L}^{B}\chi^{A}(\mathbf{x}_1)\boldsymbol{\Phi}(\mathbf{x}_1,\mathbf{x}_2)\mathbf{L}^{B}\chi^{A}(\mathbf{x}_2)\boldsymbol{\Phi}(\mathbf{x}_2,\mathbf{x}_3)\mathbf{L}^{B}\chi^{A}(\mathbf{x}_3)\boldsymbol{\varepsilon}_{\infty}d\mathbf{x}_2 d\mathbf{x}_3 + \ldots\end{aligned} \tag{13.93}$$

Take the ensemble average:

$$\begin{aligned}\langle\mathbf{p}\rangle = \langle\chi^{A}(\mathbf{x}_1)\rangle\mathbf{L}^{B}\boldsymbol{\varepsilon}_{\infty} + \int \langle\chi^{A}(\mathbf{x}_1)\chi^{A}(\mathbf{x}_2)\rangle\mathbf{L}^{B}\boldsymbol{\Phi}(\mathbf{x}_1,\mathbf{x}_2)\mathbf{L}^{B}\boldsymbol{\varepsilon}_{\infty}d\mathbf{x}_2 \ldots \\ + \int\int \langle\chi^{A}(\mathbf{x}_1)\chi^{A}(\mathbf{x}_2)\chi^{A}(\mathbf{x}_3)\rangle\mathbf{L}^{B}\boldsymbol{\Phi}(\mathbf{x}_1,\mathbf{x}_2)\mathbf{L}^{B}\boldsymbol{\Phi}(\mathbf{x}_2,\mathbf{x}_3)\mathbf{L}^{B}\boldsymbol{\varepsilon}_{\infty}d\mathbf{x}_2 d\mathbf{x}_3 + \ldots\end{aligned} \tag{13.94}$$

Now use relations (13.86) and (13.87), and the fact that the isotropic Green's function depends only on the vector $\mathbf{r}_{12}$ between points $\mathbf{x}_1$ and $\mathbf{x}_2$:

$$\begin{aligned}\langle\mathbf{p}\rangle = P(A)\mathbf{L}^{B}\boldsymbol{\varepsilon}_{\infty} + \int P(A,A|\mathbf{r}_{12})\mathbf{L}^{B}\boldsymbol{\Phi}(\mathbf{r}_{12})\mathbf{L}^{B}\boldsymbol{\varepsilon}_{\infty}d\mathbf{r}_{12} \ldots \\ + \int\int P(A,A,A|\mathbf{r}_{12},\mathbf{r}_{23})\mathbf{L}^{B}\boldsymbol{\Phi}(\mathbf{r}_{12})\mathbf{L}^{B}\boldsymbol{\Phi}(\mathbf{r}_{23})\mathbf{L}^{B}\boldsymbol{\varepsilon}_{\infty}d\mathbf{r}_{12}d\mathbf{r}_{23} + \ldots\end{aligned} \tag{13.95}$$

Using successive substitutions (i.e., solve for the first case of $\boldsymbol{\varepsilon}_{\infty}$ on the right side, then substitute this back in for the other remaining cases; repeat this process indefinitely), one may solve this equation for $\boldsymbol{\varepsilon}_{\infty}$:

$$\begin{aligned}\boldsymbol{\varepsilon}_{\infty} = \frac{(\mathbf{L}^{B})^{-1}}{P(A)}\langle\mathbf{p}\rangle - \int \frac{P(A,A|\mathbf{r}_{12})}{P(A)^{2}}(\mathbf{L}^{B})^{-1}\mathbf{L}^{B}\boldsymbol{\Phi}(\mathbf{r}_{12})\mathbf{L}^{B}(\mathbf{L}^{B})^{-1}\langle\mathbf{p}\rangle d\mathbf{r}_{12} \ldots \\ -\int\int\left(\begin{aligned}&\frac{P(A,A,A|\mathbf{r}_{12},\mathbf{r}_{23})}{P(A)^{2}}(\mathbf{L}^{B})^{-1}\mathbf{L}^{B}\boldsymbol{\Phi}(\mathbf{r}_{12})\mathbf{L}^{B}\boldsymbol{\Phi}(\mathbf{r}_{23})\mathbf{L}^{B}(\mathbf{L}^{B})^{-1}\langle\mathbf{p}\rangle \ldots \\ &-\frac{P(A,A|\mathbf{r}_{12})P(A,A|\mathbf{r}_{23})}{P(A)^{3}}(\mathbf{L}^{B})^{-1}\boldsymbol{\Phi}(\mathbf{r}_{12})\mathbf{L}^{B}(\mathbf{L}^{B})^{-1}\mathbf{L}^{B}\boldsymbol{\Phi}(\mathbf{r}_{23})\mathbf{L}^{B}(\mathbf{L}^{B})^{-1}\langle\mathbf{p}\rangle\end{aligned}\right)d\mathbf{r}_{12}d\mathbf{r}_{23} + \ldots\end{aligned} \tag{13.96}$$

Taking the ensemble average of (13.82) over S samples,

$$\frac{1}{S}\sum^{S}\mathbf{f}(\mathbf{x}_1) = \boldsymbol{\varepsilon}_\infty + \frac{1}{S}\sum^{S}\int \boldsymbol{\Phi}(\mathbf{x}_1,\mathbf{x}_2)\mathbf{p}(\mathbf{x}_2)d\mathbf{x}_2 \tag{13.97}$$

By the ergodic assumption, the left side of Equation 13.97 is independent of $\mathbf{x}_1$. Also, $\boldsymbol{\Phi}(\mathbf{x}_1,\mathbf{x}_2)$ is clearly independent of the particular sample, so

$$\langle\mathbf{f}\rangle = \boldsymbol{\varepsilon}_\infty + \int \boldsymbol{\Phi}(\mathbf{r}_{12})\left\{\frac{1}{S}\sum^{S}\mathbf{p}(\mathbf{x}_2)\right\}d\mathbf{x}_2 \tag{13.98}$$

That is,

$$\langle\mathbf{f}\rangle = \boldsymbol{\varepsilon}_\infty + \int \boldsymbol{\Phi}(\mathbf{r}_{12})\langle\mathbf{p}\rangle d\mathbf{r}_{12} \tag{13.99}$$

Substituting into Eq. (13.98):

$$\begin{aligned}\langle\mathbf{f}\rangle = \frac{(\mathbf{L}^B)^{-1}}{P(A)}\langle\mathbf{p}\rangle + \int \boldsymbol{\Phi}(\mathbf{r}_{12})\langle\mathbf{p}\rangle d\mathbf{x}_2 - \int \frac{P(A,A|\mathbf{r}_{12})}{P(A)^2}\boldsymbol{\Phi}(\mathbf{r}_{12})\langle\mathbf{p}\rangle d\mathbf{r}_{12}\ldots \\ -\int\int\begin{pmatrix}\dfrac{P(A,A,A|\mathbf{r}_{12},\mathbf{r}_{23})}{P(A)^2}\boldsymbol{\Phi}(\mathbf{r}_{12})\mathbf{L}^B\boldsymbol{\Phi}(\mathbf{r}_{23})\langle\mathbf{p}\rangle\ldots \\ -\dfrac{P(A,A|\mathbf{r}_{12})P(A,A|\mathbf{r}_{23})}{P(A)^3}\boldsymbol{\Phi}(\mathbf{r}_{12})\mathbf{L}^B\boldsymbol{\Phi}(\mathbf{r}_{23})\langle\mathbf{p}\rangle\end{pmatrix}d\mathbf{r}_{12}d\mathbf{r}_{23} + \ldots\end{aligned} \tag{13.100}$$

Note that the second term is similar to the adjustment term for a nonspherical domain as used in some weak-contrast calculations (Kalidindi et al., 2006).

The shape independence of the integrals in Eq. (13.100) arises due to cancellation of consecutive terms at infinity, which, in turn, is due to the following identities:

$$\begin{aligned}&\lim_{|x_n-x_{n-1}|\to 0} P(A,A,..,A|\mathbf{r}_{12},\ldots,\mathbf{r}_{1n-1},\mathbf{r}_{1n}) = P(A,A,..,A|\mathbf{r}_{12},\ldots,\mathbf{r}_{1n-1}) \\ &\lim_{|x_n-x_{n-1}|\to \infty} P(A,A,..,A|\mathbf{r}_{12},\ldots,\mathbf{r}_{1n-1},\mathbf{r}_{1n}) = P(A,A,..,A|\mathbf{r}_{12},\ldots,\mathbf{r}_{1n-1})P(A)\end{aligned} \tag{13.101}$$

For example, take the terms in parenthesis of the fourth term of Eq. (13.100), and employ Eq. (13.101) for large $\mathbf{r} = |\mathbf{x}_2 - \mathbf{x}_1|$:

$$\begin{aligned}&\lim_{r\to\infty}\frac{P(A,A|\mathbf{r}_{12})}{P(A)^2} - 1 = 0 \\ &\lim_{r\to\infty}\left(\frac{P(A,A,A|\mathbf{r}_{12},\mathbf{r}_{23})}{P(A)^2} - \frac{P(A,A|\mathbf{r}_{12})P(A,A|\mathbf{r}_{23})}{P(A)^3}\right) = 0\end{aligned} \tag{13.102}$$

Note also that the fourth- and higher-order terms also vanish for small $|\mathbf{r}_{12}|$, making exclusion of the integral about $|\mathbf{r}_{12}| = 0$ unnecessary.

Using Eq. (13.88), we may now find $\mathbf{L}^e$:

$$(\mathbf{L}^e)^{-1} = \frac{(\mathbf{L}^B)^{-1}}{P(A)} + \int \mathbf{\Phi}(\mathbf{r}_{12})d\mathbf{r}_{12} - \int \frac{P(A,A|\mathbf{r}_{12})}{P(A)^2}\mathbf{\Phi}(\mathbf{r}_{12})d\mathbf{r}_{12} \ldots$$
$$- \iint \left(\begin{matrix} \dfrac{P(A,A,A|\mathbf{r}_{12},\mathbf{r}_{23})}{P(A)^2}\mathbf{\Phi}(\mathbf{r}_{12})\mathbf{L}^B\mathbf{\Phi}(\mathbf{r}_{23}) \ldots \\ \\ -\dfrac{P(A,A|\mathbf{r}_{12})P(A,A|\mathbf{r}_{23})}{P(A)^3}\mathbf{\Phi}(\mathbf{r}_{12})\mathbf{L}^B\mathbf{\Phi}(\mathbf{r}_{23}) \end{matrix} \right) d\mathbf{r}_{12}d\mathbf{r}_{23} + \ldots \tag{13.103}$$

Note that it may be convenient to multiply Eq. (13.103) on the left by $\mathbf{L}^B$ to remove the inverse tensor in the first term and to obtain a repeated $\mathbf{L}^B\mathbf{\Phi}$ term on the right. We then find $\mathbf{C}^e$ from Eq. (13.89):

$$(\mathbf{L}^e)^{-1} = \{\mathbf{I} + \mathbf{E}[\mathbf{C}^e - \mathbf{C}^R]\}[\mathbf{C}^e - \mathbf{C}^R]^{-1} \tag{13.104}$$

Hence,

$$\mathbf{C}^e = [(\mathbf{L}^e)^{-1} - \mathbf{E}]^{-1} + \mathbf{C}^R \tag{13.105}$$

13.5.2 Strong contrast for multiple phases

Let us assume that we have N phases, enumerated with the index n. We will use the summation convection such that repeated indices in the same term indicate summation. Since the indicator functions are used to pick out the correct phase at each point of the material, Eq. (13.84) now takes the following form:

$$p_{ij}(x) = \{L^n_{ijkl}\chi^n(\mathbf{x})\}f_{kl}(\mathbf{x}) \tag{13.106}$$

Note that if phase n is the reference phase, then the corresponding $\mathbf{L}^n$ is 0.

Then Eq. (13.93) becomes

$$\mathbf{p}(x_1) = \mathbf{L}^n\chi^n(\mathbf{x}_1)\boldsymbol{\varepsilon}_\infty + \int \mathbf{L}^{n'}\chi^{n'}(\mathbf{x}_1)\mathbf{\Phi}(\mathbf{x}_1,\mathbf{x}_2)\mathbf{L}^{n''}\chi^{n''}(\mathbf{x}_2)\boldsymbol{\varepsilon}_\infty d\mathbf{x}_2 \ldots$$
$$+ \iint \mathbf{L}^m\chi^m(\mathbf{x}_1)\mathbf{\Phi}(\mathbf{x}_1,\mathbf{x}_2)\mathbf{L}^{m'}\chi^{m'}(\mathbf{x}_2)\mathbf{\Phi}(\mathbf{x}_2,\mathbf{x}_3)\mathbf{L}^{m''}\chi^{m''}(\mathbf{x}_3)\boldsymbol{\varepsilon}_\infty d\mathbf{x}_2 d\mathbf{x}_3 + \ldots \tag{13.107}$$

which leads to a new Eq. (13.94):

$$\langle\mathbf{p}\rangle = \mathbf{L}^n\langle\chi^n(\mathbf{x}_1)\rangle\boldsymbol{\varepsilon}_\infty + \int \langle\chi^{n'}(\mathbf{x}_1)\chi^{n''}(\mathbf{x}_2)\rangle\mathbf{L}^{n'}\mathbf{\Phi}(\mathbf{x}_1,\mathbf{x}_2)\mathbf{L}^{n''}\boldsymbol{\varepsilon}_\infty d\mathbf{x}_2 \ldots$$
$$+ \iint \langle\chi^m(\mathbf{x}_1)\chi^m(\mathbf{x}_2)\chi^{m'}(\mathbf{x}_3)\rangle\mathbf{L}^{m'}\mathbf{\Phi}(\mathbf{x}_1,\mathbf{x}_2)\mathbf{L}^{m''}\mathbf{\Phi}(\mathbf{x}_2,\mathbf{x}_3)\mathbf{L}^{m''}\boldsymbol{\varepsilon}_\infty d\mathbf{x}_2 d\mathbf{x}_3 + \ldots \tag{13.108}$$

To make the notation a little more compact and gain the added benefits of the summation convention, let us write the probability that a randomly placed vector $\mathbf{r}_{12} = (\mathbf{x}_2 - \mathbf{x}_1)$ starts at a location with phase n and ends at a location with phase n' as $P^{nn'}(\mathbf{r}_{12})$. Then

$$\begin{aligned}\langle \mathbf{p} \rangle = \overline{\mathbf{L}}\boldsymbol{\varepsilon}_\infty + \int P^{nn'}(\mathbf{r}_{12})\mathbf{L}^n\boldsymbol{\Phi}(\mathbf{r}_{12})\mathbf{L}^{n'}\boldsymbol{\varepsilon}_\infty d\mathbf{r}_{12}\ldots \\ + \iint P^{mm'm''}(\mathbf{r}_{12},\mathbf{r}_{23})\mathbf{L}^m\boldsymbol{\Phi}(\mathbf{r}_{12})\mathbf{L}^{m'}\boldsymbol{\Phi}(\mathbf{r}_{23})\mathbf{L}^{m''}\boldsymbol{\varepsilon}_\infty d\mathbf{r}_{12}d\mathbf{r}_{23} + \ldots\end{aligned} \tag{13.109}$$

Now solving for $\boldsymbol{\varepsilon}_\infty$:

$$\begin{aligned}\boldsymbol{\varepsilon}_\infty = (\overline{\mathbf{L}})^{-1}\langle \mathbf{p} \rangle - \int P^{nn'}(\mathbf{r}_{12})(\overline{\mathbf{L}})^{-1}\mathbf{L}^n\boldsymbol{\Phi}(\mathbf{r}_{12})\mathbf{L}^{n'}(\overline{\mathbf{L}})^{-1}\langle \mathbf{p} \rangle d\mathbf{r}_{12}\ldots \\ - \iint \begin{pmatrix} P^{mm'm''}(\mathbf{r}_{12},\mathbf{r}_{23})(\overline{\mathbf{L}})^{-1}\mathbf{L}^m\boldsymbol{\Phi}(\mathbf{r}_{12})\mathbf{L}^{m''}\boldsymbol{\Phi}(\mathbf{r}_{23})\mathbf{L}^{m''}(\overline{\mathbf{L}})^{-1} - \ldots \\ P^{nn'}(\mathbf{r}_{12})P^{mm'}(\mathbf{r}_{23})(\overline{\mathbf{L}})^{-1}\mathbf{L}^n\boldsymbol{\Phi}(\mathbf{r}_{12})\mathbf{L}^{n'}(\overline{\mathbf{L}})^{-1}\mathbf{L}^m\boldsymbol{\Phi}(\mathbf{r}_{23})\mathbf{L}^{m'}(\overline{\mathbf{L}})^{-1} \end{pmatrix}\langle \mathbf{p} \rangle d\mathbf{r}_{12}d\mathbf{r}_{23} + \ldots\end{aligned} \tag{13.110}$$

and hence

$$\begin{aligned}(\mathbf{L}^e)^{-1} = (\overline{\mathbf{L}})^{-1} + \int \boldsymbol{\Phi}(\mathbf{r}_{12})d\mathbf{r}_{12} - \int P^{nn'}(\mathbf{r}_{12})(\overline{\mathbf{L}})^{-1}\mathbf{L}^n\boldsymbol{\Phi}(\mathbf{r}_{12})\mathbf{L}^{n'}(\overline{\mathbf{L}})^{-1}d\mathbf{r}_{12}\ldots \\ - \iint \begin{pmatrix} P^{mm'm''}(\mathbf{r}_{12},\mathbf{r}_{23})(\overline{\mathbf{L}})^{-1}\mathbf{L}^m\boldsymbol{\Phi}(\mathbf{r}_{12})\mathbf{L}^{m''}\boldsymbol{\Phi}(\mathbf{r}_{23})\mathbf{L}^{m'''}(\overline{\mathbf{L}})^{-1} - \ldots \\ P^{nn'}(\mathbf{r}_{12})P^{mm'}(\mathbf{r}_{23})(\overline{\mathbf{L}})^{-1}\mathbf{L}^n\boldsymbol{\Phi}(\mathbf{r}_{12})\mathbf{L}^{n'}(\overline{\mathbf{L}})^{-1}\mathbf{L}^m\boldsymbol{\Phi}(\mathbf{r}_{23})\mathbf{L}^{m'}(\overline{\mathbf{L}})^{-1} \end{pmatrix} d\mathbf{r}_{12}d\mathbf{r}_{23} + \ldots\end{aligned} \tag{13.111}$$

and

$$\mathbf{C}^e = [(\mathbf{L}^e)^{-1} - \mathbf{E}]^{-1} + \mathbf{C}^R \tag{13.112}$$

as before. As for the two-phase case, we still have absolutely convergent integrals (assuming no long-distance correlation; that is, $\boldsymbol{\Phi}$ is a constant at large distances) since, for example,

$$\begin{aligned}(P^{nn'}(\mathbf{r}_{12})\ (\overline{\mathbf{L}})^{-1}\mathbf{L}^n\boldsymbol{\Phi}\mathbf{L}^{n'}(\overline{\mathbf{L}})^{-1} - \boldsymbol{\Phi}) &= ((\overline{\mathbf{L}})^{-1}\mathbf{L}^n P^n\boldsymbol{\Phi}\mathbf{L}^{n'}(\overline{\mathbf{L}})^{-1}P^{n'} - \boldsymbol{\Phi}) \\ &= ((\overline{\mathbf{L}})^{-1}(\mathbf{L}^n P^n)\boldsymbol{\Phi}(\mathbf{L}^{n'}P^{n'})(\overline{\mathbf{L}})^{-1} - \boldsymbol{\Phi}) \\ &= ((\overline{\mathbf{L}})^{-1}(\overline{\mathbf{L}})\boldsymbol{\Phi}(\overline{\mathbf{L}})(\overline{\mathbf{L}})^{-1} - \boldsymbol{\Phi}) = 0\end{aligned} \tag{13.113}$$

where P^n is the volume fraction of phase n.

13.5.3 Two isotropic phases: Macroscopically isotropic/anisotropic

To validate the new formulation of the strong-contrast model it was run with two isotropic phases of various contrasts for both a randomly dispersed mixture and one with fibers of the higher-modulus phase. Figure 13.6 shows the rectangular model used for the material in both cases. The models were

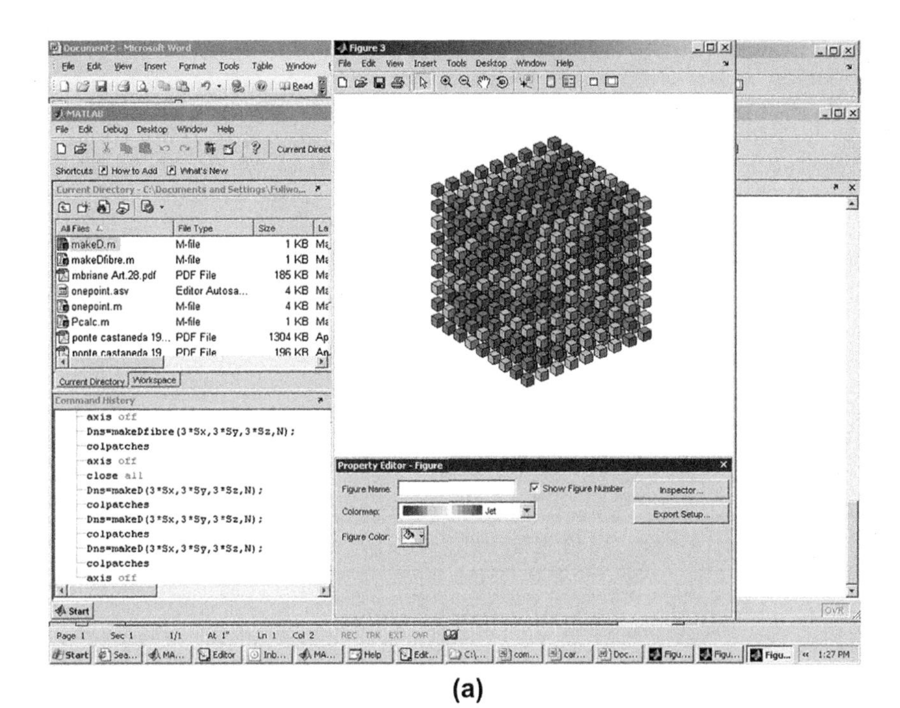

(a)

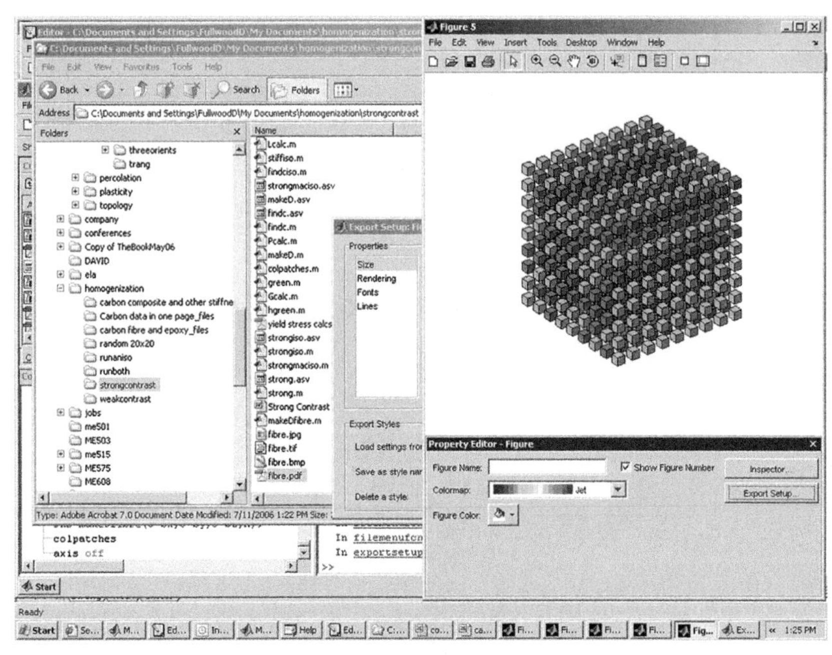

(b)

FIGURE 13.6

(a) Schematic of randomly dispersed material and (b) fiber-aligned material in $\mathbf{e}_1$ direction for two isotropic phases.

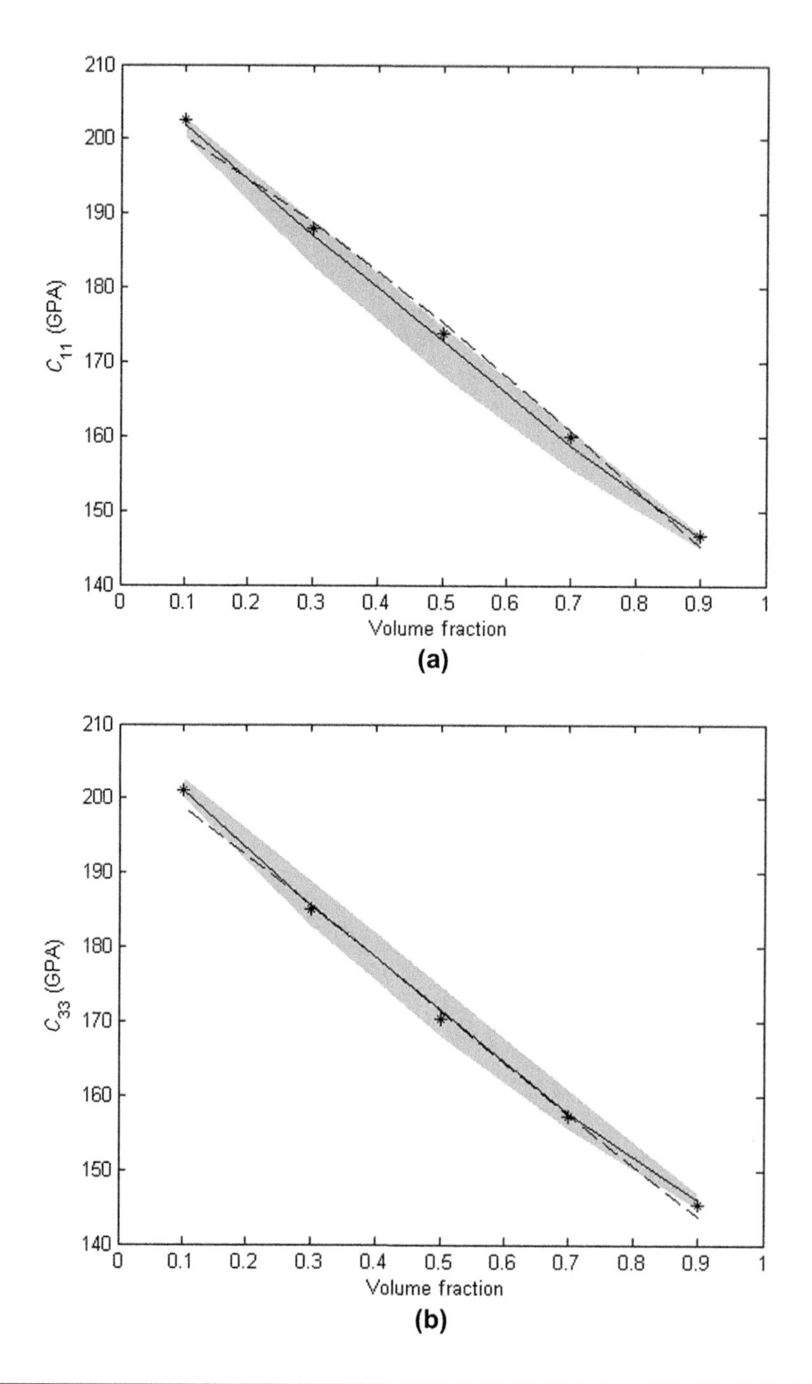

FIGURE 13.7

Results for fiber-arranged isotropic phases. The one-point bounds are *shaded*; the strong- and weak-contrast results are shown with a *solid* and a *dashed line*, respectively; the FE results are *asterisks*.

compared with previously published data (Kalidindi et al., 2006). The fiber results are shown in Figure 13.7, and they demonstrate good agreement with the finite-element calculations previously published.

13.5.4 Two anisotropic phases: Macroscopically anisotropic

A strongly anisotropic example (carbon fiber in epoxy resin) was subsequently used to test the abilities of the model to cope with highly anisotropic phases and anisotropic effective properties. The results for C_{33} (perpendicular to the fiber direction) are shown in Figure 13.8.

The results are reasonable and far superior to the weak-contrast results, which are not even sensible values. However, the results in the C_{11} direction were not as stable and indicate that for high anisotropy more work may be required in the numerical derivation to ensure accurate results, or perhaps higher-order terms are required.

13.5.5 Multiphase anisotropic property closure

Finally, we wish to demonstrate the full utility of the new formulation by generating a multiphase property closure of anisotropic material. For the sake of this exercise a single-phase polycrystalline material is assumed; the global properties will be determined by the orientation of the anisotropic crystallites throughout the sample (i.e., the state space of interest is the orientation space). The strengths of the method become apparent, as shown in Figure 13.9.

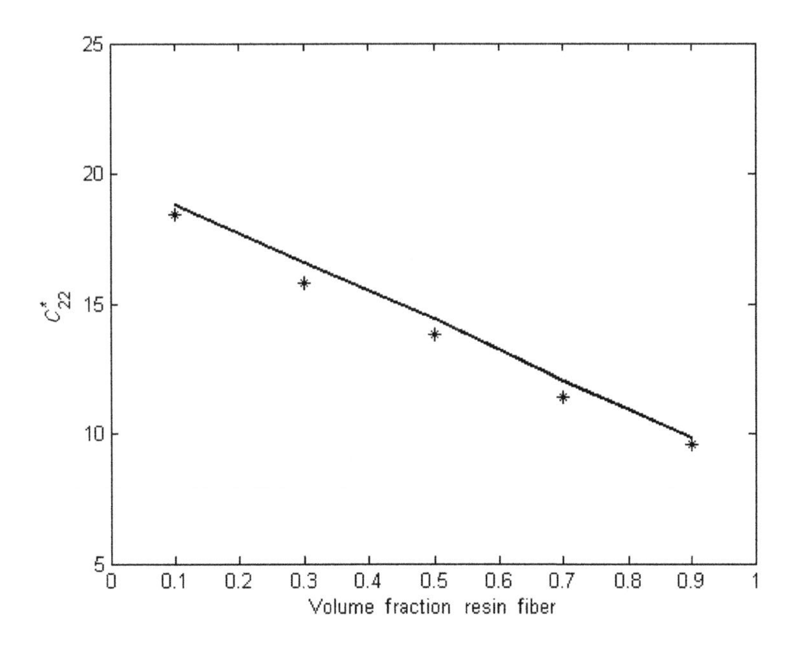

FIGURE 13.8

C^*_{22} for carbon fiber in resin. The upper bound is given by the *straight line*, the calculated figures are *asterisks*.

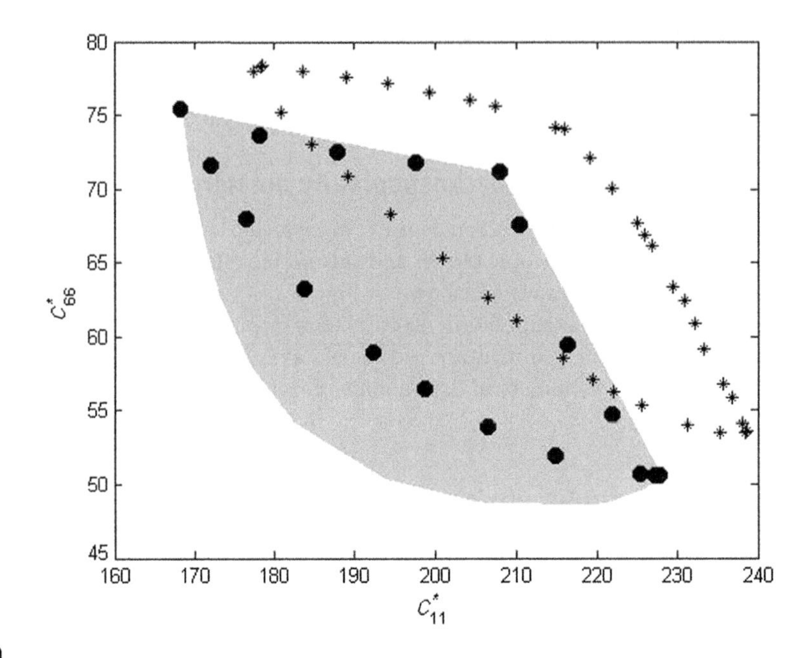

FIGURE 13.9

One-point rigorous bounds (*shaded*), weak-contrast closure (*asterisks*), and strong-contrast closure (*circles*) for copper polycrystal.

It may be seen from Figure 13.9 that the strong-contrast solution lies completely within the rigorous one-point bounds; the weak-contrast solution gives approximately the correct shape, but is translated from the correct position.

In summary, the strong-contrast formulation for elasticity perturbation homogenization relations has been extended to an arbitrary number of phases, with full anisotropy. The method has been validated against previously published data for isotropic phases and multiple anisotropic phases. For very high anisotropy the results are understandably dependent on the reference tensor, and improvements may be needed in terms of numerical accuracy (e.g., during numerical integration of Green's function) and perhaps the addition of the third-order terms in the series. Further study is required in this area.

SUMMARY

In this chapter homogenization relations were developed that utilize the n-point structure descriptors defined in Chapter 12. These relations are based on perturbation expansions, and the so-called weak- and strong-contrast expansions were both described (the former being the most commonly used). With both the higher-order structure metrics and the higher-order homogenization relations in place, Chapter 14 incorporates both concepts into a higher-order microstructure-sensitive design framework via the relevant structure and property design spaces (hulls and closures), as for first-order design.

Exercises

Section 13.2

13.1. Take the definition of $\widetilde{H}_{ss't}$ as given in Eq. (13.31). Prove that we may write (Eq. 13.32)

$$\widetilde{H}_{ss't} = h(s_1, s_1', t_1)h(s_2, s_2', t_2)h(s_3, s_3', t_3)$$

for some function h (the function does not need to be defined at this point).

13.2. Using the result from the previous exercise, prove that (Eq. 13.10)

$$h(s_i, s_i', t_i) = \begin{cases} \delta/2 & \text{if} \quad t_i = s_i' - s_i + 2^P \text{or } t_i = s_i' - s_i + 2^P + 1 \\ 0 & \text{otherwise} \end{cases}$$

where $\delta = D/2^P$ is the cell dimension for a cubic sample region of length D. The cells in $\Psi(\Omega)$ are chosen to have the same dimensions as the cells in Ω, and $\Psi(\Omega)$ is eight times as large as Ω rom its definition

CHAPTER

Second-Order Hull, Property Closure, and Design

14

CHAPTER OUTLINE

The object of microstructure-sensitive design (MSD) is to provide a searchable atlas of effective (global) material properties that can be directly related to the microstructure. Using such a tool, a designer would be able to optimize the properties of a given component by choosing the appropriate microstructure. An interim step in arriving at such an atlas is to determine the closure of potential properties, and related microstructures.

This goal has historically involved the utilization of bounding theories operating on first-order (volume-fraction) descriptions of the microstructure (Chapter 4). Following this approach, one may arrive at rigorous bounds for the material properties, but there is no guarantee that a given value of a property within these bounds is physically realizable, and a material (microstructure) design that achieves a particular property value is not determined beyond specification of the texture.

In Chapter 13 we obtained homogenization relations that allow a second-order description of material properties to be manipulated in a manner that is consistent with exploration of the property closure. The corresponding second-order microstructure hull contains the set of all possible two-point correlation functions for describing structure. In this chapter we present the hull in its discrete Fourier transform representation. This results in a searchable design space for microstructure that can be combined with the second-order homogenization relations of Chapter 13 to generate second-order property closures as the basis for inverse design.

However, one of the main issues relating to this approach is the lack of familiarity with two-point correlations as a microstructure description, together with the potential complexity of the descriptions. By way of contrast, one-point descriptions (e.g., in the form of pole figures) are well recognized and understood. Hence, the perceived need to reconstruct real microstructures from a given set of optimized two-point statistics to understand the implications of the microstructure description in a real situation. Various reconstruction algorithms exist to aid in this step, but the efficient reconstruction of realistic polycrystalline materials from a set of two-point statistics is still some way off. In this chapter we demonstrate a more direct link between the microstructure function and the resultant property closures that retains many of the benefits of MSD (e.g., splitting the structure and property parameters in the model) but avoids the reconstruction.

Determination of a property closure will be formulated in such a manner that standard optimization techniques can be employed. We then take two pareto-front optimization techniques and generalize them in such a way that they can be effectively employed to find the full property closure. The results are related to one-point (first-order) closures and discussed in this context. We then use the property closures to resolve (at the second order) the inverse design problem for homogenous and heterogeneous design.

The main drawback with this approach is the high dimensionality of the structure search space. We demonstrate the method for a very limited example, but a more realistic application will likely require a more efficient spectral decomposition of the microstructure, which will lead to some of the same issues associated with reconstructing from the two-point statistics, that is, the problem of arriving at a detailed realistic structure.

14.1 HULL OF TWO-POINT CORRELATIONS

Having established the interdependencies in the DFT of a given set of two-point correlations (see Section 12.3), we now turn our attention to identifying the complete set of theoretically feasible sets of two-point correlations for a given material system. This problem has been tackled in various guises in the past. For example, using the Wiener-Khinchin theorem, it has been demonstrated (Cohen, 1998; Torquato, 2002) that a semi-positive definite two-point correlation in 1D space relates to a real random process. Readers are referred to the work of Torquato (2006; Uche et al., 2006) and Gokhale (2005) for other important results obtained to date using such approaches.

Jiao et al. (2007) have recently proposed a novel approach to exploring a certain subspace of the two-point correlation space described here. The space explored by these authors is called the S_2 space and captures only the auto-correlations of eigen-microstructures. In particular, for composites with more than two distinct local states, the S_2 space will be missing the important phase information that is only present in the cross-correlations (i.e., the θ_k^n described in Section 12.3).

In Chapter 9 we demonstrated the advantages of visualizing in Fourier spaces the set of all theoretically feasible one-point statistics in a given material system; this design space is referred to as the *microstructure hull*. In these constructs, the one-point statistics of any microstructure are visualized as a point in the Fourier space of which the coordinates are identified by the Fourier coefficients of the selected one-point statistics. More specifically, it was demonstrated that the space corresponding to the set of all theoretically feasible one-point statistics in a given material system is compact and convex. These constructs have been found to be of tremendous value in designing components for optimized performance characteristics.

It is particularly emphasized here that, in material systems with a continuous local state space, the hulls for one-point statistics are most easily established in the Fourier space. This is because the values of the distribution (probability density function) used to define the one-point statistics could be unbounded for many microstructures. For example, in describing the one-point statistics in polycrystalline microstructures, the values taken by the orientation distribution function (ODF) have to be positive, but are otherwise unbounded. In fact, the ODF for a single-crystal microstructure would be a Dirac-delta function. However, the Fourier coefficients corresponding to any theoretically feasible texture occupy a bounded region, making it very convenient to visualize the complete set of all theoretically possible one-point statistics in a given material system.

In the current approach, because of the use of discrete sampling of the microstructure in both the spatial domain and the local state space, we could theoretically delineate hulls for two-point correlations either in the vector space of f_t^{np} or in the spectral space of F_k^{np}. The discrete sampling employed in defining f_t^{np} is, of course, tantamount to the use of a primitive Fourier basis via the indicator functions or characteristic functions (see Section 3.1), and therefore both descriptions can be interpreted as Fourier representations. However, as shown in Section 12.3, the use of DFTs allows us to identify many of the redundancies that exist in the set of F_k^{np} much more easily than in f_t^{np}. Therefore, in this work, we focus exclusively on the delineation of the hulls of two-point correlations in the DFT space.

The concept of eigen-microstructures described earlier, where all $m_s^n = D_s^n/N$ values are either 0 or 1, is central to the construction of the microstructure hulls. Let $\hat{F}_k^{np}$ denote the DFT of the two-point statistics of an eigen-microstructure of a material system of interest. Let $\hat{H}$ denote the set of all theoretically feasible $\hat{F}_k^{np}$. Then the convex region defined by the elements of $\hat{H}$ represents the hull of two-point correlations sought here. Equations (12.63) through (12.67) explicitly define the hull in a $(2N-3)\lceil (S+1)/2 \rceil$ -dimensional space (where $\lceil x \rceil$ denotes the "ceiling" function, which returns the smallest integer $\geq x$). The $(2N-3)$ term comes from the fact that $(N-1)$ correlations are necessary to describe an N local state material system (see Section 12.3) and the observation that the Fourier terms in cross-correlations are complex valued and real valued in autocorrelations. The $\lceil (S+1)/2 \rceil$ term comes from Eq. (12.67) and the symmetry of each Fourier transform. It is anticipated that the eigen-microstructures will occupy an extremely small portion of the points in the hull.

The points in the hull that are not occupied by eigen-microstructures can be interpreted in two different ways. One option is to interpret them as belonging to noneigen-microstructures. However, our investigations have revealed that it may not be possible to assign a (single) specific noneigen-microstructure, of spatial extent S, to every point in the hull. It is important to note that the inability to generate a single instantiation is the by-product of the discreteness in space. It is expected that, in the continuous limit, a single noneigen-instantiation could be found for all points in the hull.

The second option is to interpret these points as the ensemble-averaged two-point statistics over multiple instantiations of the microstructure drawn either from a single sample or from multiple samples with nominally the same processing history. We prefer the second interpretation for the following two main reasons:

1. Defining a representative microstructure in terms of the ensemble average of statistics from multiple instantiations (rather than from a single instantiation) assures us that the microstructure statistics for the sample are captured as accurately as possible.
2. Ensemble averaging also assures us that the hull is convex and compact, which in turn makes it possible to search for microstructures with optimal combinations of properties using the available computationally efficient quadratic programming algorithms (Boggs and Tolle, 2000).

All of the concepts described previously will be clarified through a very simple case study in the next section.

14.1.1 Hull of 1D two-phase microstructures

As a simple case study that is mainly intended for clarification of the various concepts presented earlier, we investigate a set of hypothetical 1D two-phase microstructures extracted from a larger 2D sample shown in Figure 14.1. These represent multiple instantiations of eigen-microstructures extracted from a hypothetical sample. The two distinct phases present in these microstructures are shown in black and white. Furthermore, the spatial domain in these microstructures was divided into only 10 bins as shown in the figure. Since there are only two phases, we need only one two-point correlation to represent the microstructure. Let F_k^{11} denote the DFTs selected for representation in a two-point hull. Since S has been selected as 10 in this simple case study, the relevant DFT space here is six-dimensional (taking advantage of Eq. (12.67) and the fact that the autocorrelation coefficients F_k^{11} are all real). This is the main reason for selecting a highly simplified idealized microstructure for this case study.

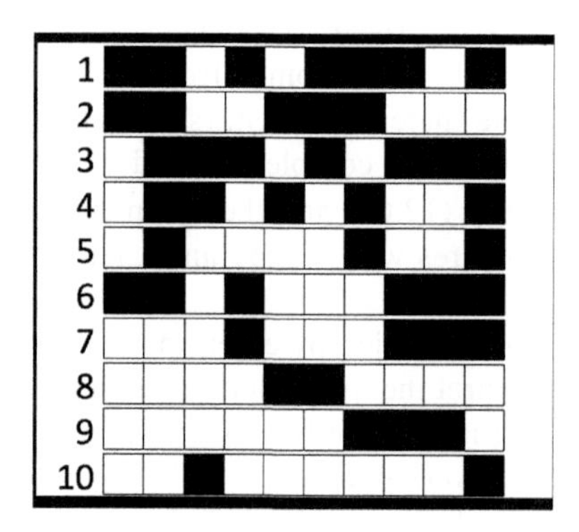

FIGURE 14.1

A set of 10 1D two-phase samples extracted from a very large hypothetical sample. It is expected that the large sample can be represented as an ensemble average of these smaller samples.

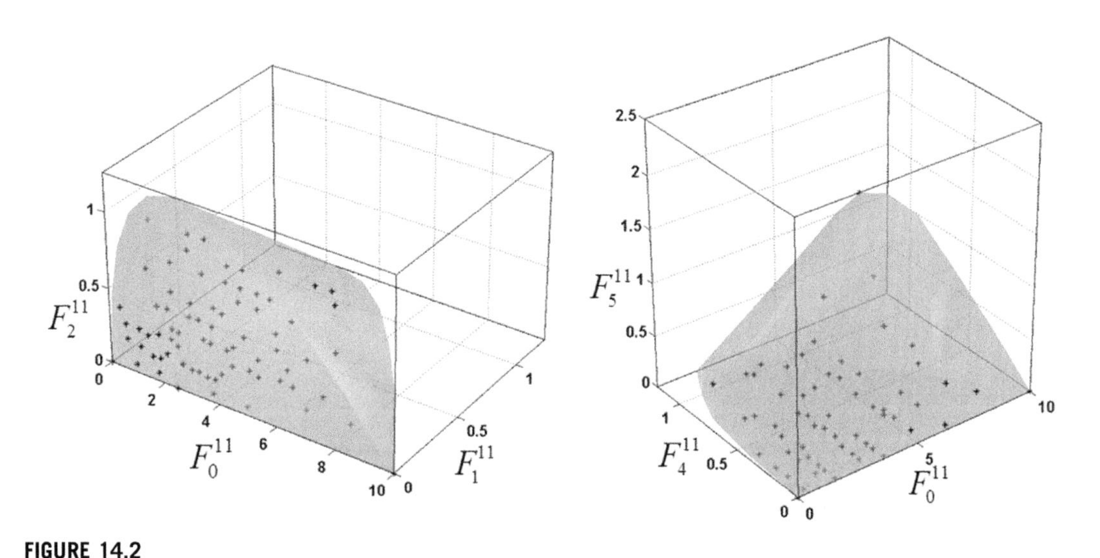

FIGURE 14.2

Projections of the computed two-point correlation hull for the selected 1D material system. The starred points inside the hull represent the set of all possible eigen-microstructures in this material system.

As one can see, the dimensionality of the space of the two-point hulls described in this work grows approximately proportionally with the product *SN*. However, it is only possible to view projections of this hull in at most a selected 3D subspace. Although the mathematical framework presented before can be employed on microstructures with much larger values of both *S* and *N*, the visualization of their two-point hulls in 3D subspaces is not particularly insightful. Nevertheless, readers might be interested in knowing that we have successfully used the mathematical framework presented here on microstructures with *S* values of about 2,000,000 and *N* values of about 500 (Fullwood et al., 2008c).

Figure 14.2 shows the projections of the computed two-point hull for the selected 1D material system in the $F_0^{11}, F_1^{11}, F_2^{11}$ subspace and the $F_0^{11}, F_4^{11}, F_5^{11}$ subspace. The starred points inside the hull in the figure denote the set of all possible eigen-microstructures in this material system. Although one might expect a total of 2^{10} eigen-microstructures (based on $S = 10$ and $N = 2$), the number of eigen-microstructures shown is substantially less. Two factors can help explain this discrepancy. First, a very large number of eigen-microstructures share the exact same representation in the two-point hull shown in Figure 14.2. In fact, in the DFT representation two-point correlations are invariant under a translation and/or an inversion of the microstructure. It is easy to see from the definitions in Eq. (12.43) that the microstructure data sets m_s^n and $m_{\pm s+a}^n$ would produce identical two-point correlations, where *a* denotes a translation of the microstructure data set by an integer number of grid points on the spatial domain of the microstructure. Second, the projection of the 6D hull into a 3D subspace causes some of the distinct points in the 6D space to occupy the same location in the smaller 3D space.

The eigen-microstructures shown in Figure 14.1 have been identified as black squares in Figure 14.3. These are a subset of the starred points shown in Figure 14.2. Note that eigen-microstructures 1 and 3 are

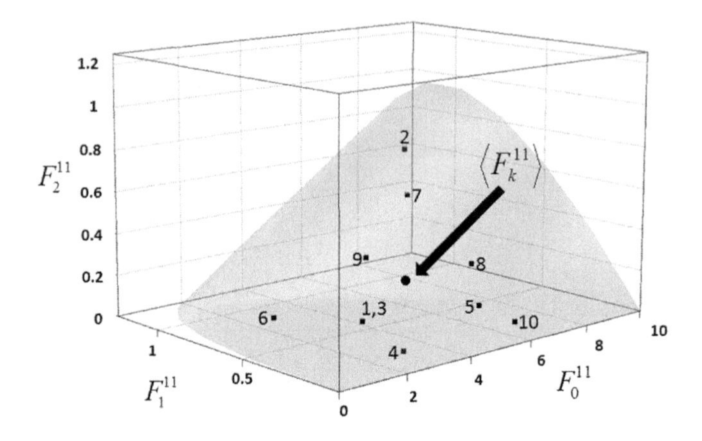

FIGURE 14.3

The projection of the ensemble of microstructures from Figure 14.1 into the two-point correlation hull. The average statistics for the ensemble are denoted by the black circle.

related to each other by a translation and therefore occupy the same location in Figure 14.3. The ensemble-averaged two-point statistics, denoted $\langle F_k^{11} \rangle$, for the entire set of microstructures shown in Figure 14.1 are shown as a filled circle in Figure 14.3. The ensemble-averaged statistics in this case did not correspond to any of the possible eigen-microstructures for the selected material system. Given the relatively small number of points in the hull occupied by the eigen-microstructures, it should be expected that there is only an exceedingly low chance that the ensemble-averaged two-point statistics would correspond to an eigen-microstructure. As discussed earlier, all of the points in the hull that do not belong to an eigen-microstructure can be interpreted as representing the ensemble-averaged two-point statistics of a set of microstructures extracted from one or more samples with nominally the same processing history.

One method of visualizing the points in the two-point hull is through reconstruction techniques. Recent advances in reconstruction techniques using phase-retrieval algorithms have resulted in rapid restoration of microstructures from a given set of two-point statistics. The procedures used for the reconstructions have been described in papers (Fullwood et al., 2008b; Fullwood et al., 2008c) and have been highly successful in noneigen-reconstructions and exact reconstructions of eigen-microstructures (up to an arbitrary translation or inversion). However, efforts to reconstruct a noneigen-microstructure corresponding to the ensemble-averaged statistics in Figure 14.3 have resulted in only limited success.

In general, reconstructions from ensemble-averaged statistics result in multiple solutions where the statistics of the reconstructed microstructures are close to the ensemble-averaged statistics, but the normalized error is still outside the bounds of what would be considered a "successful" reconstruction (Jiao et al., 2007; Fullwood et al., 2008c). This leads us to believe that there are points in the two-point hull that would not correspond to any single noneigen-microstructure (defined at the adopted spatial resolution). An example of the reconstructed noneigen-microstructure that comes closest to capturing the ensemble-averaged statistics is shown in Figure 14.4. This inability to

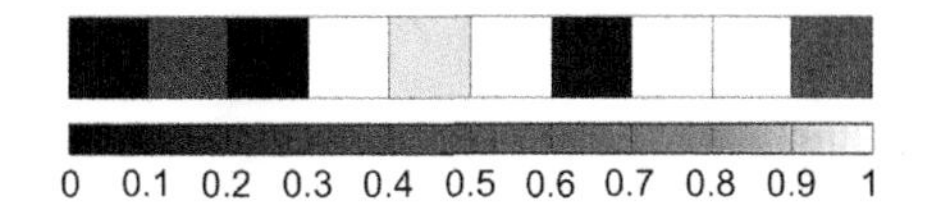

FIGURE 14.4

Best possible reconstruction of a noneigen-microstructure corresponding to the ensemble-averaged statistics shown in Figure 14.3. The statistics of the reconstructed microstructure differ from the ensemble-averaged statistics by approximately 1%.

capture the ensemble-averaged statistics is a direct result of the limited size of the spatial discretization used in this example.

Given the extremely limited spatial discretization (only 10 spatial points in the sample), it would be impossible to build a microstructure for each point in the convex hull. In principle, if one is willing to refine the spatial grid indefinitely, it will become possible to find a noneigen-microstructure with the statistics of the ensembled structures to a predefined approximation. Any noneigen-microstructure will eventually become an eigen-microstructure (to a given approximation) as the spatial grid on which it is discretized is made finer. Alternatively, the noneigen-microstructures can be thought of as eigen-microstructures on a coarse spatial grid.

Instead of the previous approach, we prefer to represent the ensemble-averaged two-point statistics for the set of microstructures shown in Figure 14.3 using a weighted set of representative volume elements (RVEs). For this purpose we define a Euclidian distance between any two points in the two-point correlation hull, normalized by the size of the two-point hull (i.e., the largest distance of the hull vertices from the origin), and use it to identify a set of eigen-microstructures that are close to the ensemble-averaged statistics of the given set of microstructures. Figure 14.5 describes examples of how a weighted set of eigen-microstructures can approach the ensemble-averaged two-point statistics. As expected, the more microstructures we can use in the RVE set, the closer we can approximate the ensemble-averaged statistics.

14.2 SECOND-ORDER PROPERTY CLOSURE

The concept of property closure was introduced in Chapter 9. We now extend the previous results to the second-order closure, obtained directly from the primitive basis representation of the microstructure function. For illustrative purposes, assume that we have two properties, P and Q, both of which can be written in the form of Eq. (13.47) (dropping the hats on the N and J tensors):

$$P = {}^1J^n_s D^n_s - {}^1\aleph^{nn'}_{ss'} D^n_s D^{n'}_{s'} \quad \text{and} \quad Q = {}^2J^n_s D^n_s - {}^2\aleph^{nn'}_{ss'} D^n_s D^{n'}_{s'} \tag{14.1}$$

with the constraints that $\sum_{n=1}^{N} D^n_s = N$ for each s. Then we can search over all feasible microstructures (defined by D^n_s subject to the constraints), with each one giving a point in PQ-space.

Such a search, however, is no trivial undertaking. We illustrate this in the following example by choosing a large number of random microstructure designs (using a Monte Carlo–type approach) and plotting them on a previously calculated property closure.

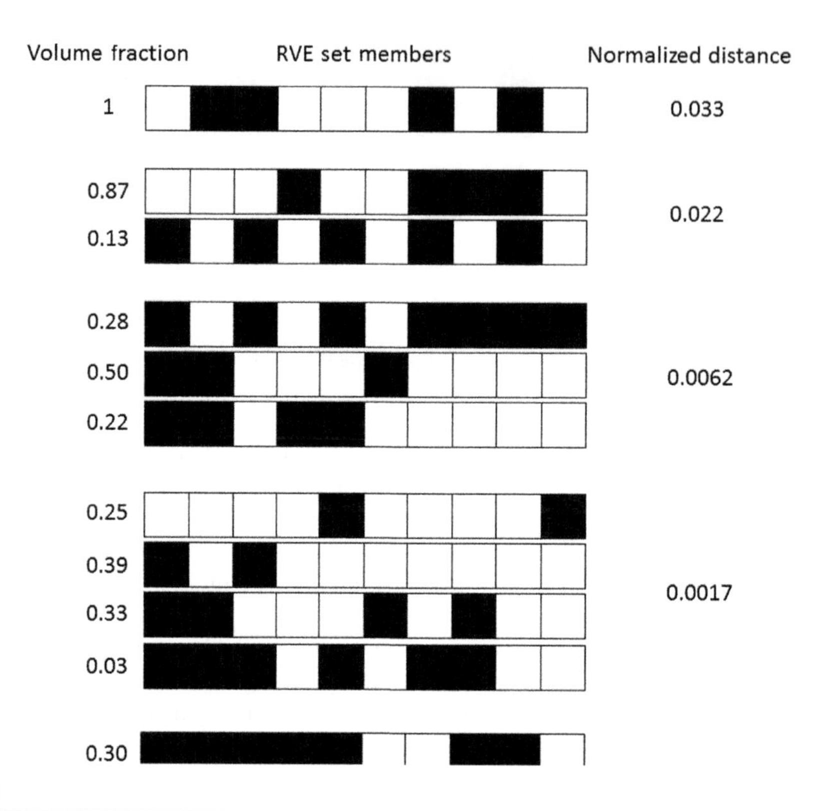

FIGURE 14.5

RVE for a large hypothetical sample represented as the weighted fractions of five eigen-microstructures.

Example 14.1

Take the material defined in Example 13.3, in a 4 × 4 × 4 sample with four cells (orientations) in the fundamental zone. Plot the points in the property closure for C_{11} and C_{66} corresponding to 10,000 randomly chosen microstructures.

Solution

The matrices for the properties C_{11} and C_{66} are in a similar manner to those found in Example 13.3. To find the random microstructures we take each value of s for D_s^n in turn. Since $N = 4$ we choose four random numbers that add up to 4 by choosing the first number between 0 and 4; then choose the second number between this number and 4; then take the third number between the sum of these numbers and 4; then the fourth number is 4 minus the sum of these numbers. We randomly place these numbers in the four available positions of D_s^n for the given s. Continue this process for each s. Then find the properties C_{11} and C_{66} given by this microstructure.

The results of 10,000 points are plotted in Figure 14.6 over the full closure found using methods described in Chapter 15.

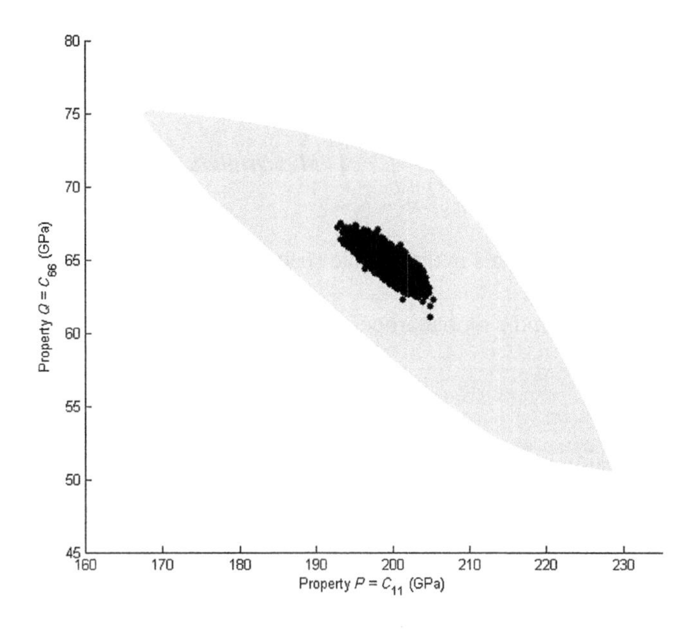

FIGURE 14.6

Properties relating to 10,000 random microstructures plotted inside the relevant property closure.

It is clear that finding microstructures that correspond to properties near the border of the closure is difficult. Methods are introduced in the next section that facilitate the search for points on the boundary of the closure. It will become apparent that once these points are found it is straightforward to find microstructures corresponding to any point in the closure.

14.3 PARETO-FRONT TECHNIQUES ON THE PROPERTY CLOSURE

In this section we briefly review relevant material given earlier and arrive at a practical formulation that can be immediately related to optimization techniques.

14.3.1 Problem formulation

Recall that the material sample space, Ω, has been divided into S cells and the state space, H, has been divided into N cells. The microstructure function is defined such that $M(x, h)dh$ is the volume fraction of material (dV/V) in an infinitesimally small neighborhood of the sample point x that has orientation lying within the infinitesimal neighborhood dh of state h. Within each cell, ω_s ($s \leq S$), of the sample we defined a discrete variable D_s^n that approximates the microstructure function. D_s^n is related to the probability distribution ${}_sf(h)$ in cell s, and is given the $N\times$ (i.e., volume fraction of material in cell s of the sample space with orientation lying within cell $n(n \leq N$ i.e., of state space H). In fact, D_s^n is normalized such that $\sum_{n=1}^{N} D_s^n = N$, and consequently $\sum_{s=1}^{S}\sum_{n=1}^{N} D_s^n = NS$:

$$D_s^n = \frac{N}{V_s} \int\limits_{\mathbf{x} \in \omega_s} \int\limits_{h \in \omega_n} M(\mathbf{x}, h) dh d\mathbf{x} \tag{14.2}$$

where V_s is the volume of cell number s in the sample space, and similarly n refers to the cell in H (see Eqs. (5.45), (5.46), and (4.29)).

The second-order equation for the global property, C^*, in terms of a particular microstructure D_s^n is (Eq. 13.47):

$$C^* = J_s^n D_s^n - \aleph_{ss'}^{nn'} D_s^n D_{s'}^{n'} \tag{14.3}$$

For the sake of clarity in using optimization techniques, we will combine the two indices for D_s^n into a single index: D^α, $\alpha = (n-1)S + s$ ($1 \leq \alpha \leq NS$).

Let us also take a specific example, in that the global property we require is the fourth-order stiffness tensor, C_{ijkl}^*. We also choose an isotropic reference tensor, C^r, in our calculation of the second-order terms since the Green's function in this case was solved in Chapter 13 and can be easily manipulated. The final form for the effective global stiffness tensor can be written as

$$C_{ijkl}^* = J_{ijkl}^\alpha D^\alpha - \aleph_{ijkl}^{\alpha\alpha'} D^\alpha D^{\alpha'} \tag{14.4}$$

where $ijkl$ emphasizes that we are considering one component of C^* at a time. We will henceforth omit the subscript for brevity, and refer to a particular property (e.g., C_{ijkl}^*) using the letter P or Q.

Clearly, Eq. (14.3) can be written in matrix form as

$$\mathbf{P} = \mathbf{j}^T\mathbf{D} - \mathbf{D}^T\aleph\mathbf{D} \tag{14.5}$$

where T refers to a matrix–vector transpose. The constraints are that, for each s, $D^\alpha \geq 0$ and $\sum_{n=1}^{N} D^{(n-1)S+s} = N$. Equation (14.4) now has the form of a typical quadratic surface and can be solved using standard quadratic programming.

14.3.2 Bi-objective optimization

As discussed earlier, the objective of this section is to use optimization techniques to find the property closure for a number of properties. We illustrate the solution technique using two properties, but the methodology could readily be extended to higher dimensions.

Take a pair of physical properties, P and Q. If the property closure of feasible properties were strictly convex and closed in PQ-space, then there would exist unique points representing maximum and minimum values of P and Q. In a traditional design approach one would be looking for a microstructure design that in some sense optimized both P and Q.

If it were the case that the designer wanted to minimize both P and Q, then the pareto-front of points between the global minimum values of P and Q would be found, and the designer would choose a value along this front. The (pareto) points represent "nondominated" values, in the sense that any microstructure that decreases the value of P must increase the value of Q and vice versa.

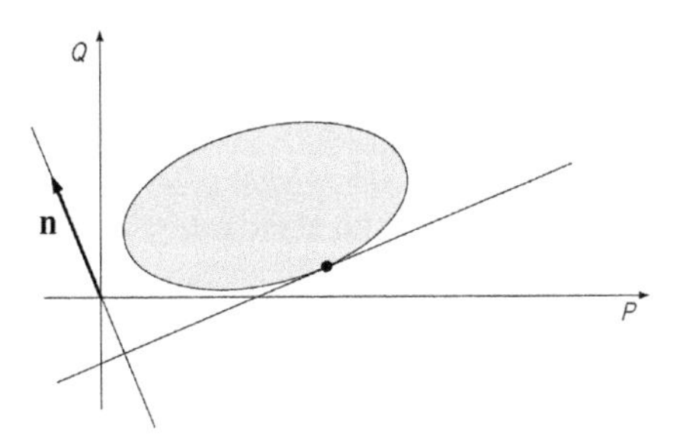

FIGURE 14.7

Point found on the boundary of the property closure for an arbitrary **n**.

If the objective were changed to minimizing P and maximizing Q, a further quadrant of the property closure would be found, and so on, to find all four quadrants.

14.3.3 Generalized weighted sum and adaptive NBI methods

A more elegant way of following this process is to generalize the weighted sum method used to find a set of evenly spaced pareto-front points (e.g., Kim and de Weck, 2005). In a typical approach one defines a function:

$$Y = \lambda_1 P + \lambda_2 Q \tag{14.6}$$

such that $\lambda_1 + \lambda_2 = 1$, and both are positive. One then minimizes Y, subject to the same constraints on D required for feasible P and Q.

To adapt this approach in a manner that is easy to visualize, we modify the definition of the weights, λ_i. Let $\mathbf{n} = [\lambda_1, \lambda_2]^T$ such that $\lambda_1^2 + \lambda_2^2 = 1$; that is, $\mathbf{n}$ is the unit length. Then for a point (P,Q) in PQ-space, Y represents the dot product of $\mathbf{n}$ with the vector defined by $[P, Q]^T$; in other words, the projection of the vector $[P, Q]^T$ in the direction $\mathbf{n}$. Thus, minimizing Y gives the point on the closure that has the minimum projection in direction $\mathbf{n}$ (Figure 14.7).

If the direction of $\mathbf{n}$ is varied to cover the whole unit circle, every point on the strictly convex closure can be found. We refer to this method as a generalized weighted sum (GWS) approach.

Example 14.2

Demonstrate the GWS approach using the material from Example 13.3, with a $4 \times 4 \times 4$ sample. The two properties to be calculated are the stiffness matrix parameters $P = C_{11}$ and $Q = C_{66}$.

Solution

The matrices for P and Q are found in the same manner as in Example 13.3. Hence, we arrive at the following two equations (Eq. 14.4):

$$P = \mathbf{J}_P^T\mathbf{D} - \mathbf{D}_P^T\aleph\mathbf{D}_P \text{ and } Q = \mathbf{J}_Q^T\mathbf{D} - \mathbf{D}_Q^T\aleph\mathbf{D}_Q$$

Let us search over 25 evenly spaced points on the unit circle; hence,

$$\mathbf{n}_k = \begin{bmatrix} \cos(2\pi k/25) \\ \sin(2\pi k/25) \end{bmatrix}$$

The constraints on the microstructure variables, D, may be implemented as

$$-I(NS,NS)^{*}D \leq zeros(NS,1) \text{ and } [I(S,S),\ldots,I(S,S)]^{*}D = ones(S,1)^{*}N$$

where $I(S,S)$ is the $S \times S$ identity matrix, $zeros(NS,1)$ is a vector of NS zeros, and so forth. The negative version of the positivity constraint is used to be consistent with standard $\leq$ constraints (e.g., in MATLAB).

Then for each $\mathbf{n}_k$ we wish to minimize,

$$\mathbf{n}_k \cdot \begin{bmatrix} P \\ Q \end{bmatrix} = \cos(2\pi k/25)P + \sin(2\pi k/25)Q$$

where P and Q are defined as before, subject to the given constraints. Since this is a quadratic programming problem (quadratic function with linear constraints), one might use the `quadprog` function in MATLAB, for example, to find the minimum each time.

For this exercise we have implemented the GWS algorithm using the MATLAB quadratic programming solver for an initial number of evenly spaced directions for $\mathbf{n}$ in the unit circle. Further points might then be obtained by identifying badly spaced points and choosing extra directions for $\mathbf{n}$ between the previous values.

In this example we have stopped the algorithm after several iterations to demonstrate that points congregate around areas of high curvature, and points on "flat" regions of the closure are more difficult to find (Figure 14.8).

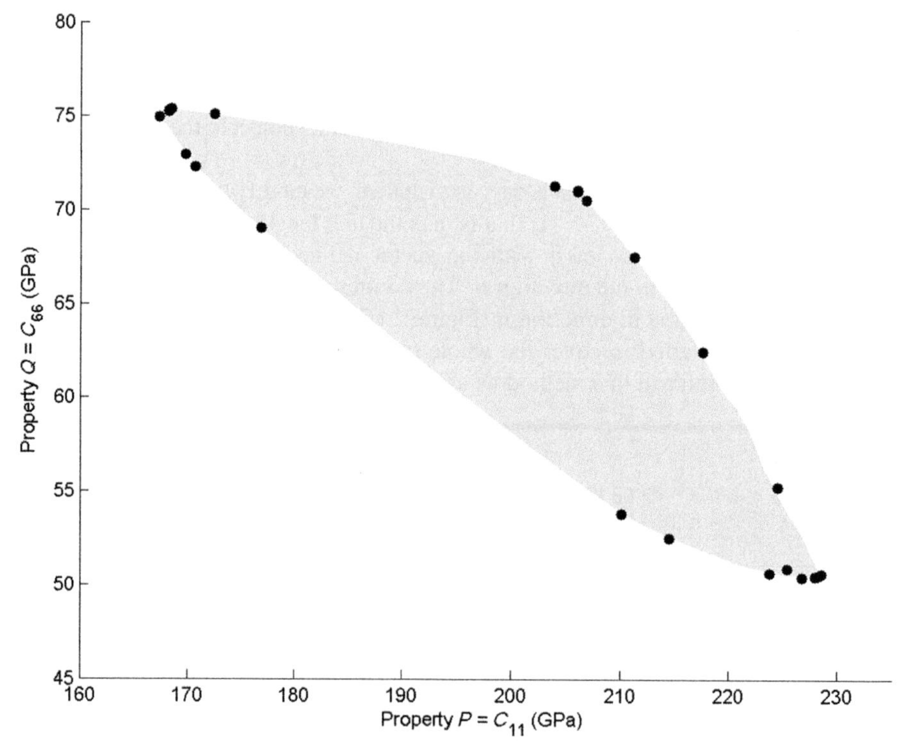

FIGURE 14.8

Points on the boundary of the property closure for C_{11} and C_{66} found using the GWS approach.

Several authors (e.g., Koski, 1985; Das and Dennis, 1997; Messac and Mattson, 2002) have discussed the two disadvantages of the weighted sum method highlighted by this example:

1. The resultant points may not be evenly spaced.
2. Points on regions of the closure that are not strictly convex will not necessarily be found.

In practice, we do not expect that the closure will be strictly convex, or even convex, in which case a significant modification to our approach is required. Two methods that can find points in the concave region of the closure are the normal boundary intersection (NBI) method (Das and Dennis, 1998) and the adaptive weighted sum (AWS) method (Kim and de Weck, 2005). Both of these methods will find points in the concave region of the closure. However, the AWS method will only find nondominated (pareto) points. From a designer's viewpoint, this may be an advantage, as pareto points may be the only design points of interest, and the AWS method should be chosen. However, since we set out to find all points on the closure we have chosen the NBI method.

One disadvantage with both of these methods is that they introduce nonlinear constraints into the optimization routing, thus barring quadratic programming as a potential solution method. Up until this point, all constraints have been linear, but once a constraint is placed in the *PQ*-space (*P* and *Q* being quadratic functions of the microstructure) this linear property is lost. Of course, it is no theoretical hardship to move to a nonlinear programming method such as sequential quadratic programming (SQP). The issue is the time required to solve the problem. For a relatively small problem in a real microstructure there might be tens of thousands of variables, and keeping the problem as simple as possible is important.

For the typical previous example we recorded the time taken to solve 100 optimization steps (the same problem each time for this exercise) using 100 different directions of **n**. The times are plotted against each other for the QP and the SQP algorithms (Figure 14.9). One may see that the SQP

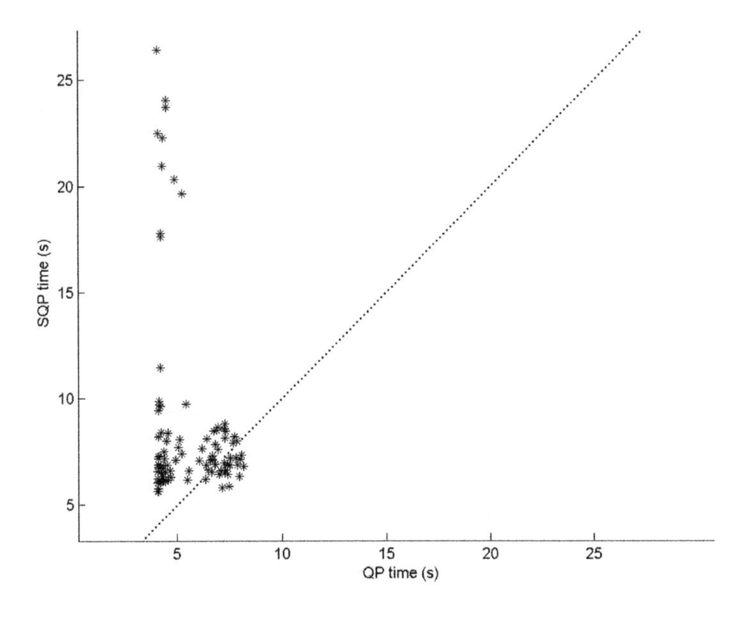

FIGURE 14.9

Times to solve 100 identical problems using QP and SQP algorithms in MATLAB. Points above the dotted line indicate that the SQP algorithm took longer.

algorithm generally takes longer to solve the problem (an average of 55% longer for this problem), and in some cases it takes six times as long. These results are for a reasonably small and well-behaved problem; we would expect the difference to be more pronounced as the problem size increases and perhaps becomes less well behaved.

The *NBI method* is set up to solve a traditional pareto-front minimization problem where only one quadrant of the property closure is being sought. A chord is drawn between the point that minimizes P and the point that minimizes Q. A line normal to this chord is found, and the minimal point in the property closure is found along this line.

This scheme does not suit our requirements for a number of reasons, and we define an adaptive NBI routine (ANBI) as follows. First, a number of evenly spread vectors, $\mathbf{n}_k$, are defined in the unit circle. J is minimized with respect to each of these vectors to find a series of points on the closure. We define a maximum tolerance on the "distance" between points on the front. This is done with respect to each parameter by taking the difference between the maximum and minimum values for P and dividing by some factor (e.g., five in our examples), and similarly for Q. Thus, the step in the P direction and the step in the Q direction must both lie within the respective tolerances. Starting at the first point on the closure we check the "distance" between subsequent points. If it exceeds the tolerance we take a vector $\mathbf{n}$ halfway between the two values of $\mathbf{n}_k$ used to find these points. We minimize J for this value of $\mathbf{n}$, still using quadratic programming; that is, up to this point we are using the GWS method. If no new point is found, we switch to ANBI for these points.

For the ANBI step we first draw a line between the points and find the unit normal to this line that points into the property closure. We then draw the bisecting line (i.e., in the direction of the normal and halfway between the points; see Figure 14.10). We define J using the normal, $\mathbf{n}$, that is

$$\left(J = \mathbf{n} \cdot \begin{bmatrix} P \\ Q \end{bmatrix}\right)$$

and minimize J using SQP with the added constraint that the points (P,Q) must lie on the bisecting line. A schematic is given in Figure 14.11. In this manner, we use linear constraints wherever possible, and introduce nonlinear constraints only where the boundary is not strictly convex or is almost flat.

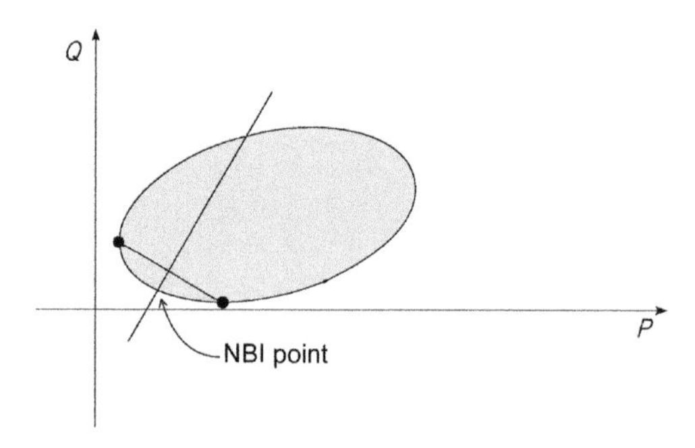

FIGURE 14.10

Traditional NBI approach. A chord is drawn between the minimum points for P and Q, and normal lines are used to constrain the search for boundary points.

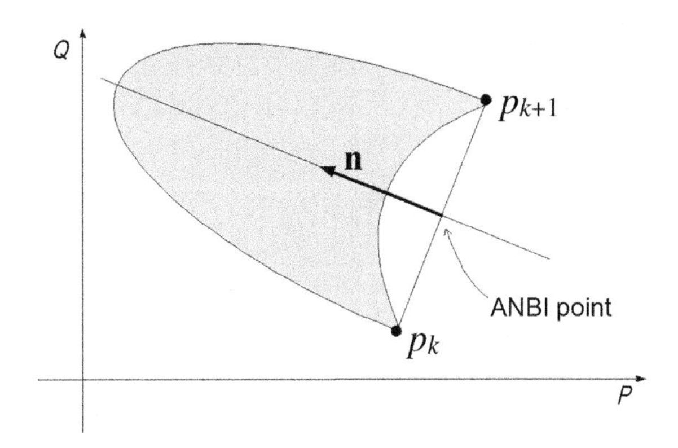

FIGURE 14.11

Schematic for finding a point on a concave region of the property closure between points p_k and p_{k+1} that have already been found.

Example 14.3

Repeat Example 14.2 using the ANBI method.

Solution

We implement the method described before using the QP and SQP solvers in MATLAB (`quadprog` and `fmincon`, respectively). The extra constraints used during the ANBI routine are given by defining the perpendicular line previously described as a function called from `fmincon`. Points in the flat region are now easily found, as demonstrated by Figure 14.12.

Note that the points are not perfectly spaced. This is due to the fact that the points found using the GWS method are not as evenly spread as those found using the ANBI method. This will always be the case since the ANBI method puts constraints on the spacing in *PQ*-space, while the GWS method uses the original linear constraints with a modification to the function being optimized, and hence cannot accurately predict the spacing in *PQ*-space.

To emphasize this fact, we demonstrate with a set of points found using only the ANBI method after the original seven points were found using the GWS method. The result is a much more even spread (Figure 14.13). This approach takes significantly more computer time, however, due to the nonlinear constraints (the difference is more noticeable the larger the number of variables).

14.3.4 Improvements to the method

Two improvements to the previous method are mentioned. The first is simply to attempt multiple values of **n** in the GWS approach before resorting to the ANBI method. A judgment must be made in terms of the extra time required for the ANBI method, and thus a limit on initial GWS trials is imposed.

For example, an example with fairly flat sides was formulated using only two orientations. The closure was found using only the ANBI method (after finding the first nine points using GWS). A total of 21 ANBI points were required to meet the constraint on the distance between points.

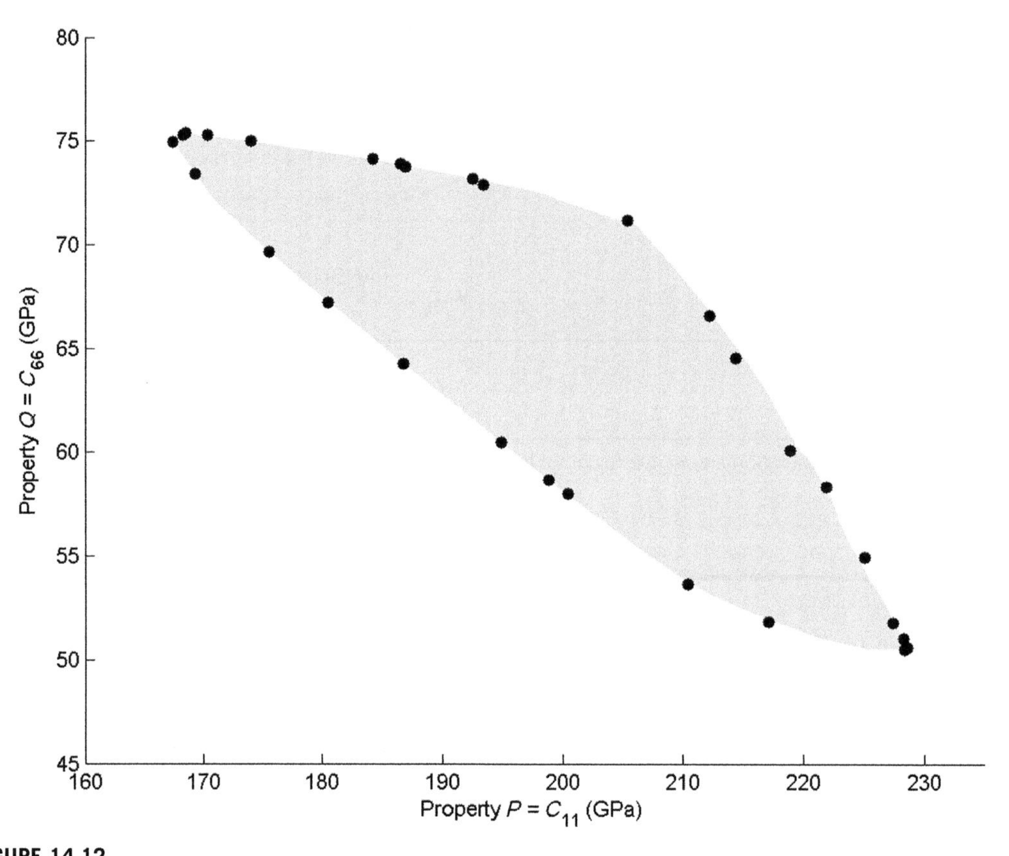

FIGURE 14.12

Points found using the GWS and ANBI methods for C_{11} and C_{66} property closures.

The same problem was repeated using GWS for the first attempt at finding a point between each pair of existing points. The number of ANBI points was reduced to 11 (note that the flatness of the surfaces resulted in a high number of ANBI points). When multiple attempts at the GWS methods were allowed (in this case, up to five attempts using different values of **n**), the number of ANBI points was reduced to 7.

The second, and more sophisticated, improvement is inspired by the AWS method (Kim and de Weck, 2005). The constraints for the AWS method are formulated in the property space, resulting in the requirement for SQP programming. The idea is to constrain the properties to values that lie between the two points that have already been found. Since we wish to formulate the constraints in the microstructure space, we instead put constraints on the ODF. The ODF is the first-order description of the microstructure and determines what fraction of each orientation is present in the design. Since we are working in a discretized space, we may define the relevant discrete ODF as

$$f^n = \frac{1}{S}\sum_{s=1}^{S} D_s^n \tag{14.7}$$

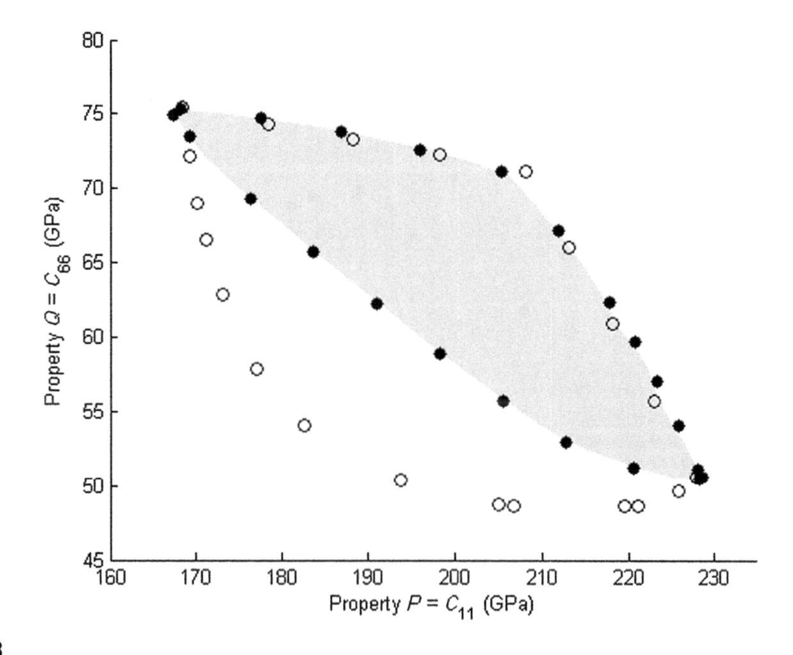

FIGURE 14.13

Points on the property closures for C_{11} and C_{66} found using only the ANBI method (after the original seven points were found using GWS). The hollow circles are from the first-order closure for comparison.

Since the properties are continuous functions of the ODF, it should be possible to force the optimization routine to find a point between the current two points on the closure by excluding an open set around the ODFs at the current points. For example, if we have points x_i and x_{i+1} on the closure and we wish to find a point between them, we constrain each f^n to lie between $(0.1f^n(x_i) + 0.9f^n(x_{i+1}))$ and $(0.9f^n(x_i) + 0.1f^n(x_{i+1}))$.

Since this constraint is linear the GWS QP algorithm can still be used. For the example tested this resulted in no ANBI points; that is, no SQP programming was needed. However, it should be borne in mind with this formulation that the constraint may result in a point slightly off the closure, and hence if it is used repeatedly over a long stretch of the closure, then small errors may accumulate.

Example 14.4

Repeat Example 14.3 using the constrained ODF method rather than the ANBI method.

Solution

In this example, seven initial directions for **n** were chosen, and the GWS method was used to find the initial points corresponding to these directions. Subsequent points were then found by constraining the ODF (as given by Eq. 14.6) for the next point to lie halfway between the ODFs for the points on either side. The search direction was also taken to be halfway between the points on either side. The resulting closure was obtained without any requirement for ANBI; however, the ODF constraint pulls the resultant points slightly off the actual boundary of the closure, as can be seen by comparing Figure 14.14 with the figures from the previous examples.

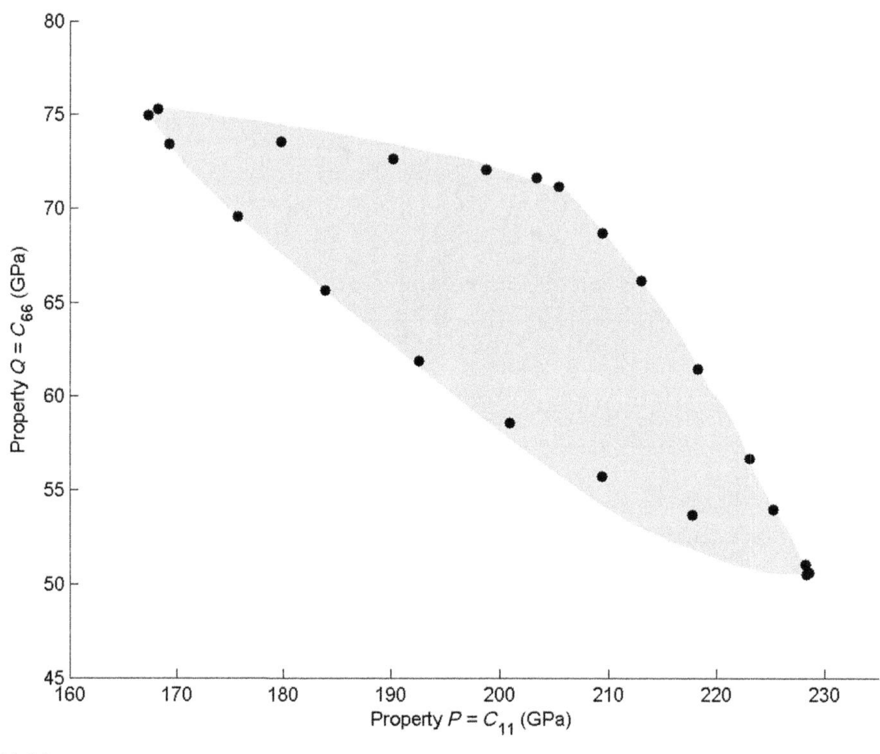

FIGURE 14.14

Property closure for C_{11} and C_{66} using the ODF constraint instead of ANBI. The resultant points are slightly off the actual boundary, especially on the top edge of the closure.

14.3.5 Accuracy of the purely linear solution

The equations for the properties, as developed earlier, involve a first-order and a second-order term. As one might expect in this type of expansion of terms, the first-order term dominates the problem. In the equation $P = \mathbf{J}^T\mathbf{D} - \mathbf{D}^T\aleph\mathbf{D}$ for the examples we have used, the coefficients of the matrix $\aleph$ are of the order of 25 times smaller than the coefficients of $\mathbf{J}$.

If we consider only the problem $P_L = \mathbf{J}^T\mathbf{D}$, the solution is much simpler than the quadratic case. Since the equation is now linear, and the feasible region is bounded by straight lines (i.e., the constraints are linear), the optimum points for P_L must lie at the vertices of the feasible region. In essence, this means that each cell of the sample contains only one phase; that is, the resultant microstructures are eigen-microstructures as defined in Section 9.1.1.

The immediate drawback with this approach is that $\mathbf{J}$ has only N distinct elements—these being the values of property P associated with each cell in the fundamental zone, divided by NS. Thus, a maximum of N points may be found on the closure, and these will simply be the microstructures that have the same orientation in every real cell. Hence, the approach is so far of no real importance.

However, if we now reduce the dimensions of the space using the technique in Section 13.2.3, the number of distinct values in vector **B** (the transformed **J**) increases dramatically (up to $N(S-1)$), and we may find a large number of points that are near the boundary of the closure. We demonstrate with an example.

Example 14.5

Demonstrate the linear approximation to the closure using Example 14.4 (derived from Example 13.3).

Solution

To find general points on the property closure, we proceed using the GWS method. Define $J = \lambda_1 P' + \lambda_2 Q'$ where P' and Q' are the linear parts of the property definition. Now minimize J for a series of $\mathbf{n} = [\lambda_1, \lambda_2]^T$ in the unit circle, as before. Once the optimal microstructures (i.e., vectors **x**) have been determined, the corresponding values of P and Q are determined using the original quadratic equation. The resulting points lie almost exactly on the closure for PQ, as illustrated in Figure 14.15.

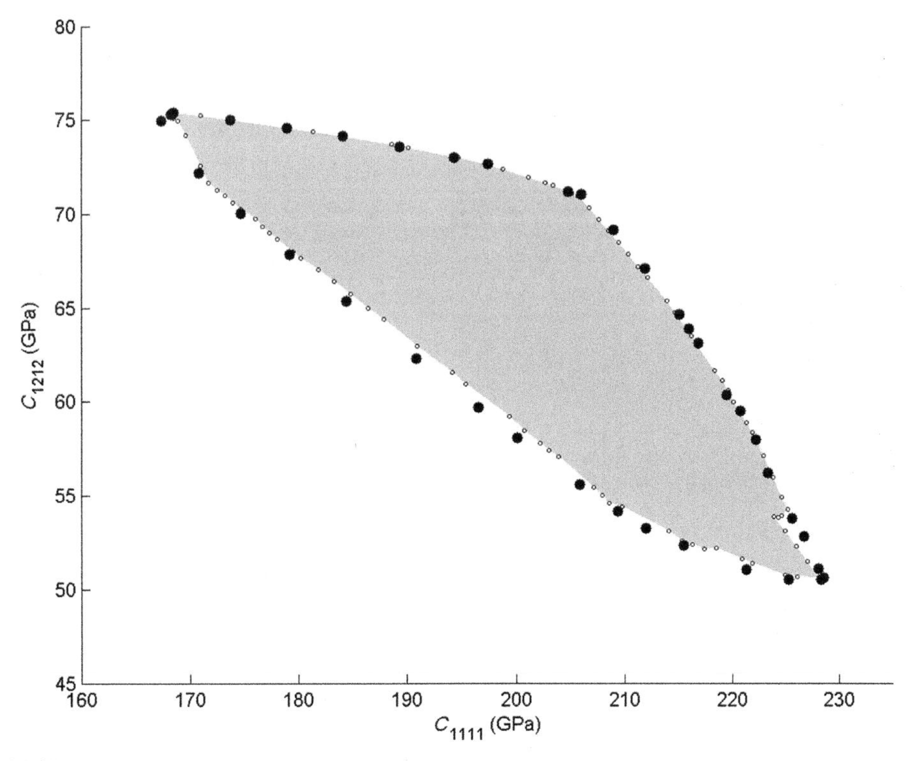

FIGURE 14.15

Second-order property closures (shaded area) for C_{1111} and C_{1212} using the linearization method. The small hollow circles give the boundary points found using the linearization technique; the solid markers give the points found using ANBI/GWS.

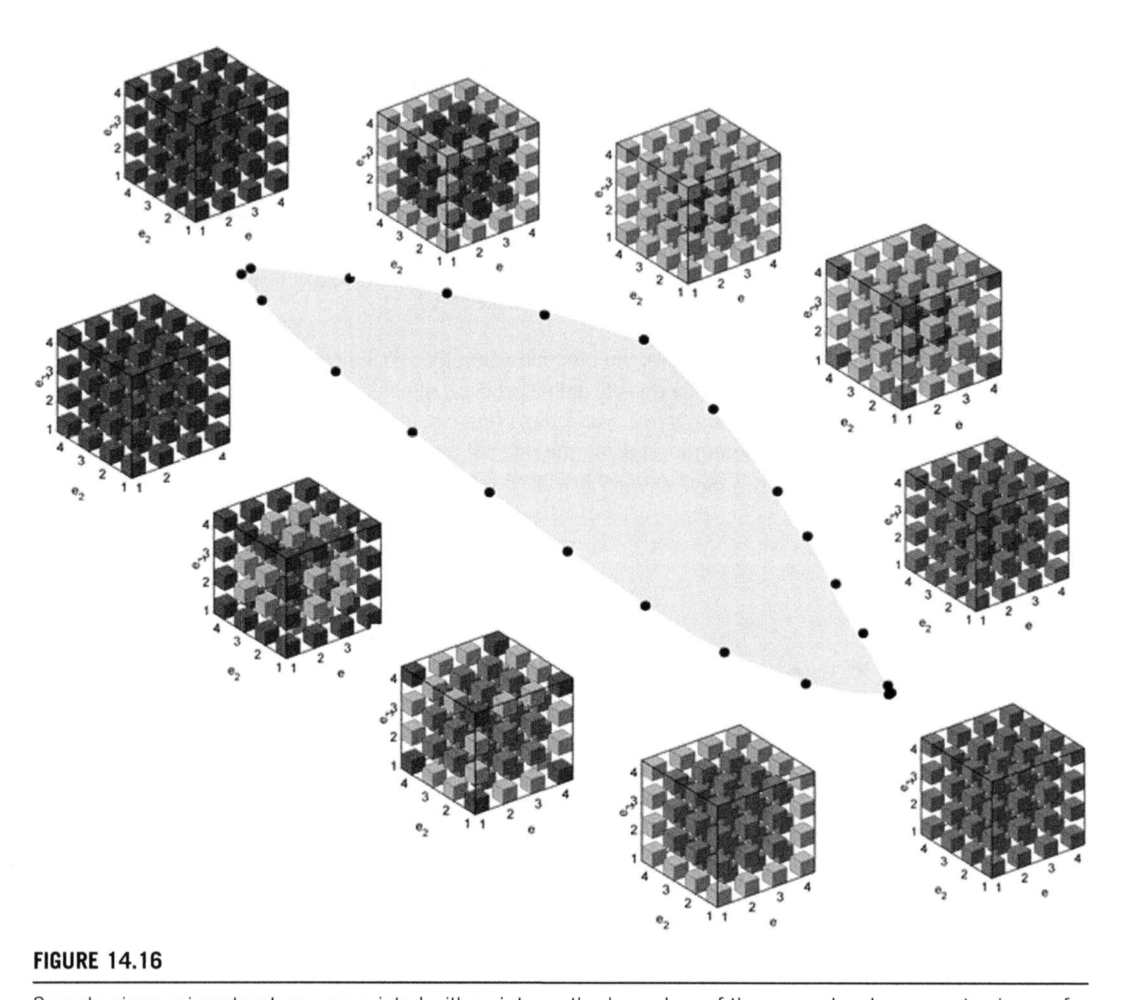

FIGURE 14.16

Sample eigen-microstructures associated with points on the boundary of the second-order property closure for material in 4 × 4 × 4 cells. Each represents a different orientation.

In Figure 14.16 we visualize the mapping between the property closure and the microstructure hull. Since we have found only eigen-microstructures on the closure, we have a single orientation in each cell of the 4 × 4 × 4 sample. Each of the four orientations (cells in the fundamental zone) is illustrated by a different gray shade. We have chosen the more symmetrical microstructures from around the boundary; in-between structures evolve from one to the next without necessarily remaining symmetrical.

While it is not expected that the linear optimization approach will result in the exact closure for larger problems, this approach may give a first approximation. Apart from giving an approximate closure the points could also be used as the starting point for a genetic algorithm.

14.3.6 Genetic algorithm approach

Since the methods being developed for property closures are intended for use on large problems, an alternative approach to quadratic or other gradient-based programming is to use genetic algorithms (GAs). This approach is good for large numbers of variables where an approximate answer is sought, rather than a single optimum point.

Again, the traditional pareto-front search can be adapted to find the whole property closure using an approach related to that given before. In this case, however, we search each quadrant (i.e., between the four optimal points for P and Q; clearly these points are not necessarily distinct, as would be the case if $N = 2$, for example), rather than producing the full closure in a single search. This is because the GA approach automatically spreads out the points along the closure for the searches in each of the four quadrants.

We adopt the maximin approach (Balling, 2000), which assigns a fitness to each "design" (or point in the PQ-space) in such a manner that the optimization naturally moves points toward the boundary of the closure while also spreading them out. Although there are many variations that can be made on the theme, we describe our approach briefly to highlight the main points of the method.

First, the PQ-space is searched using a QP algorithm to find microstructures that optimize P and Q (maximum and minimum for each). We define a generation size, G, and choose a unit normal vector $\mathbf{n}$ in the direction $[1,-1]$ (thus we are minimizing P and maximizing Q; note that whether or not $\mathbf{n}$ is unit length is irrelevant).

A "design" is a microstructure of NS variables. The genes are the microstructure variables that are associated with each cell in real space; therefore, a gene is of length N. Gene i includes $\{D^i, D^{S+i}, D^{2S+i}, \ldots, D^{(N-1)S+i}\}$, where i takes values from 1 to S.

We randomly choose $(G - 2)$ microstructures within the linear constraints, and add the microstructures that minimize P and maximize Q to give a total of G. We then choose a tournament size, T, and randomly select T of the G designs. We choose the fittest design in this tournament as our first parent based on the maximin fitness ranking. For the ith design, this is defined as

$$f(i) = \max_{j \neq i}(\min((P_i - P_j), (Q_i - Q_j))) \tag{14.8}$$

where P_i is the value of property P for the ith design, and so on.

We similarly find a second parent. The two children from these parents are defined starting from the parents themselves, and it is then decided whether to cross over their "genes" based on a crossover probability. If they are to be crossed over, a random crossover point, c, is defined, and the first child is composed of the first c genes from the first parent and the remainder from the second parent (and vice versa for the second child).

A mutation probability is defined, and if mutation is decided for a given child, then a random number is again chosen for each gene. If the number is below the mutation probability, the gene is switched to a different value randomly chosen from the feasible region of microstructure orientations (i.e., such that the original constraints are satisfied).

To ensure that values are spread out across the full quadrant of the closure, one must periodically reinject the microstructures corresponding to the minimum value of P and the maximum value of Q.

This is done at the mutation stage, where periodically two of the children are replaced by the "optimum" number of microstructures.

The fitness of the parents and children are now found with respect to the whole group, and the fittest G are chosen to go on to the next generation (i.e., elitism is used). Once the desired number of generations is reached, the process is repeated with $\mathbf{n}$ in the directions [1,1], [−1,1], [−1,−1] to find all four quadrants.

Example 14.6

Demonstrate the results from 50 generations (of 100 parents in each generation) in each quadrant using the same material as Example 14.1 (based on Example 13.3). Plot the actual closure found using the ANBI method to check the accuracy of the method.

Solution

It can be seen from Figure 14.17 that the results are reasonable for even this modest number of generations. The time taken for the calculation is significantly less (around 25% in this case) than that for the ANBI method (which found 42 points in the comparison case).

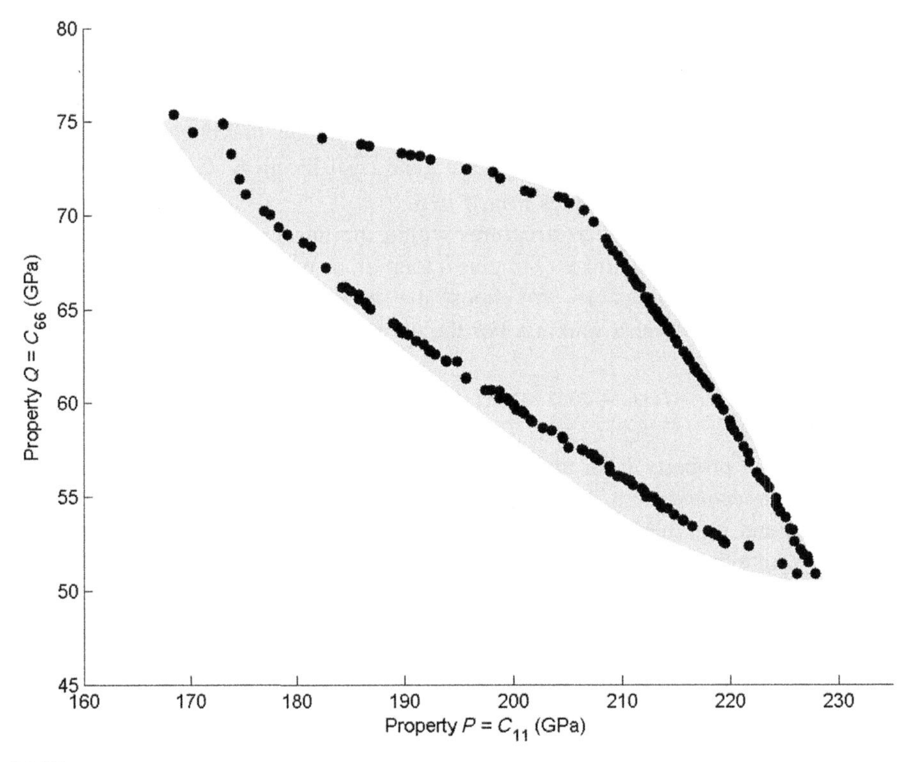

FIGURE 14.17

Genetic algorithm results (points) versus ANBI method (shaded) for property closure for C_{11} and C_{66}.

14.4 SECOND-ORDER DESIGN

As explained earlier, the motivation for finding the property closure is to allow a designer to search this space for the optimal property combination required for a design. The closure may be passed to the designer as a data set of points on the boundary, and the designer can interpolate between these points for the full boundary. Alternatively, for the 2D case, polynomials may be fitted to the points making up the boundary. In either case the designer requires a map back to the microstructure hull to enable the optimal microstructure to be specified once the optimal properties have been chosen.

If the optimal properties are on the boundary of the closure, and the data set of microstructures on the boundary is being used, then it may be adequate to interpolate between these microstructures (using convex combinations) to obtain the desired optimal microstructure. However, if the optimal property set is not on the boundary of the closure, then it will not be adequate to use convex combinations of known microstructures to obtain the required optimal microstructure design. Linear interpolation between points on the boundary may produce significant errors. We demonstrate a method of mapping between the closure and the microstructure hull using an example.

Example 14.7

Determine the optimal microstructure for a plate with a hole that is under tension. Use the same material as used in Example 14.2.

Solution

The example is similar to that in Kalidindi et al. (2004), but simplified so as not to distract from the goal of illustrating the MSD method. The geometry is that of a plate made of anisotropic material with a hole in it. The goal is to reduce the stress concentration at the edge of the hole by choice of optimal microstructure. The result in Kalidindi et al. (2004) was an optimal texture; the second-order approach seeks an actual microstructure design. The equations for stress (which is circumferential in direction) at an angle θ from the x direction, around the hole, are taken from Lekhnitskii (1968):

$$\sigma_{\theta\theta} = \sigma_\infty E_\theta \left(\sqrt{E_x/E_y}\cos^2(\theta) + (1+N)\sin^2(\theta) \right) \Big/ E_x$$

$$1/E_\theta = \sin^4(\theta)/E_x + \left(1/G_{xy} - 2\nu_{xy}/E_x \right)\sin^2(\theta)\cos^2(\theta) + \cos^4(\theta)/E_y$$

$$N = \sqrt{E_x/G_{xy} - 2\nu_{xy} + 2\sqrt{E_x/E_y}}$$

$$E_x = 1/S_{11}, E_y = 1/S_{22}, G_{xy} = 1/S_{66}, \nu_{xy} = -S_{12}/S_{11}$$

It is assumed that C_{11} and C_{66} are the most critical values of the stiffness tensor in our problem; hence, these are the coefficients that are allowed to vary during the optimization. Using the same approach as in the previous examples in this chapter, the other values for **C** are simply taken to be the average of the values for the four orientations being considered.

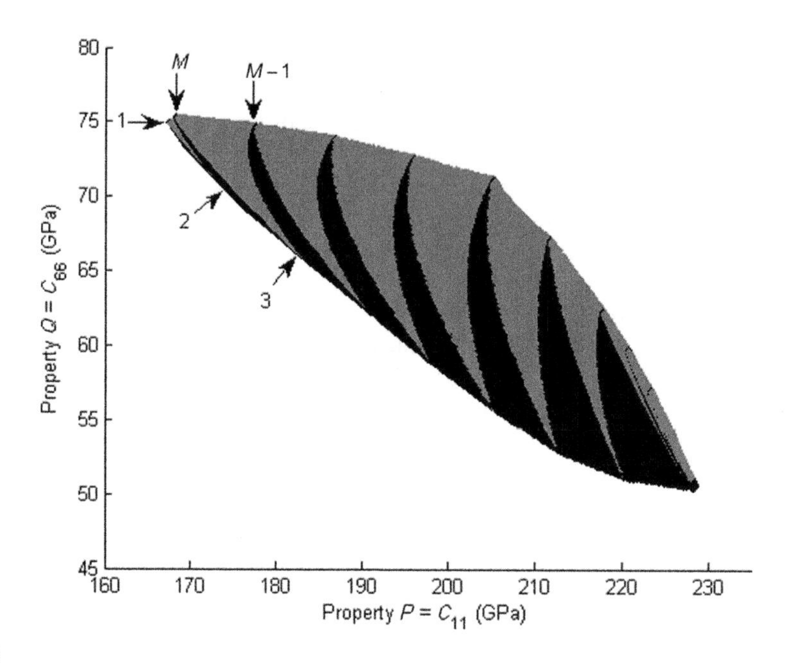

FIGURE 14.18

Areas found (alternately shaded black or gray) for convex combinations of each triple of points on the boundary of the property closure for C_{11} and C_{66}.

The closure found using the ANBI method is used. It is assumed that a designer wishes to find a microstructure that minimizes the stress concentration factor, K. Thus, the designer wishes to search the property closure for the ideal point in property space, and map that point back to the microstructure hull. There are several ways that such a mapping might be accomplished, and different methods will suit different applications. If the optimal property is assumed to be on the boundary of the closure, then the mapping is straightforward using convex combinations of the microstructures found around this border. An illustration of these microstructures for our example was shown in Figure 14.16.

A more general approach is demonstrated that allows the optimal point to be in the center of the closure. The output from the ANBI routine is a sequence of points in PQ-space that encircle the closure in an anticlockwise direction. Convex combinations of the microstructures associated with these points have properties that fill the property closure. If M points have been identified on the periphery of the closure, then first take the points $\{M,1,2\}$ (Figure 14.18) and find the minimum K for any convex combination of these microstructures (working out P and Q for each combination—Eq. (14.4)—and using quadratic programming). Repeat this exercise for the points $\{M,2,3\}$, $\{M-1,M,3\}$, and so forth, until all points have been used.

This sequence of steps takes minimum computer effort and time since the points on the boundary are already identified, and the algorithm is only optimizing over three coefficients (that define the convex combination) at each step. Then take the minimum of these optimum points and identify the related microstructure. This will be the global minimum for all points within the closure. The search is illustrated in Figure 14.18 with each convex combination forming a deformed triangle in the property closure.

For the example problem, the optimum microstructure is illustrated in Figure 14.19. It comes from the area of the top left corner of the closure. Four shades of gray are used to represent the four orientations. The dominant orientation in each cell is represented.

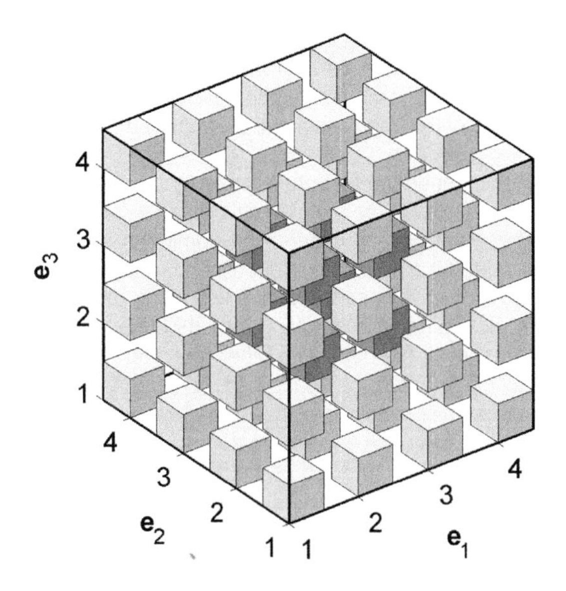

FIGURE 14.19

Representation of optimal microstructure for 4 × 4 × 4 sample. Each shade represents one of the four available orientations; only two orientations are represented in this design.

Example 14.8

In this example we demonstrate the use of the spectral expansion given in Eq. (13.63) to design a two-phase structure with a desired set of properties. The phases are assumed to have Lamé constants of $\mu = 35$ and 100 and $\lambda = 70$ and 200 GPa, respectively. The desired property combination lies near the edge of the first-order property closure and is given by $C_{11} = 220$ GPa, $C_{22} = 200$ GPa, and $C_{55} = 55$ GPa. The microstructure is assumed to occupy a cubic space with 19^3 bins.

Solution

While we could design using either the two-point statistics ($F_k^{jj'}$ in Eq. 13.63) or the DFTs of the microstructure function, related by $F_k^{jj'} = \frac{1}{S}\tilde{M}_k^j M_k^{j'}$, we choose, once again, to design in the microstructure function space (or in this case, the DFT of the microstructure function space). This saves us having to reconstruct the resultant two-point statistics to arrive at a representative microstructure. If the microstructure function itself were used as the design space, it would require $(N - 1) * S$ degrees of freedom to fully span the space, where N is the number of possible states and S is the number of elements in the model. For the relatively coarse resolution presented here, that is 6859 degrees of freedom, an unreasonably high number to map and search.

In Fourier space one may ignore the majority of the terms in $F_k^{jj'}$ (or, similarly, the majority of the M_k^j terms) of Eq. (13.63) with little influence on the accuracy; the ignored terms will be set to 0. Two methods of reducing the number of terms of the second-order representation of the microstructure were tested. The first was the removal of just the high-frequency terms of the Fourier transform of the microstructure function in Fourier space. However, this did not converge rapidly since high-frequency terms can still have significant influence in certain structures.

The second is the selection of only the dominant frequencies of the same space. The selected-terms method introduces more degrees of freedom per M_k^j term because the frequencies (k) of the terms are allowed to vary. However, using selected terms spans the space of possible microstructures with far fewer degrees of freedom overall because of the improved accuracy achieved by focusing on dominant (rather than low-frequency) terms.

Since there are only two phases, once the first phase is assigned, the second phase will be determined automatically. Eq. (13.63) was used to evaluate the material properties of a given point, or microstructure, in the reduced design space. These design parameters (the dominant spectral terms and their associated frequencies) were then optimized to best match a set of desired elastic properties.

The design space was limited to the five linearly independent M_k^1 terms (all other terms are set to 0) of the Fourier transform of the microstructure function of a $19 \times 19 \times 19$ cube. One of these terms was the volume-fraction term, representing 1 degree of freedom. The other four had arbitrary frequency and magnitude, having 5 degrees of freedom each. In total, the space had 21 degrees of freedom: 12 discrete (the DFT frequency terms) and 9 continuous (the magnitude of the volume-fraction term and the magnitude and phase angle of the other four terms).

To optimize these variables to best fit the desired properties, a least mean squares solution was sought. Because any change of volume fraction requires a recalculation of Green's function, the volume fraction was first optimized by itself using a gradient-based method, using the average of the first-order upper and lower bounds of the stiffness matrix as an estimate of the properties. The resulting stiff-phase volume fraction was 35%.

To optimize the remaining 20 variables to best fit the desired properties, a simulated annealing algorithm was used. Simulated annealing is a stochastic optimization method where a single design is randomly perturbed. Improved designs are accepted outright, but worse designs have a probability of being accepted. This probability decreases exponentially with each iteration. The simulated annealing was performed with 150 iterations, with an initial probability of accepting a poor design of 0.7 and a final probability of 0.0005. Once a good design was found, the algorithm was run again from that point to refine the design.

Two constraints were applied. The first related to the symmetry of the Fourier space, that is, $M_k^j = \tilde{M}_{S-k}^j$. The second was to force the parameters to correspond to an eigen-microstructure. An eigen-microstructure is one where only a single local state is present at each position; that is, at a given position, $\mathbf{x}$, $m(\mathbf{x}, h)$ takes the value 1 for a certain h, and 0 for all other states. In the case of a two-phase composite either one phase or the other is present at any given point.

Eigen-microstructures are desirable because they provide a deterministic model that can be readily evaluated and modeled using finite element analysis (FEA). To apply the eigen-microstructure constraint, for a given trial set of coefficients the symmetry condition may be applied; the inverse transform may then be taken to give the microstructure function. This function may then be thresholded to arrive at a suitable eigen-microstructure for analysis.

The optimized design is shown in Figure 14.20. It consists of discontinuous lamina of the stiff phase with a normal somewhere between the y and z directions. The stiffest direction is in the fiber direction.

FEA analysis was performed on the final design to validate the results of the search algorithm. A comparison of the predicted and FEA properties of this design are shown in Table 14.1. The properties derived from the method were within 0.5% to 1.5% of the desired properties in the normal direction but were 16.5% off in the shear direction.

The FEA results showed the error of the second-order design framework to be only 6% in the shear direction and less than 0.1% in the normal directions. The addition of more terms will most likely improve results without making the design space unsearchable.

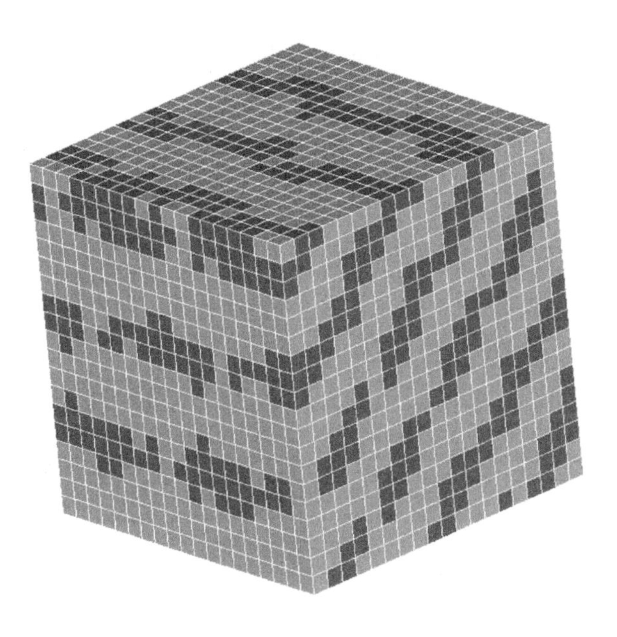

FIGURE 14.20

Optimized microstructure produced by the reduced second-order homogenization method with darker areas representing the stiffer phase.

Table 14.1 Optimized Microstructure Design Resulting Properties

Component	Desired Value	Second-Order Properties	FEA Properties
C_{11}	220	219	219.7
C_{22}	200	197.1	201.8
C_{55}	55	45.9	51.8

SUMMARY

This chapter presented the hull of two-point correlation functions in a discrete Fourier transform space. The resulting space of possible microstructures can be searched by restricting consideration to only dominant terms in the space and allowing the frequencies of these terms to vary. Such an approach is not developed in this text, partially because the resultant set of two-point statistics would then need to be reconstructed to provide a real microstructure instantiation that designers could deal with (sets of two-point statistics are not yet routinely used or interpreted by engineers). Hence, linkages to property closures were described in terms of the microstructure function so that design points chosen from such closures could be mapped directly back to actual microstructures. A small ($4 \times 4 \times 4$) microstructure design problem was illustrated, followed by a more extensive example using spectral methods to design a two-phase microstructure for desired stiffness coefficients. These examples demonstrate the potential for the method, and larger design spaces can be readily accommodated. Work is under way by the authors and many others to develop these ideas into a practical design framework for higher-order MSDPO.

In Chapter 15 we introduce some ideas that look promising for implementation of higher-order MSDPO in a processing environment.

CHAPTER

Higher-Order Models of Deformation Processing

15

CHAPTER OUTLINE

15.1 HIGHER-ORDER MODEL OF VISCO-PLASTICITY

In this chapter we introduce texture evolution theory that utilizes the primitive basis approach to model deformation over time and space. We use several of the techniques that were introduced in previous chapters; hence, this chapter will provide a useful mechanism for reviewing earlier work.

We do not explicitly discretize the sample or orientation space for the development of the following equations, but one should keep in mind that the aim is to track the evolution of texture in individual cells of both of these spaces during plastic deformation. We first reintroduce a few parameters of interest. The velocity gradient is given by

$$L_{ij} = v_{i,j} \tag{15.1}$$

Then, for rigid plasticity (ignoring elasticity),

$$\mathbf{L} = \mathbf{D}^P + \mathbf{W}^p + \mathbf{W}^* \tag{15.2}$$

where $\mathbf{D}^p$ is the plastic deformation gradient:

$$\mathbf{D}^p = \sum_{\alpha} \frac{\dot{\gamma}^\alpha}{2}(\mathbf{m}^\alpha \otimes \mathbf{n}^\alpha + \mathbf{n}^\alpha \otimes \mathbf{m}^\alpha) = \frac{1}{2}(\mathbf{L} + \mathbf{L}^T) = \frac{1}{2}\left(\mathbf{L}^p + (\mathbf{L}^p)^T\right) \tag{15.3}$$

and $\mathbf{W}^p$ is the plastic spin tensor:

$$\mathbf{W}^p = \sum_{\alpha} \frac{\dot{\gamma}^\alpha}{2}(\mathbf{m}^\alpha \otimes \mathbf{n}^\alpha - \mathbf{n}^\alpha \otimes \mathbf{m}^\alpha) = \frac{1}{2}\left(\mathbf{L}^p - (\mathbf{L}^p)^T\right) \tag{15.4}$$

with slip plane normals $\mathbf{n}^\alpha$ and slip directions $\mathbf{m}^\alpha$; $\dot{\gamma}^\alpha$ denotes the effective shear rate on this slip system.

Implicit in Eqs. (15.3) and (15.4) is the definition of $\mathbf{L}^p$ as the plastic velocity gradient:

$$\mathbf{L}^p = \sum_{\alpha} \dot{\gamma}^{\alpha} \mathbf{m}^{\alpha} \otimes \mathbf{n}^{\alpha} \tag{15.5}$$

In the absence of elastic deformations, the difference between the *applied velocity gradient tensor*, $\mathbf{L}$, and that accommodated by the slip processes, $\mathbf{L}^p$, needs to be accommodated by the *rigid-body spin*, $\mathbf{W}^*$ (see Section 7.6.1). This rigid-body spin causes a lattice rotation, $\mathbf{R}^*$. Therefore, the relationship between the lattice rotation tensor and the plastic velocity gradient tensor can be expressed as

$$\mathbf{W}^* = \dot{\mathbf{R}}^* (\mathbf{R}^*)^T = \mathbf{L} - \mathbf{L}^p \tag{15.6}$$

We next distinguish between the applied, macroscopic homogeneous velocity gradient, $\overline{\mathbf{L}}$, and the local velocity gradient, $\mathbf{L}$. Assume that the sample space is divided into cells, ω_s, and that we wish to determine the evolution of the texture within each cell. We need to determine the local velocity to calculate the amount of material entering and leaving the cell, and we need the local lattice spin, $\mathbf{W}^*$.

The approach is to apply a global $\overline{\mathbf{L}}$; we then need to calculate the local v_i and $\mathbf{W}^*$. To calculate the $\mathbf{W}^*$ we require the local $\mathbf{L}$. We obtain this by using the homogenization relation that links the local $\mathbf{L}$ to the global $\mathbf{L}$. Following Adams (1989), let $\mathbf{N}$ be the secant modulus; then

$$\sigma_{ij}^D = N_{ijkl} \varepsilon_{kl}^D \tag{15.7}$$

where D stands for the deviatoric stress and strain. Also, the creep compliance is given by

$$\mathbf{M} = \sum_{\alpha} \left(\dot{\gamma}^0 \frac{\left| \boldsymbol{\sigma}^D \cdot \mathbf{P}^{\alpha} \right|^{\frac{1}{m}-1}}{(s^{\alpha})^{\frac{1}{m}}} (\mathbf{P}^{\alpha} \otimes \mathbf{P}^{\alpha}) \right) \tag{15.8}$$

and the secant modulus $\mathbf{N} = \mathbf{M}^{-1}$ (see also the alternative form in Adams et al. (1989)).

For rate-independent plasticity, the slip resistance, S^a, is exactly the same as the critical resolved shear stress (see Section 7.6.1), and m is the rate sensitivity parameter in the slip law:

$$\tau^{(k)} / \tau_0^{(k)} = \left(\gamma^{(k)} / \gamma_0 \right)^m \tag{15.9}$$

Furthermore, if $n = 1/m$, then $\mathbf{N}(\lambda\varepsilon) = \lambda^{(1-n)/n} \mathbf{N}(\varepsilon)$, which reduces the number of times $\mathbf{N}$ must be calculated.

The Cauchy stress is arrived at from the deviatoric stress by adding the hydrostatic pressure:

$$T_{ij} = N_{ijkl} \varepsilon_{kl} - p\delta_{ij} = N_{ijkl} L_{kl} - p\delta_{ij} \tag{15.10}$$

The equilibrium equation requires that

$$T_{ij,j} = \left(N_{ijkl} L_{kl} \right)_{,j} - p_{,i} = 0 \tag{15.11}$$

Now define a reference secant modulus,

$$N^R_{ijkl} = \int_{g \in FZ} f(h) N_{ijkl}\left(\overline{\mathbf{L}}, \tau_0^*, g\right) dg \tag{15.12}$$

where it is assumed that the reference stress τ_0^* may be taken as a constant across the material. A polarized modulus is then given as

$$\tilde{N}_{ijkl}\left(\mathbf{L}, \tau_0^*, g\right) = N_{ijkl}\left(\mathbf{L}, \tau_0^*, g\right) - N^R_{ijkl} \tag{15.13}$$

Then we can write the equilibrium equation as

$$N^R_{ijkl} L_{kl,j} - p_{,i} + \left[\tilde{N}_{ijkl}(\mathbf{L}, g) L_{kl}\right], i = N^R_{ijkl} L_{kl,j} - p_{,i} + f_i = 0 \tag{15.14}$$

where f_i is considered a fictitious body force. This produces the three differential equations. The fourth equation is given by incompressibility:

$$L_{ii} = 0 \tag{15.15}$$

Now, using the Green's function approach to solve these equations, with Green's functions G and H,

$$N^R_{ijkl} G_{km,lj}(x - x') - H_{m,i}(x - x') + \delta_{im} \delta(x - x') = 0 \tag{15.16}$$

$$G_{im,i}\left(x - x'\right) = 0 \tag{15.17}$$

leads to solutions for v and p:

$$v_n(x) = \overline{v}_n(x) + \int_V G_{ki}(x - x') f_i(x') \mathrm{d}x'^3 \tag{15.18}$$

$$p(x) = \overline{p}(x) + \int_V H_i(x - x') f_i(x') dx'^3 \tag{15.19}$$

One might rearrange the equation for **V**, taking the derivative with respect to x and integrating by parts to obtain

$$L_{ik}(x) = \overline{L}_{ik}(x) + \int_V G_{ij,kl}(x - x') \tilde{N}_{jlrs}(\mathbf{L}(x'), g(x')) L_{rs}(x') dx'^3 \tag{15.20}$$

The corresponding Fourier transform of the partial differential equations yields

$$N^R_{ijkl} k_j k_1 \tilde{G}_{km}(k) - i k_i \tilde{H}_m(k) + \delta_{im} = 0 \tag{15.21}$$

$$k_k \tilde{G}_{km}(k) = 0 \tag{15.22}$$

Example 15.1

Find the Green's function, G, assuming isotropic $\mathbf{N}^R$.

Solution

If we assume that $\mathbf{N}^R$ is isotropic then it can be written as (Molinari et al., 1987)

$$N^R_{ijkl} = \frac{\mu}{2}\left(\delta_{ik}\delta_{jl} + \delta_{il}\delta_{jk}\right) \tag{15.23}$$

Then we can write (15.21) as

$$\frac{\mu^r}{2}(\delta_{ik}\delta_{jl} + \delta_{il}\delta_{jk})k_j k_1 \tilde{G}_{km}(k) - ik_i\tilde{H}_m(k) + \delta_{im} = 0 \tag{15.24}$$

Substituting Eq. (15.22) into Eq. (15.24), we obtain

$$\left(-\frac{\mu^r}{2}\right)k_i k_k \tilde{G}_{km}(k) - \frac{\mu^r}{2}|k|^2\tilde{G}_{im}(k) + ik_i\tilde{H}_m(k) + \delta_{im} = 0 \tag{15.25}$$

Taking Eq. (15.25) $\times\, k_i$ and summing over repeated indices, and changing the repeated i's to ks in the second term, we get

$$\left(-\frac{\mu^r}{2}\right)|k|^2 k_k \tilde{G}_{km}(k) - \frac{\mu^r}{2}|k|^2 k_k \tilde{G}_{km}(k) + i\cdot|k|^2\tilde{H}_m(k) + k_m = 0$$

That is,

$$-\mu^r|k|^2 k_k\tilde{G}_{km}(k) + i\cdot|k|^2\tilde{H}_m(k) + k_m = 0 \tag{15.26}$$

Thus,

$$\tilde{G}_{km}(k) = \frac{i\cdot|k|^2\tilde{H}_m(k) + k_m}{\mu^r|k|^2 k_k} \tag{15.27}$$

Substituting Eq. (15.27) into Eq. (15.25) and setting the first term to 0 (Eq. 15.22), we get

$$-\frac{\mu^r}{2}|k|^2\cdot\frac{i\cdot|k|^2\tilde{H}_m(k) + k_m}{\mu^r|k|^2 k_k} + ik_i\tilde{H}_m(k) + \delta_{im} = 0$$

$$\frac{i\cdot|k|^2\tilde{H}_m(k) + k_m}{2k_k} + ik_i\tilde{H}_m(k) + \delta_{im} = 0 \tag{15.28}$$

$$\tilde{H}_m\left(\frac{-i|k|^2}{2k_k} + ik_i\right) - \frac{k_m}{2k_k} + \delta_{im} = 0$$

Taking Eq. (15.28) $\times k_i$, we get

$$\tilde{H}_m\left(\frac{-i|k|^2 k_i}{2k_k}+i|k|^2\right)-\frac{k_m k_i}{2k_k}+k_m = 0$$
$$\tilde{H}_m\left(\frac{-i|k|^2}{2}+i|k|^2\right)+\frac{k_m}{2} = 0 \tag{15.29}$$

$$\tilde{H}_m = \frac{-k_m}{i|k|^2} = \frac{ik_m}{k^2} \tag{15.30}$$

Substituting Eq. (15.30) into Eq. (15.25) with Eq. (15.22), we get

$$-\frac{\mu^r}{2}|k|^2\tilde{G}_{im}(k)+i\cdot k_i\frac{i\cdot k_m}{k^2}+\ \delta_{im} = 0$$
$$-\frac{\mu^r}{2}|k|^2\tilde{G}_{im}(k)-\frac{k_i k_m}{k^2}+\delta_{im} = 0 \tag{15.31}$$

$$\tilde{G}_{im} = \frac{2\delta_{im}}{\mu^r k^2}-\frac{2(k_i k_m)}{\mu^r k^4} \tag{15.32}$$

We now demonstrate how to find the inverse Fourier transform for $\tilde{Z} = \frac{1}{k^2}$. In 3D space the inverse Fourier transform is given by

$$Z(\mathbf{x}) = \frac{1}{(2\pi)^3}\iiint_{R^3} e^{i\mathbf{k}.\mathbf{x}}\tilde{Z}(\mathbf{k})dv \tag{15.33}$$

where x is the position vector in R^3. Changing to spherical coordinates (note that k indicates the length of vector k, and hence this is the radial variable),

$$\left.\begin{aligned} k_1 &= k\cos\theta\sin\phi \\ k_2 &= k\sin\theta\sin\phi \\ k_3 &= k\cos\theta \end{aligned}\right\} dv = k^2\sin\phi d\phi d\theta dk \tag{15.34}$$

It can be shown (Champeney, 1973) that for a spherically symmetric function (letting r denote the radial variable in real space),

$$\frac{1}{(2\pi)^3}\iiint_{R^3} e^{i\,\mathbf{k}.\mathbf{x}}\tilde{Z}(\mathbf{k})k^2\sin\phi d\phi d\theta dk = \frac{1}{(2\pi)^3}\int_0^\infty \tilde{Z}(\mathbf{k})\frac{\sin kr}{kr}4\pi k^2 dk \tag{15.35}$$

Hence,

$$Z(r) = \frac{4\pi}{(2\pi)^3}\int_0^\infty \frac{\sin kr}{k^3 r}k^2 dk = \frac{1}{2\pi^2}\frac{\pi}{2r} = \frac{1}{4\pi r} \tag{15.36}$$

To determine the inverse Fourier transform for the remainder of Eq. (15.32) we utilize the often quoted Green's function for elasticity:

$$\tilde{J} = \frac{\delta_{km}}{\mu|k|^2} - \frac{(\lambda+\mu)k_k k_m}{\mu(\lambda+2\mu)|k|^4}, \quad J = \frac{(\lambda+3\mu)\delta_{km}}{8\pi\mu(\lambda+2\mu)|x|} + \frac{(\lambda+\mu)x_k x_m}{8\pi\mu(\lambda+2\mu)|x|^3} \tag{15.37}$$

where $|\mathbf{x}| = r$. Using the previous result for $\tilde{Z} = \frac{1}{k^2}$ it is easy to determine the inverse transform for the second term in Eq. (15.32):

$$\tilde{Y} = \frac{k_k k_m}{|k|^4}, \quad Y = \frac{\delta_{km}}{8\pi|x|} - \frac{x_k x_m}{8\pi|x|^2} \tag{15.38}$$

This leads us to

$$G(\tilde{x}) = \frac{2\delta_{im}}{4\pi\mu^r|x|} - \frac{2}{\mu^r}\left(\frac{\delta_{im}}{8\pi|x|} - \frac{x_i x_m}{8\pi|x|^3}\right) \tag{15.39}$$

Note that this assumes isotropic $\mathbf{N}$. To ensure that this is the case, one may define a new reference tensor that is isotropic, and as close as possible to the original $\mathbf{N}^R$. Another alternative is to use the formulation in Adams et al. (1989).

We now employ an iterative scheme to recover **L** from Eq. (15.20). Write the equation in reduced form as follows:

$$L_{ik}(x) = \overline{L}_{ik}(x) + \mathrm{G}_{ij,kl}(x-x')^* \tilde{N}_{jlrs}(x')L_{rs}(x') \tag{15.40}$$

Then, inserting this equation into itself,

$$L_{ik}(x) = \overline{L}_{ik}(x) + \mathrm{G}_{ij,kl}(x-x')^* \tilde{N}_{jlrs}(x')\left(\overline{L}_{rs}(x) + \mathrm{G}_{rm,sn}(x-x')^* \tilde{N}_{mnpq}(x')L_{pq}(x')\right) \tag{15.41}$$

This process may be repeated over and over. If we assume that the polarization terms are small compared to the reference values, then higher-order terms will decrease in value. This equation is analogous to similar equations for elasticity in Chapter 13, such as Eq. (13.11). The theory could be developed in the same way as elasticity to incorporate higher-order correlation functions. We do not fully develop in that direction in this chapter. Instead, we use the results to model the evolution of texture with time. For the following, we only consider the first two terms on the right side of Eq. (15.41). Hence,

$$L_{ik}(x) = \overline{L}_{ik}(x) + \mathrm{G}_{ij,kl}(x-x')^* \tilde{N}_{jlrs}(x')\overline{L}_{rs}(x') \tag{15.42}$$

Note that there is a singularity in Green's function at $x = x'$. One way to deal with this is to take the average value of the function in a small neighborhood of x for the value at x (see Adams et al., 1989). Also note that since we are actually looking for average values of **L** within each cell of the sample, this scheme is likely to suit our requirements. We must be careful to take account of the nonspherical shape of the sample region if we use the analytical values for Green's function (see Section 13.1).

The velocity, **v**, is obtained by integrating **L** according to Eq. (15.1). To recover the lattice rotation we require $\mathbf{L}^p$. This is obtained by recovering $\mathbf{D}^p$, and therefore $\dot{\gamma}^\alpha$, from Eq. (15.3). Then, using Eq. (15.5), we obtain $\mathbf{L}^p$. Finally, we obtain $\mathbf{W}^*$ from Eq. (15.6).

Once $\mathbf{W}^*$ is available, the actual rotation rate may be obtained (Adams and Field, 1991) (these results are based on linearization of the equations; see Morawiec, 2004):

$$\begin{aligned}\dot{\varphi}_1 &= W_{23}^* \frac{\sin\varphi_2}{\sin\Phi} - W_{13}^* \frac{\cos\varphi_2}{\sin\Phi} \\ \dot{\Phi} &= W_{23}^* \cos\varphi_2 + W_{13}^* \sin\varphi_2 \\ \dot{\varphi}_2 &= -\dot{\phi}_1 \cos\Phi + W_{12}^*\end{aligned} \tag{15.43}$$

We use a different form of these equations in Section 15.2 as it becomes apparent that it is easier to work with Rodriguez angles in the calculations.

15.2 TIME- AND SPACE-DEPENDENT MODELING OF TEXTURE EVOLUTION

As originally stated, we are interested in the evolution of the orientation distribution function across a sample; hence, we must employ the microstructure function, which contains information about both space and orientation distribution. Recall the definition of the microstructure function $M(x, g)$:

$$M(x, g)dg = dV/V \tag{15.44}$$

This function tells us that the volume fraction density of material lying in the neighborhood of position x has orientation g. Consider a region $\Re$ of the material surrounded by surface ζ. The volume of material at time t lying within region $\Re$, and having orientation lying within invariant measure dg of g, is

$$\iiint_{\Re} M(x, g)dgdV \tag{15.45}$$

Over an increment of time δt, during which deformation is taking place, the orientation g at a given point or at a given particle in $\Re$ will (in general) change, and some of the particles will travel across the surface ζ, thereby transporting particles of various orientation into and out of $\Re$. The rate of change in the volume of material associated with invariant measure dg of g is given by the notation

$$\frac{D}{Dt} \iiint_{\Re} M(x, g)dgdV \tag{15.46}$$

The rate of increase of this orientation component is equal to the sum of the rate of increase of this component associated with the particles instantaneously within $\Re$, plus the net rate of influx of this component through ζ into $\Re$. Thus,

$$\frac{D}{Dt} \iiint_{\Re} M(x, g)dV = \iiint_{\Re} \frac{\partial M(x, g)}{\partial t} dV + \iint_{\zeta} M(x, g)\mathbf{v}\cdot\hat{n}dS \tag{15.47}$$

where $\hat{n}dS$ is an infinitesimal patch of the surface ζ of measure dS and outward normal $\hat{n}$, and $\mathbf{v} = \mathbf{v}(x)$ is the velocity of the current material particle at position x. By applying the divergence theorem to the surface integral in relation (15.47), we obtain

$$\frac{D}{Dt}\iiint_{\Re} M(x,g)dV = \iiint_{\Re}\left\{\frac{\partial M(x,g)}{\partial t} + div(M(x,g)\mathbf{v})\right\}dV \tag{15.48}$$

It is useful to look at the integrand of relation (15.48) more carefully, by examining it in index form. Note that $M(x,g)\mathbf{v}$ is a vector of magnitude equal to the velocity of the material point at position x, weighted by the volume fraction density of orientation g at that position. The velocity depends only on position x and not on orientation g. In component form the integrand of (15.48) can be written as

$$\frac{\partial M(x,g)}{\partial t} + div(M(x,g)\mathbf{v}) = \frac{\partial M(x,g)}{\partial t} + M(x,g)\frac{\partial v_j}{\partial x_j} + v_j\frac{\partial M(x,g)}{\partial x_j} \tag{15.49}$$

Our next task is to evaluate the time derivative of the texture function (i.e., the first term on the right side of relation (15.49)). Assuming that via some model the lattice rotation rate field $\varpi = \varpi(x,g)$ is known, then the orientation flow field at position x is $M(x,g)\varpi(x,g)$. It is to be understood here that the dependence of ϖ on position x is related through the deformation model to the local velocity gradient tensor at that position. Given that crystallite orientations are neither created nor destroyed, a principle of conservation of orientations can be expressed in the form of a continuity relation (Clement, 1982; Morawiec, 2004):

$$\frac{\partial M(x,g)}{\partial t} = -div(M(x,g)\varpi(x,g)) \tag{15.50}$$

The divergence in relation (15.50) is a divergence with respect to orientation parameters associated with g, and they must be evaluated in accordance with the specialized geometry of the orientation space. The simplest form of the term on the right side of (15.50) is in terms of the Rodriguez parameters, r^1, r^2, r^3, which are conveniently related to the axis $(\hat{n})$ and angle (θ) parameters, for a rotation θ about $\hat{n}$, by

$$r^i = n_i \tan(\theta/2) \tag{15.51}$$

The axis and angle parameters are themselves related to Bunge's orientation parameters (passive direction cosines) via the relation

$$g_{ij} = \delta_{ij}\cos\theta + \varepsilon_{ijk}n_k\sin\theta + n_i n_j\left(1 - \cos\theta\right) \tag{15.52}$$

In terms of the Rodriguez vector the divergence term in (15.50) has the form

$$\begin{aligned} div(M(x,r)\varpi(x,r)) &= \frac{\partial(M(x,r)\varpi(x,r))^j}{\partial r^j} - 4\frac{(M(x,r)\varpi(x,r))^j r^j}{(1 + r^k r^k)} \\ &= M(x,r)\frac{\partial\varpi^j(x,r)}{\partial r^j} + \varpi^j(x,r)\frac{\partial M(x,r)}{\partial r^j} - 4\frac{M(x,r)\varpi^j(x,r)r^j}{(1 + r^k r^k)} \end{aligned} \tag{15.53}$$

Thus, in terms of the Rodriguez parameterization of lattice orientation, relation (15.48) becomes

$$\frac{D}{Dt}\iiint\limits_{\Re} M(x,r)dV = \iiint\limits_{\Re}\left\{\frac{\partial M(x,r)}{\partial t} + div(M(x,r)\mathbf{v})\right\}dV \tag{15.54}$$

And, incorporating the results of relations (15.49) through (15.53), this becomes

$$\begin{aligned}\frac{D}{Dt}\iiint\limits_{\Re} M(x,r)dV = \iiint\limits_{\Re}\Bigg\{ & M(x,r)\frac{\partial v_j}{\partial x_j} + v_j\frac{\partial M(x,r)}{\partial x_j} \\ & - M(x,r)\frac{\partial \varpi^j(x,r)}{\partial r^j} - \varpi^j(x,r)\frac{\partial M(x,r)}{\partial r^j} \\ & + 4\frac{M(x,r)\varpi^j(x,r)r^j}{(1+r^k r^k)}\Bigg\}dV\end{aligned} \tag{15.55}$$

At this point in the derivation, readers should note that $\partial v_i/\partial x_j = L_{ij}$ is the well-known velocity gradient tensor, and in incompressible media, $L_{jj} = 0$. It follows that relation (15.55) simplifies to

$$\begin{aligned}\frac{D}{Dt}\iiint\limits_{\Re} M(x,r)dV = \iiint\limits_{\Re}\Bigg\{ & v_j\frac{\partial M(x,r)}{\partial x_j} - M(x,r)\frac{\partial \varpi^j(x,r)}{\partial r^j} \\ & - \varpi^j(x,r)\frac{\partial M(x,r)}{\partial r^j} + 4\frac{M(x,r)\varpi^j(x,r)r^j}{(1+r^k r^k)}\Bigg\}dV\end{aligned} \tag{15.56}$$

The lattice rotation rate vector, $\varpi(x,r)$, is related to the local velocity gradient tensor in a particular way. In terms of the time rate of change of the (passive) parameters of Bunge, the relationship is

$$\frac{dg_{ij}}{dt} = \delta_{ij} - \varepsilon_{ijk}\varpi^k \tag{15.57}$$

In terms of the deformation model, assume that the local velocity gradient tensor is $L(x)$. We have shown how to recover the lattice spin, $\mathbf{W}^*$. The components of the lattice rotation rate vector are known to be related to $\mathbf{W}^*$ by the expression

$$\varpi^i = \varepsilon_{ijk}W^*_{jk}/2 \tag{15.58}$$

Therefore, having obtained $\mathbf{v}$ and $\varpi(x,r)$ we may now use Eq. (15.56) to determine the evolution of the microstructure function in small cells represented by the region $\Re$ in the integrals.

SUMMARY

In this chapter readers were exposed to a theoretical framework for capturing the higher-order details of microstructure evolution using higher-order composite theories. It is hoped that readers can appreciate the MSDPO framework's power in addressing this highly challenging problem. It should be recognized that process design even using the first-order composite theories is itself highly challenging. Therefore, it is not surprising that process design using higher-order composite theories is substantially more difficult. This chapter showed the essential foundations for addressing higher-order process design. However, it should be clear to readers that additional development is needed before we can obtain practical solutions to the process design problem using higher-order theories.

CHAPTER

Electron Backscatter Diffraction Microscopy and Basic Stereology 16

CHAPTER OUTLINE

This chapter focuses on microstructure characterization to determine the structure metrics (e.g., grain size and crystal orientation) used in the preceding chapters. Traditional electron backscatter diffraction methods are described, and recent advances in high-resolution techniques are briefly overviewed. Some stereology basics used to infer 3D information from 2D data are also mentioned.

16.1 INTRODUCTION

Traditionally, microstructure has referred to a structure that can be viewed in a microscope. Thus, features such as grain size or shape are visually critical. In some imaging modes, different phases are also visually evident due to higher or lower intensity than the surrounding matrix. However, if we focus purely on the visual aspects of the microstructure (2D shape and morphology of the constituents), we miss aspects of the structure that are critical to the material properties. Two key aspects of the microstructure that are not visually evident in basic microscopic imaging are the chemical composition and crystallographic structure and orientation of the constituent grains. The focus of modern microscopy is microstructure at the scale that is revealed in the scanning electron microscope (SEM).

Spatial variation in chemical composition can be measured using X-ray energy dispersive spectroscopy (EDS). The crystallographic structure of the constituents is best revealed using electron backscatter diffraction (EBSD). The combination of EBSD and EDS is that which is most critical to the microstructure design concepts presented in this book. In many cases, EDS is mainly helpful in distinguishing phase identity. This is especially important when two or more phases have similar crystallographic characteristics, leading to ambiguous phase identification by EBSD alone. The main focus of this chapter is on the aspects of EBSD that are most important for MSDPO, but readers are reminded that EDS may be important for distinguishing phase in some materials.

16.2 PATTERN FORMATION

An EBSD pattern is formed when a stationary electron beam is focused on a highly tilted (approximately 70°) sample mounted in the SEM as shown in Figure 16.1. The interaction of the incident electron beam with the sample creates an interaction or diffraction volume within the sample. The volume acts essentially as an electron source with electrons being scattered in all possible directions. As the scattered electrons interact with crystallographic planes within the crystal lattice, those satisfying Bragg's law are coherently diffracted out of the sample along the crystal planes. As these electrons collide with a phosphor screen placed within the SEM chamber a pattern is formed. The pattern can be imaged using a low-light camera (typically a charge-coupled device, or CCD). Such patterns provide considerable information about the crystal lattice structure within the interaction volume. Here, our focus is on retrieval of lattice orientation from EBSD patterns, but readers should be aware that other features, such as local elastic distortion, may also be recovered.

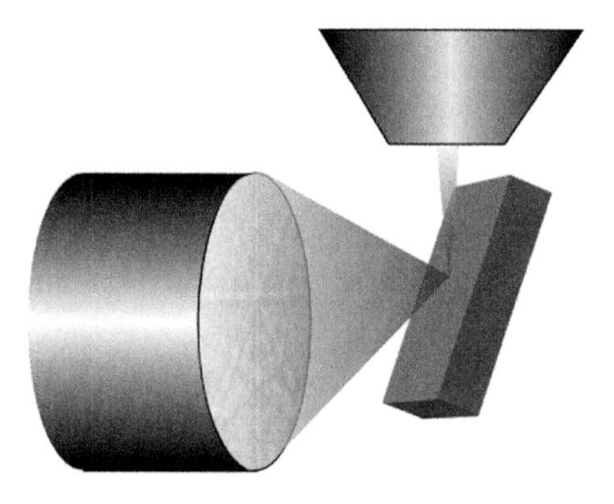

FIGURE 16.1

Schematic of EBSD. *Courtesy of EDAX/TSL.*

Bragg's law is given by the following equation:

$$\lambda = 2d_{hkl} \sin \theta \tag{16.1}$$

where λ is the wavelength of the incident radiation (electrons in this case), d_{hkl} is the interplanar spacing or *d*-spacing associated with the (*hkl*) plane, and θ is the angle between the incident wave and the scattered wave. The implications of Bragg's law on the patterns can be observed in the patterns. The first is the wavelength. The wavelength is a function of the energy of the electron beam. The impact of the SEM's accelerating voltage is evident in the patterns shown in Figure 16.2. Increasing the voltage decreases the width of the bands. The width of the bands is also a function of the *d*-spacing of the atomic plane. As d_{hkl} increases $\sin \theta$ must decrease to maintain Bragg's law (θ decreases). The width of the bands in the pattern is proportional to $\sin \theta$. Thus, thinner bands are associated with planes of higher *d*-spacing, which are typically the lower index planes as shown in Figure 16.3.

To a first-order, kinematic approximation, the band intensity associated with a given plane is given by the square of the structure factor:

$$F_{hkl} = \sum_i f_i(\theta) e^{-2\pi i(hx_i + ky_i + lz_i)} \tag{16.2}$$

where F_{hkl} is the structure factor associated with a given plane, $f_i(\theta)$ is the atomic scattering factor, and x_i, y_i, and z_i are the positions of given atoms within the unit cell. One of the important implications of Eq. (16.2) is the effect of the atomic scattering factor. The atomic scattering factor is a function of the atomic density: The higher the density, the higher the scattering factor. This effect is evident in Figure 16.4.

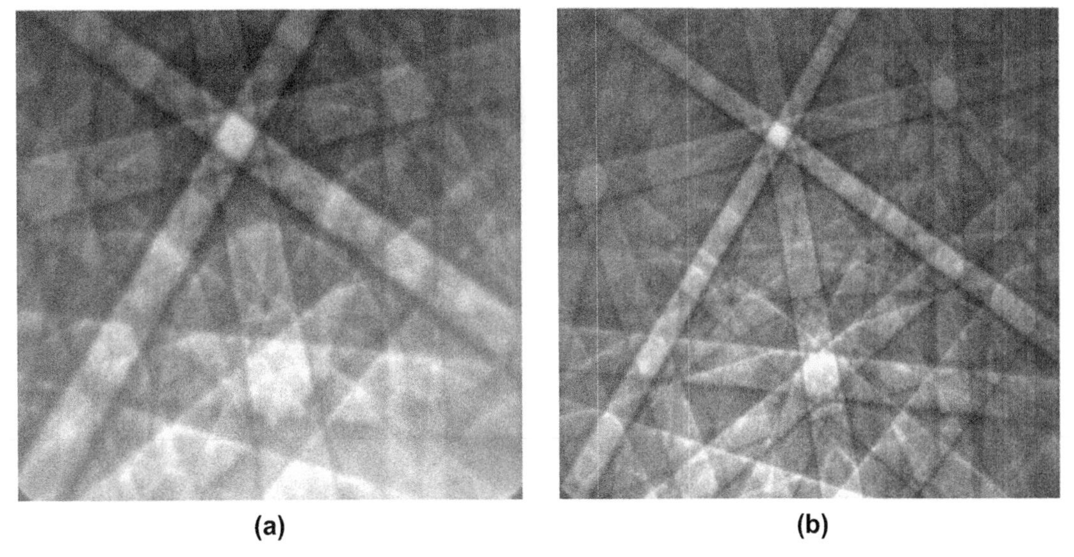

FIGURE 16.2

EBSD patterns from single-crystal silicon obtained at accelerating voltages of (a) 10 kV and (b) 30 kV. *Courtesy of EDAX/TSL.*

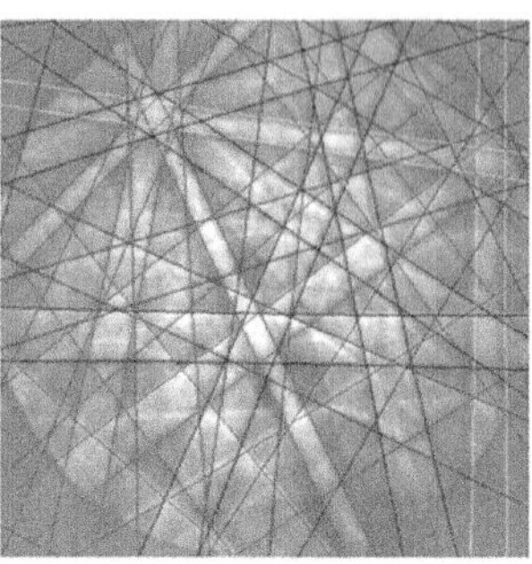

Color	(*hkl*)	*d*-spacing
	111	2.31
	200	2
	220	1.41
	311	1.21
	331	0.92
	042	0.89

FIGURE 16.3

EBSD pattern from nickel showing the relation between band widths and *d*-spacing. *Courtesy of EDAX/TSL.*

Besides the species of atoms present in the unit cell, the quality of the diffraction pattern is a function of the crystal lattice state within the diffraction volume. For instance, if the beam is located near a boundary, then the resulting pattern is a superposition of the patterns from the two individual lattices on either side of the boundary as shown in Figure 16.5. This same idea can be extended to deformed materials. If the crystal lattice within the diffraction volume is plastically deformed then the volume will essentially contain a superposition of several crystal lattices slightly misoriented from each other. In addition, the dislocations present associate with local elastic distortions that disturb the perfection of the lattice. These effects lead to a more diffuse diffraction pattern.

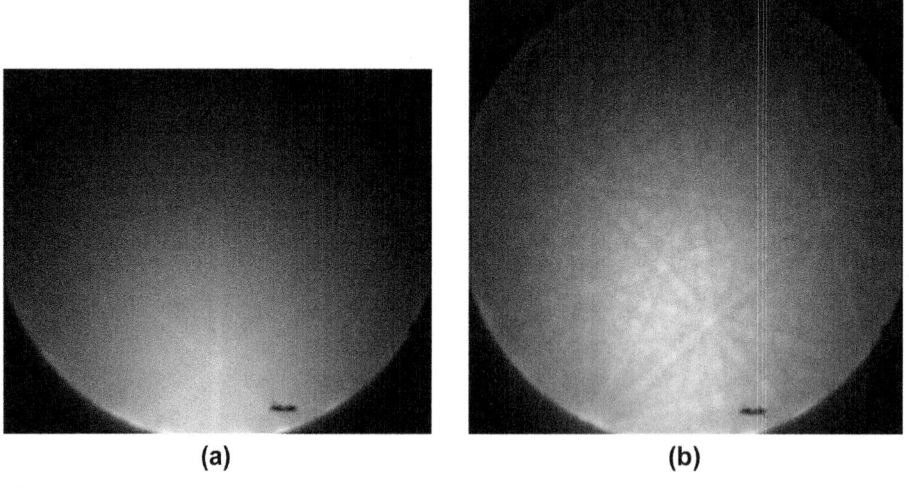

FIGURE 16.4

EBSD patterns from (a) aluminum and (b) tantalum patterns obtained at identical operating conditions without any image processing. *Courtesy of EDAX/TSL.*

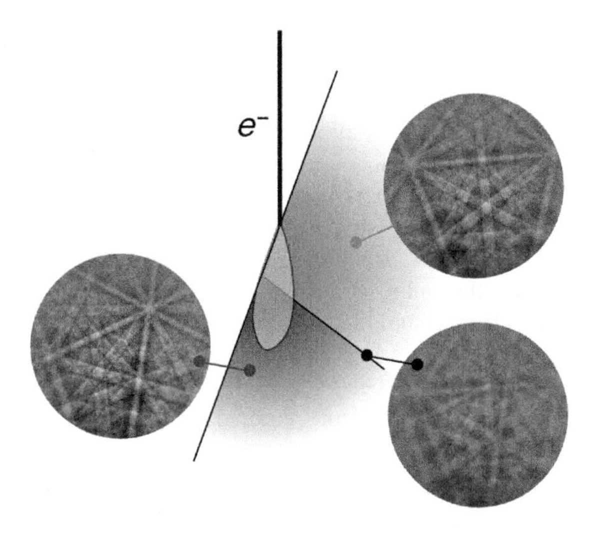

FIGURE 16.5

Schematic showing superposition of EBSD patterns at a grain boundary. *Courtesy of EDAX/TSL.*

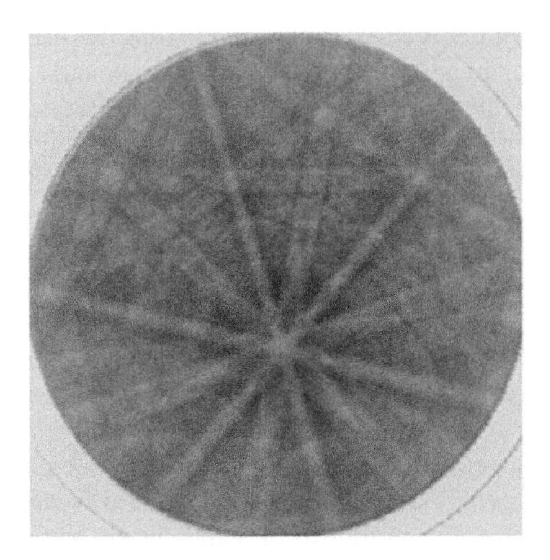

FIGURE 16.6

EBSD patterns from a recrystallized grain and a deformed grain in partially recrystallized low-carbon steel. *Courtesy of EDAX/TSL.*

This effect is shown in Figure 16.6. Elastic strains distort the crystal lattice, producing a change in the lattice parameters. The distortion manifests itself in the patterns as a shift in the zone axes along with a change in the width of the diffraction bands. In the uniform dilation case, the changes will only occur in the bandwidths. These changes to the pattern are on the order of 1 part in 1000 and are thus difficult, although not impossible, to detect. In addition, the elastically "bent" planes result in a range of solutions to Bragg's law resulting in more diffuse scattering.

The spatial resolution of the EBSD technique is a function of the interaction volume. This is a function of the materials as well as the accelerating voltage of the SEM. Higher resolutions can be achieved using lower voltages, but there is a cutoff at approximately 5 keV. At voltages less than 5 keV it is difficult to obtain EBSD patterns, although changes in the phosphor detector may enable EBSD at lower energies. The greatest impact on the spatial resolution is the electron beam itself. To achieve the optimal resolution the goal is to have as much current as possible in as narrow a beam as possible. Thus, thermal field–emitter microscopes provide the best spatial resolution over tungsten-filament SEMs. However, microscopes that use immersion lenses to achieve higher resolutions produce magnetic fields within the microscope chamber that have a negative impact on EBSD pattern formation. As electrons travel from the sample to the phosphor screen they pass through the magnetic field. Thus, they will follow a curved rather than straight path, resulting in distorted patterns. Grains as small as 10 nm have been successfully imaged using EBSD. However, these grains are situated at the tail end of a distribution centered on a mean grain size closer to 100 nm in diameter (Nowell and Field, 1998).

16.3 AUTOMATED INDEXING

The crystal lattice orientation within the diffracting volume (relative to a reference coordinate system associated with the sample) can be determined from the diffraction patterns. Traditionally, this was done by the user identifying the symmetry associated with given zone axes in the pattern (the zone axes are the intersection between bands in the crystal and these correlate to the intersections between planes in the lattice), thereby identifying the corresponding indices; that is, the [001] axis in a cubic material would exhibit fourfold symmetry. If these indices for two or three axes can be identified, then the orientation of the crystal lattice can be determined. This requires a good working knowledge of the crystallography and recognition of the symmetry in the pattern. However, indexing a pattern (i.e., determining the orientation of the diffracting lattice) is an automated procedure in today's modern systems (Wright, 2000). It is essentially a two-step process. The first step is to automatically detect the position of the bands in the pattern. The second step is to determine the orientation based on the geometry of the bands using a voting scheme.

The most common technique employed to detect the bands in the pattern is the Hough transform (first proposed by Krieger-Lassen et al., 1992). This transform is an image-processing approach modified to find the bands. In the Hough transform, bands in the pattern are transformed to peaks, reducing the band detection problem to a peak-finding problem. The key to the Hough transformation is the following equation:

$$\rho = x \cos \theta + y \sin \theta \tag{16.3}$$

Here, x and y are the coordinates (column and row) of a pixel in the EBSD pattern image, and ρ and θ are the coordinates of lines that pass through the pixel. Thus, a pixel in the image space becomes a sinusoidal curve in the transform space as shown in Figure 16.7.

An accumulator array is constructed containing bins, typically at steps of 1° in θ, and ρ is divided up into approximately 100 steps. The intensity of the pixel x, y (ranging from 0 to 255 for 8-bit grayscale images) is added to each bin along the sinusoidal curve. This process is repeated for all

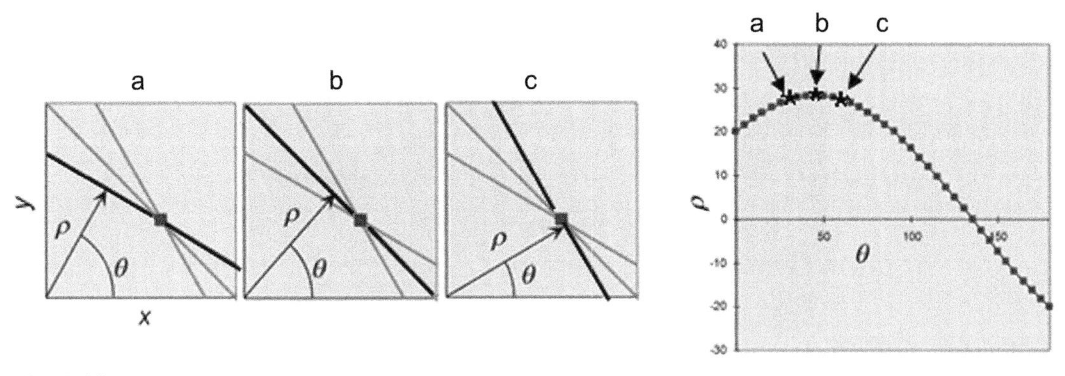

FIGURE 16.7

Schematic of transform from image space to Hough space. *Courtesy of EDAX/TSL.*

pixels in the image. An example is shown in Figure 16.8. Close inspection of the Hough transform shows that the individual sinusoidal curves can be recognized, each one passing through the strong intensity peak in the Hough transform. Thus, the line of white dots is reduced to a peak in the Hough transform.

Figure 16.9 shows the Hough transform for an EBSD pattern. Note that the profile of the peak for constant θ has a distinct footprint—namely, dark at the edges of the band and light at the center. This characteristic is used to aid in peak detection. A so-called "butterfly" mask is used to enhance the peaks with the characteristic butterfly shape. This helps distinguish true peaks from noise in the pattern.

Once the bands have been found the next step is to determine the orientation of the crystal lattice from the geometrical arrangement of the bands. Initially, a look-up table of all possible interplanar angles is constructed. This step presumes that the crystal structure of the material phase is known. An example for a face-centered cubic material is shown in Table 16.1.

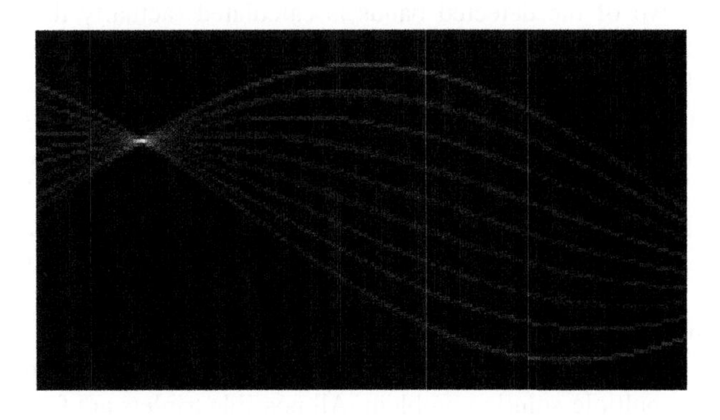

FIGURE 16.8

Hough transform of a set of discrete pixels along a line. *Courtesy of EDAX/TSL.*

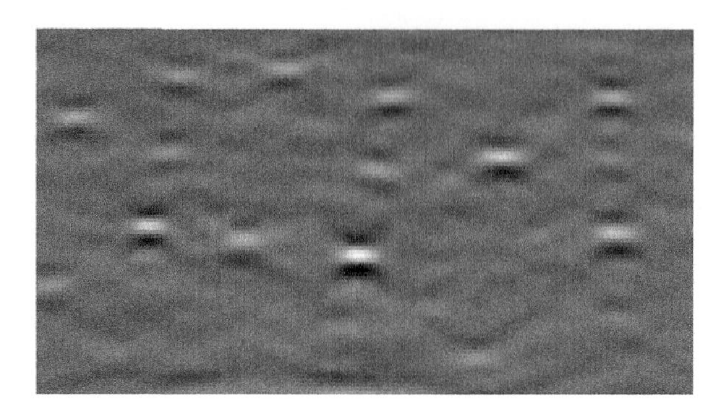

FIGURE 16.9

Hough transform of an EBSD pattern from titanium. *Courtesy of EDAX/TSL.*

Table 16.1 Interplanar Angles for a Face-Centered Cubic Material

Interplanar Angle	$(hkl)_1$	$(hkl)_2$	Interplanar Angle	$(hkl)_1$	$(hkl)_2$
25.2	200	311	64.8	220	$3\bar{1}1$
29.5	111	311	70.5	111	$11\bar{1}$
31.5	220	311	72.5	200	131
35.1	311	$31\bar{1}$	80	111	$3\overline{11}$
35.3	111	220	84.8	311	$1\bar{3}1$
45	200	220	90	111	220
50.5	311	$3\overline{11}$	90	200	020
54.7	111	200	90	200	022
58.5	111	$31\bar{1}$	90	220	$1\bar{1}3$
60	220	202	90	220	$2\bar{2}0$
63	311	$13\bar{1}$			

The angle between two of the detected bands is calculated (actually the projected angle) and compared against the look-up table. Figure 16.10 shows an example. The angle measured between band A and band B in this instance is 63.9°. If a tolerance window of ±3° is assumed, then band A and band B could both be (311) types, or band A could be a (220) type and band B a (311) type, or vice versa. Which (*hkl*) are associated with each specific band could possibly be determined by examining the width of the bands. However, this is unreliable as the patterns are often fairly diffuse, particularly when trying to collect them as quickly as possible for scanning. The choices of indices can be narrowed considerably using three bands. A logic table can be easily formed to identify the bands as shown in Figure 16.11.

Even with three bands, multiple solutions may be identified as shown in Figure 16.12. (The number of possible solutions will be a function of the tolerance window.) However, a voting scheme has been devised to overcome the multiple-solution problem. All possible triplets are formed from the detected set of bands. For each triplet, all possible solutions are found. The solution that appears most often is identified as the most likely solution. A sample of the voting results for a face-centered cubic pattern is

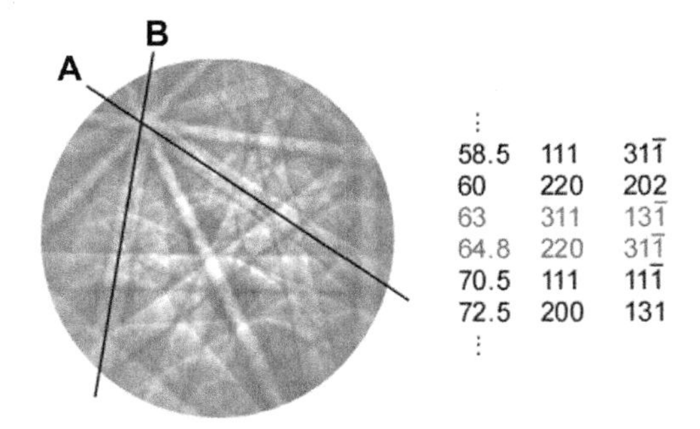

⋮		
58.5	111	$31\bar{1}$
60	220	202
63	311	$13\bar{1}$
64.8	220	$31\bar{1}$
70.5	111	$11\bar{1}$
72.5	200	131
⋮		

FIGURE 16.10

Possible (*hkl*) pairs for the bands indicated. *Courtesy of EDAX/TSL.*

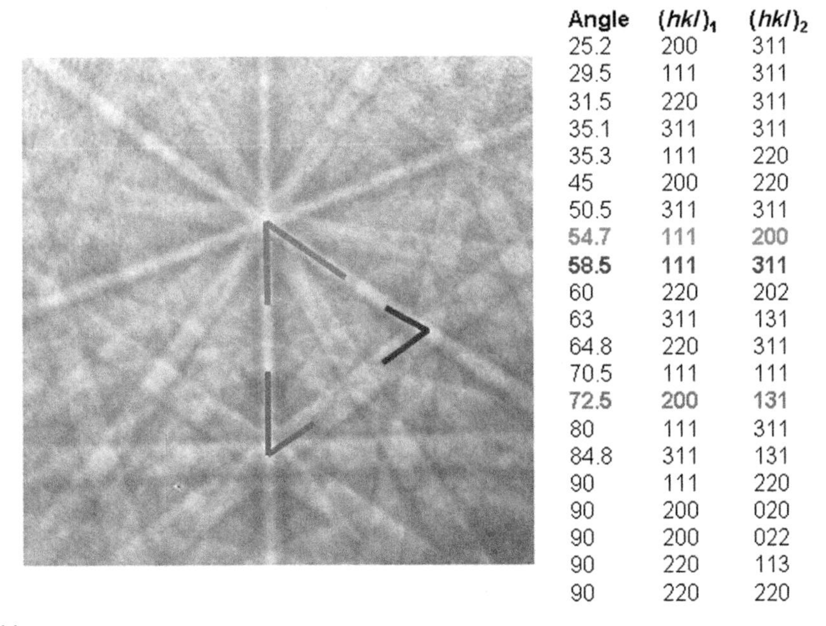

Angle	$(hkl)_1$	$(hkl)_2$
25.2	200	311
29.5	111	311
31.5	220	311
35.1	311	311
35.3	111	220
45	200	220
50.5	311	311
54.7	111	200
58.5	**111**	**311**
60	220	202
63	311	131
64.8	220	311
70.5	111	111
72.5	200	131
80	111	311
84.8	311	131
90	111	220
90	200	020
90	200	022
90	220	113
90	220	220

FIGURE 16.11

Schematic of using a triple to find a single indexing solution. *Courtesy of EDAX/TSL.*

shown in Figure 16.13. In this example, five bands were found by the Hough transform. This means that 10 triplets can be formed. Three possible solutions were found using the 10 triplets. Solution 1 received a vote from each triplet for a total of 10, whereas the other two solutions each received only one vote. Thus, solution 1 is assumed to be the correct solution. This voting approach is particularly robust in dealing with rogue bands identified by the Hough transform.

When the sample contains multiple phases, a phase differentiation process is applied to find the best solution. This works well when the phases are structurally quite different. However, when they are similar, identifying the correct phase can be difficult. If, however, the chemical compositions of the

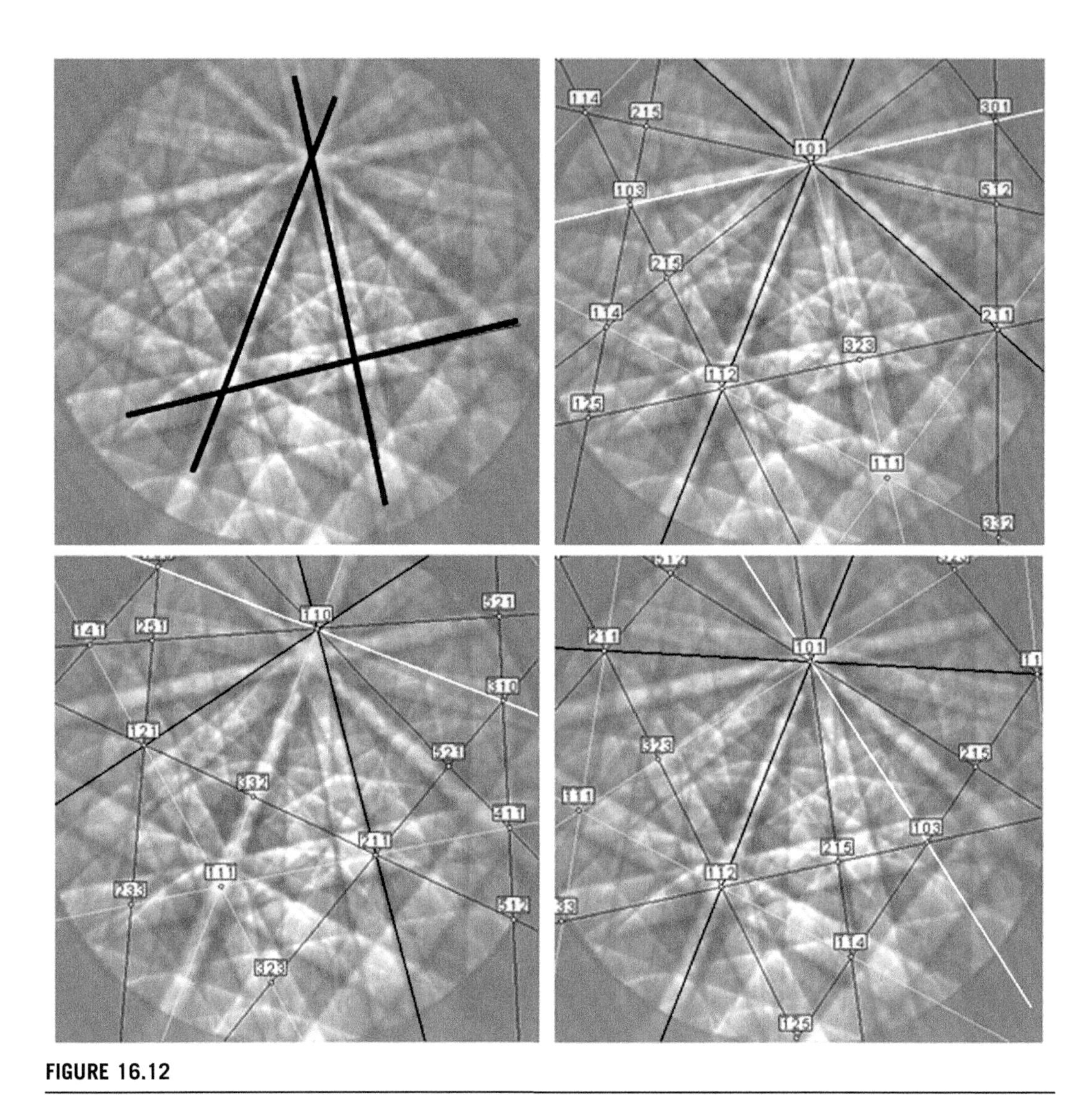

FIGURE 16.12

Three potential indexing solutions for the three bands highlighted. *Courtesy of EDAX/TSL.*

two phases differ from each other, EDS data can be used to differentiate between the phases (Nowell and Wright, 2004).

When the orientations can be reliably identified without any operator intervention, then a scan over a defined area can be performed, a technique termed Orientation Imaging Microscopy™ (OIM™). As the orientation measurements are collected on regular grids, various orientation or other parameters associated with the individual patterns can be used to construct 2D maps. Examples are shown in Figure 16.14. These maps provide a visualization of the spatial arrangement of

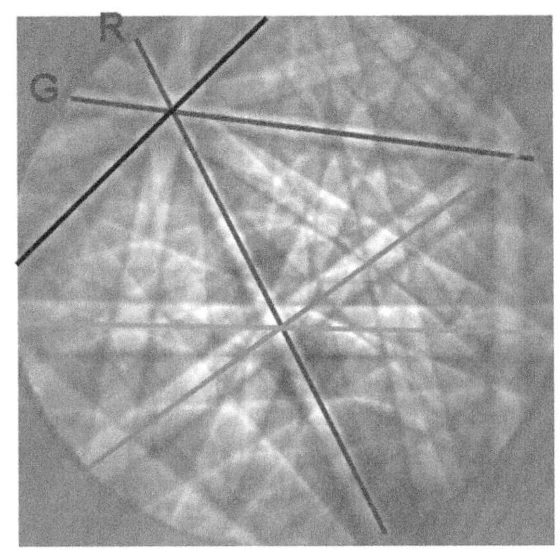

Triplet	Solution 1(V_1)	Solution 2(V_2)	Solution 3(V_3)
R, G, Y	✓		
R, G, **B**	✓		
R, G, M	✓		
R, Y, **B**	✓		
R, Y, M	✓	✓	
R, **B**, M	✓		
G, Y, **B**	✓		
G, Y, M	✓		
G, **B**, M	✓		
Y, **B**, M	✓		✓
Total	10	1	1

FIGURE 16.13

Schematic of the triplet voting procedure used to identify the most probable indexing procedure. *Courtesy of EDAX/TSL.*

FIGURE 16.14

OIM maps from nickel. The shades in the map on the *left* reflect the crystal direction aligned with the sample normal varying continuously between crystal directions [001], [111] and [110]. The shade in the grayscale map on the right is linearly scaled with the quality of the corresponding EBSD pattern at each point in the OIM scan. The boundaries are colored according to misorientation. While difficult to distinguish in the grayscale image, the shades correspond with misorientations in the ranges: 5–15°, 15–30°, 30–45°, and greater than 45°. Readers might note some evidence of polishing scratches in the figure on the right. *Courtesy of EDAX/TSL.*

crystallographic orientation within the microstructure as well as other features that can be derived from the patterns.

One limitation of the technique arises from materials exhibiting pseudo-symmetry (Nowell and Wright, 2005). For example, consider lead zirconium titanate (PZT). PZT is only slightly tetragonal—the c-to-a ratio is only 1.03. While it is possible to see this difference when comparing individual patterns, it is difficult for the automated indexing routines to differentiate the c-axis from the a-axis in a single pattern. One way to overcome the problem is to run a sample as if it were simply cubic. However, this misses some of the salient distinguishing features of the microstructure. Various clean-up methods have also been proposed to overcome these problems.

To improve the band detection, various image-processing procedures can be applied to the patterns prior to processing them via the Hough transform. The most common of these is background correction. The Hough transform does not perform well when the incoming patterns exhibit significant intensity gradients. However, intensity gradients are relatively common in the EBSD patterns. Background correction is a helpful procedure for reducing these gradients. A background pattern can be subtracted or divided from the incoming pattern as shown in Figure 16.15. The background pattern can be formed one of two ways: statically or dynamically. A static background pattern can be formed by collecting a pattern while the beam is rastered. The resulting pattern is essentially the sum of all patterns in the scan area. If the scan area is composed of many grains then the resulting pattern will not show any features except the overall gradient of the pattern. A dynamic

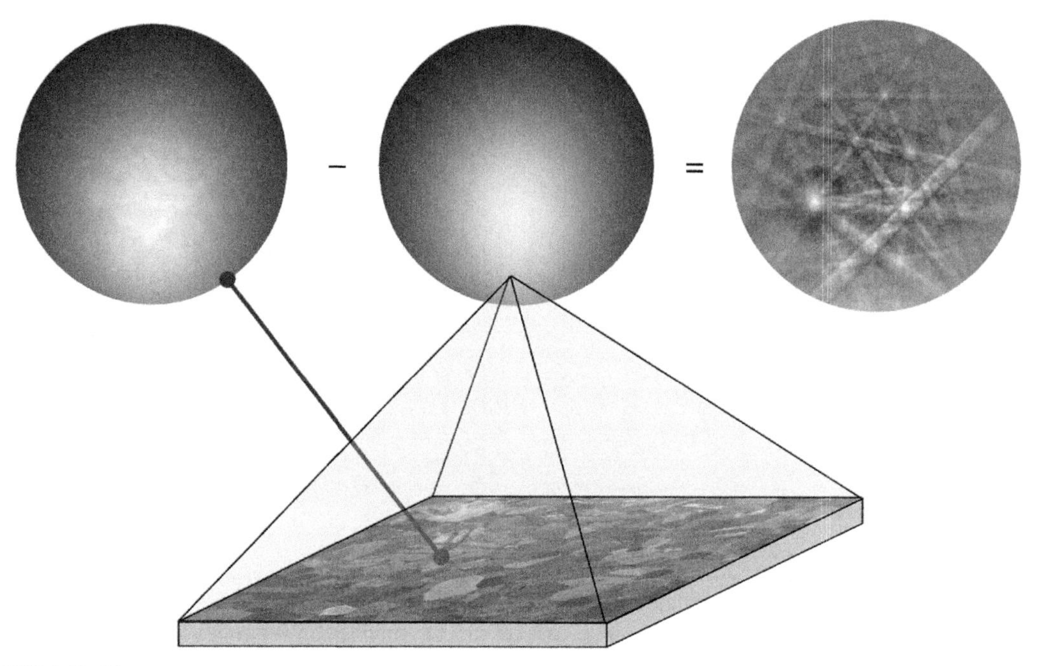

FIGURE 16.15

Schematic showing background correction of an EBSD using a background pattern formed by a summation pattern from many grains. *Courtesy of EDAX/TSL.*

background pattern can be formed from the incoming pattern itself by smoothing out the pattern using various smoothing operations. One disadvantage of this approach is the computational time required to perform the smoothing operations.

The angular resolution of the automated OIM technique should be considered in two different ways: first, the absolute angular resolution and, second, the relative angular resolution. The absolute resolution is dependent on many factors, that is, the calibration of the system, the mounting of the sample in the microscope, sectioning of the sample, among others. However, it is possible to measure orientations to within 1° or 2° resolution routinely with careful operation. The point-to-point misorientations within a scan can be measured to much higher resolution. While 1° is sometimes quoted with careful calibration of the system and optimized band detection, it is possible to exceed 0.3° relative angular resolution.

16.4 PHASE IDENTIFICATION

While we have focused to this point on orientation determination via EBSD, it is also possible to use EBSD to identify phases in multiphase materials. The approach is straightforward. First, measure the chemical composition at a given point in the microstructure typically using EDS. It is not necessary to identify precisely the composition of the elements present (this is difficult at the high tilts required for EBSD), just which elements are present. This information is used as a filter against a database of known phases to form a candidate list of phases. Once the phases are extracted from the database, an attempt is made to index a pattern from the same point using the salient structure parameters (symmetry, lattice parameters, and reflecting planes) for each phase in the candidate list. The indexing solution that best fits the pattern identifies the phase.

Some work has proceeded toward identifying complete unknowns from fundamental analysis of an EBSD pattern; however, such work is still in its infancy.

16.5 ORIENTATION ANALYSIS

While the OIM maps are visually pleasing, the real power of the OIM technique lies in the quantitative nature of the orientation data collected. At each point in an OIM scan an orientation is determined and recorded. This orientation is given in relation to a coordinate system based on the diffraction geometry as shown schematically in Figure 16.16.

While it is common to represent crystallographic orientation in terms of a plane and direction—that is, {*hkl*}<*uvw*> (the {*hkl*} represents the crystal plane parallel to the surface of the sample and <*uvw*> is the crystal direction parallel to the "1" direction of the sample—RD in this case)—such a description is not numerically practical. For example, {*hkl*}<*uvw*> is given in integers. This limits the continuous nature of the orientation measurements to a finite set of discrete orientations. For example, consider the orientation {21 23 22}<34 −32 1> would be reduced simply to {111}<1 $\bar{1}$ 0>, which is 2.1° away from the original orientation. In this book the main approach is to use the set of three Bunge–Euler angles (i.e., φ_1, Φ, and φ_2).

As described in Chapter 2, the set of physically distinct Euler angles forms a bounded space of 0° to 360° in φ_1 and φ_2 and 0° to 180° in Φ. Readers are reminded of the weakness of Euler angles,

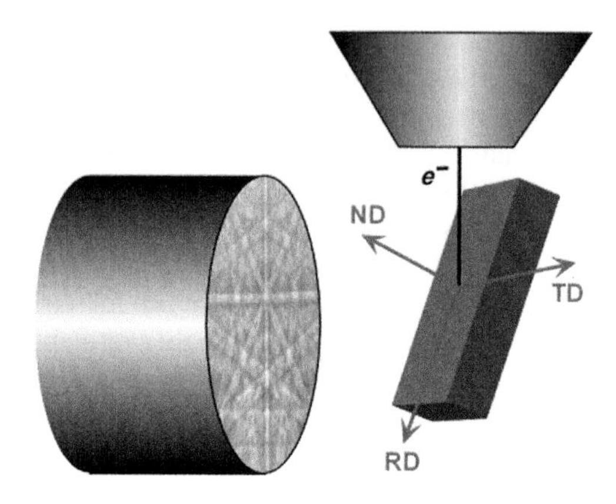

FIGURE 16.16

EBSD reference system. *Courtesy of EDAX/TSL.*

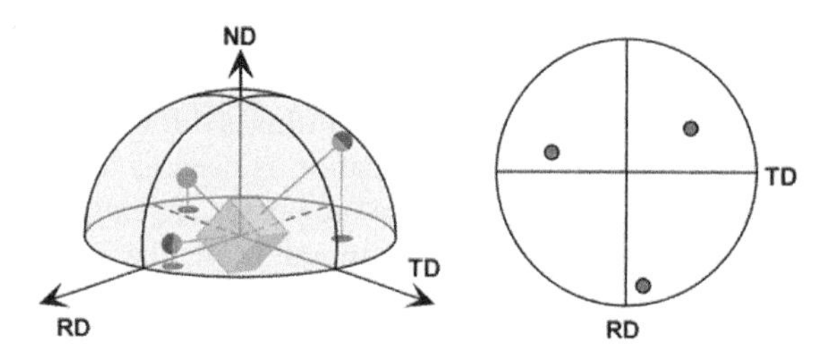

FIGURE 16.17

Schematic illustration of an (001) pole figure.

which is that as the Φ angle approaches 0 the other two angles are not well defined. In the computer codes, the orientations are generally represented by orientation matrices, which are the direction cosines $g_{ij}^{s \rightarrow c}$. Recall that it is relatively straightforward to convert between these different descriptions of orientation.

Orientations can be plotted in a variety of ways. Three common approaches are using pole figures, inverse pole figures, and sections through Euler space. A pole figure shows the normal to a specified plane (or pole) projected onto a 2D section. For example, consider the {001} poles of an arbitrarily oriented cube as shown in Figure 16.17. The intersections of the poles with a hemisphere are projected onto a plane (generally only the poles in the upper hemisphere are shown). Thus, the pole figure shows the orientation of a given pole with respect to the sample axes.

The inverse pole figure shows the orientation of a given sample axis with respect to the crystal coordinate frame. Alternatively, an inverse pole figure can be thought of as showing which crystal direction is parallel to a given pole direction. Orientations can also be plotted relative to their location in Euler space. This is most commonly represented as a series of sections through Euler space. An example showing all three different ways of presenting orientation data is shown in Figure 16.18.

Such discrete plots can be helpful, but when many points are included it becomes difficult to interpret. Consider, for example, the pole figure shown in Figure 16.19. The large cluster at the center

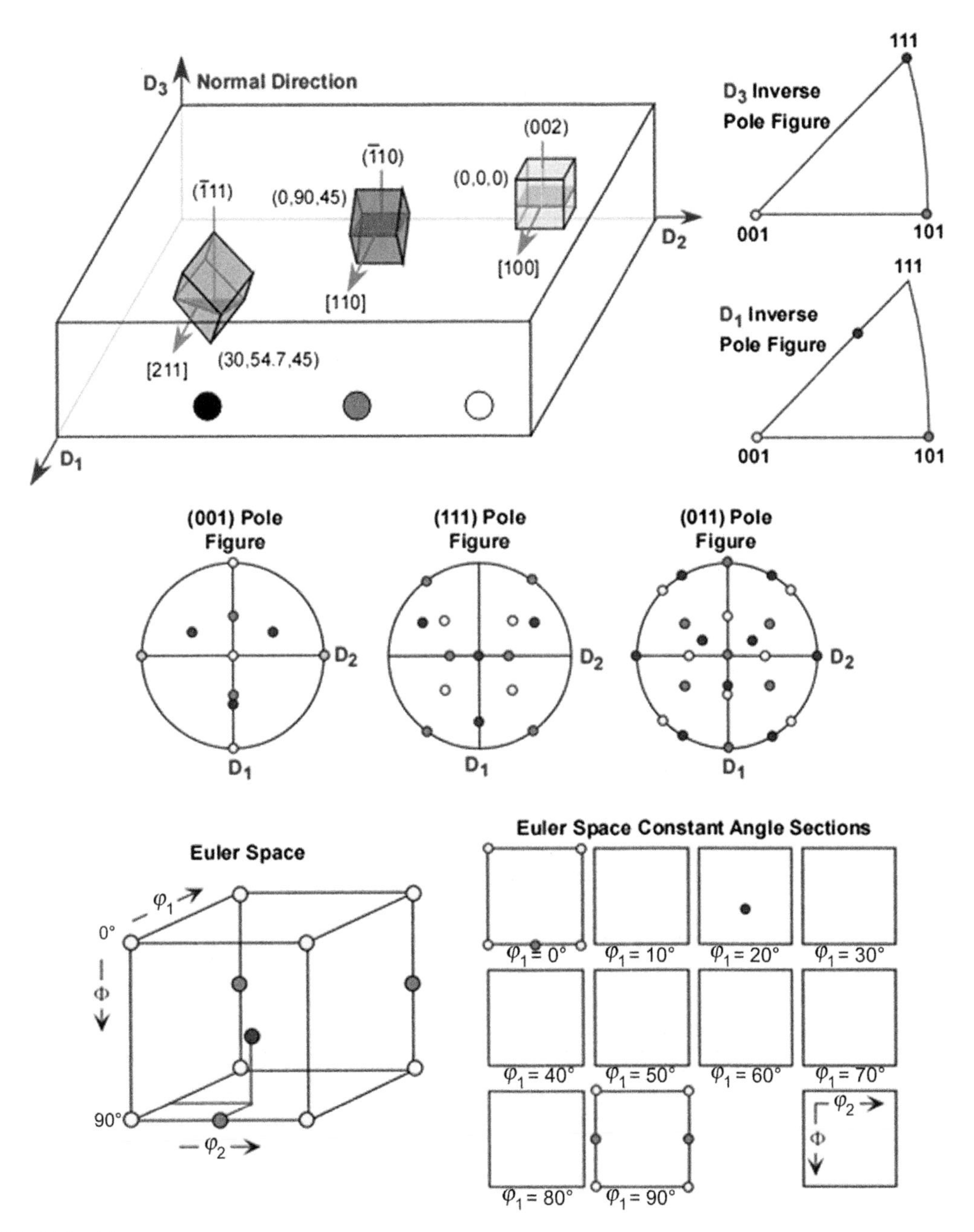

FIGURE 16.18

Inverse pole figures, pole figures, and Euler space 3D plot and constant angle sections showing three different crystallographic orientations. The three small cubes/orientations are represented by a *black dot*, a *gray dot*, and a *white dot*, respectively (left to right). Only a portion of Euler space is shown. Multiple points appear for each orientation because of crystal symmetry.

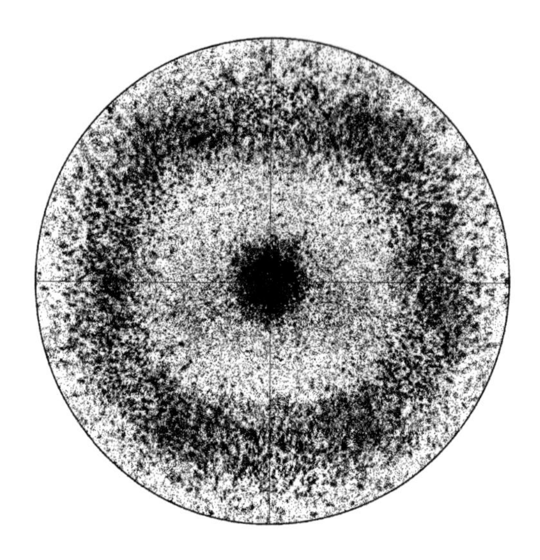

FIGURE 16.19

This is a (111) pole figure for an aluminum thin film. *Courtesy of EDAX/TSL.*

shows a preference for the (111) poles of a copper thin film sample to be aligned with the sample normal. However, it is not possible to quantitatively describe this preference. The tools of texture analysis are well suited to the statistical analysis of orientation distributions. The quantitative description of crystallographic texture was the subject of Chapter 4.

16.6 HIGH-RESOLUTION EBSD

The local state space information extraction techniques discussed thus far in this chapter use traditional methods that have broadly followed a Hough transform approach since the automation of EBSD methods in the early 1990s. However, more recent developments have led to high-resolution EBSD (HREBSD) that significantly extends the microstructure characterization capabilities reviewed thus far (see, for example, Troost et al., 1993; Wilkinson et al., 2006; Landon et al., 2008; Kacher et al., 2009; Villert et al., 2009). The basic idea of HREBSD is to determine the distortion that must be applied to a reference structure to arrive at the local structure under consideration by comparing the EBSD patterns of the two structures using convolution techniques (briefly mentioned in Section 3.3). The reference pattern can either be taken from an EBSD measurement on the same sample/material (Wilkinson et al., 2006), or can be simulated by a variety of techniques (Winkelmann, 2007; Kacher et al., 2009).

Regions of interest (ROIs) are defined in the two images, and fast Fourier transform (FFT) convolutions are applied to determine the shift, $\mathbf{q}$, that must be made to the reference ROI to line up with the ROI of the image being analyzed (Figures 16.20 and 16.21). Typically the EBSD patterns are saved for later analysis to reduce the required microscope time. The relationship between the elastic distortion and the pattern shift seen on the phosphor is represented in Figure 16.21 and is approximately given by (Wilkinson et al., 2006)

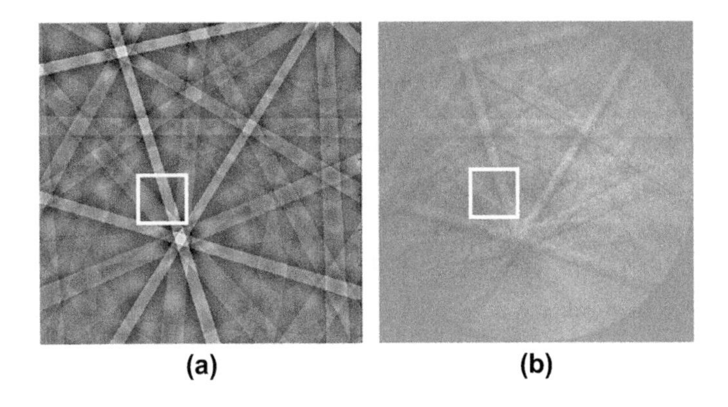

FIGURE 16.20

ROI highlighted in (a) a reference simulated pattern and (b) a real EBSD pattern with the corresponding ROI marked for comparison via convolution. *Courtesy of EDAX/TSL.*

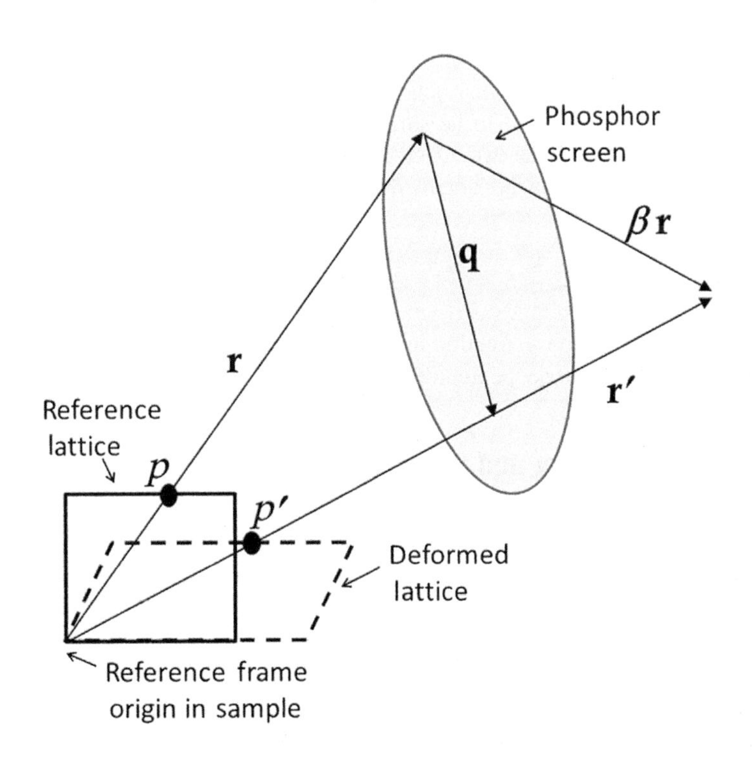

FIGURE 16.21

Deformation geometry and projection of shifts onto the phosphor screen.

$$\mathbf{q} = \beta\mathbf{r} - (\beta\mathbf{r}\cdot\hat{\mathbf{r}})\hat{\mathbf{r}} \tag{16.4}$$

where $\mathbf{r}$ is the vector from the sample origin to the center of the ROI on the phosphor, and $\hat{\mathbf{r}}$ is the unit vector in the same direction; β is once again the elastic distortion, or displacement gradient. This equation makes some reasonable approximations, including assuming that the vector from the sample to the ROI is normal to the phosphor surface. If no such approximations are made a more exact equation arises (Kacher, 2009):

$$\mathbf{q} = \beta\mathbf{r} - (\beta\mathbf{r}\cdot\hat{\mathbf{r}}')\hat{\mathbf{r}}' + (\mathbf{q}\cdot\hat{\mathbf{r}}')\hat{\mathbf{r}}' \tag{16.5}$$

where $\mathbf{r}'$ is the vector formed from applying the lattice deformation to $\mathbf{r}$ and goes from the interaction volume through the center of the ROI in the deformed pattern (Figure 16.21).

The usual solution strategy for finding β aims to remove the last term from Eq. (16.4), reducing the number of independent simultaneous equations from three to two. By measuring values of $\mathbf{q}$ and $\mathbf{r}$ for at least four ROIs for Eq. (16.4) or (16.5), the off-diagonal elements of β can be uniquely determined, and the differences of the diagonal terms can be resolved. To arrive at the full distortion tensor a further constraint is required. This is generally obtained by assuming a traction-free boundary condition, consistent with the presence of the free surface of the sample:

$$\sigma_{ij}n_j = C_{ijkl}\varepsilon_{kl}n_j = 0 \tag{16.6}$$

where C_{ijkl} contains the elastic constants, $\bar{\varepsilon}$ is the elastic strain, and $\bar{n}$ is the normal to the surface. It is generally assumed that the average surface normal is aligned with the 3-axis of the sample reference frame, introducing a potential for error that is not considered here.

The reference structure is assumed to be sufficiently close to the real structure that only small deformations are obtained between the two. Once the full distortion tensor has been determined, strain and rotation tensors ε_{ij} and ω_{ij} can be obtained by splitting β into its respective symmetric and asymmetric parts:

$$\varepsilon = \frac{1}{2}(\beta + \beta^T) \quad \text{and} \quad \omega = \frac{1}{2}(\beta - \beta^T) \tag{16.7}$$

Options for further information extraction are discussed next after briefly discussing the critical issue of the measurement of microscope geometry.

16.6.1 Microscope geometry and the pattern center

Electron backscatter diffraction patterns are typically captured as images from a flat phosphor screen that is conveniently situated relative to the electron beam and the material sample in the SEM. To interpret the Kikuchi bands on a given EBSD pattern in terms of atomic geometry in the material, a reference frame for the image is required. This frame is generally specified in terms of a pattern center (PC). For a given image and related sample, the PC, (X^*, Y^*, Z^*), provides the position on the phosphor screen, (X^*, Y^*), from which a normal vector would intersect the interaction volume in the sample, and the distance, (Z^*), from this position on the phosphor screen to the interaction volume.

A common commercial EBSD software, OIM (EDAX, 2010), approximates pattern center parameters during a calibration exercise, and this value is used for subsequent sample analysis. Bands on a Kikuchi pattern are identified (either manually or using the Hough transform). An approximate

PC is assumed, and the bands are subsequently indexed. A tuning process is then used that adjusts the PC position in 0.1 pixel steps to optimize the fit to the indexed bands.

Several alternative approaches to pattern center determination are available (Engler and Randle, 2010), including screen moving (Carpenter et al., 2007), shadow casting (Venables and Bin Jaya, 1977), iterative fitting (Krieger and Lassen, 1999), and strain minimization (Kacher, 2009; Kacher et al., 2010). They all offer accuracy ranging roughly from 1 to 0.2% of the phosphor width (which is typically 37 mm) for different PC components. This resolution is adequate for standard EBSD applications where a given lattice orientation is assumed to be accurate only to 0.5° (Wright, 1993); that is, the resultant error is in line with traditional EBSD resolution. Figure 16.22 illustrates the potential errors in strain and orientation measurements from an incorrect pattern center.

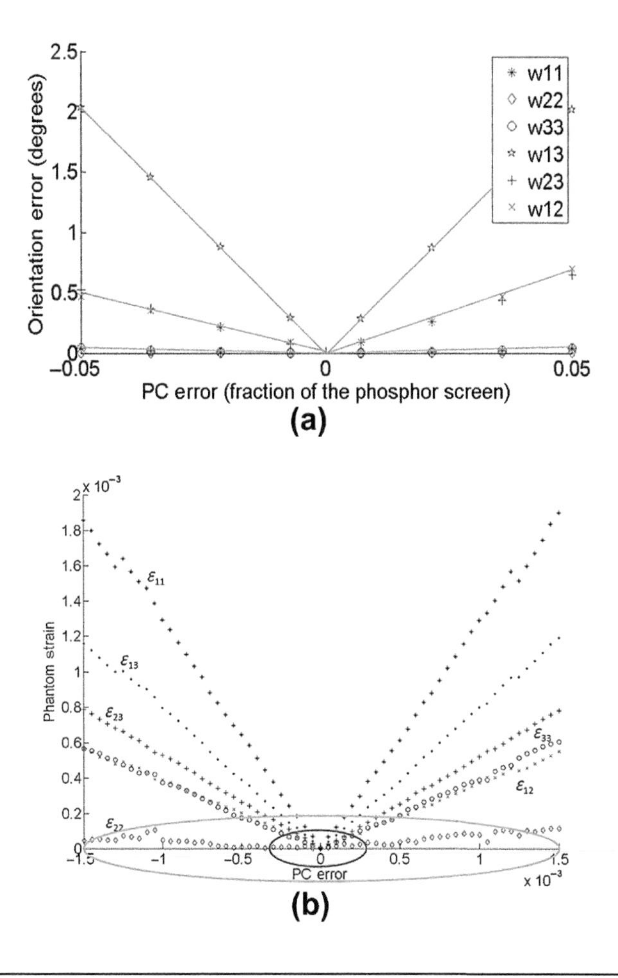

FIGURE 16.22

Expected error (a) in rotation matrix components (Kacher, 2009) and (b) in strain components (Basinger et al., 2011) versus pattern center error. The elipeses indicate typical resolution for standard EBSD (the larger elipese) and HREBSD.

More recently, work has been undertaken to improve the PC resolution for high-resolution EBSD techniques, where the PC error can potentially dominate the overall error in orientation and strain measurements. An image-processing technique that considers parallelism of bands in the EBSD image, after mapping back onto a sphere using the assumed PC position, provided resolution of approximately 0.03% of the phosphor width in various tests (Basinger et al., 2011).

16.6.2 Orientation measurement

Assuming only small distortions between the reference and sample pattern, the rotation matrix, given by $R = I + \omega$, provides the relative orientation change between the reference pattern and the measured pattern. If the reference pattern is taken from a nearby point in the same polycrystalline grain, then the resolution of the relative orientation measurement is shown to be of the order of 0.006° (Wilkinson et al., 2006). If the simulated pattern method is used, the absolute orientation is then obtained by combining the rotation matrix of the simulated pattern with that of the relative orientation. The resultant resolution (using simple Bragg's law–simulated patterns for calculation speed) is of the order of 0.04° (Kacher et al., 2009).

Figure 16.23 illustrates the variability/noise associated with traditional EBSD and HREBSD for a well-behaved material, germanium (Ge). The increase in smoothness of the data by using HREBSD

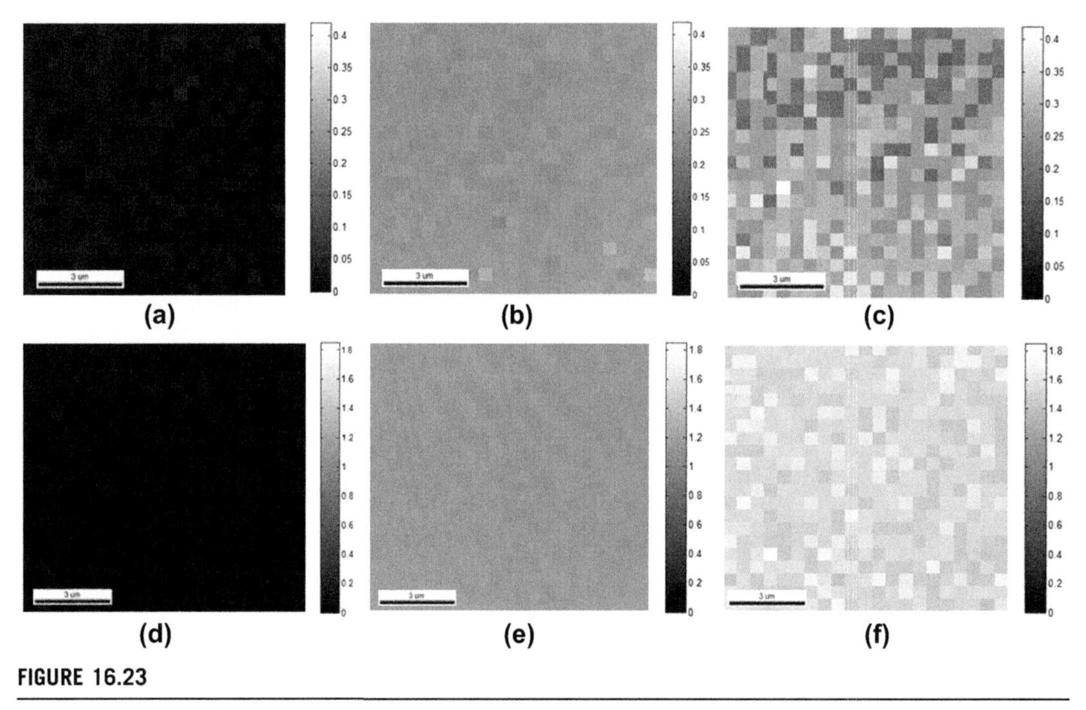

FIGURE 16.23

Misorientation maps (in degrees) from the average orientation of the PC-calibrated HREBSD scan (simulated reference patterns used in all HR-EBSD runs). For germanium: (a) HREBSD scan with PC calibration, (b) HREBSD scan without PC calibration, and (c) OIM scan; and for silicon: (d) HREBSD scan with PC calibration, (e) HREBSD scan without PC calibration, and (f) OIM scan.

is dramatic; furthermore, application of the correct pattern center leads to both accurate and smooth orientation measurements.

16.6.3 Strain measurement

The image convolution method, on which HREBSD is based, was first implemented with the goal of extracting strain measurements from EBSD patterns (Troost et al., 1993). If only relative strain is required, a typical application of the method will use a reference pattern taken from a relaxed area in the center of a grain and convolve this pattern with other patterns obtained from across the grain to arrive at a map of relative strain (via Eq. 16.7). The accuracy of the resultant map is reported to be approximately $1e^{-4}$ in the various relative strain components (Wilkinson et al., 2006).

However, if a measure of absolute strain is required, a simulated pattern may be used as the reference pattern, resulting in a typical accuracy of $7e^{-4}$ (Kacher et al., 2009). This accuracy is greatly influenced by the accuracy of pattern center determination. Figure 16.22 illustrates the magnitude of "phantom strains" associated with pattern center errors of varying size. The effect was highlighted more recently in a study of "strain-free" Ge and silicon (Si). If the pattern center determined via standard EBSD software was used in the HREBSD strain measurement, the error was typically an order of magnitude higher than that achieved with the more sophisticated pattern center algorithm introduced in Basinger et al. (2011) and described before (Basinger et al., in press).

The pattern center issue is also relevant to the measurement of relative strain via the real–reference pattern approach. If a reference pattern is taken from a position that is a significant distance from the measurement point, then an adjustment must be made for a change in pattern centers between the images or phantom strains will arise (Villert et al., 2009; Britton and Wilkinson, 2011; Basinger et al., in press).

Figure 16.24 demonstrates the determination of a stress field in a sample of ultrasonically consolidated (UC) nickel (thin sheets of nickel are bonded by UC) (Gardner et al., 2010). The compression was applied to the sample in the horizontal direction, and an area of poor bonding was chosen for analysis using the simulated-pattern HREBSD method described in Kacher et al. (2009). The traction-free regions on either side of the unconsolidated areas (the dark horizontal stripes) are evident.

16.6.4 Geometrically necessary dislocation measurement

Dislocation content of a material is an important local state for input into various crystal plasticity models and is typically quoted in terms of dislocation density, ρ, the line length of dislocations per unit volume. In 1953, Nye introduced the dislocation density tensor, α, which combines the slip activity on a multitude of slip systems into a single tensor:

$$\alpha_{ij} = \sum_m \rho^{(m)} b_i^{(m)} v_j^{(m)} \tag{16.8}$$

where (m) indicates the dislocation type, and b and v represent the particular Burgers vector and line direction, respectively, for the given type. Note that this equation is only strictly accurate when the lattice rotations due to deformation are additive (e.g., for infinitesimal rotations); however, an analogous equation may be used for arbitrary deformations (Cermelli and Gurtin, 2001).

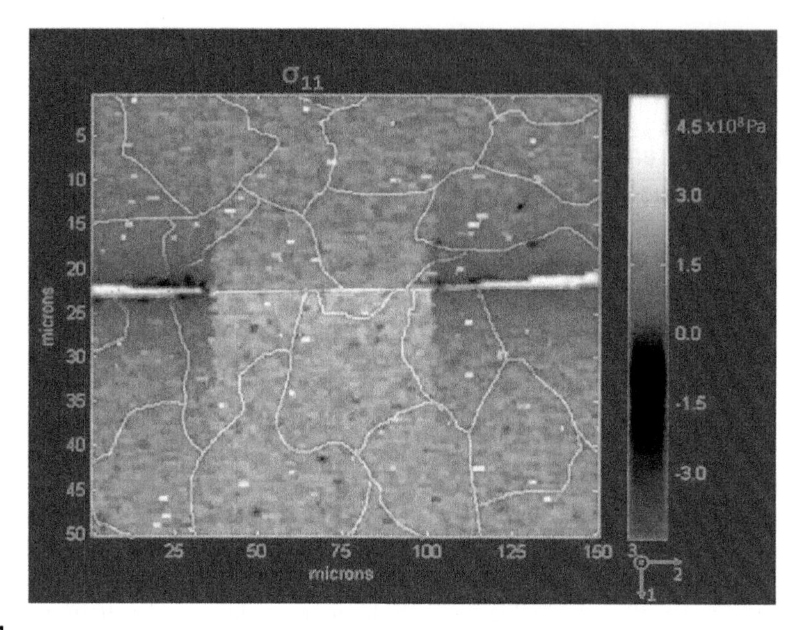

FIGURE 16.24

Stress field in an ultrasonically consolidated nickel sample (horizontal layers) with a consolidated central region and with unconsolidated regions on either side (showing up as *white regions*), as reported in Gardner et al. (2010). The sample is under compression in the horizontal (2) direction.

The Nye tensor forms the basis of continuum dislocation theory. By considering the geometrical distortion caused by the presence of geometrically necessary dislocations (GNDs), the Nye tensor may be written in terms of the plastic distortion, β^P: $\alpha = -curl\beta^P = -\nabla \times \beta^P$; in other words, $\alpha_{lj} = -\in_{nmj}\beta^P_{\text{ln},m}$ ($\in$ is the permutation tensor; note that Einstein's summation convention holds on repeated indices). The total distortion is given as $\beta^T = \beta^P + \beta$, where $\beta^T_{ij} = u^T_{i,j}$ for some total displacement field, u^T_i, and β is the elastic distortion (displacement gradient). Since the total displacement field must be continuous and differentiable to maintain a connected body (Kroner, 1958), we have that $curl\beta^T = curl\ (grad\ \ u) = 0$. Hence,

$$\alpha = \nabla \times \beta \tag{16.9}$$

Equation (16.9) is known as the fundamental equation of continuum dislocation theory. Since not all gradients of the distortion tensor are available from a surface characterization technique, such as EBSD, dislocation information (GND content) is typically recovered from EBSD data using approximate relationships between lattice curvature and the Nye dislocation tensor, α:

$$\alpha_{ij} = \in_{nmj} \beta_{in,m} = \in_{nmj} \omega_{in,m} + \in_{nmj} \varepsilon_{in,m} = \kappa_{ji} - \kappa_{kk}\,\delta_{ij} + \in_{nmj} \varepsilon_{in,m} \approx \kappa_{ji} - \kappa_{kk}\,\delta_{ij} \tag{16.10}$$

where κ refers to the lattice curvature, and δ is the usual Kronecker-delta function. The curvature is obtained from the local orientation gradient. If the surface normal direction is taken to be x_3 then we are unable to measure the curvatures $\kappa_{13}, \kappa_{23}, \kappa_{33}$. Hence, only five components of α are accessible using this approach: α_{12}, α_{21}, and α_{i3} (Pantleon, 2008).

An equivalent formulation to the one just presented may be written in terms of the rotation matrix, g, that captures the orientation change between adjacent scan points (see, for example, Brewer et al., 2009):

$$\alpha_{ij} = \in_{imn} g_{jn,m} \tag{16.11}$$

With traditional EBSD techniques it is natural to quantify GND content via the measurable quantity of orientation gradient, or curvature, in Eq. (16.10). However, in HREBSD the derived tensor is the elastic distortion, β (Eq. 16.4), and it is more natural to apply Eq. (16.9) directly. The expanded form of this equation may be written as

$$\alpha_{ij} \equiv \begin{bmatrix} (\tilde{\beta}_{12,3} - \beta_{13,2}) & (\beta_{13,1} - \tilde{\beta}_{11,3}) & (\beta_{11,2} - \beta_{12,1}) \\ (\tilde{\beta}_{22,3} - \beta_{23,2}) & (\beta_{23,1} - \tilde{\beta}_{21,3}) & (\beta_{21,2} - \beta_{22,1}) \\ (\tilde{\beta}_{32,3} - \beta_{33,2}) & (\beta_{33,1} - \tilde{\beta}_{31,3}) & (\beta_{31,2} - \beta_{32,1}) \end{bmatrix} \tag{16.12}$$

The unavailable components are marked by a tilde, indicating that the right column of the Nye tensor is fully accessible using EBSD. Other components are partially accessible; for example, one common approximation is made by expanding the components of the distortion tensor: $\beta_{ij} = \varepsilon_{ij} + \omega_{ij}$, $\varepsilon_{ij} = \varepsilon_{ji}$, $\omega_{ij} = -\omega_{ji}$. Then

$$\alpha_{11} - \alpha_{22} = \varepsilon_{12,3} + \varepsilon_{21,3} - \beta_{13,2} - \beta_{23,1} \approx -\beta_{13,2} - \beta_{23,1} \tag{16.13}$$

That is, if the strain gradient components are assumed to be negligible, then we can determine this difference. Similarly, since $\omega_{11} = \omega_{22} = 0$ (the rotation tensor being anti-symmetric), if we once again ignore some of the strain gradient terms we can readily deduce that

$$\begin{aligned} \alpha_{12} &= \beta_{13,1} - \varepsilon_{11,3} \approx \beta_{13,1} \\ \alpha_{21} &= \beta_{23,2} - \varepsilon_{22,3} \approx \beta_{23,2} \end{aligned} \tag{16.14}$$

which are measurable. These estimates for α_{i3}, α_{12}, α_{21}, and $\alpha_{11} - \alpha_{22}$ are potentially more accurate than those given in Eq. (16.10) since not all elastic terms need to be dropped.

An estimate for the overall dislocation density is typically then obtained in one of two ways. The first is to solve Eq. (16.8) for the various contributing values of $\rho^{(m)}$ and then sum these to obtain a total density (El-Dasher et al., 2003; Field et al., 2010). Since there are an inadequate number of equations to uniquely solve for these parameters, other assumptions must be made. The other method is to estimate the dislocation content from the L^1 norm of the Nye tensor (El-Dasher et al., 2003), where the L^1 norm is defined as

$$||\alpha||_1 = \sum_i \sum_j |\alpha_{ij}| \tag{16.15}$$

The L^1 norm of the Nye tensor and the approximate bulk dislocation density are then related by

$$\rho \approx \frac{1}{b} \|\alpha\|_1 \tag{16.16}$$

where b is the average Burgers vector length. The validity of this approximation is verified using simulated dislocation fields in Ruggles and Fullwood (in press).

Since not all Nye tensor components are available, a further approximation must be made by scaling up the values obtained from the available components. For example, the absolute value of the third column of the Nye tensor may be summed and multiplied by 3 for a crude estimate of the total dislocation density. If the approximates for α_{12} and α_{21} are included in the approximation, then the best scaling factor turns out to be 30/14 (Ruggles and Fullwood, in press). Similarly, if all available betas are used, as indicated in Eq. (16.12), then the sum should be scaled by 30/20.

The resolution of the method is dependent on the step size used for the orientation gradient and the accuracy of the orientation data. Recommendations for step size choice from two viewpoints are represented in Figure 16.25 (Adams and Kacher, 2009; Kysar et al., 2010; Field et al., 2012). An illustration of the application of this methodology is shown in Figure 16.26, where GND content is measured in a deformed single-crystal copper sample using both EBSD and X-ray microdiffraction (Field et al., 2010). The absolute values of measured GND content are similar, but the structure seen by the two techniques is quite different. This may be due to noise in the EBSD approach or to the fact that the EBSD technique is measuring the GND content at the surface while the X-ray method is measuring it deeper in the sample.

16.7 STEREOLOGY: VOLUME FRACTIONS ESTIMATION

Stereology, as it applies to our interests, is the study of how one may recover information about 3D features of microstructure from 2D data. The need for stereology is pervasive in microstructure characterization, for the principal reason that most materials are opaque to the probing radiation that is used to study their structure. For example, scanning and transmission electron microscopes are widely used to study microstructure of both soft and hard materials. In the study of the microstructure of crystalline materials by EBSD, it is typical that electrons in the range of 5 to 30 keV penetrate only a small distance into the material. Interaction volumes are typically of dimension 20 to 100 nm, depending on accelerating voltage, atomic number, and other factors.

For this reason only the near-surface region of the sample is readily accessible, and when scanning is employed in characterizations, the data obtained has 2D character. This characterization extends to optical microscopy and other microprobe instruments. By way of contrast, imaging by neutron diffraction and high-energy electron beams, such as those produced by state-of-the-art synchrotron light sources, can have interaction volumes on the order of centimeters or more, providing true 3D characterizations of microstructure. Few instruments of this character exist, however, and these are not readily available to most materials scientists, as they are very expensive to build and operate.

Our purpose here is to provide a brief introduction to that component of stereology that is directly connected to characterization of materials by EBSD, using the scanning methods that are known as

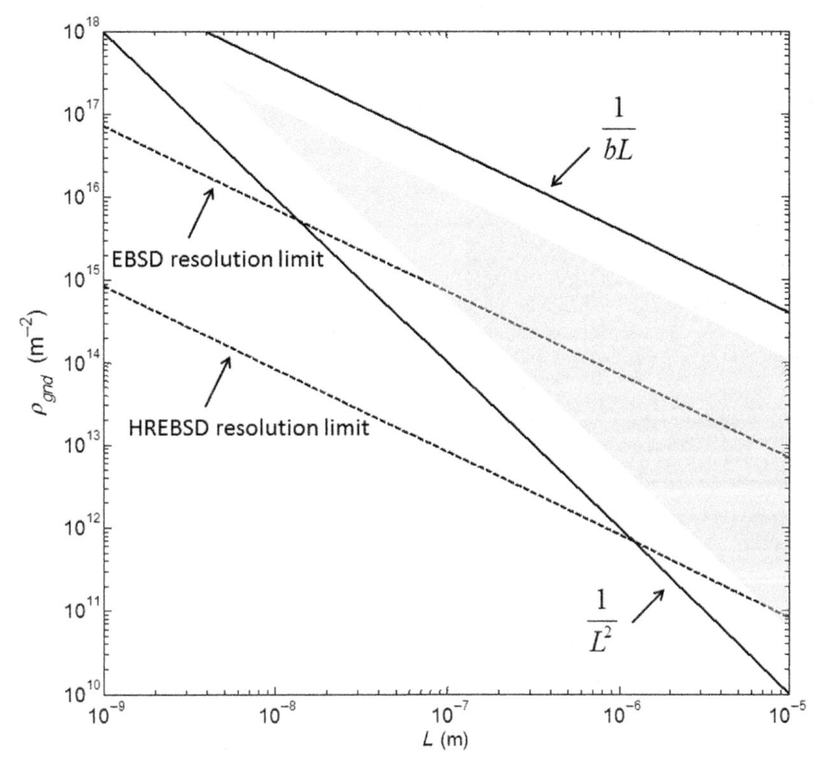

FIGURE 16.25

GND resolution versus step size (L). The *dashed lines* indicate lower bounds of GND resolution for an assumed EBSD orientation resolution of 0.5° and HREBSD resolution of 0.006°, respectively, according to Adams and Kacher (2009). The *solid lines* indicate a lower bound, on resolution relating to a single dislocation within the volume bounded by a step, and an upper bound, relating to variations in the plastic deformation field. The *shaded area* is the recommended characterization region according to Kysar et al. (2010).

OIM. These methods are in widespread use worldwide and are particularly useful in recovering statistical correlation functions that lie at the heart of many material structure descriptions. The basic relations linking 3D volume fractions to 2D area fractions (Delesse, 1847), linear fractions (Rosiwal, 1898), and point fractions (Thomsen, 1930; Glagolev, 1934) are reviewed.

In this section we assume that the microstructures being examined are statistically homogeneous. A specimen belonging to the ensemble of microstructures occupies region Ω and has volume $V(\Omega)$. For simplicity we assume that the microstructure comprises two phases, α and β. Our interest is focused on the volume fraction of β, which is defined by

$$V_\beta = \frac{V(\beta)}{V(\alpha) + V(\beta)} = \frac{V(\beta)}{V(\Omega)} \tag{16.17}$$

These definitions are illustrated in Figure 16.27.

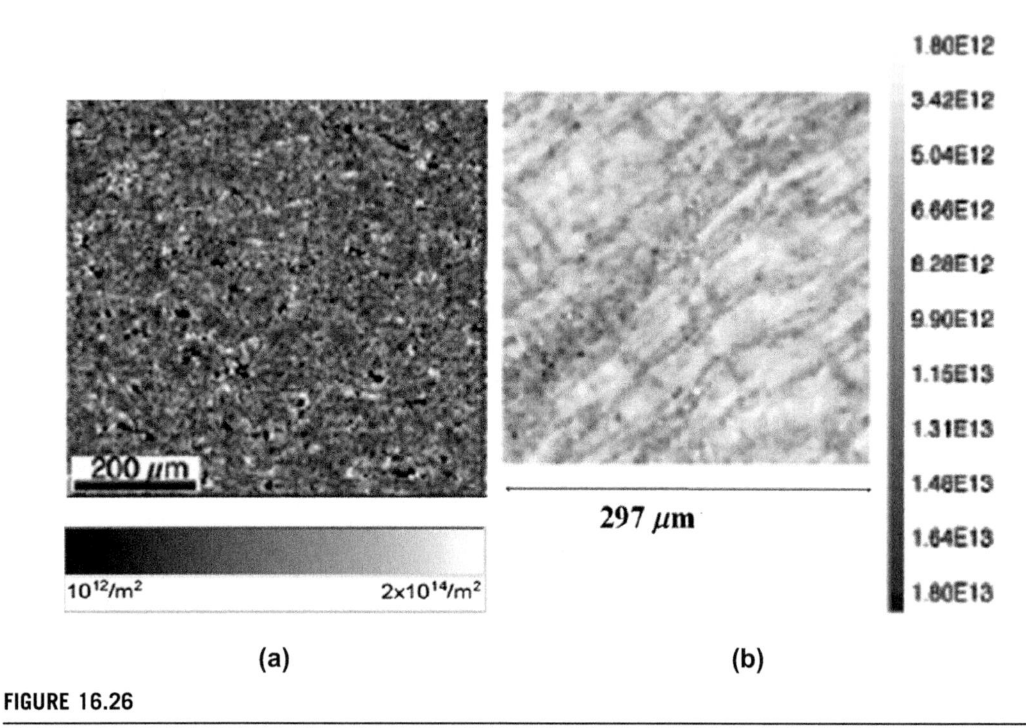

FIGURE 16.26

GND content in deformed single-crystal copper as measured by (a) EBSD and (b) X-ray microdiffraction. *Source:* Field et al. (2010). Used with permission.

16.7.1 Area fractions and Delesse's principle

Suppose that $d\beta$ is an infinitesimal volume element of β in Ω.[1] Let $T = T_2$ be a planar 2D *test section* that forms a profile of area $A(d\beta \cap T)$ with $d\beta$. The total area of intersection of T with β is found by integration:

$$A(\beta \cap T) = \int_{\beta} A(d\beta \cap T) \tag{16.18}$$

Stated in words, the total profile area of phase β with test section T is estimated by summing the profile of each infinitesimal element $d\beta$ with T. In the limit that $d\beta \rightarrow 0$, this is the integral expressed in (16.18).

Now consider not a fixed test section, T, but a random one. This is accomplished by random placement of T in a particular sample. We ask what is the expectation value of $A(d\beta \cap T)$ obtained from many randomly placed test sections; this we denote as $E(A(d\beta \cap T))$. If our sample is a representative volume element of the microstructure, this expectation is equivalent to the ensemble average of

[1] We think of $d\beta$ as occupying a specific subregion $d\omega \subset \Omega$, which can be filled by either phase α or phase β but not both. If $d\omega$ is occupied by phase β, then $d\beta = d\omega$; if it is occupied by phase α, then $d\beta = 0$ (and $d\alpha = d\omega$).

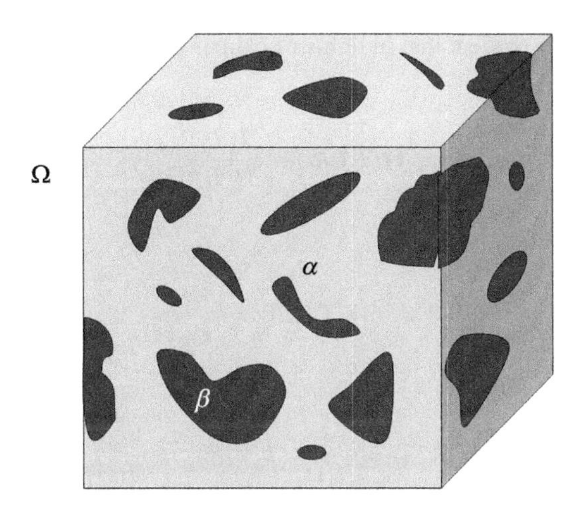

FIGURE 16.27

Typical two-phase microstructure comprising phases α and β.

$A(d\beta \cap T)$, which is obtained from fixed test sectioning over the large number of elements of the ensemble. Thus,

$$E(A(d\beta \cap T)) = \langle A(d\beta \cap T) \rangle \tag{16.19}$$

Incorporating (16.19) into (16.18), and noting that integration commutes with arithmetic averaging, we obtain

$$\langle A(\beta \cap T) \rangle = \left\langle \int_{\beta} A(d\beta \cap T) \right\rangle = \int_{\beta} \langle A(d\beta \cap T) \rangle \tag{16.20}$$

Note that the integrand in (16.20) can be expressed in terms of the probability that T hits $d\beta$, $\Pr(T \uparrow d\beta)$:

$$\langle A(\beta \cap T) \rangle = \int_{\beta} \langle A(d\beta \cap T | T \uparrow d\beta) \Pr(T \uparrow d\beta) \rangle \tag{16.21}$$

Since the ensemble itself is statistically homogeneous, the probability $\Pr(T \uparrow d\beta)$ is entirely independent of the location of the test section T. Thus,

$$\langle A(\beta \cap T) \rangle = \int_{\beta} \langle A(d\beta \cap T | T \uparrow d\beta) \rangle \Pr(T \uparrow d\beta) \tag{16.22}$$

Next, consider the *caliper length* of each element $d\beta$. Let $\hat{n}_T$ be the unit normal vector perpendicular to the test section T. Project each point belonging to $d\beta$ onto a line passing through $\hat{n}_T$ (arbitrary origin). Define the caliper length of $d\beta$ on T, $H(d\beta)$, to be the length of the projection of $d\beta$ onto the line passing through $\hat{n}_T$. It is evident that in terms of caliper lengths the probability $\Pr(T \uparrow d\beta)$ can be expressed as a ratio of caliper lengths:

$$\Pr(T \uparrow d\beta) = \frac{H(d\beta)}{H(\Omega)} \tag{16.23}$$

Furthermore, it is expected that

$$\langle A(d\beta \cap T | T \uparrow d\beta) \rangle = \frac{V(d\beta)}{H(d\beta)} \tag{16.24}$$

And, hence, it follows that

$$\langle A(\beta \cap T) \rangle = \frac{1}{H(\Omega)} \int_{\beta} V(d\beta) = \frac{V(\beta)}{H(\Omega)} \tag{16.25}$$

The same analysis leads to

$$\langle A(\Omega \cap T) \rangle = \frac{V(\Omega)}{H(\Omega)} \tag{16.26}$$

Thus, by combining (16.25) and (16.26) we retrieve the result of Delesse (1847) that the expectation of the area fraction observed in test sections of the microstructure is equal to the volume fraction:

$$\langle A_A \rangle = \frac{\langle A(\beta \cap T) \rangle}{\langle A(\Omega \cap T) \rangle} = \frac{V(\beta)}{V(\Omega)} = V_V \tag{16.27}$$

16.7.2 Line fractions and Rosiwal's principle

If we now change from a test section to a *test line* ($T = T_1$) and inquire as to the intersection of the test line with a volume element of phase β, $d\beta$, the line segment of intersection is denoted $L(d\beta \cap T)$. The total intersection in all of β is then

$$L(\beta \cap T) = \int_{\beta} L(d\beta \cap T) \tag{16.28}$$

Now consider not a single, fixed test line T but a random test line. In statistically homogeneous microstructures we expect

$$E(L(\beta \cap T)) = \langle L(\beta \cap T) \rangle = \left\langle \int_{\beta} L(d\beta \cap T) \right\rangle = \int_{\beta} \langle L(d\beta \cap T) \rangle \tag{16.29}$$

However, this can be expressed in terms of the probability that test line T hits the volume element $d\beta$:

$$\langle L(\beta \cap T)\rangle = \int_{\beta} \langle L(d\beta \cap T|T \uparrow d\beta)\text{Pr}(T \uparrow d\beta\rangle = \int_{\beta} \langle L(d\beta \cap T|T \uparrow d\beta)\rangle \text{Pr}(T \uparrow d\beta) \tag{16.30}$$

The expression on the right in (16.30) is a consequence of the observation that the probability that test line T hits the volume element $d\beta$ will be the same for all placement locations within Ω.

Next, project each point of $d\beta$ onto a plane that is perpendicular to the test line, $\hat{n}_T$. Let the area of projection be $A_p(d\beta)$. It is evident that

$$\text{Pr}(T \cap d\beta) = \frac{A_p(d\beta)}{A_p(\Omega)} \tag{16.31}$$

It can easily be shown that

$$\langle L(d\beta \cap T|T \uparrow d\beta)\rangle = \frac{V(d\beta)}{A_p(d\beta)} \tag{16.32}$$

and hence

$$\langle L(\beta \cap T)\rangle = \frac{1}{A_p(\Omega)} \int_{\beta} V(d\beta) = \frac{V(\beta)}{A_p(\Omega)} \tag{16.33}$$

The same analysis gives

$$\langle L(\Omega \cap T)\rangle = \frac{V(\Omega)}{A_p(\Omega)} \tag{16.34}$$

Combining (16.33) and (16.34) to obtain the expectation of linear fraction, $\langle L_L\rangle$, we have

$$\langle L_L\rangle = \frac{\langle L(\beta \cap T)\rangle}{\langle L(\Omega \cap T)\rangle} = \frac{V(\beta)}{V(\Omega)} = V_V \tag{16.35}$$

Relation (16.35) is *Rosiwal's principle*. In words, it states that the expected linear fraction of test line T, randomly placed in the microstructure, to intersect phase β is equivalent to the volume fraction of phase β that occurs in Ω.

16.7.3 Point counting and the principle of Thomsen and Glagolev

The procedure followed to obtain the principles of Delesse and Rosiwal can be extended to *test points* $T = T_0$. Let the intercept of test point T with a volume element $d\beta$ be $P(d\beta \cap T)$, which has value either 0 or 1. Integrating, we obtain

$$P(\beta \cap T) = \int_{\beta} P(d\beta \cap T) \in \{0, 1\} \tag{16.36}$$

Now consider a random set of test points and the expectation of intersection, which we again equate to the ensemble average in statistically homogeneous microstructures:

$$E(P(\beta \cap T)) = \langle P(\beta \cap T) \rangle = \int_{\beta} \langle P(d\beta \cap T) \rangle \in [0,1] \tag{16.37}$$

Relating this to the probability that T hits $d\beta$, the expression becomes

$$\langle P(\beta \cap T) \rangle = \int_{\beta} \langle P(d\beta \cap T | T \uparrow d\beta) \rangle \Pr(\langle T \rangle \uparrow d\beta) \tag{16.38}$$

Clearly,

$$\Pr(T \uparrow d\beta) = \frac{V(\beta)}{V(\Omega)} \tag{16.39}$$

and

$$\langle P(\beta \cap T) \rangle = \frac{1}{V(\Omega)} \int_{\beta} V(d\beta) = \frac{V(\beta)}{V(\Omega)} \tag{16.40}$$

By the same arguments, it must be that

$$\langle P(\Omega \cap T) \rangle = \frac{1}{V(\Omega)} \int_{\Omega} V(d\Omega) = \frac{V(\Omega)}{V(\Omega)} = 1 \tag{16.41}$$

Combining these we have the final expression:

$$\langle P_P \rangle = \frac{\langle P(\beta \cap T) \rangle}{\langle P(\Omega \cap T) \rangle} = \frac{V(\beta)}{V(\Omega)} = V_V \tag{16.42}$$

This construction is attributed to Thomsen (1930) and Glagolev (1934). In words, the expectation for randomly placed points in Ω to fall in phase β is equal to the volume fraction of Ω occupied by phase β.

By way of summary, the basic stereology of volume fractions equates the expectation of point fractions $\langle P_P \rangle$ with that of line fractions $\langle L_L \rangle$ and area fractions $\langle A_A \rangle$:

$$V_V = \langle A_A \rangle = \langle L_L \rangle = \langle P_P \rangle \tag{16.43}$$

SUMMARY

This chapter overviewed microscopy-based structure characterization techniques that feed microstructure data into the framework described in the rest of the book. While many other data extraction techniques exist (including, for example, X-ray–based methods and transmission electron microscopy), it is hoped that a reasonably detailed description of just one of these methods will provide readers with an intuitive link to "viewing" real structures that they wish to improve.

Exercises

Section 16.1

(short essay). Write a one-page essay about sample preparation for OIM samples. Describe mechanical and chemical techniques. Focus on samples containing two or more distinct material phases.

Section 16.6

(short project). Local elastic strain of the crystal lattice gives rise to changes in the six lattice parameters required to describe the crystal lattice. Consider the problem of measuring the local elastic strain of the lattice using measured EBSD patterns. Suppose a reference EBSD pattern for a (strain-free) lattice were available for an orientation of interest. Describe how you could use image correlation techniques between the reference and strained EBSD patterns to recover an estimate of the local elastic strain.

APPENDIX

1 Symmetry Point Operators

In the following, the reference axes $\mathbf{e}_1$, $\mathbf{e}_2$, $\mathbf{e}_3$ are labeled x, y, z, and the lattice basis vectors for the crystal are labeled $\mathbf{a}_i$.

NONHEXAGONAL LATTICES

Symbol	Transformation*	Description
E	x,y,z	Identity
C_{2x}	$x,\bar{y},\bar{z}$	180° rotation about $\mathbf{a}_1$
C_{2y}	$\bar{x},y,\bar{z}$	180° rotation about $\mathbf{a}_2$
C_{2z}	$\bar{x},\bar{y},z$	180° rotation about $\mathbf{a}_3$
C_{2a}	$y,x,\bar{z}$	180° rotation about $\mathbf{a}_1 + \mathbf{a}_2$
C_{2b}	$\bar{y},\bar{x},\bar{z}$	180° rotation about $\mathbf{a}_1 - \mathbf{a}_2$
C_{2c}	$z,\bar{y},x$	180° rotation about $\mathbf{a}_1 + \mathbf{a}_3$
C_{2d}	$\bar{x},z,y$	180° rotation about $\mathbf{a}_2 + \mathbf{a}_3$
C_{2e}	$\bar{z},\bar{y},\bar{x}$	180° rotation about $\mathbf{a}_1 - \mathbf{a}_3$
C_{2f}	$\bar{x},\bar{z},\bar{y}$	180° rotation about $\mathbf{a}_2 - \mathbf{a}_3$
C_{31}^{+}	z,x,y	120° rotation about $\mathbf{a}_1 + \mathbf{a}_2 - \mathbf{a}_3$
C_{23}^{+}	$\bar{z},x,\bar{y}$	120° rotation about $-\mathbf{a}_1 - \mathbf{a}_2 + \mathbf{a}_3$
C_{33}^{+}	$\bar{z},\bar{x},y$	120° rotation about $\mathbf{a}_1 - \mathbf{a}_2 - \mathbf{a}_3$
C_{34}^{+}	$z,\bar{x},\bar{y}$	120° rotation about $-\mathbf{a}1 + \mathbf{a}_2 - \mathbf{a}_3$
C_{31}^{-}	y,z,x	240° rotation about $\mathbf{a}_1 + \mathbf{a}_2 + \mathbf{a}_3$
C_{32}^{-}	$y,\bar{z},\bar{x}$	240° rotation about $-\mathbf{a}_1 - \mathbf{a}_2 + \mathbf{a}_3$
C_{33}^{-}	$\bar{y},z,\bar{x}$	240° rotation about $\mathbf{a}_1 - \mathbf{a}_2 - \mathbf{a}_3$
C_{34}^{-}	$\bar{y},\bar{z},x$	240° rotation about $-\mathbf{a}_1 + \mathbf{a}_2 - \mathbf{a}_3$
C_{4x}^{+}	$x,\bar{z},y$	90° rotation about $\mathbf{a}_1$
C_{4y}^{+}	$z,y,\bar{x}$	90° rotation about $\mathbf{a}_2$
C_{4z}^{+}	$\bar{y},x,z$	90° rotation about $\mathbf{a}_3$

Note: This appendix is taken from H. Stokes, Solid State Physics, *3rd ed., Provo, UT: Brigham Young University. Courtesy of Dr. Stokes.*
Note, for example, transformation $\bar{y},\bar{x},z$ indicates $x \to -y, y \to -x, z \to z$, etc.

(continued)

Symbol	Transformation*	Description
C_{4x}^-	$x,z,\bar{y}$	270° rotation about $\mathbf{a}_1$
C_{4y}^-	$\bar{z},y,x$	270° rotation about $\mathbf{a}_2$
C_{4z}^-	$y,\bar{x},z$	270° rotation about $\mathbf{a}_3$
I	$\bar{x},\bar{y},\bar{z}$	Inversion
σ_x	$\bar{x},y,z$	Reflection in plane ⊥ to $\mathbf{a}_1$
σ_y	$x,\bar{y},z$	Reflection in plane ⊥ to $\mathbf{a}_2$
σ_z	$x,y,\bar{z}$	Reflection in plane ⊥ to $\mathbf{a}_3$
σ_{da}	$\bar{y},\bar{x},z$	Reflection in plane ⊥ to $\mathbf{a}_1 + \mathbf{a}_2$
σ_{db}	y,x,z	Reflection in plane ⊥ to $\mathbf{a}_1 - \mathbf{a}_2$
σ_{dc}	$\bar{z},y,\bar{x}$	Reflection in plane ⊥ to $\mathbf{a}_1 + \mathbf{a}_3$
σ_{dd}	$x,\bar{z},y$	Reflection in plane ⊥ to $\mathbf{a}_2 + \mathbf{a}_3$
σ_{de}	z,y,x	Reflection in plane ⊥ to $\mathbf{a}_1 - \mathbf{a}_3$
σ_{df}	x,z,y	Reflection in plane ⊥ to $\mathbf{a}_2 - \mathbf{a}_3$
S_{61}^+	$\bar{y},\bar{z},\bar{x}$	60° rotation about $\mathbf{a}_1 + \mathbf{a}_2 + \mathbf{a}_3$ followed by reflection in plane ⊥ to that axis
S_{62}^+	$\bar{y},z,x$	60° rotation about $-\mathbf{a}_1 - \mathbf{a}_2 + \mathbf{a}_3$ followed by reflection in plane ⊥ to that axis
S_{63}^+	$y,\bar{z},x$	60° rotation about $\mathbf{a}_1 - \mathbf{a}_2 - \mathbf{a}_3$ followed by reflection in plane ⊥ to that axis
S_{64}^+	$y,z,\bar{x}$	60° rotation about $-\mathbf{a}_1 + \mathbf{a}_2 - \mathbf{a}_3$ followed by reflection in plane ⊥ to that axis
S_{61}^-	$\bar{z},\bar{x},\bar{y}$	300° rotation about $\mathbf{a}_1 + \mathbf{a}_2 + \mathbf{a}_3$ followed by reflection in plane ⊥ to that axis
S_{62}^-	$z,\bar{x},y$	300° rotation about $-\mathbf{a}_1 - \mathbf{a}_2 + \mathbf{a}_3$ followed by reflection in plane ⊥ to that axis
S_{63}^-	$z,x,\bar{y}$	300° rotation about $\mathbf{a}_1 - \mathbf{a}_2 - \mathbf{a}_3$ followed by reflection in plane ⊥ to that axis
S_{64}^-	$\bar{z},x,y$	300° rotation about $-\mathbf{a}_1 + \mathbf{a}_2 - \mathbf{a}_3$ followed by reflection in plane ⊥ to that axis
S_{4x}^+	$\bar{x},\bar{z},y$	90° rotation about $\mathbf{a}_1$ followed by reflection in plane ⊥ to that axis
S_{4y}^+	$z,\bar{y},\bar{x}$	90° rotation about $\mathbf{a}_2$ followed by reflection in plane ⊥ to that axis
S_{4z}^+	$\bar{y},x,\bar{z}$	90° rotation about $\mathbf{a}_3$ followed by reflection in plane ⊥ to that axis
S_{4x}^-	$\bar{x},z,\bar{y}$	270° rotation about $\mathbf{a}_1$ followed by reflection in plane ⊥ to that axis
S_{4y}^-	$\bar{z},\bar{y},x$	270°rotation about $\mathbf{a}_2$ followed by reflection in plane ⊥ to that axis
S_{4z}^-	$y,\bar{x},\bar{z}$	270° rotation about $\mathbf{a}_3$ followed by reflection in plane ⊥ to that axis

HEXAGONAL LATTICES

Symbol	Transformation*	Description
E	x,y,z	Identity
C_2	$\bar{x},\bar{y},z$	180° rotation about $\mathbf{a}_3$
C_3^+	$\bar{y},x-y,z$	120° rotation about $\mathbf{a}_3$
C_3^-	$y-x,\bar{x},z$	240° rotation about $\mathbf{a}_3$
C_6^+	$x-y,x,z$	60° rotation about $\mathbf{a}_3$
C_6^-	$y,y-x,z$	300° rotation about $\mathbf{a}_3$
C'_{21}	$y-x,y,\bar{z}$	180° rotation about $\mathbf{a}_1 + 2\mathbf{a}_2$
C'_{22}	$x,x-y,\bar{z}$	180° rotation about $2\mathbf{a}_1 + \mathbf{a}_2$
C'_{23}	$\bar{y},\bar{x},\bar{z}$	180° rotation about $\mathbf{a}_1 - \mathbf{a}_2$
C''_{21}	$x-y,\bar{y},\bar{z}$	180° rotation about $\mathbf{a}_1$
C'_{22}	$\bar{x},y-x,\bar{z}$	180° rotation about $\mathbf{a}_2$
C''_{23}	$y,x,\bar{z}$	180° rotation about $\mathbf{a}_1 + \mathbf{a}_2$
I	$\bar{x},\bar{y},\bar{z}$	Inversion
σ_h	$x,y,\bar{z}$	Reflection in plane ⊥ to $\mathbf{a}_3$
S_3^+	$\bar{y},x-y,\bar{z}$	120° rotation about $\mathbf{a}_3$ followed by reflection in plane ⊥ to that axis
S_3^-	$y-x,\bar{x},\bar{z}$	240° rotation about $\mathbf{a}_3$ followed by reflection in plane ⊥ to that axis
S_6^+	$x-y,x,\bar{z}$	60° rotation about $\mathbf{a}_3$ followed by reflection in plane ⊥ to that axis
S_6^-	$y,y-x,\bar{z}$	300° rotation about $\mathbf{a}_3$ followed by reflection in plane ⊥ to that axis
σ_{d1}	$x-y,\bar{y},z$	Reflection in plane ⊥ to $\mathbf{a}_1 + 2\mathbf{a}_2$
σ_{d2}	$\bar{x},y-x,z$	Reflection in plane ⊥ to $2\mathbf{a}_1 + \mathbf{a}_2$
σ_{d3}	y,x,z	Reflection in plane ⊥ to $\mathbf{a}_1 - \mathbf{a}_2$
σ_{v1}	$y-x,y,z$	Reflection in plane ⊥ to $\mathbf{a}_1$
σ_{v2}	$x,x-y,z$	Reflection in plane ⊥ to $\mathbf{a}_2$
σ_{v3}	$\bar{y},\bar{x},z$	Reflection in plane ⊥ to $\mathbf{a}_1 + \mathbf{a}_2$

BRAVAIS LATTICES

Conditions on the basis vectors:

- $\mathbf{a}_1$, $\mathbf{a}_2$, $\mathbf{a}_3$ are the magnitudes of the basis vectors $\mathbf{a}_1$, $\mathbf{a}_2$, $\mathbf{a}_3$, respectively
- α is the angle between $\mathbf{a}_2$ and $\mathbf{a}_3$
- β is the angle between $\mathbf{a}_3$ and $\mathbf{a}_1$
- γ is the angle between $\mathbf{a}_1$ and $\mathbf{a}_3$

Crystal System	Condition
Triclinic	none
Monoclinic	$\alpha = \beta = 90^\circ$
Orthorhombic	$\alpha = \beta = \gamma = 90^\circ$
Tetragonal	$\mathbf{a}_1 = \mathbf{a}_2$, $\alpha = \beta = \gamma = 90^\circ$
Hexagonal	$\mathbf{a}_1 = \mathbf{a}_2$, $\alpha = \beta = 90^\circ$ $\gamma = 120^\circ$
Cubic	$\mathbf{a}_1 = \mathbf{a}_2 = \mathbf{a}_3$, $\alpha = \beta = \gamma = 90^\circ$

Centered Lattices

Type of Centering	Symbol	Centered Point ($\mathbf{a}_1$, $\mathbf{a}_2$, $\mathbf{a}_3$)
Primitive	P	None
Base centered	A	$(0, 1/2, 1/2)$
	B	$(1/2, 0, 1/2)$
	C	$(1/2, 1/2, 0)$
Face centered	F	$(0, 1/2, 1/2)$, $(1/2, 0, 1/2)$, $(1/2, 1/2, 0)$
Body centered	I	$(1/2, 1/2, 1/2)$
Rhombohedral centered	R	$(2/3, 1/3, 1/3)$, $(1/3, 2/3, 2/3)$

Schematics of the Bravais Lattices

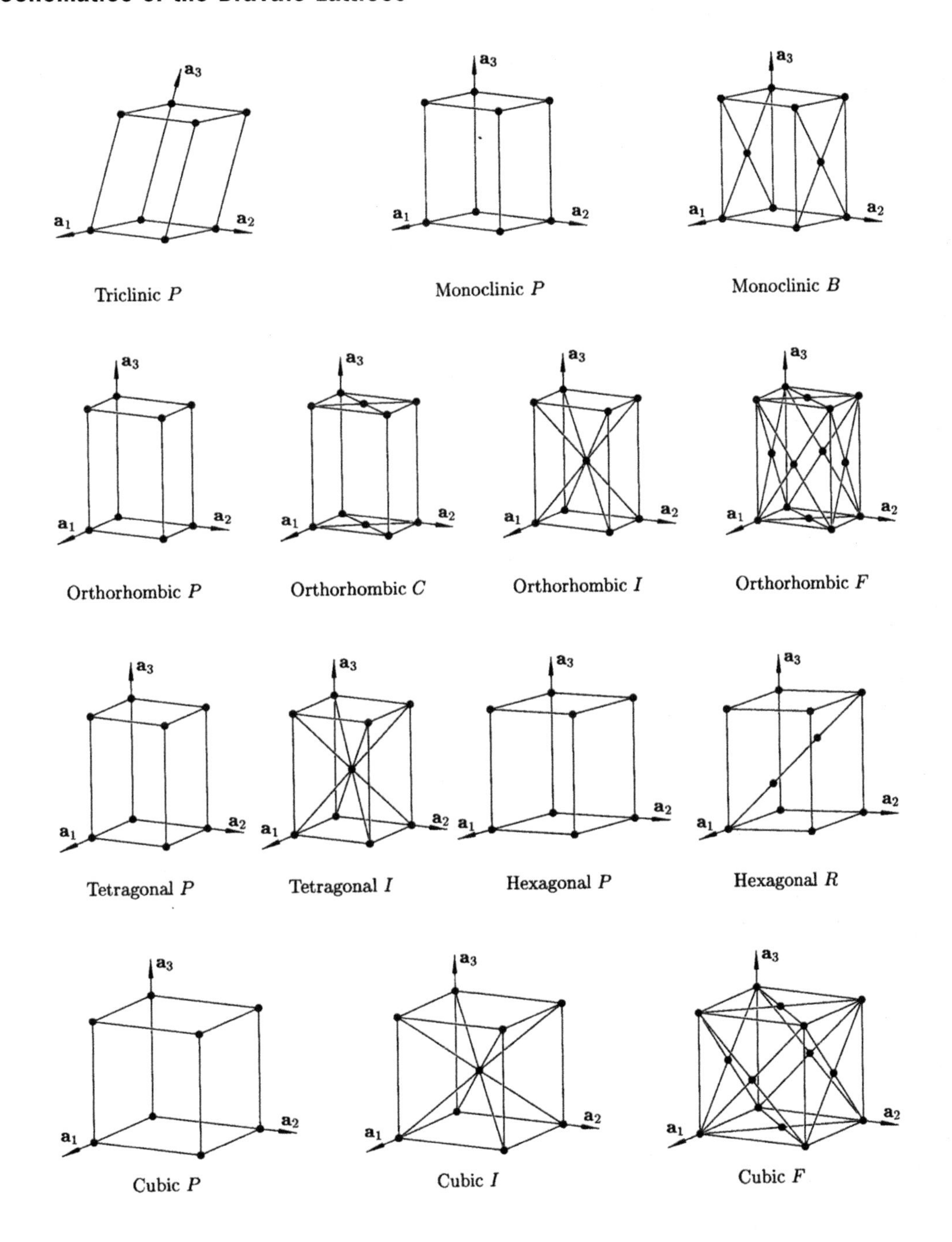

Space Groups	Class		Point Operators
Monoclinic			
3–5	C_2	2	E, C_{2z}
6–9	C_s	m	E, σ_z
10–15	C_{2h}	$2/m$	E, C_{2z}, I, σ_z
Orthorhombic			
16–24	D_2	222	$E, C_{2x}, C_{2y}, C_{2z}$
25–46	C_{2v}	$mm2$	$E, C_{2x}, \sigma_x, \sigma_y$
47–74	D_{2h}	mmm	$E, C_{2x}, C_{2y}, C_{2z}, I, \sigma_x, \sigma_y, \sigma_z$
Hexagonal			
143–146	C_3	3	$E, C_3^{\pm}$
147–148	S_6	$\overline{3}$	$E, C_3^{\pm}, I, S_6^{\pm}$
149–155	D_3	32	(1) $E, C_3^{\pm}, C_{21}, C_{22}, C_{23}$ (2) $E, C_3^{\pm}, C''_{21}, C''_{22}, C''_{23}$
	C_{3v}	$3m$	(1) $E, C_3^{\pm}, \sigma_{d1}, \sigma_{d2}, \sigma_{d3}$ (2) $E, C_3^{\pm}, \sigma_{v1}, \sigma_{v2}, \sigma_{v3}$
	D_{3d}	$\overline{6}$	(1) $E, C_3^{\pm}, C_{21}, C_{22}, C_{23}, I, S_6^{\pm}, \sigma_{d1}, \sigma_{d2}, \sigma_{d3}$ (2) $E, C_3^{\pm}, C''_{21}, C''_{22}, C''_{23}, I, S_6^{\pm}, \sigma_{v1}, \sigma_{v2}, \sigma_{v3}$
	C_6	6	$E, C_6^{\pm}, C_3^{\pm}, C_2$
	C_{3h}	$\overline{6}$	$E, S_3^{\pm}, C_3^{\pm}, \sigma_h$
	C_{6h}	$6/m$	$E, C_6^{\pm}, C_3^{\pm}, C_2, I, S_3^{\pm}, S_6^{\pm}, \sigma_h$
	D_6	622	$E, C_6^{\pm}, C_3^{\pm}, C_2, C_{21}, C_{22}, C_{23}, C''_{21}, C''_{22}, C''_{23}$
	C_{6v}	$6mm$	$E, C_6^{\pm}, C_3^{\pm}, C_2, \sigma_{d1}, \sigma_{d2}, \sigma_{d3}, \sigma_{v1}, \sigma_{v2}, \sigma_{v3}$
	D_{3h}	$\overline{6}m2$	(1) $E, C_3^{\pm}, C_{21}, C_{22}, C_{23}, \sigma_h, S_3^{\pm}, \sigma_{v1}, \sigma_{v2}, \sigma_{v3}$ (2) $E, C_3^{\pm}, C''_{21}, C''_{22}, C''_{23}, \sigma_h, S_3^{\pm}, \sigma_{d1}, \sigma_{d2}, \sigma_{d3}$
	D_{6h}	$6/mmm$	$E, C_6^{\pm}, C_3^{\pm}, C_2, C_{21}, C_{22}, C_{23}, C''_{21}, C''_{22}, C''_{23}, I, S_6^{\pm}, S_3^{\pm}, \sigma_h, \sigma_{d1}, \sigma_{d2}, \sigma_{d3}, \sigma_{v1}, \sigma_{v2}, \sigma_{v3}$
Cubic			
195–199	T	23	$E, C_{31}^{\pm}, C_{32}^{\pm}, C_{33}^{\pm}, C_{34}^{\pm}, C_{2x}, C_{2y}, C_{2z}$
200–206	T_h	$m\overline{3}$	$E, C_{31}^{\pm}, C_{32}^{\pm}, C_{33}^{\pm}, C_{34}^{\pm}, C_{2x}, C_{2y}, C_{2z}, I, S_{61}^{\pm}, S_{62}^{\pm}, S_{63}^{\pm}, S_{64}^{\pm}, \sigma_x, \sigma_y, \sigma_z$
207–214	O	432	$E, C_{31}^{\pm}, C_{32}^{\pm}, C_{33}^{\pm}, C_{34}^{\pm}, C_{2x}, C_{2y}, C_{2z}, C_{2a}, C_{2b}, C_{2c}, C_{2d}, C_{2e}, C_{2f}, C_{4x}^{\pm}, C_{4y}^{\pm}, C_{4z}^{\pm}$
215–220	T_d	$\overline{4}3m$	$E, C_{31}^{\pm}, C_{32}^{\pm}, C_{33}^{\pm}, C_{34}^{\pm}, C_{2x}, C_{2y}, C_{2z}, \sigma_{da}, \sigma_{db}, \sigma_{dc}, \sigma_{dd}, \sigma_{de}, \sigma_{df}, S_{4x}^{\pm}, S_{4y}^{\pm}, S_{4z}^{\pm}$
221–230	O_h	$m\overline{3}m$	$E, C_{31}^{\pm}, C_{32}^{\pm}, C_{33}^{\pm}, C_{34}^{\pm}, C_{2x}, C_{2y}, C_{2z}, C_{2a}, C_{2b}, C_{2c}, C_{2d}, C_{2e}, C_{2f}, C_{4x}^{\pm}, C_{4y}^{\pm}, C_{4z}^{\pm},$ $I, S_{61}^{\pm}, S_{62}^{\pm}, S_{63}^{\pm}, S_{64}^{\pm}, \sigma_x, \sigma_y, \sigma_z, \sigma_{da}, \sigma_{db}, \sigma_{dc}, \sigma_{dd}, \sigma_{de}, \sigma_{df}, S_{4x}^{\pm}, S_{4y}^{\pm}, S_{4z}^{\pm}$

Space Groups	Class		Point Operators
Triclinic			
1	C_1	1	E
2	C_i	$\overline{1}$	E, I
Tetragonal			
75–80	C_4	4	$E, C_{4z}^{\pm}, C_{2z}$
81–82	S_4	$\overline{4}$	$E, S_{4z}^{\pm}, C_{2z}$
83–88	C_{4h}	$4/m$	$E, C_{4z}^{\pm}, C_{2z}, I, S_{4z}^{\pm}, \sigma_z$
89–98	D_4	422	$E, C_{4z}^{\pm}, C_{2z}, C_{2x}, C_{2y}, C_{2a}, C_{2b}$
99–110	C_{4v}	$4mm$	$E, C_{4z}^{\pm}, C_{2z}, \sigma_x, \sigma_y, \sigma_{da}, \sigma_{db}$
111–122	D_{2d}	$\overline{4}2m$	(1) $E, S_{4z}^{\pm}, C_{2z}, C_{2x}, C_{2y}, \sigma_{da}, \sigma_{db}$ (2) $E, S_{4z}^{\pm}, C_{2z}, C_{2a}, C_{2b}, \sigma_x, \sigma_y$
123–142	D_{4h}	$4/mmm$	$E, C_{4z}^{\pm}, C_{2z}, C_{2x}, C_{2y}, C_{2a}, C_{2b}, I, S_{4z}^{\pm}, \sigma_z, \sigma_x, \sigma_y, \sigma_{da}, \sigma_{db}$

CRYSTAL STRUCTURES

The following list shows the crystal structures of various elements and compounds. The lattice parameters, **a**, in units of Å are also given.

Elements						Compounds with Sodium Chloride Structure					
With bcc Lattice											
Li	3.5	Ba	5.02	Mo	3.14	LiF	4.02	RbI	7.32	BaTe	6.99
Na	4.3	V	3.04	W	3.15	LiCl	5.14	CsF	6	MnO	4.43
K	5.2	Nb	3.3	Fe	2.86	LiBr	5.49	MgO	4.2	MnS	5.21
Rb	5.59	Ta	3.32	Eu	4.38	LiI	6	MgS	5.19	MnSe	5.45
Cs	6.5	Cr	2.87			NaF	4.61	MgSe	5.45	FeO	4.28
With fcc Lattice						NaC1	5.63	CaO	4.8	CoO	4.25
Ca	5.36	Pd	3.87	Al	4.04	NaBr	5.96	CaS	5.68	NiO	4.17
Sr	6.08	Pt	3.9	Pb	4.93	NaI	6.46	CaSe	5.91	AgF	4.92
Ac	5.31	Cu	3.61	Ce	5.12	KF	5.36	CaTe	6.34	AgCl	5.54
Rh	3.8	Ag	4.07	Yb	5.48	KC1	6.27	SrO	5.13	AgBr	5.76
Jr	3.82	Au	4.07	Th	5.08	KBr	6.58	SrS	6.01	CdO	4.7
Ni	3.52					KI	7.05	SrSe	6.23	SnTe	6.28
With Diamond Structure						RbF	5.63	SrTe	6.65	PbS	5.93
C	3.56	Ge	5.65	Sn	6.46	RbCl	6.53	BaO	5.53	PbSe	6.14
Si	5.42					RbBr	6.85	BaSe	6.59	PbTe	6.44

(continued)

Elements						Compounds with Sodium Chloride Structure					
Compounds with Cesium Chloride Structure						**Compounds with Zincblende Structure**					
CsCl	4.11	TlI	4.18	CuPd	2.99	BeS	4.86	ZnTe	6.09	GaP	5.44
CsBr	4.28	TlSb	3.85	AgZn	3.16	CuC1	5.41	CdTe	6.46	GaAs	5.64
CsI	4.56	Tlbi	3.9	AuZn	3.15	CuBr	5.68	AlP	5.45	GaSh	6.09
T1C1	3.84	CuZn	2.95	AlNi	2.82	CuI	6.05	AlAs	5.63	InSb	6.45
TlBr	3.97					ZnS	5.42	A1Sb	6.1	SnSb	6.13
						ZnSe	5.66				

Tables of Spherical Harmonic Functions

2

Table A2.1 Symmetry Coefficients $A_l^{m\,\mu}$ for Cubic Symmetry, [001] Parallel to the *z*-axis (for use in Eq. 6.2.2)

l	*μ*	*m*	*B*
4	1	0	0.3046972
4	1	1	0.36418281
6	1	0	−0.1410474
6	1	1	0.52775103
8	1	0	0.28646862
8	1	1	0.21545346
8	1	2	0.32826995
9	1	0	0
9	1	1	−0.47483629
9	1	2	0.3046972
10	1	0	−0.16413497
10	1	1	0.33078546
10	1	2	0.39371345
12	1	0	0.26141975
12	1	1	0.27266871
12	1	2	0.0327746
12	1	3	0.32589402
12	2	0	0.09298802
12	2	1	−0.23773812
12	2	2	0.49446631
12	2	3	0
13	1	0	0
13	1	1	−0.44370187
13	1	2	−0.12875807
13	1	3	0.32382078
14	1	0	−0.17557309
14	1	1	0.25821932
14	1	2	0.27709173
14	1	3	0.3364536
15	1	0	0
15	1	1	0.16910989
15	1	2	−0.4570458
15	1	3	0.28429012
16	1	0	0.24370673
16	1	1	0.29873515
16	1	2	0.06447688
16	1	3	0.00377
16	1	4	0.32574495
16	2	0	0.12039646
16	2	1	−0.25330128
16	2	2	0.23950998
16	2	3	0.40962508
16	2	4	0
17	1	0	0
17	1	1	−0.41566875
17	1	2	−0.19764468
17	1	3	−0.02159338
17	1	4	0.32557592
18	1	0	−0.16914245
18	1	1	0.1701734
18	1	2	0.34598142
18	1	3	0.07433932
18	1	4	0.32696037
18	2	0	0.06901768
18	2	1	0.16006562
18	2	2	−0.24743528
18	2	3	0.47110273
18	2	4	0
19	1	0	0
19	1	1	0.21950324
19	1	2	−0.36729917
19	1	3	−0.18180666
19	1	4	0.31963396
20	1	0	0.23067026
20	1	1	0.31151832
20	1	2	0.09287682
20	1	3	0.01089683
20	1	4	0.00037564
20	1	5	0.32573563
20	2	0	0.1361542
20	2	1	−0.25048007
20	2	2	0.12882081
20	2	3	0.28642879
20	2	4	0.34620433
20	2	5	0
21	1	0	0
21	1	1	−0.38704231
21	1	2	−0.24503942
21	1	3	−0.04858614
21	1	4	−0.00282954
21	1	5	0.32572258
21	2	0	0
21	2	1	−0.08431638
21	2	2	0.21737195
21	2	3	−0.44006168
21	2	4	0.26513353
21	2	5	0
22	1	0	−0.1610956
22	1	1	0.10244188
22	1	2	0.36285175
22	1	3	0.13377513
22	1	4	0.01314399
22	1	5	0.32585583
22	2	0	−0.09620055
22	2	1	0.20244115
22	2	2	−0.22389483
22	2	3	0.17928946
22	2	4	0.42017231
22	2	5	0

Source: *Generated using Pospiech and Jura (1975); see also Bunge (1993, Table 15.2.1).*

Table A2.2 Generalized Legendre Functions, $P_l^{mn}(\Phi)$, for $l = 1, 2, 3$; $m = 0, 1, 2$; $n = 0, 1, 2$ (for use in Eq. 3.4.5)

Φ	$l = 1, m = 0, n = 0$								
0	1.0000	**5**	0.9962	**10**	0.9848	**15**	0.9659	**20**	0.9397
30	0.8660	**35**	0.8192	**40**	0.7660	**45**	0.7071	**50**	0.6428
60	0.5000	**65**	0.4226	**70**	0.3420	**75**	0.2588	**80**	0.1737
90	0.0000	**95**	–0.0872	**100**	–0.1737	**105**	–0.2588	**110**	–0.3420
120	–0.5000	**125**	–0.5736	**130**	–0.6428	**135**	–0.7071	**140**	–0.7660
150	–0.8660	**155**	–0.9063	**160**	–0.9397	**165**	–0.9659	**170**	–0.9848

Φ	$l = 1, m = 1, n = 0$								
0	0.0000	**5**	–0.0616	**10**	–0.1228	**15**	–0.1830	**20**	–0.2418
30	–0.3536	**35**	–0.4056	**40**	–0.4545	**45**	–0.5000	**50**	–0.5417
60	–0.6124	**65**	–0.6409	**70**	–0.6645	**75**	–0.6830	**80**	–0.6964
90	–0.7071	**95**	–0.7044	**100**	–0.6964	**105**	–0.6830	**110**	–0.6645
120	–0.6124	**125**	–0.5792	**130**	–0.5417	**135**	–0.5000	**140**	–0.4545
150	–0.3536	**155**	–0.2988	**160**	–0.2418	**165**	–0.1830	**170**	–0.1228

Φ	$l = 1, m = 1, n = 1$								
0	1.0000	**5**	0.9981	**10**	0.9924	**15**	0.9830	**20**	0.9699
30	0.9330	**35**	0.9096	**40**	0.8830	**45**	0.8536	**50**	0.8214
60	0.7500	**65**	0.7113	**70**	0.6710	**75**	0.6294	**80**	0.5868
90	0.5000	**95**	0.4564	**100**	0.4132	**105**	0.3706	**110**	0.3290
120	0.2500	**125**	0.2132	**130**	0.1786	**135**	0.1465	**140**	0.1170
150	0.0670	**155**	0.0468	**160**	0.0302	**165**	0.0170	**170**	0.0076

Φ	$l = 2, m = 0, n = 0$								
0	1.0000	**5**	0.9886	**10**	0.9548	**15**	0.8995	**20**	0.8245
30	0.6250	**35**	0.5065	**40**	0.3802	**45**	0.2500	**50**	0.1198
60	–0.1250	**65**	–0.2321	**70**	–0.3245	**75**	–0.3995	**80**	–0.4548
90	–0.5000	**95**	–0.4886	**100**	–0.4548	**105**	–0.3995	**110**	–0.3245
120	–0.1250	**125**	–0.0065	**130**	0.1198	**135**	0.2500	**140**	0.3802
150	0.6250	**155**	0.7321	**160**	0.8245	**165**	0.8995	**170**	0.9548

Φ	$l = 2, m = 1, n = 0$								
0	0.0000	**5**	–0.1063	**10**	–0.2094	**15**	–0.3062	**20**	–0.3936
30	–0.5303	**35**	–0.5754	**40**	–0.6031	**45**	–0.6124	**50**	–0.6031
60	–0.5303	**65**	–0.4691	**70**	–0.3936	**75**	–0.3062	**80**	–0.2094
90	0.0000	**95**	0.1063	**100**	0.2094	**105**	0.3062	**110**	0.3936
120	0.5303	**125**	0.5754	**130**	0.6031	**135**	0.6124	**140**	0.6031
150	0.5303	**155**	0.4691	**160**	0.3936	**165**	0.3062	**170**	0.2094

Φ — *l* = 2, *m* = 1, *n* = 1

0	1.0000	5	0.9905	10	0.9623	15	0.9160	20	0.8529
30	0.6830	35	0.5806	40	0.4699	45	0.3536	50	0.2346
60	0.0000	65	–0.1101	70	–0.2120	75	–0.3036	80	–0.3830
90	–0.5000	95	–0.5360	100	–0.5567	105	–0.5624	110	–0.5540
120	–0.5000	125	–0.4578	130	–0.4082	135	–0.3536	140	–0.2962
150	–0.1830	155	–0.1318	160	–0.0868	165	–0.0500	170	–0.0226

Φ — *l* = 2, *m* = 2, *n* = 0

0	0.0000	5	–0.0047	10	–0.0185	15	–0.0410	20	–0.0716
30	–0.1531	35	–0.2015	40	–0.2530	45	–0.3062	50	–0.3594
60	–0.4593	65	–0.5030	70	–0.5407	75	–0.5714	80	–0.5939
90	–0.6124	95	–0.6077	100	–0.5939	105	–0.5714	110	–0.5407
120	–0.4593	125	–0.4109	130	–0.3594	135	–0.3062	140	–0.2530
150	–0.1531	155	–0.1094	160	–0.0716	165	–0.0410	170	–0.0185

Φ — *l* = 2, *m* = 2, *n* = 1

0	0.0000	5	–0.0870	10	–0.1723	15	–0.2544	20	–0.3317
30	–0.4665	35	–0.5217	40	–0.5676	45	–0.6036	50	–0.6292
60	–0.6495	65	–0.6447	70	–0.6305	75	–0.6080	80	–0.5779
90	–0.5000	95	–0.4547	100	–0.4069	105	–0.3580	110	–0.3092
120	–0.2165	125	–0.1747	130	–0.1368	135	–0.1036	140	–0.0752
150	–0.0335	155	–0.0198	160	–0.0103	165	–0.0044	170	–0.0013

Φ — *l* = 2, *m* = 2, *n* = 2

0	1.0000	5	0.9962	10	0.9849	15	0.9662	20	0.9406
30	0.8705	35	0.8273	40	0.7797	45	0.7286	50	0.6747
60	0.5625	65	0.5060	70	0.4503	75	0.3962	80	0.3444
90	0.2500	95	0.2083	100	0.1707	105	0.1373	110	0.1082
120	0.0625	125	0.0455	130	0.0319	135	0.0214	140	0.0137
150	0.0045	155	0.0022	160	0.0009	165	0.0003	170	0.0001

Φ — *l* = 3, *m* = 0, *n* = 0

0	1.0000	5	0.9773	10	0.9106	15	0.8042	20	0.6649
30	0.3248	35	0.1454	40	–0.0252	45	–0.1768	50	–0.3002
60	–0.4375	65	–0.4452	70	–0.4130	75	–0.3449	80	–0.2474
90	0.0000	95	0.1291	100	0.2474	105	0.3449	110	0.4130
120	0.4375	125	0.3886	130	0.3002	135	0.1768	140	0.0252
150	–0.3248	155	–0.5016	160	–0.6649	165	–0.8042	170	–0.9106

Table A2.3 Cubic Surface Spherical Harmonics $\dot{k}_l^m(\Phi\beta)$ for $l = 4, 6, 8$; $m = 1$ (as calculated by Eq. 3.5.4)

Φ →

β ↓

$l = 4$ $m = 1$

	52.5	55.0	57.5	60.0	62.5	65.0	67.5	70.0
0.0	−0.1075	−0.0699	−0.0240	0.0291	0.0878	0.1507	0.2160	0.2820
5	−0.1171	−0.0808	−0.0361	0.0156	0.0731	0.1346	0.1987	0.2635
10	−0.1449	−0.1121	−0.0711	−0.0231	0.0306	0.0885	0.1490	0.2104
15	−0.1875	−0.1601	−0.1248	−0.0824	−0.0344	0.0179	0.0728	0.1289
20	−0.2397	−0.2191	0–0.195	−0.1552	−0.1142	−0.0688	−0.0206	0.0290
25	−0.2953	−0.2817	−0.2605	−0.2326	−0.1990	−0.1611	−0.1200	−0.0773
30	−0.3475	−0.3407	−0.3263	−0.3054	−0.2788	−0.2477	−0.2134	−0.1772
35	−0.3901	−0.3887	−0.3799	−0.3647	−0.3438	−0.3184	−0.2896	−0.2586
40	−0.4179	−0.4200	−0.4149	−0.4034	−0.3862	−0.3645	−0.3393	−0.3118
45	−0.4275	−0.4309	−0.4271	−0.4168	−0.4010	−0.3805	−0.3566	−0.3302

$l = 4$ $m = 1$

	72.5	75.0	77.5	80.0	82.5	85.0	87.5	90.0
0.0	0.3469	0.4091	0.4668	0.5185	0.5629	0.5987	0.6249	0.6410
5	0.3274	0.3885	0.4453	0.4962	0.5398	0.5750	0.6009	0.6167
10	0.2710	0.3291	0.3831	0.4317	0.4733	0.5070	0.5317	0.5468
15	0.1845	0.2381	0.2880	0.3329	0.3715	0.4027	0.4256	0.4397
20	0.0785	0.1264	0.1713	0.2117	0.2465	0.2748	0.2955	0.3083
25	−0.0343	0.0076	0.0470	0.0827	0.1136	0.1386	0.1571	0.1684
30	−0.1403	−0.1040	−0.0697	−0.0385	−0.0114	0.0107	0.0270	0.0370
35	−0.2267	−0.1950	−0.1649	−0.1373	−0.1132	−0.0936	−0.0790	−0.0701
40	−0.2831	−0.2544	−0.2270	−0.2017	−0.1797	−0.1616	−0.1482	−0.1400
45	−0.3027	−0.2751	−0.2485	−0.2241	−0.2028	−0.1853	−0.1723	−0.1643

$l = 6$ $m = 1$

	52.5	55.0	57.5	60.0	62.5	65.0	67.5	70.0
0.0	0.5211	0.4772	0.4236	0.3616	0.2930	0.2196	0.1432	0.0661
5	0.4864	0.4436	0.3919	0.3329	0.2681	0.1993	0.1283	0.0570
10	0.3865	0.3466	0.3007	0.2502	0.1965	0.1411	0.0853	0.0306
15	0.2335	0.1981	0.1610	0.1235	0.0868	0.0519	0.0195	−0.0098
20	0.0458	0.0159	−0.0104	−0.0319	−0.0478	−0.0576	−0.0613	−0.0594
25	−0.1540	−0.1780	−0.1928	−0.1973	−0.1910	−0.1741	−0.1473	−0.1122
30	−0.3417	−0.3602	−0.3642	−0.3527	−0.3256	−0.2835	−0.2281	−0.1618
35	−0.4948	−0.5087	−0.5039	−0.4794	−0.4353	−0.3728	−0.2940	−0.2022
40	−0.5947	−0.6056	−0.5951	−0.5621	−0.5069	−0.4310	−0.3370	−0.2286
45	−0.6294	−0.6393	−0.6268	−0.5908	−0.5318	−0.4512	−0.3519	−0.2377

Φ →

β ↓

$l = 6$ $m = 1$

	72.5	75.0	77.5	80.0	82.5	85.0	87.5	90.0
0.0	−0.0098	−0.0824	−0.1498	−0.2103	−0.2621	−0.3039	−0.3346	−0.3533
5	−0.0129	−0.0795	−0.1410	−0.1960	−0.2430	−0.2808	−0.3085	−0.3255
10	−0.0219	−0.0710	−0.1156	−0.1548	−0.1880	−0.2144	−0.2336	−0.2453
15	−0.0357	−0.0579	−0.0766	−0.0918	−0.1037	−0.1126	−0.1188	−0.1224
20	−0.0526	−0.0420	−0.0288	−0.0144	−0.0004	0.0122	0.0220	0.0283
25	−0.0706	−0.0250	0.0221	0.0679	0.1096	0.1450	0.1719	0.1886
30	−0.0875	−0.0090	0.0699	0.1452	0.2130	0.2698	0.3127	0.3393
35	−0.1013	0.0040	0.1089	0.2082	0.2973	0.3716	0.4275	0.4622
40	−0.1103	0.0125	0.1343	0.2494	0.3523	0.4380	0.5024	0.5424
45	−0.1134	0.0155	0.1432	0.2637	0.3714	0.4610	0.5284	0.5702

$l = 8$ $m = 1$

	52.5	55.0	57.5	60.0	62.5	65.0	67.5	70.0
0.0	0.2822	0.1699	0.0526	−0.0557	−0.1414	−0.1935	−0.2041	−0.1697
5	0.2654	0.1524	0.0335	−0.0775	−0.1673	−0.2249	−0.2423	−0.2160
10	0.2223	0.1090	−0.0127	−0.1297	−0.2292	−0.3002	−0.3350	−0.3296
15	0.1718	0.0623	−0.0589	−0.1795	−0.2876	−0.3728	−0.4273	−0.4466
20	0.1357	0.0377	−0.0747	−0.1906	−0.2993	−0.3910	−0.4581	−0.4959
25	0.1283	0.0512	−0.0420	−0.1420	−0.2391	−0.3246	−0.3913	−0.4345
30	0.1507	0.1011	0.0350	−0.0395	−0.1143	−0.1816	−0.2349	−0.2698
35	0.1899	0.1684	0.1311	0.0847	0.0363	−0.0072	−0.0402	−0.0587
40	0.2257	0.2253	0.2100	0.1855	0.1584	0.1346	0.1194	0.1163
45	0.2400	0.2475	0.2403	0.2242	0.2051	0.1890	0.1808	0.1839

$l = 8$ $m = 1$

	72.5	75.0	77.5	80.0	82.5	85.0	87.5	90.0
0.0	−0.0916	0.0241	0.1676	0.3255	0.4831	0.6251	0.7379	0.8102
5	−0.1470	−0.0409	0.0928	0.2413	0.3903	0.5250	0.6322	0.7012
10	−0.2841	−0.2033	−0.0955	0.0279	0.1540	0.2693	0.3617	0.4214
15	−0.4300	−0.3810	−0.3062	−0.2152	−0.1190	−0.0292	0.0438	0.0912
20	−0.5030	−0.4814	−0.4361	−0.3748	−0.3066	−0.2409	−0.1866	−0.1509
25	−0.4524	−0.4459	−0.4188	−0.3771	−0.3280	−0.2794	−0.2385	−0.2114
30	−0.2842	−0.2788	−0.2563	−0.2216	−0.1806	−0.1400	−0.1057	−0.0830
35	−0.0607	−0.0467	−0.0191	0.0181	0.0596	0.0995	0.1327	0.1545
40	0.1270	0.1512	0.1866	0.2293	0.2745	0.3168	0.3513	0.3738
45	0.1999	0.2285	0.2674	0.3128	0.3599	0.4036	0.4389	0.4618

Table A2.4 Cubic-Orthorhombic Generalized Harmonics, $\dot{\ddot{T}}_l^{\mu n}(\phi_1 \Phi \phi_2)$, for $l = 4, 6$; $\mu = 1$; $n = 2$ (as calculated by Eq. 3.4.5)

φ_1 $l = 4$ $\mu = 1$ $n = 2$ *Cross Section:* $\phi_2 = 0$

Φ	0	10	20	30	40	50	60	70	80	90
0	0.0000	0.0000	0.0000	0.0000	0.0000	0.0000	0.0000	0.0000	0.0000	0.0000
10	−0.0999	−0.0939	−0.0765	−0.0499	−0.0173	0.0173	0.0499	0.0765	0.0939	0.0999
20	−0.3528	−0.3315	−0.2703	−0.1764	−0.0613	0.0613	0.1764	0.2703	0.3315	0.3528
30	−0.6404	−0.6018	−0.4906	−0.3202	−0.1112	0.1112	0.3202	0.4906	0.6018	0.6404
40	−0.8282	−0.7782	−0.6344	−0.4141	−0.1438	0.1438	0.4141	0.6344	0.7782	0.8282
50	−0.8282	−0.7782	−0.6344	−0.4141	−0.1438	0.1438	0.4141	0.6344	0.7782	0.8282
60	−0.6404	−0.6018	−0.4906	−0.3202	−0.1112	0.1112	0.3202	0.4906	0.6018	0.6404
70	−0.3528	−0.3315	−0.2703	−0.1764	−0.0613	0.0613	0.1764	0.2703	0.3315	0.3528
80	−0.0999	−0.0939	−0.0765	−0.0499	−0.0173	0.0173	0.0499	0.0765	0.0939	0.0999
90	0.0000	0.0000	0.0000	0.0000	0.0000	0.0000	0.0000	0.0000	0.0000	0.0000

$l = 4$ $\mu = 1$ $n = 2$ *Cross Section:* $\phi_2 = 10$

	0	10	20	30	40	50	60	70	80	90
0	0.0000	0.0000	0.0000	0.0000	0.0000	0.0000	0.0000	0.0000	0.0000	0.0000
10	−0.0940	−0.0827	−0.0615	−0.0329	−0.0003	0.0324	0.0611	0.0825	0.0939	0.0940
20	−0.3308	−0.2902	−0.2146	−0.1132	0.0020	0.1169	0.2177	0.2922	0.3315	0.3308
30	−0.5967	−0.5201	−0.3807	−0.1955	0.0134	0.2207	0.4013	0.5335	0.6014	0.5967
40	−0.7627	−0.6573	−0.4726	−0.2309	0.0387	0.3035	0.5318	0.6959	0.7761	0.7627
50	−0.7453	−0.6296	−0.4379	−0.1934	0.0745	0.3333	0.5520	0.7040	0.7712	0.7453
60	−0.5468	−0.4434	−0.2866	−0.0951	0.1078	0.2977	0.4517	0.5512	0.5842	0.5468
70	−0.2543	−0.1823	−0.0882	0.0164	0.1191	0.2074	0.2707	0.3014	0.2957	0.2543
80	−0.0001	0.0315	0.0593	0.0800	0.0910	0.0911	0.0801	0.0595	0.0317	0.0001
90	0.0999	0.0939	0.0765	0.0499	0.0173	−0.0173	−0.0499	−0.0765	−0.0939	−0.0999

$l = 4$ $\mu = 1$ $n = 2$ *Cross Section:* $\phi_2 = 20$

	0	10	20	30	40	50	60	70	80	90
0	0.0000	0.0000	0.0000	0.0000	0.0000	0.0000	0.0000	0.0000	0.0000	0.0000
10	−0.0789	−0.0656	−0.0444	−0.0178	0.0109	0.0383	0.0611	0.0765	0.0827	0.0789
20	−0.2751	−0.2269	−0.1513	−0.0575	0.0433	0.1388	0.2176	0.2702	0.2901	0.2751
30	−0.4861	−0.3945	−0.2553	−0.0854	0.0949	0.2637	0.4007	0.4894	0.5190	0.4861
40	−0.5968	−0.4698	−0.2861	−0.0679	0.1585	0.3658	0.5289	0.6283	0.6519	0.5968
50	−0.5356	−0.3948	−0.2064	0.0069	0.2194	0.4054	0.5425	0.6142	0.6118	0.5356
60	−0.3097	−0.1831	−0.0345	0.1183	0.2568	0.3643	0.4279	0.4399	0.3989	0.3097
70	−0.0048	0.0823	0.1596	0.2175	0.2493	0.2510	0.2224	0.1670	0.0914	0.0048
80	0.2526	0.2858	0.2845	0.2490	0.1833	0.0956	−0.0037	−0.1025	−0.1889	−0.2526
90	0.3528	0.3315	0.2703	0.1764	0.0613	−0.0613	−0.1764	−0.2703	−0.3315	−0.3528

φ_1 →, Φ ↓

$l = 4$ $\mu = 1$ $n = 2$ Cross Section: $\phi_2 = 30$

	0	10	20	30	40	50	60	70	80	90
0	0.0000	0.0000	0.0000	0.0000	0.0000	0.0000	0.0000	0.0000	0.0000	0.0000
10	−0.0618	−0.0506	−0.0333	−0.0119	0.0109	0.0324	0.0499	0.0615	0.0656	0.0618
20	−0.2118	−0.1712	−0.1100	−0.0355	0.0433	0.1168	0.1763	0.2145	0.2268	0.2118
30	−0.3602	−0.2838	−0.1731	−0.0415	0.0951	0.2202	0.3188	0.3789	0.3933	0.3602
40	−0.4083	−0.3036	−0.1623	−0.0014	0.1596	0.3014	0.4068	0.4632	0.4637	0.4083
50	−0.2971	−0.1837	−0.0483	0.0930	0.2231	0.3263	0.3901	0.4069	0.3746	0.2971
60	−0.0400	0.0572	0.1476	0.2202	0.2662	0.2801	0.2602	0.2089	0.1325	0.0400
70	0.2789	0.3384	0.3572	0.3329	0.2684	0.1715	0.0540	−0.0701	−0.1857	−0.2789
80	0.5400	0.5500	0.4937	0.3778	0.2164	0.0289	−0.1621	−0.3336	−0.4648	−0.5400
90	0.6404	0.6018	0.4906	0.3202	0.1112	−0.1112	−0.3202	−0.4906	−0.6018	−0.6404

$l = 4$ $\mu = 1$ $n = 2$ Cross Section: $\phi_2 = 40$

	0	10	20	30	40	50	60	70	80	90
0	0.0000	0.0000	0.0000	0.0000	0.0000	0.0000	0.0000	0.0000	0.0000	0.0000
10	−0.0507	−0.0447	−0.0333	−0.0178	−0.0003	0.0173	0.0329	0.0444	0.0506	0.0507
20	−0.1704	−0.1491	−0.1099	−0.0574	0.0020	0.0612	0.1130	0.1512	0.1711	0.1704
30	−0.2781	−0.2397	−0.1724	−0.0843	0.0140	0.1106	0.1938	0.2537	0.2830	0.2781
40	−0.2852	−0.2364	−0.1591	−0.0625	0.0415	0.1406	0.2227	0.2779	0.2996	0.2852
50	−0.1414	−0.0952	−0.0375	0.0247	0.0839	0.1330	0.1661	0.1791	0.1705	0.1414
60	0.1360	0.1652	0.1746	0.1628	0.1315	0.0842	0.0269	−0.0338	−0.0903	−0.1360
70	0.4640	0.4662	0.4122	0.3084	0.1674	0.0063	−0.1556	−0.2988	−0.4059	−0.4640
80	0.7275	0.7005	0.5889	0.4064	0.1748	−0.0779	−0.3212	−0.5257	−0.6668	−0.7275
90	0.8282	0.7782	0.6344	0.4141	0.1438	−0.1438	−0.4141	−0.6344	−0.7782	−0.8282

$l = 4$ $\mu = 1$ $n = 2$ Cross Section: $\phi_2 = 50$

	0	10	20	30	40	50	60	70	80	90
0	0.0000	0.0000	0.0000	0.0000	0.0000	0.0000	0.0000	0.0000	0.0000	0.0000
10	−0.0507	−0.0506	−0.0444	−0.0329	−0.0173	0.0003	0.0178	0.0333	0.0447	0.0507
20	−0.1704	−0.1711	−0.1512	−0.1130	−0.0612	−0.0020	0.0574	0.1099	0.1491	0.1704
30	−0.2781	−0.2830	−0.2537	−0.1938	−0.1106	−0.0140	0.0843	0.1724	0.2397	0.2781
40	−0.2852	−0.2996	−0.2779	−0.2227	−0.1406	−0.0415	0.0625	0.1591	0.2364	0.2852
50	−0.1414	−0.1705	−0.1791	−0.1661	−0.1330	−0.0839	−0.0247	0.0375	0.0952	0.1414
60	0.1360	0.0903	0.0338	−0.0269	−0.0842	−0.1315	−0.1628	−0.1746	−0.1652	−0.1360
70	0.4640	0.4059	0.2988	0.1556	−0.0063	−0.1674	−0.3084	−0.4122	−0.4662	−0.4640
80	0.7275	0.6668	0.5257	0.3212	0.0779	−0.1748	−0.4064	−0.5889	−0.7005	−0.7275
90	0.8282	0.7782	0.6344	0.4141	0.1438	−0.1438	−0.4141	−0.6344	−0.7782	−0.8282

φ_1 →, Φ ↓

$l = 4$ $\mu = 1$ $n = 2$ Cross Section: $\phi_2 = 60$

	0	10	20	30	40	50	60	70	80	90
0	0.0000	0.0000	0.0000	0.0000	0.0000	0.0000	0.0000	0.0000	0.0000	0.0000
10	−0.0618	−0.0656	−0.0615	−0.0499	−0.0324	−0.0109	0.0119	0.0333	0.0506	0.0618
20	−0.2118	−0.2268	−0.2145	−0.1763	−0.1168	−0.0433	0.0355	0.1100	0.1712	0.2118
30	−0.3602	−0.3933	−0.3789	−0.3188	−0.2202	−0.0951	0.0415	0.1731	0.2838	0.3602
40	−0.4083	−0.4637	−0.4632	−0.4068	−0.3014	−0.1596	0.0014	0.1623	0.3036	0.4083
50	−0.2971	−0.3746	−0.4069	−0.3901	−0.3263	−0.2231	−0.0930	0.0483	0.1837	0.2971
60	−0.0400	−0.1325	−0.2089	−0.2602	−0.2801	−0.2662	−0.2202	−0.1476	−0.0572	0.0400
70	0.2789	0.1857	0.0701	−0.0540	−0.1715	−0.2684	−0.3329	−0.3572	−0.3384	−0.2789
80	0.5400	0.4648	0.3336	0.1621	−0.0289	−0.2164	−0.3778	−0.4937	−0.5500	−0.5400
90	0.6404	0.6018	0.4906	0.3202	0.1112	−0.1112	−0.3202	−0.4906	−0.6018	−0.6404

(continued)

Table A2.4—(*continued*)

l = 4 μ = 1 *n* = 2 Cross Section: ϕ_2 = 70

	0	10	20	30	40	50	60	70	80	90
0	0.0000	0.0000	0.0000	0.0000	0.0000	0.0000	0.0000	0.0000	0.0000	0.0000
10	−0.0789	−0.0827	−0.0765	−0.0611	−0.0383	−0.0109	0.0178	0.0444	0.0656	0.0789
20	−0.2751	−0.2901	−0.2702	−0.2176	−0.1388	−0.0433	0.0575	0.1513	0.2269	0.2751
30	−0.4861	−0.5190	−0.4894	−0.4007	−0.2637	−0.0949	0.0854	0.2553	0.3945	0.4861
40	−0.5968	−0.6519	−0.6283	−0.5289	−0.3658	−0.1585	0.0679	0.2861	0.4698	0.5968
50	−0.5356	−0.6118	−0.6142	−0.5425	−0.4054	−0.2194	−0.0069	0.2064	0.3948	0.5356
60	−0.3097	−0.3989	−0.4399	−0.4279	−0.3643	−0.2568	−0.1183	0.0345	0.1831	0.3097
70	−0.0048	−0.0914	−0.1670	−0.2224	−0.2510	−0.2493	−0.2175	−0.1596	−0.0823	0.0048
80	0.2526	0.1889	0.1025	0.0037	−0.0956	−0.1833	−0.2490	−0.2845	−0.2858	−0.2526
90	0.3528	0.3315	0.2703	0.1764	0.0613	−0.0613	−0.1764	−0.2703	−0.3315	−0.3528

l = 4 μ = 1 *n* = 2 Cross Section: ϕ_2 = 80

	0	10	20	30	40	50	60	70	80	90
0	0.0000	0.0000	0.0000	0.0000	0.0000	0.0000	0.0000	0.0000	0.0000	0.0000
10	−0.0940	−0.0939	−0.0825	−0.0611	−0.0324	0.0003	0.0329	0.0615	0.0827	0.0940
20	−0.3308	−0.3315	−0.2922	−0.2177	−0.1169	−0.0020	0.1132	0.2146	0.2902	0.3308
30	−0.5967	−0.6014	−0.5335	−0.4013	−0.2207	−0.0134	0.1955	0.3807	0.5201	0.5967
40	−0.7627	−0.7761	−0.6959	−0.5318	−0.3035	−0.0387	0.2309	0.4726	0.6573	0.7627
50	−0.7453	−0.7712	−0.7040	−0.5520	−0.3333	−0.0745	0.1934	0.4379	0.6296	0.7453
60	−0.5468	−0.5842	−0.5512	−0.4517	−0.2977	−0.1078	0.0951	0.2866	0.4434	0.5468
70	−0.2543	−0.2957	−0.3014	−0.2707	−0.2074	−0.1191	−0.0164	0.0882	0.1823	0.2543
80	−0.0001	−0.0317	−0.0595	−0.0801	−0.0911	−0.0910	−0.0800	−0.0593	−0.0315	0.0001
90	0.0999	0.0939	0.0765	0.0499	0.0173	−0.0173	−0.0499	−0.0765	−0.0939	−0.0999

φ_1 → ; Φ ↓

l = 4 μ = 1 *n* = 2 Cross Section: ϕ_2 = 90

	0	10	20	30	40	50	60	70	80	90
0	0.0000	0.0000	0.0000	0.0000	0.0000	0.0000	0.0000	0.0000	0.0000	0.0000
10	−0.0999	−0.0939	−0.0765	−0.0499	−0.0173	0.0173	0.0499	0.0765	0.0939	0.0999
20	−0.3528	−0.3315	−0.2703	−0.1764	−0.0613	0.0613	0.1764	0.2703	0.3315	0.3528
30	−0.6404	−0.6018	−0.4906	−0.3202	−0.1112	0.1112	0.3202	0.4906	0.6018	0.6404
40	−0.8282	−0.7782	−0.6344	−0.4141	−0.1438	0.1438	0.4141	0.6344	0.7782	0.8282
50	−0.8282	−0.7782	−0.6344	−0.4141	−0.1438	0.1438	0.4141	0.6344	0.7782	0.8282
60	−0.6404	−0.6018	−0.4906	−0.3202	−0.1112	0.1112	0.3202	0.4906	0.6018	0.6404
70	−0.3528	−0.3315	−0.2703	−0.1764	−0.0613	0.0613	0.1764	0.2703	0.3315	0.3528
80	−0.0999	−0.0939	−0.0765	−0.0499	−0.0173	0.0173	0.0499	0.0765	0.0939	0.0999
90	0.0000	0.0000	0.0000	0.0000	0.0000	0.0000	0.0000	0.0000	0.0000	0.0000

l = 6 $\mu = 1$ *n* = 4 Cross Section: $\phi_2 = 0$

	0	10	20	30	40	50	60	70	80	90
0	0.93541	0.71657	0.16243	−0.4677	−0.879	−0.879	−0.4677	0.16243	0.71657	0.93541
10	0.7576	0.58036	0.13156	−0.3788	−0.7119	−0.7119	−0.3788	0.13156	0.58036	0.7576
20	0.30737	0.23546	0.05337	−0.1537	−0.2888	−0.2888	−0.1537	0.05337	0.23546	0.30737
30	−0.2046	−0.1568	−0.0355	0.10231	0.19228	0.19228	0.10231	−0.0355	−0.1568	−0.2046
40	−0.5388	−0.4127	−0.0936	0.2694	0.50631	0.50631	0.2694	−0.0936	−0.4127	−0.5388
50	−0.5388	−0.4127	−0.0936	0.2694	0.50631	0.50631	0.2694	−0.0936	−0.4127	−0.5388
60	−0.2046	−0.1568	−0.0355	0.10231	0.19228	0.19228	0.10231	−0.0355	−0.1568	−0.2046
70	0.30737	0.23546	0.05337	−0.1537	−0.2888	−0.2888	−0.1537	0.05337	0.23546	0.30737
80	0.7576	0.58036	0.13156	−0.3788	−0.7119	−0.7119	−0.3788	0.13156	0.58036	0.7576
90	0.93541	0.71657	0.16243	−0.4677	−0.879	−0.879	−0.4677	0.16243	0.71657	0.93541

l = 6 $\mu = 1$ *n* = 4 Cross Section: $\phi_2 = 45$

	0	10	20	30	40	50	60	70	80	90
0	−0.9354	−0.7166	−0.1624	0.46771	0.879	0.879	0.46771	−0.1624	−0.7166	−0.9354
10	−0.7607	−0.5827	−0.1321	0.38034	0.71481	0.71481	0.38034	−0.1321	−0.5827	−0.7607
20	−0.3492	−0.2675	−0.0606	0.1746	0.32813	0.32813	0.1746	−0.0606	−0.2675	−0.3492
30	0.04567	0.03499	0.00793	−0.0228	−0.0429	−0.0429	−0.0228	0.00793	0.03499	0.04567
40	0.21213	0.1625	0.03684	−0.1061	−0.1993	−0.1993	−0.1061	0.03684	0.1625	0.21213
50	0.11059	0.08471	0.0192	−0.0553	−0.1039	−0.1039	−0.0553	0.0192	0.08471	0.11059
60	−0.1407	−0.1078	−0.0244	0.07034	0.13219	0.13219	0.07034	−0.0244	−0.1078	−0.1407
70	−0.3858	−0.2955	−0.067	0.1929	0.36253	0.36253	0.1929	−0.067	−0.2955	−0.3858
80	−0.5371	−0.4114	−0.0933	0.26855	0.50471	0.50471	0.26855	−0.0933	−0.4114	−0.5371
90	−0.5846	−0.4479	−0.1015	0.29232	0.54938	0.54938	0.29232	−0.1015	−0.4479	−0.5846

l = 6 $\mu = 1$ *n* = 4 Cross Section: $\phi_2 = 90$

	0	10	20	30	40	50	60	70	80	90
0	0.93541	0.71657	0.16243	−0.4677	−0.879	−0.879	−0.4677	0.16243	0.71657	0.93541
10	0.7576	0.58036	0.13156	−0.3788	−0.7119	−0.7119	−0.3788	0.13156	0.58036	0.7576
20	0.30737	0.23546	0.05337	−0.1537	−0.2888	−0.2888	−0.1537	0.05337	0.23546	0.30737
30	−0.2046	−0.1568	−0.0355	0.10231	0.19228	0.19228	0.10231	−0.0355	−0.1568	−0.2046
40	−0.5388	−0.4127	−0.0936	0.2694	0.50631	0.50631	0.2694	−0.0936	−0.4127	−0.5388
50	−0.5388	−0.4127	−0.0936	0.2694	0.50631	0.50631	0.2694	−0.0936	−0.4127	−0.5388
60	−0.2046	−0.1568	−0.0355	0.10231	0.19228	0.19228	0.10231	−0.0355	−0.1568	−0.2046
70	0.30737	0.23546	0.05337	−0.1537	−0.2888	−0.2888	−0.1537	0.05337	0.23546	0.30737
80	0.7576	0.58036	0.13156	−0.3788	−0.7119	−0.7119	−0.3788	0.13156	0.58036	0.7576
90	0.93541	0.71657	0.16243	−0.4677	−0.879	−0.879	−0.4677	0.16243	0.71657	0.93541

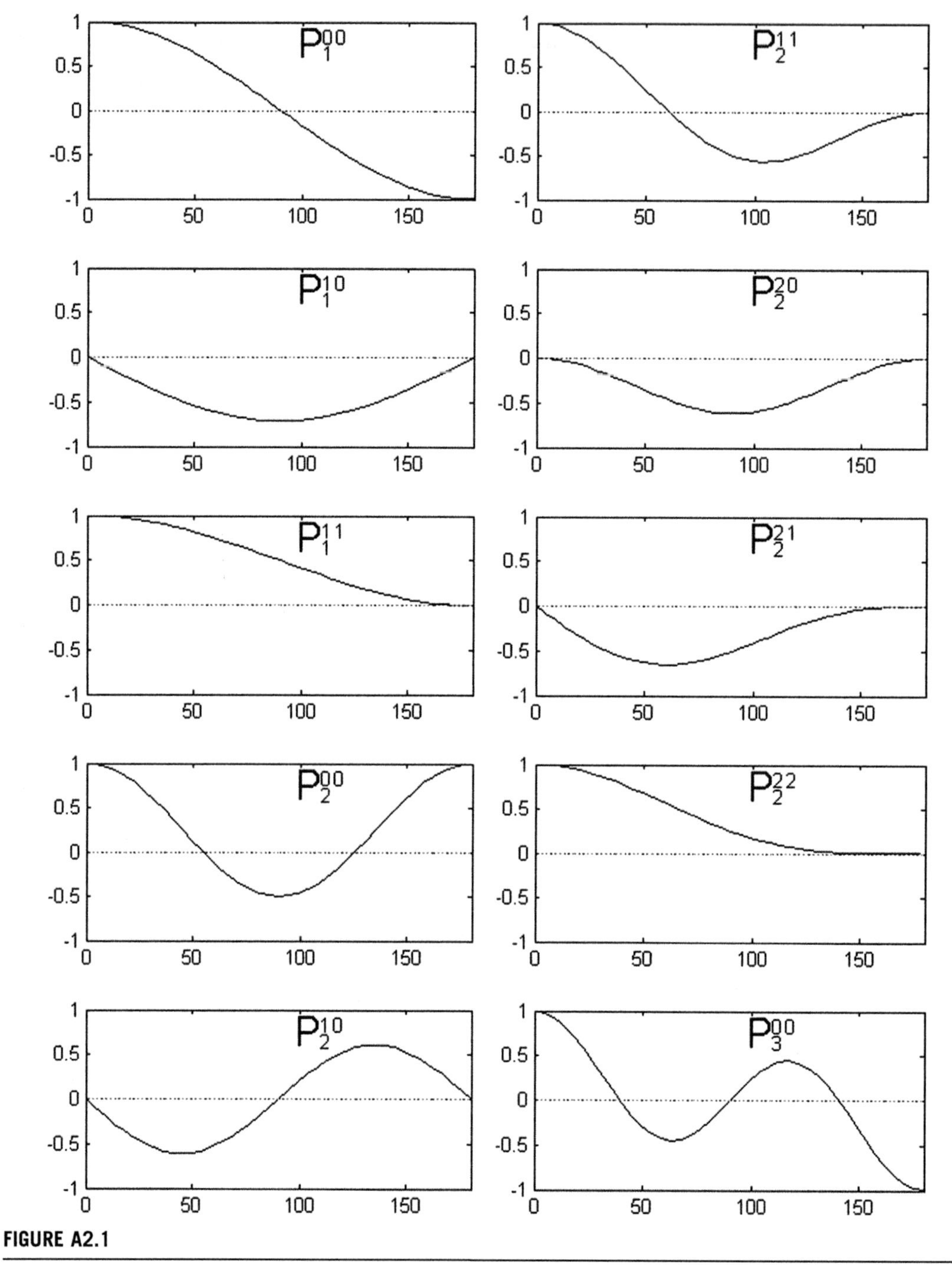

FIGURE A2.1

Generalized Legendre functions, $P_l^{mn}(\Phi)$ for $l = 1, 2, 3$; $m = 0, 1, 2$; $n = 0, 1, 2$.

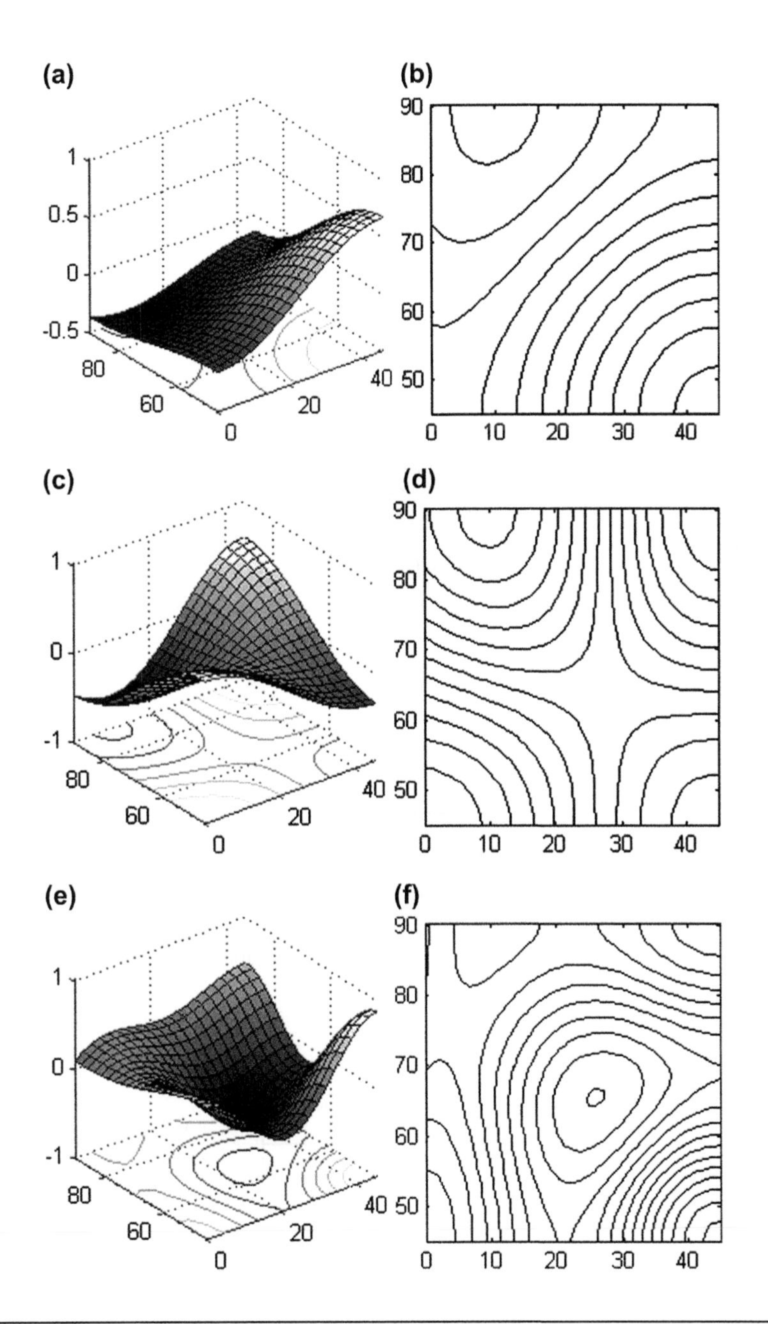

FIGURE A2.2

Cubic surface spherical harmonics $\dot{k}_l^{\mu}(\Phi\beta)$ for $l = 4, 6, 8$; $\mu = 1$: (a) $\dot{k}_l^{\mu}(\Phi\beta)$, $l = 4$, $\mu = 1$; contour plot interval is 0.1, $0 < \Phi < 45$, $45 < \beta < 90$. (b) $\dot{k}_l^{\mu}(\Phi\beta)$, $l = 6$, $\mu = 1$; contour plot interval is 0.1, $0 < \Phi < 45$, $45 < \beta < 90$. (c) $\dot{k}_l^{\mu}(\Phi\beta)$, $l = 8$, $\mu = 1$; contour plot interval is 0.1, $0 < \Phi < 45$, $45 < \beta < 90$.

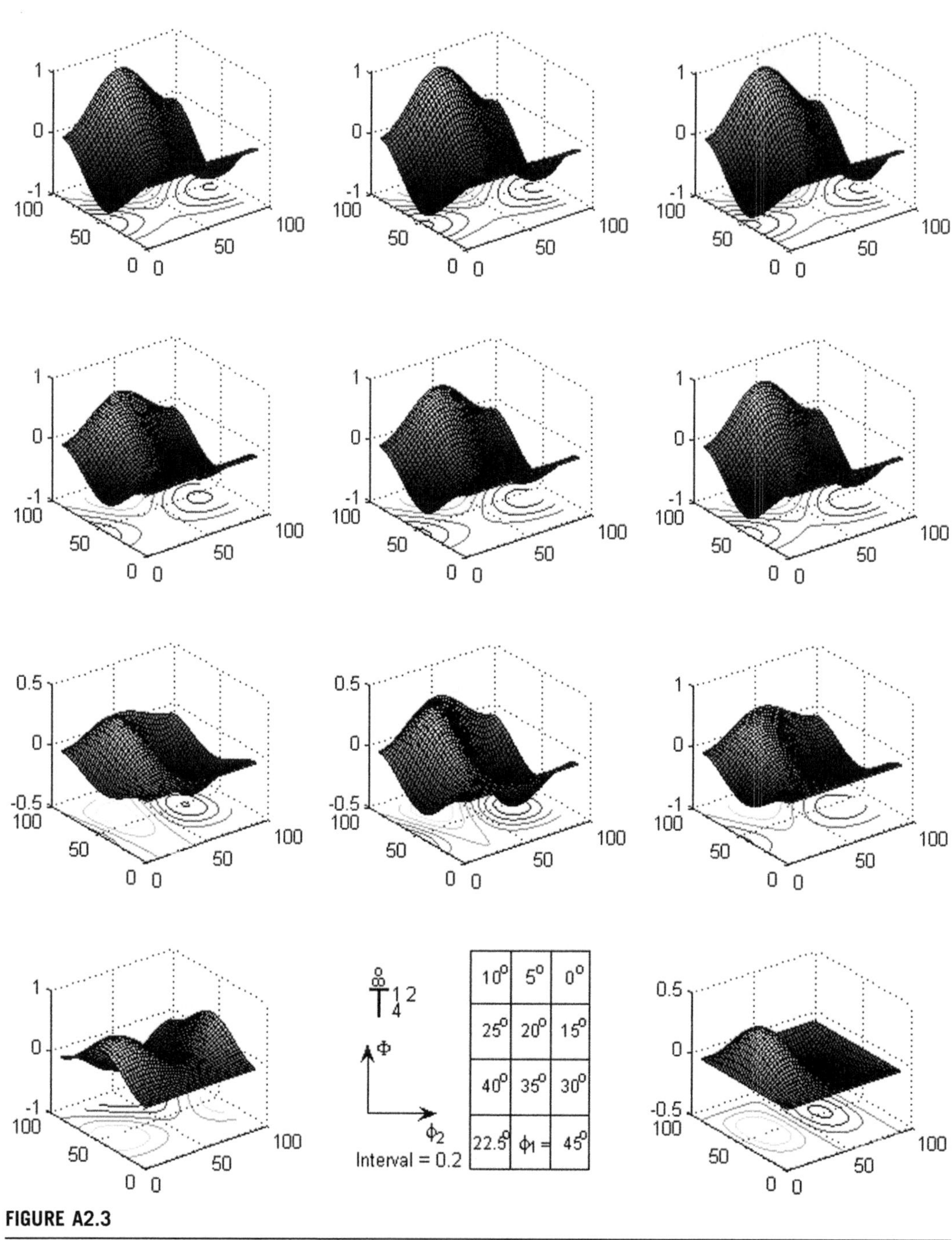

FIGURE A2.3

Cubic-orthorhombic generalized harmonics, $\overset{\therefore}{T}{}_l^{\mu n}(\phi_1\Phi\phi_2)$ for $l = 4, 6$; $\mu = 1$; $n = 2$.

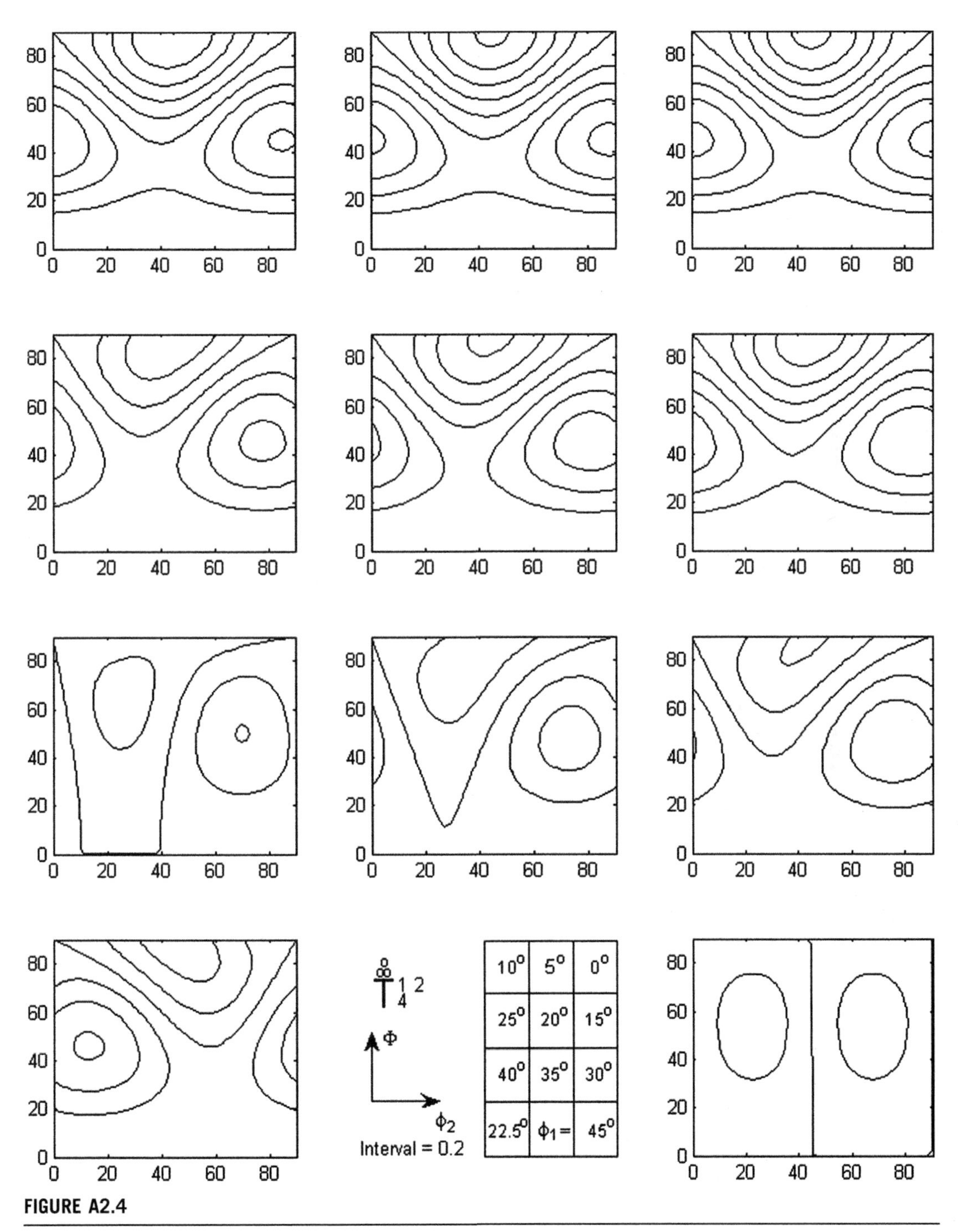

FIGURE A2.4

Generalized spherical harmonic function for $l = 4$, $\mu = 1$, $n = 2$; contour interval is 0.2.

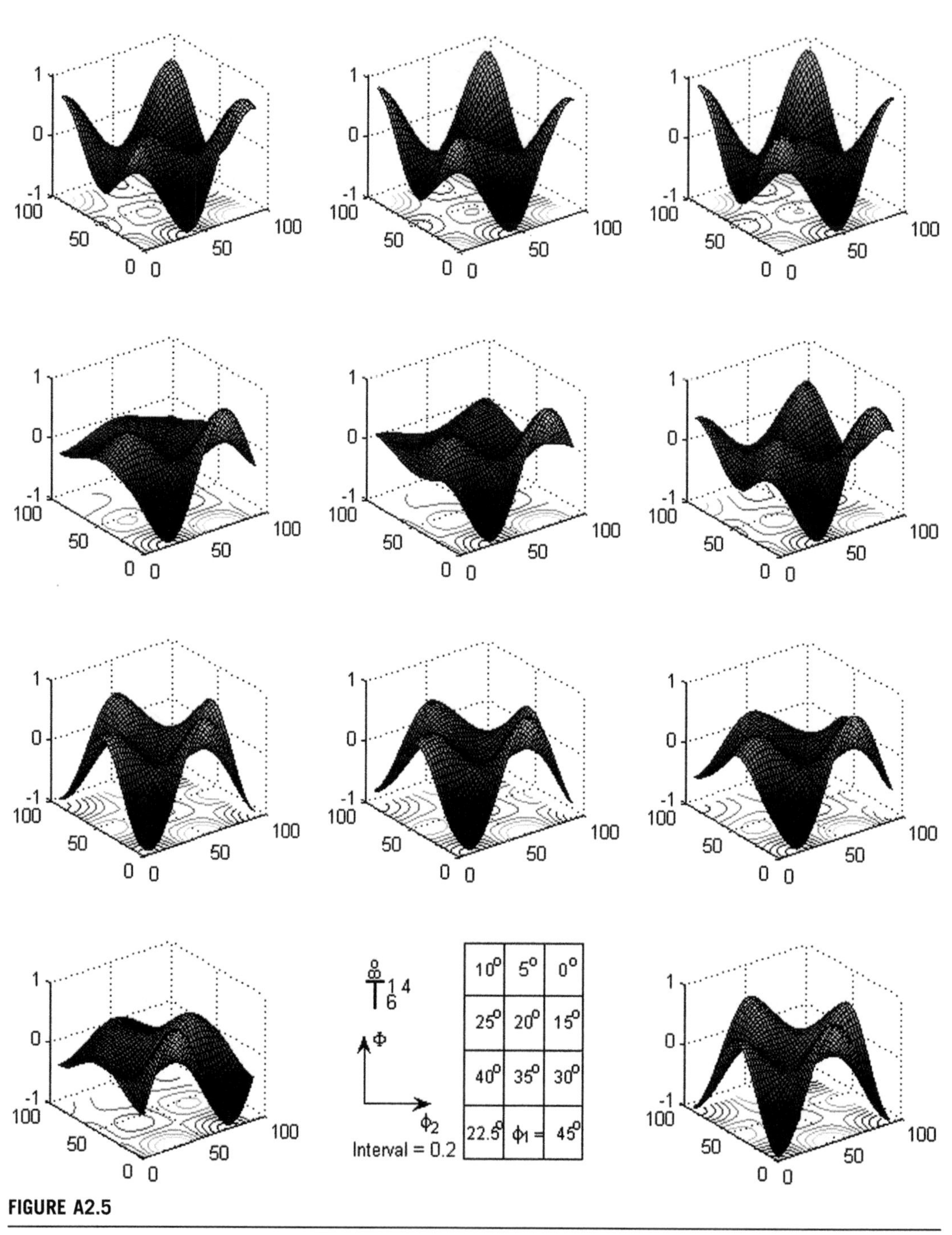

FIGURE A2.5

Generalized spherical harmonic function for $l = 6$, $\mu = 1$, $n = 4$.

FIGURE A2.6

Generalized spherical harmonic function for $l = 6$, $\mu = 1$, $n = 4$; contour interval is 0.2.

References

Adams, B.L., Canova, G.R., Molinari, A., 1989. A statistical formulation of viscoplastic behavior in heterogeneous polycrystals. Textures and Microstructures 11, 57.

Adams, B.L., Field, D.P., 1991. Statistical theory of creep in polycrystalline materials. Acta Metallurgica et Materialia 39 (10), 2405–2417.

Adams, B.L., Gao, X., Kalidindi, S.R., 2005. Finite approximations to the second-order properties closure in single phase polycrystals. Acta Materialia 53 (13), 3563–3577.

Adams, B.L., Henrie, A., Henrie, B., Lyon, M., Kalidindi, S.R., Garmestani, H., 2001. Microstructure-sensitive design of a compliant beam. Journal of the Mechanics and Physics of Solids 49 (8), 1639–1663.

Adams, B.L., Kacher, J., 2009. EBSD-based microscopy: Resolution of dislocation density. Computers, Materials and Continua 14 (3), 185–196.

Adams, B.L., Wright, S.I., Kunze, K., 1993. Orientation imaging: The emergence of a new microscopy. Metallurgical Transactions A (Physical Metallurgy and Materials Science) 24A (4), 819–831.

Arnold, S.M., Saleeb, A.F., Al-Zoubi, N.R., 2001. Deformation and life analysis of composite flywheel disk and multi-disk systems. NASA/TM (2001–210578).

Asaro, R.J., 1983a. Crystal plasticity. Journal of Applied Mechanics 50 (4b), 921–934.

Asaro, R.J., 1983b. Micromechanics of crystals and polycrystals. Advances in Applied Mechanics 23, 1–115.

Asaro, R.J., Needleman, A., 1985. Texture development and strain hardening in rate dependent polycrystals. Acta Metallurgica et Materialia 33 (6), 923–953.

Asaro, R.J., Rice, J.R., 1977. Strain localization in ductile single crystals. Journal of the Mechanics and Physics of Solids 25 (5), 309–338.

Balling, R.J., 2000. Pareto sets in decision-based design. Journal of Engineering Valuation and Cost Analysis 3 (2), 189–198.

Basinger, J., Fullwood, D., Kacher, J., Adams, B., 2011. Pattern center determination in EBSD microscopy. Microscopy and Microanalysis 17, 330–340.

Basinger, J. A., C. J. Gardner, Fullwood, D.T., Adams, B.L. Pattern center sensitivity for simulated reference patterns in high resolution electron backscatter diffraction. Microscopy and Microanalysis.

Bassani, J.L., Wu, T.Y., 1991. Latent hardening in single crystals, II. Analytical characterization and prediction. Proceedings of the Royal Society of London, Series A 435, 21–41.

Boggs, P.T., Tolle, J.W., 2000. Sequential quadratic programming for large-scale nonlinear optimization. Journal of Computational and Applied Mathematics 124 (1–2), 123–137.

Bracewell, R.N., 1978. The Fourier Transform and Its Applications. McGraw-Hill.

Brewer, L.N., Field, D.P., Merriman, C.C., 2009. Mapping and Assessing Plastic Deformation Using EBSD. Electron Backscatter Diffraction in Materials Science. In: Schwartz, A.J., Kumar, M., Adams, B.L., Field, D.P. (Eds.), 2nd ed.. Springer-Verlag, pp. 251–262.

Britton, T.B., Wilkinson, A.J., 2011. Measurement of residual elastic strain and lattice rotations with high resolution electron backscatter diffraction. Ultramicroscopy 111 (8), 1395–1404.

Brown, W.F., 1955. Solid mixture permittivities. The Journal of Chemical Physics 33 (8), 1514–1517.

Bunge, H.-J., 1993. Texture Analysis in Materials Science: Mathematical Methods. Cuvillier-Verlag.

Bunge, H., Esling, C., 1984. Texture development by plastic deformation. Scripta Metall 18, 191–195.

Caflisch, R.E., 1998. Monte Carlo and quasi–Monte Carlo methods. Acta Numerica 7, 1–49.

Caratheodory, C., 1911. Ueber den Variabilitätsbereich der Fourierschen Konstanten von positiven harmonischen Functionen. Rend. Circ. Mat. Palermo 32, 193–217.

Carpenter, D.A., Pugh, J.L., Richardson, G.D., Mooney, L.R., 2007. Determination of pattern centre in EBSD using the moving-screen technique. Journal of Microscopy 227 (September), 246–247.

Cermelli, P., Gurtin, M.E., 2001. On the characterization of geometrically necessary dislocations in finite plasticity. Journal of the Mechanics and Physics of Solids 49 (7), 1539–1568.

Chamis, C., 1989. Mechanics of composite materials: Past, present, and future. Journal of Composites Technology and Research 11 (1), 3–14.

Champeney, D.C., 1973. Fourier Transforms and Their Physical Applications. Academic Press.

Clement, A., 1982. Prediction of deformation texture using a physical principle of conservation. Materials Science and Engineering 55, 203–210.

Cohen, L., 1998. Generalization of the Wiener-Khinchin theorem. Signal Processing Letters, IEEE 5 (11), 292–294.

Das, I., Dennis, J.E., 1997. A closer look at drawbacks of minimizing weighted sums of objectives for Pareto set generation in multicriteria optimization problems. Struct. Optim. 14, 63–69.

Das, I., Dennis, J.E., 1998. "Normal-boundary intersection: A new method for generating Pareto optimal points in multicriteria optimization problems." SIAM J. Optim 8, 631–657.

Delesse, M.A., 1847. Procédé méchanique pour déterminer la composition des roches. Comptes Rendus Acad. Sci. (Paris) 25, 544–545.

EDAX-TSL, 2010. Manual for Orientation Imaging Microscopy (OIM™), version 6.0.

El-Dasher, B.S., Adams, B.L., Rollett, A.D., 2003. Viewpoint: Experimental recovery of geometrically necessary dislocation density in polycrystals. Scripta Materialia 48 (2), 141–145.

Engler, O., Randle, V., 2010. Introduction to Texture Analysis: Macrotexture, Microtexture, and Orientation Mapping. CRC Press/Taylor and Francis Group.

Eshelby, J.D., 1957. The determination of the elastic field of an ellipsoidal inclusion, and related problems. Proc. R. Soc. Lond. A 241, 376–396.

Field, D.P., Magid, K.R., Mastorakos, I.N., Florando, J.N., Lassila, D.H., Morris Jr., J.W., 2010. Mesoscale strain measurement in deformed crystals: A comparison of X-ray microdiffraction with electron backscatter diffraction. Philosophical Magazine 90 (11), 1451–1464.

Field, D.P., Merriman, C.C., Allain-Bonasso, N., Wagner, F., 2012. Quantification of dislocation structure heterogeneity in deformed polycrystals by EBSD. Model. Simul. Mater. Sci. Eng. 20, 1–12.

Frisch, H.L., Stillinger, F.H., 1963. Contribution to the statistical geometric basis of radiation scattering. The Journal of Chemical Physics 38 (9), 2200–2207.

Fullwood, D.T., Adams, B.L., Kalidindi, S.R., 2008a. A strong contrast homogenization formulation for multi-phase anisotropic materials. J. Mech. Phys. Solids 56 (6), 2287–2297.

Fullwood, D.T., Kalidindi, S.R., Adams, B.L., Ahmadi, S., 2009. A discrete Fourier transform framework for localization relations. Computers, Materials and Continua 299 (1), 1–14.

Fullwood, D.T., Kalidindi, S.R., Niezgoda, S.R., Fast, T., Hampson, N., 2008b. Gradient-based microstructure reconstructions from distributions using fast Fourier transforms. Materials Science and Engineering A 494, 68–72.

Fullwood, D.T., Niezgoda, S.R., Kalidindi, S.R., 2008c. Microstructure reconstructions from 2-point statistics using phase recovery algorithms. Acta Materialia 56 (5), 942–948.

Gardner, C.J., Adams, B.L., Basinger, J., Fullwood, D.T., 2010. EBSD-based continuum dislocation microscopy. International Journal of Plasticity 26, 1234–1247.

Gelfand, I., Minlos, R., Shapiro, ZYa, 1963. Representations of the Rotation and Lorentz Groups and Their Applications. Pergamon Press.

Glagolev, A.A., 1934. On the geometrical methods of quantitative minerologic analysis of rocks. Transactions Inst. Econ. Min. (Moscow) 59, 1–47.

Gokhale, A.M., Tewari, A., Garmestani, H., 2005. Constraints on microstructural two-point correlation functions. Scripta Materialia 53 (8), 989–993.

Hamermesh, M., 1962. Group Theory and Its Application to Physical Problems. Addison Wesley.

Hashin, Z., Shtrikman, S., 1963. A variational approach to the theory of the elastic behaviour of multiphase materials. Journal of the Mechanics and Physics of Solids 11 (2), 127–140.

Herakovich, C.T., 1998. Mechanics of Fibrous Composites. John Wiley & Sons.

Hill, R., 1948. A theory of the yielding and plastic flow of anisotropic metals. Proceedings of the Royal Society of London. Series A, Mathematical and Physical Sciences 193 (1033), 281–297.

Hill, R., 1952. The elastic behavior of a crystalline aggregate. Proceedings of the Royal Society of London. Series A, Mathematical and Physical Sciences 65, 349–354.

Hill, R., 1965. A self-consistent mechanics of composite materials. J. Mech. Phys. Solids 13, 213–222.

Hill, R., 1966. Generalized constitutive relations for incremental deformation of metal crystals by multislip. J. Mech. Phys.Solids 14, 95–102.

Hill, R., Rice, J.R., 1972. Constitutive analysis of elastic-plastic crystals at arbitrary strain. Journal of the Mechanics and Physics of Solids 20 (6), 401–413.

Houskamp, J.R., Proust, G., Kalidindi, S.R., 2007. Integration of microstructure sensitive design with finite element methods: Elastic–plastic case studies in FCC polycrystals. Int. J. Multiscale Eng. 5, 261–272.

Hutchinson, J.W., 1976. Bounds and self-consistent estimates for creep of polycrystalline materials. Proceedings of the Royal Society of London. Series A, Mathematical and Physical Sciences (348), 101–126.

Jiao, Y., Stillinger, F.H., Torquato, S., 2007. Modeling heterogeneous materials via two-point correlation functions: Basic principles. Physical Review E (Statistical, Nonlinear, and Soft Matter Physics 76 (3), 031110–031115.

Kacher, J., 2009. High resolution cross-correlation-based texture analysis using kinematically simulated EBSD patterns. Brigham Young University, Provo, UT.

Kacher, J., Basinger, J., Adams, B.L., Fullwood, D.T., 2010. Reply to comment by Maurice, et al. in response to "Bragg's law diffraction simulations for electron backscatter diffraction analysis. Ultramicroscopy 110 (7), 760–762.

Kacher, J., Landon, C., Adams, B.L., Fullwood, D., 2009. Bragg's Law Diffraction Simulations for Electron Backscatter Diffraction Analysis. Ultramicroscopy 109 (9), 1148–1156.

Kalidindi, S.R., Anand, L., 1993. Large deformation simple compression of a copper single-crystal. Metallurgical Transactions A 24 (4), 989–992.

Kalidindi, S.R., Binci, M., Fullwood, D., Adams, B.L., 2006. Elastic properties closures using second-order homogenization theories: Case studies in composites of two isotropic constituents. Acta Materialia 54 (11), 3117–3126.

Kalidindi, S.R., Bronkhorst, C.A., Anand, L., 1992. Crystallographic texture evolution in bulk deformation processing of FCC metals. Journal of the Mechanics and Physics of Solids 40 (3), 537–569.

Kalidindi, S.R., Houskamp, J.R., Lyons, M., Adams, B.L., 2004. Microstructure sensitive design of an orthotropic plate subjected to tensile load. International Journal of Plasticity 20 (8–9), 1561–1575.

Kerner, E.J., 1956. The elastic and thermo-elastic properties of composite media. Proc. Phys. Soc. Ser. B 69 (8), 808–813.

Kim, I.Y., de Weck, O.L., 2005. Adaptive weighted-sum method for bi-objective optimization: Pareto front generation. Struct. Multidisc. Optim. 29, 149–158.

Kocks, U.F., Argon, A.S., Ashby, M.F., 1975. Thermodynamics and kinetics of slip. Prog. Mat. Sci. 19, 1–291.

Koski, J., 1985. Defectiveness of weighting method in multicriterion optimization of structures. Commun. Appl. Numer. Methods 1, 333–337.

Kregers, A., Teters, G., 1979. Use of averaging methods to determine the viscoelastic properties of spatially reinforced composites. Mechanics of Composite Materials 4, 617–625.

Krieger-Lassen, N.C., Conradsen, K., Jensen, D.J., 1992. Image processing procedures for analysis of electron back scattering patterns. Scanning Microscopy 6, 115–121.

Krieger, N.C., Lassen, 1999. Source point calibration from an arbitrary electron backscattering pattern. Journal of Microscopy 195, 204–211.

Kroner, E., 1958. Continuum theory of dislocations and self-stresses. Ergebnisse der Angewandten Mathematik 5, 1327–1347.

Kröner, E., 1977. Bounds for effective elastic moduli of disordered materials. J. Mech. Phys. Solids 25, 137–155.

Kröner, E., 1986. Statistical modeling: Modeling small deformation in polycrystals. In: Gittus, J., Zarka, J. (Eds.). Elsevier.

Kysar, J.W., Saito, Y., Oztop, M.S., Lee, D., Huh, W.T., 2010. Experimental lower bounds on geometrically necessary dislocation density. International Journal of Plasticity 26, 1097–1123.

Landon, C., Adams, B., Kacher, J., 2008. High resolution methods for characterizing mesoscale dislocation structures. Journal of Engineering Materials and Technology 130 (2), 40–45.

Lee, E.H., 1969. Elastic-Plastic Deformation at Finite Strains. Journal of Applied Mechanics 36, 1–6.

Lekhnitskii, S.G., 1968. Anisotropic Plates. Gordon and Breach.

Lyon, M., Adams, B.L., 2004. Gradient-based non-linear microstructure design. Journal of the Mechanics and Physics of Solids 52, 2569–2586.

Mathworks, 2007. Matlab. The Mathworks, Inc..

Messac, A., Mattson, C.A., 2002. Generating well-distributed sets of Pareto points for engineering design using physical programming. Optim. Eng. 3, 431–450.

Midha, A., Norton, T.W., Howell, L.L., 1994. On the nomenclature, classification, and abstractions of compliant mechanisms. Journal of Mechanical Design, Trans. ASME (116), 270–279.

Milton, G.W., 2002. The Theory of Composites. Cambridge University Press.

Molinari, A., Canova, G.R., Ahzi, S., 1987. Self-consistent approach of the large deformation polycrystal viscoplasticity. Acta Metallurgica et Materialia 35 (12), 2983–2994.

Morawiec, A., 2004. Orientations and Rotations—Computations in Crystallographic Textures. Springer-Verlag.

Mori, T., Tanaka, K., 1973. Average stress in matrix and average elastic energy of materials with misfitting inclusions. Acta Metall 21, 571–574.

Nemat-Nasser, S., Hori, M., 1999. Micromechanics: Overall Properties of Heterogeneous Materials. Elsevier.

Norris, A.N., 2006. The isotropic material closest to a given anisotropic material. J. Mech. Materials Structures 1 (2), 231–246.

Nowell, M.M., Field, D.P., 1998. Texture and grain boundary structure dependence of Hillock formation in thin metal films. Mat. Res. Soc. Symp. Proc. 516, 115–120.

Nowell, M.M., Wright, S.I., 2004. Phase differentiation via combined EBSD and XEDS. Journal of Microscopy 213, 296–305.

Nowell, M.M., Wright, S.I., 2005. Orientation effects on indexing of electron backscatter diffraction patterns. Ultramicroscopy 103, 41–58.

Nye, J.F., 1953. Some geometrical relations in dislocated crystals. Acta Metallurgica 1, 153–162.

Olson, G.B., 1997. Computational Design of Hierarchically Structured Materials. Science 277 (29), 1237–1242.

Pantleon, W., 2008. Resolving the goemetrically necessary dislocation content by conventional electron backscattering diffraction. Scripta Materialia 58, 994–997.

Paul, B., 1960. Prediction of elastic constants of multiphase materials. Trans. Metall. Soc. AIME 218, 36–41.

Peirce, D., Asaro, R.J., Needleman, A., 1983. Material rate dependence and localized deformation in crystalline solids. Acta Metallurgica et Materialia 31 (12), 1951–1976.

Pospiech, J., Jura, J., 1975. Fourier coefficients of the generalized spherical function and an exemplary computer program. Kristall und Technik 10, 783–787.

Proust, G., Kalidindi, S.R., 2006. Procedures for construction of anisotropic elastic–plastic property closures for face-centered cubic polycrystals using first-order bounding relations. Journal of the Mechanics and Physics of Solids 54 (8), 1744–1762.

Reuss, A., 1929. Berechnung der Fliebgrenze von Mischkristallen auf Grund der Plastizitatsbedingung fur Einkristalle. Zeitschrift fur Angewandte Mathematik und Mechanik 9, 49–58.

Rice, J.R., 1971. Inelastic constitutive relations for solids: An internal-variable theory and its application to metal plasticity. Journal of the Mechanics and Physics of Solids 19 (6), 433–455.

Rosen, B.W., Hashin, Z., 1970. Effective thermal expansion coefficients and specific heats of composite materials. Int. J. Engng. Sci. 8, 157–173.

Rosiwal, A., 1898. Ueber geometrische Gesteinanalysen. verhandlugen K.K. des Geolische Rechsanstalt, Wein 13, 143–175.

Ruggles, T., D. Fullwood (in press). Estimation of bulk dislocation density based on known distortion gradients recovered from EBSD. Ultramicrosopy.

Sachs, G., 1928. Zur Ableitung einer Fließbedingung. Z. Ver. Deu. Ing. 72 (22), 734.

Saheli, G., Garmestani, H., Adams, B.L., 2004. Microstructure design of a two phase composite using two-point correlation functions. Journal of Computer-Aided Materials Design 11, 103–115.

Sintay, D.S., Adams, B.L., 2005. Microstructure design for a rotating disk: With application to turbine engines. IDETC/CIE, 31st Design Automation Conference, Long Beach, CA.

Suquet, P., 1987. Elements of homogenization for inelastic solid mechanics. In: Sanchez-Palencia, E., Zaoui, A. (Eds.), Homogenization Techniques for Composite Media, Lecture Notes in Physics 272. Springer-Verlag, pp. 193–278.

Taylor, G.I., 1938. Plastic strain in metals. Journal of the Institute of Metals 62, 307–324.

Thomsen, E., 1930. Quantitative microscopic analysis. Journal of Geology 38, 193–221.

Torquato, S., 1997. Effective stiffness tensor of composite media. I. Exact series expansions. J. Mech. Phys. Solids 45, 1421–1448.

Torquato, S., 2002. Random Heterogeneous Materials. Springer-Verlag.

Torquato, S., 2006. Necessary conditions on realizable two-point correlation functions of random media. Ind. Eng. Chem. Res. 45 (21), 6923–6928.

Troost, K.Z., van der Sluis, P., Gravesteijn, D.J., 1993. Microscale elastic-strain determination by backscatter Kikuchi diffraction in the scanning electron microscope. Appl. Phys. Lett. 62 (10), 1110–1112.

Troost, K.Z., van der Sluis, P., Gravesteijn, D.J., 1993. Microscale elastic-strain determination by backscatter Kikuchi diffraction in the scanning electron microscope. Appl. Phys. Lett. 62 (10), 1110–1112.

Uche, O.U., Stillinger, F.H., Torquato, S., 2006. On the realizability of pair correlation functions. Physica A: Statistical Mechanics and Its Applications 360 (1), 21–36.

Van den Boogaart, K.G., 2002. Statistics for Individual Crystallographic Orientation Measurements. Springer-Verlag.

Venables, J.A., Bin Jaya, R., 1977. Accurate microcrystallography using electron back-scattering patterns. Phil. Mag 35 (5), 1317–1332.

Villert, S., Maurice, C., Wyon, C., Fortunier, R., 2009. Accuracy assessment of elastic strain measurement by EBSD. Journal of Microscopy 233 (2), 290–301.

Voigt, W., 1910. Lehrbuch der krystallphysik. Teubner: xxiv.

Walker, J.S., 1996. Fast Fourier Transform. CRC Press.

Wikipedia, 2007. JPEG Available at. http://en.wikipedia.org/wiki/Jpeg.

Wilkinson, A.J., Meaden, G., Dingley, D.J., 2006. High-resolution elastic strain measurement from electron backscatter diffraction patterns: New levels of sensitivity. Ultramicroscopy 106, 307–313.

Willis, J.R., 1981. Variational and related methods for the overall properties of composites. Adv. Appl. Mech 21, 1–78.

Winkelmann, A., 2007. Simulation of electron backscatter diffraction patterns. Microscopy and Microanalysis 13, 930–931.

Wright, S.I., 1993. Review of automated orientation imaging microscopy (OIM). Journal of Computer-Assisted Microscopy 5 (3), 207–221.

Wright, S.I., 2000. Fundamentals of automated EBSD. In: Schwartz, A.J., Kumar, M., Adams, B.L. (Eds.), Electron Backscatter Diffraction in Materials Science. Kluwer Academic/Plenum Publishers, pp. 51–64.

Wright, S.I., Adams, B.L., Kunze, K., 1993. Application of a new automatic lattice orientation measurement technique to polycrystalline aluminum. Materials Science and Engineering A 160 (2), 229–240.

Wu, X., Kalidindi, S.R., Necker, C., Salem, A.A., 2007. Prediction of crystallographic texture evolution and anisotropic stress-strain curves during large plastic strains in high purity [alpha]-titanium using a Taylor-type crystal plasticity model. Acta Materialia 55 (2), 423–432.

Wu, X., Proust, G., Knezevic, M., Kalidindi, S.R., 2007. Elastic-plastic property closures for hexagonal close-packed polycrystalline metals using first-order bounding theories. Acta Materialia 55 (8), 2729–2737.

Xu, L.M., Fan, H., Xie, X.M., Li, C., 2008. Effective elastic property estimation for bi-continuous heterogeneous solids. Computers, Materials and Continua 7 (3), 119–127.

Index

G

H

I

K

L

M

N

O

P

T

U

V

W

Y

Printed and bound by CPI Group (UK) Ltd, Croydon, CR0 4YY

08/05/2025

01864896-0001